Current Research in Systematic Musicology

Volume 13

The series covers recent research, hot topics, and trends in Systematic Musicology. Following the highly interdisciplinary nature of the field, the publications connect different views upon musical topics and problems with the field's multiple methodology, theoretical background, and models. It fuses experimental findings, computational models, psychological and neurocognitive research, and ethnic and urban field work into an understanding of music and its features. It also supports a pro-active view on the field, suggesting hard- and software solutions, new musical instruments and instrument controls, content systems, or patents in the field of music. Its aim is to proceed in the over 100 years international and interdisciplinary tradition of Systematic Musicology by presenting current research and new ideas next to review papers and conceptual outlooks. It is open for thematic volumes, monographs, and conference proceedings.

The series therefore covers the core of Systematic Musicology, - Musical Acoustics, which covers the whole range of instrument building and improvement, Musical Signal Processing and Music Information Retrieval, models of acoustical systems, Sound and Studio Production, Room Acoustics, Soundscapes and Sound Design, Music Production software, and all aspects of music tone production. It also covers applications like the design of synthesizers, tone, rhythm, or timbre models based on sound, gaming, or streaming and distribution of music via global networks.

- Music Psychology, both in its psychoacoustic and neurocognitive as well as in its performance and action sense, which also includes musical gesture research, models and findings in music therapy, forensic music psychology as used in legal cases, neurocognitive modeling and experimental investigations of the auditory pathway, or synaesthetic and multimodal perception. It also covers ideas and basic concepts of perception and music psychology and global models of music and action.
- Music Ethnology in terms of Comparative Musicology, as the search for universals in music by comparing the music of ethnic groups and social structures, including endemic music all over the world, popular music as distributed via global media, art music of ethnic groups, or ethnographic findings in modern urban spaces. Furthermore, the series covers all neighbouring topics of Systematic Musicology.

Jakub Sawicki

Music Psychology—Balance of Relations

A Perspective of Complex Systems

Springer

Jakub Sawicki
Potsdam Institute for Climate Impact
Research (PIK)
Postdam, Germany

University of Applied Sciences
Northwestern Switzerland (FHNW)
Basel, Switzerland

Technical University of Berlin
Berlin, Germany

ISSN 2196-6966 ISSN 2196-6974 (electronic)
Current Research in Systematic Musicology
ISBN 978-3-031-97002-3 ISBN 978-3-031-97003-0 (eBook)
https://doi.org/10.1007/978-3-031-97003-0

This Springer imprint is published by the registered company Springer Nature Switzerland AG
The registered company address is: Gewerbestrasse 11, 6330 Cham, Switzerland

Jakub Sawicki

Music Psychology—Balance of Relations

A Perspective of Complex Systems

Springer

Jakub Sawicki
Potsdam Institute for Climate Impact
Research (PIK)
Postdam, Germany

University of Applied Sciences
Northwestern Switzerland (FHNW)
Basel, Switzerland

Technical University of Berlin
Berlin, Germany

ISSN 2196-6966 ISSN 2196-6974 (electronic)
Current Research in Systematic Musicology
ISBN 978-3-031-97002-3 ISBN 978-3-031-97003-0 (eBook)
https://doi.org/10.1007/978-3-031-97003-0

This Springer imprint is published by the registered company Springer Nature Switzerland AG
The registered company address is: Gewerbestrasse 11, 6330 Cham, Switzerland

Preface

This monograph examines fundamental questions in music psychology through the lens of complex systems theory, employing the principle of the balance of relations. By integrating theoretical frameworks and empirical studies, it employs the core concepts of adaptation and synchronization to model physiological and neural systems, offering predictive insights into neuronal dynamics underlying music perception. This network-based, interdisciplinary approach represents a paradigm shift from traditional localization models, which associate specific perceptual functions with discrete brain regions, toward a more holistic perspective. Music perception emerges from the interaction of the entire neural system rather than being confined to spatial or temporal localization within a neural network. Consequently, the interdisciplinary field of systematic musicology is essential for advancing the understanding of this complex phenomenon, combining physics, mathematics, biology, neuroscience and computer science. The overarching goal is to achieve a deeper comprehension of the highly complex issue of music perception from the research field of complex systems, which is able to reduce the chaotic storm of a wing beat of a butterfly to a few simple principles.

Berlin, Germany | Jakub Sawicki

Acknowledgments

My sincere appreciation goes to my academic advisers, Prof. Dr. Rolf Bader and Prof. Dr. Markus Abel, as well as Prof. Dr. Dr. h.c. Eckehard Schöll, Ph.D, whose guidance and expertise were instrumental in shaping the research direction of this monograph. The insightful feedback from my colleagues, research collaborators, and coauthors enriched the content and strengthened the arguments: Prof. Dr. Dr. h.c. mult. Jurgen Kurths, Prof. Dr. Ralph G. Andrzejak, Prof. Dr. Mehrnaz Anvari, Prof. Dr. Wolfram Barfuss, Prof. Dr. Matthias Bertsch, Prof. Dr. Igor Franović, Prof. Dr. Daniel J. Gauthier, Prof. Dr. Herwig Gerlach, Prof. Dr. Jaroslav Hlinka, Prof. Dr. Philipp Hovel, Prof. Dr. Klaus Lehnertz, Prof. Dr. Gary McPherson, Prof. Dr. Peter A. Tass, Prof. Dr. Giovanni Volpe, Prof. Dr. Anna Zakharova, Prof. Dr. Serhiy Yanchuk, PD Dr. Jost Fischer, Dr. Rico Berner, Dr. Nicola Botta, Dr. Nuria Brede, Dr. Fenja Drauschke, Dr. Isabel Fernholz, Dr. Moritz Gerster, Dr. Sebastian Goldt, Dr. Aida Hajizadeh, Dr. Omer Karin, Dr. Philipp Lorenz-Spreen, Dr. Sarah A. M. Loos, Dr. Christoph Miehl, Dr. Jan Molter, Dr. Simona Olmi, Dr. Iryna Omelchenko, Dr. Lukas Ramlow, Dr. Giulia Ruzzene, Dr. Eckhardt Schneider, Dr. Alireza Seif, Dr. Antonín Škoch, Moritz Alkofer, Lenz Hartmann, Julia M. Koulen, Andrea Löser, Thomas Löser, Marshall Mykietyshyn, Sebastian Nagies, Gabi Pilz, Markus Radke, Natalia Spitha, and Max Thiele.

I extend my gratitude to the Deutsche Forschungsgemeinschaft (DFG, German Research Foundation, project Nos. 429685422 and 440145547) for their generous support, without which this extensive research would not have been possible. I thank the Joachim Herz Foundation for funding an interdisciplinary and international workshop on “Adaptivity in nonlinear dynamical systems”, which was held from 20th to 23rd September 2022 and further thank the Potsdam Institute for Climate Impact Research for supporting and hosting the workshop. Special appreciation is extended to the Berlin University of the Arts for providing an inspiring environment, particularly during the “creative performance workshop” in 2024.

Last, I owe thanks to the scholars who paved the way in this field, your groundbreaking work laid the foundation for this exploration.

Contents

Chapter 1
Introduction

The concept of *interdisciplinarity* has become a prevalent element in the discourse surrounding scientific projects. Few research proposals are submitted without reference to interdisciplinary methods, situated within a broader context of intra-, coss-, multi-, inter-, and transdisciplinary methodologies [1]. Simultaneously, the well-known issue of overspecialization and the increasing fragmentation of scientific disciplines–where even subfields become too expansive for comprehensive mastery by individuals–highlights the perceived need for greater integrative knowledge frameworks. This fragmentation is frequently viewed as a deficit in coherent, overarching understanding, positioning interdisciplinarity as an inherently valuable approach, comparable to societal relevance or innovation in its perceived importance.

This monograph investigates fundamental questions in music psychology through the interdisciplinary framework of complex systems theory. Central to this work is the principle of the balance of relations, which emphasizes the importance of understanding entities through the dynamic relations that define them, rather than in isolation. This conceptual framework, improved within this monograph and inspired by various scientific disciplines, is widely applicable across fields such as philosophy, psychology, music, and complex systems theory, where it highlights the intricate interdependencies and emergent properties within interconnected systems. The intention behind elaborating and applying the principle of the balance of relations is to provide a structured framework and a new orientation for addressing fundamental questions in music psychology. Although the principle of the balance of relations has been proposed before, this monograph demonstrates that the applied methods enable its methodological realization on a more advanced level. Through the integration of complex systems theory and psychological approaches, relational dynamics in music psychology can be systematically investigated.

The research underlying this work has involved collaboration among musicologists, physicists, mathematicians, biologists, neuroscientists, and computer scientists, exemplifying the essence of interdisciplinary inquiry. By integrating theoretical and experimental approaches, this monograph seeks to advance the understanding of music perception through a network-based perspective. Traditional models in music

J. Sawicki, *Music Psychology—Balance of Relations*,
Current Research in Systematic Musicology 13,
https://doi.org/10.1007/978-3-031-97003-0_1

psychology, which attribute specific perceptual functions to discrete brain regions, are re-evaluated in light of network neuroscience. Rather than conceptualizing auditory processing as a linear sequence of information transfer, this approach highlights the dynamic, interconnected nature of neural systems involved in music perception. Graph theory and network-based analyses reveal that music perception engages distributed brain networks, integrating sensory, motor, cognitive, and emotional processes. A key advantage of network theory lies in its ability to capture dynamic activity patterns and interdependent neural interactions, providing a statistically rigorous framework for analyzing the complexity of music cognition. This monograph underscores the largely unexplored potential of network neuroscience in music psychology. Confidence in achieving a deeper understanding of music perception stems from the success of complex systems research, which has revealed underlying principles governing seemingly chaotic phenomena, such as the turbulence that may arise from a butterfly's wingbeat.

The aim of this introduction is to explore the principle of balance and relations (Sect. 1.1) in and between musicology and natural sciences. Music psychology in particular is an interdisciplinary act of balancing, in which the perspectives of different scientific disciplines must be related to each other in a balanced way, as specified in Sect. 1.2. In Sect. 1.3, the idea of balance of relations in dynamical systems is introduced focusing on the phenomena of adaptation and synchronization. Both phenomena can be represented by the relation of temporal changes of objects, whereby the nature of the objects can be both musical and physical. In the case of many objects, the network approach presented in Sect. 1.4 will be utilized. Section 1.5 gives the outline of this monograph.

1.1 Principle of the Balance of Relations

The paradigmatic approach of this monograph is the "principle of balance of relations", a concept expressed across numerous interdisciplinary fields, each recognizing the critical importance of relational dynamics in shaping systems. At its core, this principle emphasizes that understanding entities is not just about focusing on their individual characteristics, but rather on the relations that define and shape them. For instance, theories regarding the evolutionary origins of music suggest it emerged from emotional signaling, which facilitates and enhances the relations between individuals, thus benefiting social cohesion. Describing these relations is often as complex as describing the entities themselves, and while the definitions of relations may vary across different fields and can sometimes be difficult to grasp, their subtle changes can be far easier to perceive and have profound effects. Equilibrium states hold particular significance in this context, as they play a central role in regulating the dynamics of relations within a system. These states are especially important when small changes can lead to disproportionate outcomes, as seen in tipping points or critical states.

The principle of balance of relations is inspired by and finds applications in diverse scientific disciplines, including philosophy, psychology, sociology, medicine, and the natural sciences such as physics, chemistry, and biology. Each of these fields brings its own interpretation of the principle, but the common thread is the emphasis on the interdependencies and dynamic interactions between components, as opposed to studying isolated entities. This interdisciplinary approach shows how systems, whether biological, social, or physical, rely on relational balance to function optimally.

In philosophy, for instance, Martin Buber's relational framework is rooted in the balance of relations, focusing on the interdependence between the self and others [2]. His concept of identity reflects a dynamic equilibrium where human existence is continuously shaped by engagement, dialogue, and mutual recognition, emphasizing relational balance as central to human life. Similarly, in psychology, the systemic approach stresses that individuals are not isolated but embedded in a larger system, where maintaining balance within the system is key to its proper functioning [3]. Therapeutic interventions in this context do not impose external changes but instead provide impulses that trigger self-regulation and reorganization, facilitating the evolution of the system toward more adaptive configurations.

In the realm of applied (music) physiology, physical therapy methods, like the Alexander technique [4], the Feldenkrais method [5], or the "Resonanzlehre" developed by Lange [6] illustrate the balance of relations within a performer's interaction with their instrument and the surrounding environment. The core idea is that effortless movement leads to enhanced resonance, creating a deeper connection between the musician, the instrument, and the audience. This reflects the balance of relations within the performer's body, the instrument's response, and the acoustics of the space, with the body's center of balance playing a pivotal role in producing harmonious sound.

Holism, which contrasts with reductionism, underscores that a system's properties arise from the interactions of its components [7]. This perspective suggests that complex behaviors and emergent phenomena cannot be fully understood by analyzing the parts in isolation. Holism has influenced fields like systems biology, which views biological networks as integrated entities rather than isolated components. Similarly, in systems medicine, holistic principles guide approaches that treat the body as an interconnected whole, facilitating more comprehensive health strategies that account for the relational dynamics between different systems.

Cybernetics, another key field in understanding the balance of relations, focuses on self-regulating systems through feedback loops and circular causality. By studying how outputs influence future inputs, cybernetics explores how systems maintain stability and adaptability [8, 9]. Initially applied in engineering, biology, and information theory, cybernetics has expanded into sociology, economics, and even the arts, providing a unified framework for understanding dynamic, self-regulating systems across multiple contexts.

The tipping point concept, originating in physics and later adapted to sociology and network economics, illustrates the principle of balance of relations in systems [10]. A tipping point marks a critical threshold where small changes cause a system to transition from a stable equilibrium to an unstable one, resulting in a self-reinforcing and often irreversible transformation. This principle demonstrates how relational dynamics within a system can trigger profound changes when interdependencies reach a threshold, affecting systems in fields ranging from social behavior to opinion dynamics and demographic shifts.

Systems theory, which explores the balance of relations within networks of interdependent components, emphasizes dynamic interactions that maintain equilibrium and resilience within a system [11, 12]. Systems theory identifies universal principles that govern the stability of systems across various domains, whether they are biological, social, or technological. This approach distinguishes between active systems, which continuously adapt to changes within their environment, and passive systems, which remain static until activated. Ultimately, the goal of systems theory is to achieve "equifinality"–the idea that different pathways can lead to the same outcome while maintaining relational balance.

Synergetics, an interdisciplinary framework developed by Hermann Haken, investigates self-organization and pattern formation in open systems far from thermodynamic equilibrium. It explains how order emerges from the nonlinear interactions of subsystems, and how external factors, such as energy fluxes or environmental conditions, guide this process [13, 14]. Synergetics identifies universal principles of cooperation, or synergy, that apply across diverse fields, from physics and chemistry to biology and social dynamics. This approach underscores the balance of relations within systems, where cooperation between components leads to the spontaneous emergence of organized patterns.

Network physiology, as a field, examines how the body's organ systems interact and synchronize to maintain overall health. Unlike traditional approaches that isolate individual organs, network physiology focuses on the complex interconnections between systems [15, 16]. By studying these dynamic inter-organ communications, researchers aim to understand how disruptions in these interactions can lead to disease and how the balance of relations within the body contributes to maintaining homeostasis, the system's internal balance.

The principle of balance of relations is pervasive in discussions of complex systems, nonlinear dynamics, self-organization, collective behavior, and pattern formation (see Sect. 1.4). It resonates across a wide range of scientific disciplines, with computational methods playing an essential role in analyzing these systems. Much like the n-body problem in physics, where predicting the motions of interacting celestial bodies is complex due to the gravitational interdependence between them, the principle of balance of relations highlights the intricacies of interconnected systems. While a general closed-form solution may not always be available, this principle offers a holistic approach to describing and understanding the complexity of systems, even when their dynamics appear intractable. The ability to study and capture these complexities is crucial for developing a deeper understanding of how systems function and adapt.

1.2 Music Psychology–An Act of Balance

Music psychology is a balancing act in three ways: by nature it is highly inter- and transdisciplinary, moving within a network of different scientific and physiological fields with which it is in a productive interdependence as shown in Fig. 1.1. Secondly, music psychology has developed within musicology from a part of systematic musicology into an independent discipline and is related to the dynamics and progress of the parent disciplines. And finally, there are different research perspectives within music psychology itself, to which every researcher can find a balanced point of reference. This triple balancing act is a dynamic process per se. Albert Einstein remarked in a letter to his son in 1930: "Life is like riding a bicycle. To keep your balance you must keep moving." At the time of this quote, the term *music psychology* was coined in Germany [17] and perhaps the latter quote can therefore also be understood within music psychology as an instruction to keep an eye on the directions of scientific progress in order to keep the balance.

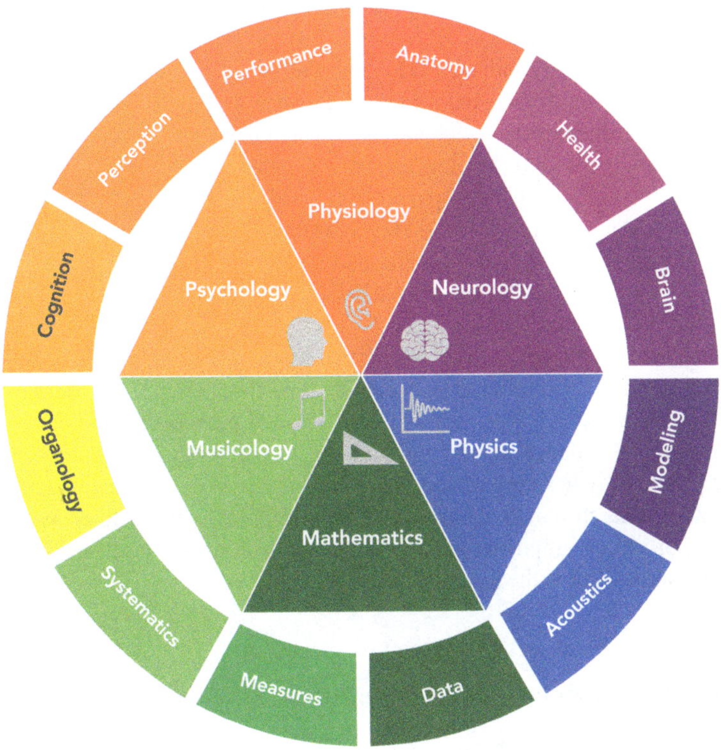

Fig. 1.1 The balance of diverse inter- and transdisciplinary connections within music psychology: At the core lie the various scientific disciplines, while the periphery encompasses their respective applications, which are examined in detail within this monograph

The question of what scientific progress is and how it differs from other areas of human activity has long been a subject of the philosophy of science and is still relevant today. A defining analogy is the comparison of science as such with the architecture of a building. Modern science is typically divided into major columns, from which further columns and arches can grow over time, reflecting the complexity and growing richness of the "building of science" [18]. Thus, music psychology as a branch of musicology has grown from a branch of systematic musicology into an independent discipline. The architecture of music psychology itself is highly interdisciplinary and diverse. It is flanked by branches of physics, mathematics, biology, medicine, psychology and neuroscience and borders on other areas of systematic musicology, such as musical acoustics, music aesthetics, music sociology, music education, music therapy and ethnomusicology.

Furthermore, music psychology is a field of research with practical relevance for many areas, including music performance, composition, education, criticism, and therapy, as well as investigations of human attitude, skill, performance, intelligence, creativity, and social behavior. As applied music psychology, the discipline thus makes an important contribution to the care of musicians, which includes aspects such as dealing with stress and stage fright, mental training, concentration practice, performance coaching and stage performance. These areas of application are based on the empirical pillars of music psychology, which deal with how music is perceived, created and responded to. Empirical topics of music psychology are the perception of music, musical socialization, musical behavior and experience as well as music production as composition, interpretation and improvisation. These pillars stand on the foundation of music psychology, which deals with the neuroscientific mechanisms and modeling of musical perception processes. Each of the three levels of this young branch of science–application, empirical research, and neuroscientific foundation–contributes to the balance of the entire building, which is flanked by a variety of scientific disciplines (see Fig. 1.1).

In this interdisciplinary field of interdependence, many methods from the various related scientific disciplines have already been transferred to the field of music psychology. However, in order not to lose the broad perspective and to be able to make a reasonable balance, it is important to focus on the fundamental mechanisms of music psychology in an interdisciplinary way. It is essential to include current and fundamental research results from neuroscientific fields in order to ensure that the foundation on which empirical music psychology studies are built and based is as solid as possible. That is the aim of this monograph.

1.3 Balance of Relations in Dynamical Systems

A dynamical system is a mathematical framework used to describe how a system evolves over time. It consists of a state space representing all possible configurations of the system, and a rule (often a function) that determines how the system's state changes. This rule or function can be deterministic, where the future state is precisely determined by the current state, or stochastic, where randomness influences the evolution. Dynamical systems can describe a wide range of phenomena, from the swinging of a simple pendulum to the dynamics of the human brain. The study of dynamical systems provides insights into processes like chaos and self-organization, with applications across disciplines like physics, mathematics, biology, medicine, neuroscience, engineering, and economics.

The "revolution of the way of thinking" brought about by Newtonian physics [19], as referred to by Immanuel Kant in his Critique of Pure Reason [20], forms the foundation for the modern description of dynamical systems. This paradigm shift, initiated by Newton's formulation of natural phenomena in terms of a *balance of relations*, laid the groundwork for differential calculus. The fundamental concept of differential calculus is the determination of the derivative of a function, which depends on a varying parameter. The derivative expresses the relation between changes in the function and changes in the parameter. By understanding this relation, it becomes possible to quantitatively describe the function itself.

In modern physics, the behavior of dynamical systems is commonly described using such differential equations, where the relation between rates of change plays a crucial role in characterizing the system's evolution over time. As with any equation, two expressions are related with an equals sign. There is a balance between the "left-hand side" and "right-hand side" of the equation and in this sense every equation is a balance. The "=" symbol, which appears in every equation and in particular in Fig. 1.2 looks like a weigh scale, was invented in 1557 by Recorde [21], who considered that nothing could be more equal than parallel straight lines with the same length.

In many cases, finding an explicit solution to a system of differential equations is infeasible due to the complexity of the system. In such instances, differential equations function as computational algorithms for computer systems. By understanding the relationships governing the rate of change within a system and providing a set of initial conditions, it becomes possible to numerically compute solutions in a stepwise manner. This approach requires substantial computational resources, which is why the advancement of nonlinear physics has been closely tied to the development of computers.

The solutions of these systems can be analyzed in terms of state or phase space, where all possible states of the system are represented, and the solution is depicted as a trajectory within this space. A nonlinear, deterministic oscillator, for example, is characterized by a closed trajectory, i.e. a self-repeating loop, within the phase space.

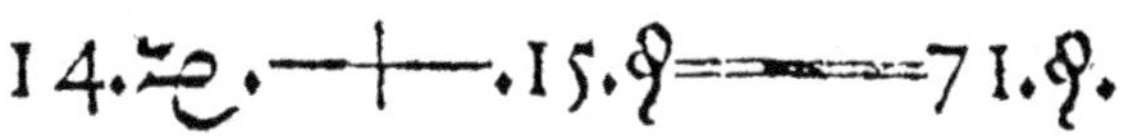

Fig. 1.2 The first use of an equals sign, equivalent to $14x + 15 = 71$ in modern notation, invented in 1557 by Recorde [21]. The "=" symbol appears like a balanced weigh scale

In Sect. 2.4.1, generic oscillator models to study the behavior of dynamical systems are introduced. One of the central concerns of nonlinear physics is the examination of such limit cycles under small perturbations. Depending on how the system responds to these disturbances, its dynamics can be classified as either stable or unstable. Additionally, when the solution of a differential system exhibits extreme sensitivity to initial conditions, the system's behavior is described as chaotic.

1.3.1 Adaptive Mechanisms

Adaptation (or adaptivity) is a dynamic characteristic pervasive across natural, socio-economic, and technological systems. For instance, adaptive couplings are observed in diverse real-world systems such as power grids, social networks, and neural networks. These couplings play a crucial role in the structure of closed-loop control mechanisms and underpin the functionality of machine learning algorithms.

Charles Darwin's principle emphasizes that survival is not determined by strength, but by the capacity to adapt to change [22]. In parallel, Stephen Hawking encapsulated the essence of intelligence by stating that "Intelligence is the ability to adapt to change" [23]. This concept is further supported by the definition of learning as "any change in a system that results in a more or less permanent increase in its capacity to adapt to its environment" [24]. Together, these insights underscore the fundamental importance of adaptivity for life itself. In simple terms, one might assert: "To live is to adapt."

Adaptation is not only vital for biological systems but also serves as a cornerstone for the development of intelligent artificial systems. From advanced control strategies in complex systems to machine learning algorithms and robotic systems, adaptive mechanisms are integral features. They allow systems to respond to dynamic environments and evolve their behavior, reflecting the omnipresence of adaptation across all living organisms.

At its core, adaptivity refers to the ability to adjust to changing conditions over time, a crucial aspect of both natural and artificial processes across numerous scientific disciplines. This capacity for adaptation is also a defining feature of the human mind, enabling creativity, the appreciation of music and visual arts, and the ability to invent. Adaptive mechanisms play a fundamental role in musical training, as well as in music-performance and musical creation. Moreover, adaptivity is the driving force behind cultural and intellectual advancements, emphasizing the deep connection between adaptability and progress in both biological and cognitive domains.

Adaptive mechanisms operate across a broad spectrum of spatial and temporal scales, ranging from the adaptation of individual neurons to the capacity of social systems to adjust to dynamic environments and process sensory inputs, extending even to the adaptation of Earth's climate system. Over recent decades, significant expertise has been developed in various scientific fields to describe and control complex systems. The growing capabilities of modern technologies, coupled with the immense challenges confronting humanity as a large-scale social system, have sparked a renewed interest in adopting an interdisciplinary approach to the study and application of adaptation [25].

1.3.2 Synchronization Phenomena

In 1665, the renowned Dutch physicist Christiaan Huygens, who had suffered from poor health throughout his life, was bedridden in his hometown of The Hague. A few years earlier, he had invented the pendulum clock, an innovation that gained him significant recognition due to its remarkable precision, with an error margin of only a few seconds per day. However, despite this success, Huygens was facing a new challenge beyond his illness: multiple sea trials had revealed that the pendulum clock was significantly less accurate when used aboard ships at sea. Anticipating potential issues in maritime environments, Huygens had already conceived the idea of constructing a double-clock system, in which one clock could serve as a backup should the other fail.

While confined to his bed, Huygens conducted an experiment by mounting two of his pendulum clocks side by side on the same wooden brace as depicted in a drawing of the set-up in Fig. 1.3. During this setup, he made a serendipitous observation, which he described as "an odd kind of sympathy": Despite disturbances to one of the pendulums, the two clocks would, after a short time, synchronize such that their pendulums swung in perfect opposition. This phenomenon, later understood as entrainment, was the first documented instance of what is now referred to as *synchronization*, a concept that would not be fully understood until centuries later.

Synchronization is a phenomenon with many manifestations in nature and has fascinated humanity for centuries. The term originates from the Greek word *synchrónos*, meaning "sharing the same time." This implies two essential conditions for synchronization: first, objects with well-defined intrinsic *time* scales, and second, a connection or interaction between these objects. Synchronization occurs when the

Fig. 1.3 An original drawing of Christiaan Huygens illustrating his experiments in 1665 with two pendulum clocks mounted next to each other on the same brace [26]

intrinsic times of individual elements become *related* through their interaction. This relationship can be established between oscillators that are coupled to one another. In the case of periodic oscillators, the intrinsic timescale is referred to as the period.

There are numerous examples of synchronization in nature, particularly in systems with many interacting elements, where synchronization patterns often seem almost magical. Examples include the coordinated flocking or swarming behavior of birds and fish, the simultaneous flashing of thousands of fireflies, or the precise coordination among musicians. Although these phenomena have been observed for centuries, the systematic study of synchronization began only in the twentieth century [27, 28].

The mechanism underlying synchronization is inherently nonlinear. The oscillators involved can be described by stable limit cycles in their state space, making them self-sustained and independent of external forcing. From an energetic perspective, these systems are dissipative, meaning they do not conserve energy but instead dissipate it over time. They will be the focus of Sect. 2.4. Equally crucial for synchronization is the coupling between oscillators. In the case of Huygens' clocks, this coupling was established through a shared supporting beam. As will be introduced in the next Sect. 1.4.1, a system of coupled oscillators can be modeled as a network, where the oscillators are represented by nodes and the interactions between them by links.

In general, synchronization requires at least two coupled nonlinear oscillators. However, there is no upper limit: complex network topologies, combined with intricate local dynamics, lead to a wide range of synchronization phenomena. A universal characteristic of synchronization, however, is the establishment of a fixed relationship between the individual time scales of the oscillators.

1.4 Relations in Complex Systems

Relations are a fundamental construct of human cognition. In general, the recognition of relations requires the ability to compare and quantify. Throughout human history, this capability has exhibited an ambivalent nature: on the one hand, it has driven progress and prosperity, while on the other, it has given rise to envy and ill-will. An intriguing aspect of relations is their extrinsic nature: "relation" as a concept exists only in reference to an observer.

As both a musician and a physicist, I am often asked about my motivation for exploring these seemingly disparate fields. However, there is, in fact, a profound connection between music and physics, both of which have been deeply influenced by the paradigm of relations. The recognition of these connections can be enriched by the cross-pollination of ideas between the two disciplines. It is no coincidence that physicists like Christiaan Huygens and Lord Rayleigh, who studied synchronization phenomena, were also dedicated musicologists. Both started to compare two objects that exhibit temporal change. These objects may be musical or physical in nature, such as a musical instrument or a pendulum. Through this comparison, they were able to better understand the underlying principles that govern both fields.

1.4.1 Network Description

In the case of two objects, it is relatively straightforward to define their relation: either a relation exists, or it does not. However, when dealing with larger systems, the task of determining the relation between all objects becomes more complex. Additionally, large systems composed of many interconnected components tend to exhibit emergent behaviors, adding an extra layer of complexity. As Aristotle famously noted, "The whole is greater than the sum of its parts" [29]. Such large interconnected systems, comprised of individual nodes with complex dynamics, are prevalent in various fields, including the natural sciences, technology, and economics.

Network science provides a robust framework for addressing these challenges, representing interconnected oscillators as nodes and their relationships as links. For instance, the human brain presents an immensely complex network of billions of nodes (neurons), interconnected by up to a quadrillion links (synapses). Interestingly, the origins of this field trace back to a simpler problem posed by the mathematician Leonhard Euler [30]. Euler investigated whether it was possible to cross all the bridges of Königsberg exactly once, returning to the starting point, which later became known as the Eulerian circuit problem. The network Euler considered had only four nodes (land masses) and seven links (bridges) as depicted in Fig. 2.2a of Sect. 2.4.2. Despite its simplicity, graph theory–the mathematical foundation of network science–provides elegant solutions to such problems. Euler demonstrated that completing an Eulerian circuit is impossible if any node in the network has an odd degree, a measure indicating the number of links connected to a node. This foundational result extends to more complex networks, where graph theory continues to offer powerful tools for analyzing intricate systems.

Today, networks have become a fundamental component of numerous fields, with applications spanning social sciences, economics, biology, chemistry, psychology, technology, mathematics, and physics [31–36]. The expansion of the World Wide Web is nearly beyond comprehension, as social networks host billions of active users daily, while search engines handle tens of thousands of queries per second, amounting to trillions of queries annually. Network scientists have developed various graph models to capture the complexities of real-world systems. These include random networks [37, 38], Erdős-Rényi networks [39, 40], small-world networks [41–53], scale-free networks [32, 54–57], hierarchical networks [58], meta-networks, and interdependent networks.

Of particular relevance today are multilayer networks, which consist of multiple types of interactions or relationships [59–63]. Multilayer networks provide a comprehensive framework for modeling complex real-world systems, such as the interactions among genes, proteins, neurons, transportation infrastructures, power grids, and social networks. A distinguishing feature of these examples is that they can be represented by networks composed of distinct layers, where nodes are connected via intra-layer links (within the same layer) and inter-layer links (across layers). A specific subclass of multilayer networks is the multiplex network, in which each layer contains an identical set of nodes. In these networks, inter-layer connections occur only between corresponding nodes across layers.

1.4.2 Network Physiology

The human body is composed of multiple parts, each characterized by its own intricate structural organization and regulatory mechanisms. These parts exhibit complex behaviors, including transient, intermittent, and nonlinear dynamics. Traditionally, medical practice has specialized in the examination of individual physiological systems: cardiologists focus on the heart and assess electrocardiogram (ECG) signals, pulmonologists evaluate lung structure and function through respiratory patterns, and neurologists investigate brain activity using magnetic resonance imaging (MRI) and electroencephalography (EEG) signals. However, the human organism functions as an integrated network, where organ systems continuously interact via various feedback mechanisms across different spatial and temporal scales to optimize and synchronize their activities. These coordinated relations and interactions are crucial for maintaining homeostasis and generating distinct physiological states, such as wakefulness and sleep, different stages of sleep, and states of consciousness or unconsciousness. Disruption of these inter-organ communications can result in dysfunction of individual systems or lead to systemic failure, manifesting as conditions like sepsis, epilepsy, pneumonia, coma, or multiple organ failure.

A fundamental aspect of the field of complex systems is to deepen the understanding of the dynamic relations and interactions between different systems. In this context, the multi-disciplinary field of network physiology is also significantly relevant. The focus of the field of network physiology is the understanding of integrated physiological functions resulting from the complex relations and interactions between different organ systems [15, 16]. The aim of this discipline is to provide a theoretical framework and a system-wide network-based approach to understand how the horizontal integration of physiological systems–each characterized by its own intricate structure and regulatory mechanisms–contributes to global behavior and distinct physiological functions at the organismal level.

The unique dynamics in physiological systems require the development of novel analytical tools and theoretical models. This challenge has attracted the attention of researchers from various disciplines, including neuroscience, physics, applied mathematics, and physiology. In recent years, research in the field of Network Physiology has expanded beyond examining individual organ interactions to explore how the dynamic output of one system is related and influenced by changes in the dynamics of other systems through a network of interconnections. Specifically, investigations have focused on the temporal synchronization and coordination among physiological systems, how these interactions evolve over time, and network communications across distinct tissues (see Sect. 4.2). Empirical studies, combined with theoretical modeling, have aimed to elucidate emergent network dynamics resulting from coupled and synchronized nonlinear oscillators.

The human organism is composed of multiple physiological and organ systems, each characterized by its own structural complexity and dynamic behavior, resulting in transient, fluctuating, and nonlinear output dynamics [64–66]. Traditional approaches in physiology and medicine are largely reductionist, focusing on individual systems to understand health and disease. However, the human body functions as an integrated network, where continuous interactions among multi-component systems are essential for generating distinct physiological states and maintaining health. These interactions occur across multiple levels and spatiotemporal scales, optimizing and synchronizing organ functions. The emerging field of Network Physiology seeks to redefine health and disease by examining the structural, dynamical, and regulatory interactions between physiological systems.

A key challenge in physical, neural, and physiological systems is understanding how global behaviors emerge from networked interactions among dynamically changing entities, often governed by nonlinear, time-varying relations. Due to the complex dynamics of organ systems, identifying and quantifying these relations is difficult [67–69]. Traditional frameworks in computational neuroscience and systems physiology, focused on neuron signaling and intra-system integration, cannot fully provide comprehensive help [70]. The field of Network Physiology has emerged to address this gap, investigating how physiological systems synchronize and integrate as a network to optimize function and maintain health.

1.5 Outline

A brief outline is now provided. This monograph is divided into three main parts: The first part (this chapter and Chap. 2) deals with the common roots of music psychology and the research field of complex systems. The second part (Chap. 3) introduces fundamental mechanisms that are of crucial importance for both research disciplines. The third part (Chaps. 4 and 5) focuses on the modeling of physiological, neuronal and perceptual processes, which gives us a basic understanding of how these processes function in the associated systems.

In the first part, this chapter and Chap. 2, the subject of music psychology and the research field of complex systems will be introduced. These two relatively young disciplines have common developmental roots and are characterized by a high degree of interdisciplinarity. Both disciplines have the concepts of *balance* of *relations* in common. The chapter explores psychoacoustic principles in Sect. 2.3 and provides an overview of complex systems theory in Sect. 2.4. In Sect. 2.5, the role of computational tools, including machine learning and neural modeling, is examined in bridging neuroscience and complex systems approaches to music. Furthermore, the chapter presents the auditory system as a complex network, detailing the transformation of sound waves into neural signals and the hierarchical processing of auditory information. In Sect. 2.6, a network-based paradigm in music psychology is introduced, moving beyond traditional localization models to emphasize the distributed and dynamic nature of neural circuits involved in music perception.

The second part, Chap. 3, is centered on two basic mechanisms: Adaptation and synchronization are essential for the functioning of many systems and many complex phenomena can be traced back to these two simple mechanisms. Before going into detail, general concepts will be discussed, along with an introduction to adaptation and synchronization in various systems, in Sect. 3.1. The following sections will proceed to multilayer systems, starting with Sect. 3.2 and considering a two layer multiplex network and analyzing cluster synchronization patterns. One way of taking control of the dynamics of a two-layer network is to implement a pace maker. In Sect. 3.3, this leads to certain partial synchronization patterns, so-called chimera states, which are quasi tamed by the pacemaker. In Sect. 3.4, an additional layer is considered . A special focus is put on relay synchronization, where two outer layers are synchronized through a remote layer in between. These results are then generalized to heterogeneities in the topology. Section 3.5 focuses on relay synchronization scenarios in a three-layer network, where the middle (relay) layer is just a single node, i.e. a hub.

In the third and final part, Chapters 4 and 5, these mechanisms will be successfully transferred to selected physiological and neuronal systems, showing in Sect. 4.1 how adaptation and synchronization can generally be applied as modeling tools. The specific areas of application range from diseases such as tumors and sepsis in Sect. 4.2, through epilepsy in Sect. 4.5, to specific sleep states, so-called unihemispheric sleep, in Sect. 4.4. The basis for the latter modeling applications is an empirical structural brain connectivity of human subjects in Sect. 4.3, where a paradigmatic model for neural dynamics maps the neuronal activity of the individual brain regions. The comparison with experimental data shows the success of this modeling approach.

The modeling expertise gained is applied to the perceptual process of music in Chap. 5. Just as for physiological and neuronal systems in Chapter 4, the two mechanisms of adaptation and synchronization play a fundamental role in Sect. 5.1. The following Sections show in detail how the perception of music determines the dynamics in the human brain. Starting with a simple tone in Sects. 5.2 and 5.3 shows the crucial role taken by the auditory cortex. The human brain also has a solution for perceiving complex musical structures, in particular their syntax. The processes required for this are modeled in Sect. 5.4.

This monograph concludes with Chap. 6, where the obtained results are summarized and discussed in Sect. 6.1. Furthermore, an outlook is provided, and future research directions are considered in Sect. 6.2.

References

1. Stember M (1991) Advancing the social sciences through the interdisciplinary enterprise. Soc Sci J 28(1):1–14. https://doi.org/10.1016/0362-3319(91)90040-B
2. Buber M (1923) Ich und du. Insel-Verlag
3. Di Nicola V (1984) Road map to Schizo-land: Mara Selvini Palazzoli and the Milan model of systemic family therapy. J Strat Syst Ther 3:50–62 (1984)
4. Alexander FM (1923) Constructive conscious control of the individual. E. P. Dutton & Co

5. Feldenkrais M, Leri D, Brousse M, Nativ M, Beringer E (2010) Higher judo: groundwork. North Atlantic Books
6. Lange T (2016) Die Resonanzlehre. IPV-Printjournal 43(11)
7. Smuts J (1926) Holism and evolution. Books for college libraries. Macmillan. https://doi.org/10.2307/2179530
8. Rosenblueth A, Wiener N, Bigelow J (1943) Behavior, purpose and teleology. Philos Sci 10(1):18–24. https://doi.org/10.1086/286788
9. Wiener N (1948) Cybernetics; or control and communication in the animal and the machine. MIT Press. https://doi.org/10.7551/mitpress/11810.001.0001
10. Grodzins M (1957) Metropolitan segregation. Sci Am 197(4):33–41. https://doi.org/10.1038/scientificamerican1057-33
11. Gorelik G (1983) Bogdanov's tektology: its nature, development and influence. Stud Sov Thought 26(1):39–57. https://doi.org/10.1007/BF00832210
12. von Bertalanffy L (1967) Robots, men, and minds: psychology in the modern world. In: Braziller G (ed) The Heinz Werner lecture series. Clark University, Worcester, Mass
13. Haken H (1983) Synergetics, an introduction, 3rd edn. Springer, Berlin. https://doi.org/10.1007/978-3-642-88338-5
14. Haken H (1987) Synergetics, 2nd edn. Springer, Berlin
15. Ivanov PC (2021) The new field of network physiology: building the human physiolome. Front Netw Physiol 1:711778. https://doi.org/10.3389/fnetp.2021.711778
16. Schöll E, Sawicki J, Berner R, Ivanov PC (2022) Editorial: adaptive networks in functional modeling of physiological systems. Front Netw Physiol 2:996784. https://doi.org/10.3389/fnetp.2022.996784
17. Kurth E (1931) Musikpsychologie. Hesse, Berlin
18. Schroeer D (1984) Das Gebäude der Wissenschaften. Vieweg+Teubner Verlag, Wiesbaden, pp 27–33. https://doi.org/10.1007/978-3-322-88798-6_4
19. Newton I (1687) Philosophiae Naturalis Principia Mathematica. Auctore Js. Newton ... Jussu Societatis Regiae ac Typis Josephi Streater ..., London. https://doi.org/10.3931/e-rara-440
20. Kant I (1787) Critik der reinen Vernunft. J. F. Hartknoch, Riga
21. Recorde R (1969) The Whetstone of Witte [by] Robert record: London, 1557. English experience, its record in early printed books published in facsimile, Da Capo. https://doi.org/10.5281/zenodo.6060870
22. Darwin C (1906) Natural selection; or the survival of the fittest. John Murray, London, England, pp 97–163. https://doi.org/10.1037/13681-004
23. Sreechinth C (2016) Stephen Hawking Wisdom: Stephen Hawking's encapsulated expressions. UB Tech
24. Simon HA (1969) The sciences of the artificial. Karl Taylor compton lectures. M.I.T. Press, Cambridge. https://doi.org/10.7551/mitpress/12107.001.0001
25. Sawicki J, Berner R, Loos SAM, Anvari M, Bader R, Barfuss W, Botta N, Brede N, Franović I, Gauthier DJ. Goldt S, Hajizadeh A, Hövel P, Karin O, Lorenz-Spreen P, Miehl C, Mölter J, Olmi S, Schöll E, Seif A, Tass PA, Volpe G, Yanchuk S, Kurths J (2023) Perspectives on adaptive dynamical systems. Chaos Interdiscip J Nonlinear Sci 33(7):071501. https://doi.org/10.1063/5.0147231
26. Huygens C (1932) Oeuvres complètes de Christiaan Huygens, vol 17. Societe Hollandaise des Sciences, La Haye. (M. Nijhoff, Ed.)
27. van der Pol B (1920) A theory of the amplitude of free and forced triode vibrations. Rad Rev 1:701
28. van der Pol B (1926) On relaxation oscillations. Philos Mag 2:978–992
29. Aristotle (1957) Metaphysica, Vol 7, 17, 1041b. Oxford University Press. https://doi.org/10.1093/actrade/9780198145134.book.1
30. Euler L (1741) Solutio problematis ad geometriam situs pertinentis. Commentarii academiae scientiarum Petropolitanae 8:128–140
31. Boccaletti S, Latora V, Moreno Y, Chavez M, Hwang DU (2006) Complex networks: structure and dynamics. Phys Rep 424(4–5):175–308. https://doi.org/10.1016/j.physrep.2005.10.009

32. Albert R, Barabási AL (2002) Statistical mechanics of complex networks. Rev Mod Phys 74:47–97. https://doi.org/10.1103/revmodphys.74.47
33. Newman M (2003) The structure and function of complex networks. SIAM Rev 45(2):167–256. https://doi.org/10.1137/s0036144503
34. Newman M, Barabási AL, Watts DJ (2006) The structure and dynamics of networks. Princeton University Press, Princeton, USA
35. Newman MEJ (2010) Networks: an introduction. Oxford University Press, Inc., New York. https://doi.org/10.1093/acprof:oso/9780199206650.001.0001
36. Boccaletti S, Pisarchik AN, del Genio CI, Amann A (2018) Synchronization: from coupled systems to complex networks. Cambridge University Press, Cambridge
37. Rapoport A (1957) Contribution to the theory of random and biased nets. Bull Math Biol 19:257–277. https://doi.org/10.1007/bf02478417
38. Solomonoff R, Rapoport A (1951) Connectivety of random nets. Bull Math Biol 13:107–117
39. Erdös P, Rényi A (1959) On random graphs. Publ Math Debrecen 6:290–297
40. Erdös P, Rényi A (1960) On the evolution of random graphs. Publ Math Inst Hung Acad Sci 5:17–61
41. Watts DJ, Strogatz SH (1998) Collective dynamics of 'small-world' networks. Nature 393:440–442. https://doi.org/10.1038/30918
42. Monasson R (1999) Diffusion, localization and dispersion relations on "small-world" lattices. Eur Phys J B 12(4):555–567. https://doi.org/10.1007/s100510050749
43. Newman M, Watts DJ (1999) Renormalization group analysis of the small-world network model. Phys Lett A 263(4–6):341–346. https://doi.org/10.1016/s0375-9601(99)00757-4
44. Adamic LA (1999) The small world web. Lecture notes in computer science, vol 1696/1999. Springer Berlin/Heidelberg. https://doi.org/10.1007/3-540-48155-9_27
45. Jeong H, Tombor B, Albert R, Oltvai ZN, Barabási AL (2000) The large-scale organization of metabolic networks. Nature 407:651. https://doi.org/10.1038/35036627
46. Barahona M, Pecora LM (2002) Synchronization in small-world systems. Phys Rev Lett 89(5):054101. https://doi.org/10.1103/physrevlett.89.054101
47. Sporns O, Tononi G, Edelman GM (2000) Theoretical neuroanatomy: relating anatomical and functional connectivity in graphs and cortical connection matrices. Cereb Cortex 10(2):127–141. https://doi.org/10.1093/cercor/10.2.127
48. Sporns O, Zwi JD (2004) The small world of the cerebral cortex. Neuroinformatics 2(2):145–162. https://doi.org/10.1385/ni:2:2:145
49. Hilgetag CC, Burns GAPC. O'Neill MA, Scannell JW, Young MP (2000) Anatomical connectivity defines the organization of clusters of cortical areas in the macaque and the cat. Philos Trans R Soc Lond Ser B 355(1393):91–110. https://doi.org/10.1098/rstb.2000.0562
50. Sporns O, Chialvo DR, Kaiser M, Hilgetag CC (2004) Organization, development and function of complex brain networks. Trends Cogn Sci 8:418. https://doi.org/10.1016/j.tics.2004.07.008
51. Bassett DS, Meyer-Lindenberg A, Achard S, Duke T, Bullmore ET (2006) Adaptive reconfiguration of fractal small-world human brain functional networks. Proc Natl Acad Sci USA 103(51):19518–19523. https://doi.org/10.1073/pnas.0604453103
52. Shefi O, Golding I, Segev R, Ben-Jacob E, Ayali A (2002) Morphological characterization of in vitro neuronal networks. Phys Rev E 66(2):021905. https://doi.org/10.1103/physreve.66.021905
53. Humphries MD, Gurney K, Prescott TJ (2006) The brainstem reticular formation is a small-world, not scale-free, network. Proc R Soc B Biol Sci 273(1585):503–511. https://doi.org/10.1098/rspb.2005.3331
54. de Solla Price DJ (1965) Networks of scientific papers. Science 149(3683):510–515. https://doi.org/10.1126/science.149.3683.510
55. Barabási AL, Albert R (1999) Emergence of scaling in random networks. Science 286:509. https://doi.org/10.1126/science.286.5439.509
56. Jeong H, Mason SP, Barabási AL, Oltvai ZN (2001) Lethality and centrality in protein networks. Nature 411(6833):41–42. https://doi.org/10.1038/35075138

57. Chen Q, Chang H, Govindan R, Jamin S, Shenker SJ, Willinger W (2002) The origin of power laws in internet topologies revisited. In: Kermani P (ed) INFOCOM 2002. Twenty-first annual joint conference of the IEEE computer and communications societies. Proceedings. IEEE, vol 2, pp 608–617. https://doi.org/10.1109/infcom.2002.1019306
58. Ravasz E, Barabási AL (2003) Hierarchical organization in complex networks. Phys Rev E 67:026112. https://doi.org/10.1103/physreve.67.026112
59. Gao J, Li D, Havlin S (2014) From a single network to a network of networks. Natl Sci Rev 1(3):346–356. https://doi.org/10.1093/nsr/nwu020
60. Boccaletti S, Bianconi G, Criado R, del Genio CI, Gómez-Gardeñes J, Romance M, Sendiña Nadal I, Wang Z, Zanin M (2014) The structure and dynamics of multilayer networks. Phys Rep 544(1):1–122. https://doi.org/10.1016/j.physrep.2014.07.001
61. Kivelä M, Arenas A, Barthélemy M, Gleeson JP, Moreno Y, Porter MA (2014) Multilayer networks. J Complex Netw 2(3):203–271. https://doi.org/10.1093/comnet/cnu016
62. Rheinwalt A, Goswami B, Boers N, Heitzig J, Marwan N, Kurths J (2014) A network of networks approach to investigate the influence of sea surface temperature variability on monsoon systems. In: EGU general assembly conference abstracts, vol 16, p 8147. Springer. https://doi.org/10.1007/978-3-319-17220-0_3
63. Wiedermann M, Donges JF, Donner RV, Kurths J (2014) Ocean-atmosphere coupling from a climate network perspective. In: EGU general assembly conference abstracts, vol 16, p 11900. EGU
64. Ivanov PC, Rosenblum MG, Peng CK, Mietus J, Havlin S, Stanley HE, Goldberger AL (1996) Scaling behaviour of heartbeat intervals obtained by wavelet-based time-series analysis. Nature 383(6598):323–327 https://doi.org/10.1038/383323a0
65. Ivanov PC, Amaral LAN, Goldberger AL, Havlin S, Rosenblum MG, Struzik ZR, Stanley HE (1999) Multifractality in human heartbeat dynamics. Nature 399(6735):461–465. https://doi.org/10.1038/20924
66. Goldberger AL, Amaral LAN, Hausdorff JM, Ivanov PC, Peng CK, Stanley HE (2002) Fractal dynamics in physiology: alterations with disease and aging. Proc Natl Acad Sci 99(suppl_1):2466–2472. https://doi.org/10.1073/pnas.012579499
67. Ivanov PC, Nunes Amaral LA, Goldberger AL, Havlin S, Rosenblum MG, Stanley HE, Struzik ZR (2001) From 1/f noise to multifractal cascades in heartbeat dynamics. Chaos Interdiscip J Nonlinear Sci 11(3):641–652. https://doi.org/10.1063/1.1395631
68. Suki B, Alencar A, Frey U, Ivanov P, Majumdar A, Stanley H, Dawson C, Krenz G (2003) Fluctuations, noise and scaling in the cardio-pulmonary system. Fluct Noise Lett 15:1–25. https://doi.org/10.1142/S0219477503001142
69. Hu K, Ivanov PC, Chen Z, Hilton MF, Stanley H, Shea SA (2004) Non-random fluctuations and multi-scale dynamics regulation of human activity. Phys A Stat Mech Appl 337(1):307–318. https://doi.org/10.1016/j.physa.2004.01.042
70. Lehnertz K, Bröhl T, Rings T (2020) The human organism as an integrated interaction network: recent conceptual and methodological challenges. Front Physiol 11. https://doi.org/10.3389/fphys.2020.598694

Chapter 2
Music Psychology and Complex Systems

In this Chapter, the intricate interplay between music psychology and the field of complex systems is analyzed. Both fields focus on emergent patterns and dynamic interactions–concepts critical for decoding the underlying mechanisms of perception, cognition, and behavior.

Section 2.1 traces the parallel historical trajectories of these domains. From ancient philosophical explorations of musical harmony to the formalization of dynamical systems and network theory, this section reveals the intellectual underpinnings that connect the study of music with the broader framework of complex systems. Section 2.2 delves into the collaborative methodologies and conceptual frameworks uniting these fields. They showcase how insights from music psychology and complex systems inform each other, from the neural modeling of auditory perception to the hierarchical and nonlinear structures inherent in musical phenomena.

Music perception involves the dynamic processing of auditory stimuli, integrating psychophysics and psychoacoustics to understand how pitch, timbre, loudness, and rhythm interact. Rather than isolated attributes, these elements form an interconnected framework shaped by neurophysiological and cultural factors. Advances in psychoacoustics highlight the complexity of auditory cognition, emphasizing the need for holistic models that capture the interplay of perceptual dimensions in music processing. This topic is explored in greater depth in Sect. 2.3, which examines key psychoacoustic principles and their relevance to music perception research.

Section 2.4 presents a comprehensive overview of the foundational principles and fundamental concepts that define complex systems theory. It examines the intrinsic properties of complex systems, such as emergence, self-organization, nonlinearity, and multiscale interactions, highlighting their relevance across scientific disciplines. It introduces and discusses scenarios involving complex dynamics on networks, as well as the dynamics of complex networks themselves, framed within the context of mathematics as a universal language, demonstrating how mathematical formalism provides a unifying framework for modeling, analyzing, and interpreting the behavior

J. Sawicki, *Music Psychology—Balance of Relations*,
Current Research in Systematic Musicology 13,
https://doi.org/10.1007/978-3-031-97003-0_2

of complex systems across diverse domains, including neuroscience, physics, and computational modeling.

Section 2.5 examines the integration of computational tools and mathematical models with neuroscience and complex systems theory, enhancing our understanding of neural dynamics. The evolution of music neuroscience is explored, from early theoretical approaches to advanced neuroimaging techniques, which map neural networks involved in music processing. Additionally, the section highlights the intersection of computational neuroscience and artificial intelligence (AI), showcasing breakthroughs in machine learning (ML), connectomics, and neural modeling. The computational modeling of tonality is analyzed, emphasizing AI-driven frameworks, including neural networks, self-organizing maps, and Bayesian generative models, which integrate music theory, cognitive psychology, and machine learning.

Finally, Sect. 2.6 examines the auditory system as a complex network that transducers sound waves into neural signals, enabling sound localization, frequency discrimination, and higher-order processing. Sound vibrations are converted into electrical impulses by cochlear hair cells and relayed through the brainstem to the auditory cortex, where frequency organization and advanced processing occur. The system comprises ascending pathways that transmit auditory information and descending pathways that modulate cochlear sensitivity and perception, facilitating adaptive auditory processing, attention, and learning. A network-based approach in music psychology provides a transformative perspective, moving beyond traditional localization models by emphasizing the dynamic, interconnected nature of neural circuits. Network neuroscience reveals that music perception engages distributed networks spanning sensory, motor, cognitive, and emotional domains, offering a rigorous framework for understanding music cognition.

2.1 Common Roots of Music Psychology and Complex Systems

Music psychology and the theory of complex systems have evolved as interdisciplinary domains that explore patterns, dynamics, and emergent phenomena, albeit in different contexts. Their historical roots reveal shared intellectual developments, shaped by parallel advancements in scientific thought, mathematics, and philosophy.

The historical roots of music psychology can be traced to ancient civilizations, where philosophers like Pythagoras investigated the mathematical relationships between musical intervals, linking sound to harmonic order [1]. Similarly, Aristotle and Plato considered music's psychological and ethical impacts, foreshadowing inquiries into its effects on human cognition and emotion [2].

The formal theory of complex systems developed significantly later, but its foundational concepts can be traced back to nineteenth-century studies of dynamical systems and networks. These early explorations laid the groundwork for understanding how individual components interact to produce collective behavior. The essence of this idea, however, can be attributed to Aristotle's philosophical insight: "The whole is greater than the sum of its parts [3]." Henri Poincaré's work on chaos theory and Ludwig Boltzmann's statistical mechanics were pivotal in framing systems that exhibit unpredictable yet patterned behavior. These mathematical tools laid the groundwork for understanding emergent phenomena, akin to the way harmonic relationships underpinned early music theories.

During the nineteenth century, Wilhelm Wundt's establishment of experimental psychology marked the beginning of systematic inquiries into music and its psychological effects. His studies explored rhythm, pitch perception, and the cognitive processes underpinning musical experiences, bridging experimental methods and theoretical reflections [4].

In the same era, complex systems theory began to coalesce as scientists tackled problems in thermodynamics and biology. The concept of "self-organization", which later became a cornerstone of complex systems, mirrored emerging ideas in music psychology, such as the Gestalt principles of perception that underscored how humans organize sensory input into coherent patterns [5].

The twentieth century witnessed a consolidation of both fields. In music psychology, Leonard Meyer introduced the theory of expectation, emphasizing the probabilistic and emergent aspects of musical perception [6]. His work paralleled Shannon's information theory, which became influential in complex systems science by providing a mathematical basis for studying communication and unpredictability in networks [7].

Concurrently, the rise of cybernetics and systems theory, as advanced by Norbert Wiener [8] and Ludwig von Bertalanffy [9], inspired approaches to understanding interrelated elements in music cognition and performance. Music began to be seen not only as a sequence of notes but as a dynamic interaction of perception, memory, and emotional resonance–concepts deeply entwined with emergent behaviors in complex systems.

The latter half of the twentieth century saw a convergence of methodologies between music psychology and complex systems theory. Advances in computational modeling enabled both fields to explore emergent phenomena with greater precision. For instance, dynamical systems models in music psychology, such as Fred Lerdahl and Ray Jackendoff's Generative Theory of Tonal Music, demonstrated how hierarchical and nonlinear structures shape musical understanding [10]. These ideas resonate with complex systems' emphasis on feedback loops and multi-level interactions.

Simultaneously, complex systems theory expanded into social and cultural domains, analyzing phenomena like synchronization, collective behavior, and adaptive networks. These concepts overlap with research in music psychology on group dynamics in musical performance, audience interaction, and the role of culture in shaping musical preferences [11].

Both music psychology and complex systems theory focus on emergent properties–phenomena that arise from the interaction of components rather than from any single element. In music, this might manifest as the perception of harmony or rhythm from individual notes, while in complex systems, it appears in the form of flocking behavior in animals or phase transitions in physical systems.

The integration of computational models, neuroscience, and data analytics further underscores their shared evolution. For instance, network theory is now used to analyze the structure of musical compositions and the neural connectivity involved in music perception, demonstrating how both fields mutually benefit from interdisciplinary insights.

Summarizing, the historical trajectory of music psychology and complex systems theory reveals a profound interconnection. Rooted in ancient philosophical inquiries, both disciplines have developed through shared paradigms, such as the study of patterns, interactions, and emergent phenomena. As they continue to evolve, their convergence promises to deepen our understanding of complex phenomena, from the neural underpinnings of music to the self-organizing principles of dynamical systems.

2.2 Discovering Interdisciplinary Pathways

Interdisciplinary research has become a defining feature of modern science, enabling the integration of diverse methodologies and perspectives to address multifaceted phenomena. This monograph seeks to illuminate one such interdisciplinary pathway, exploring the interplay between music psychology and the theory of complex systems. By leveraging tools and insights from these traditionally distinct fields, the aim is to deepen the understanding of the mechanisms underlying music perception, cognition, and production. These fields share conceptual overlaps, emphasizing the interconnectedness, adaptability, and emergent behavior inherent in dynamic systems, making their convergence a fertile ground for scientific discovery.

Music psychology investigates how individuals perceive, process, and respond to musical stimuli, encompassing cognitive, emotional, and neural dimensions. Its scope extends from understanding the intricate auditory processes in the brain to the profound psychological impact of music on human experience. However, the complexity of music perception requires an interdisciplinary approach, one that transcends the traditional boundaries of psychology and neuroscience. Complex systems theory, with its focus on the dynamic interactions and emergent properties of interconnected networks, offers a powerful framework for addressing these challenges.

The principles of complex systems theory resonate with fundamental aspects of music psychology, such as pattern recognition, synchronization, and hierarchical structures. For instance, the perception of rhythm and harmony mirrors the self-organizing principles observed in ecosystems, neural networks, and other complex systems. Music perception involves the interplay of local phenomena, like auditory processing, and global phenomena, such as emotional resonance, reflecting the multiscale dynamics central to complex systems research.

This convergence has been further enabled by the rise of computational tools and data-driven methodologies. Computational neuroscience, which integrates biology, mathematics, and computer science, serves as a critical bridge between music psychology and complex systems. By employing neural models of synchronization, adaptation, and connectivity, researchers can simulate and analyze the dynamic processes underpinning music perception. Such models highlight the coherence and variability characteristic of musical experiences, revealing how principles of complex systems can elucidate the interwoven mechanisms of perception and cognition.

Large interconnected systems, such as neural networks, transportation systems, and social networks, provide real-world examples of multilayer networks with dynamic inter- and intra-layer interactions. Understanding the synchronization and desynchronization patterns within and between these layers can have profound implications, including for computational models of brain regions associated with music perception. Insights from these systems allow us to conceptualize the highly complex and interconnected nature of the human brain, where localized neural activity integrates into broader network dynamics that support music cognition.

This monograph aims to provide a comprehensive exploration of these interdisciplinary pathways. It reflects on the shared terminology and conceptual frameworks of adaptation and synchronization across disciplines, identifying common challenges and proposing future research directions. Through the collaborative efforts of musicologists, physicists, mathematicians, biologists, neuroscientists, and computer scientists, fundamental questions in music psychology are explored using the tools of complex systems theory.

Rather than solely identifying the spatial and temporal localization of music perception in the brain, this monograph focuses on elucidating the underlying mechanisms that govern "how" and "why" music is processed. Drawing on advances in complex systems research, which has successfully explained the seemingly chaotic dynamics of physical and biological phenomena, this work aims to bring similar clarity to the intricate and multifaceted processes of music perception. This interdisciplinary approach not only deepens the understanding of music as a psychological and neurological phenomenon but also contributes to the broader study of complex networks, with implications extending to artificial intelligence, sociology, and other fields. A network-based framework represents a paradigm shift from traditional localization models in music psychology, which attribute specific perceptual functions to discrete brain regions, toward a more integrative perspective. Music perception emerges from the dynamic interactions of the entire neural system rather than from isolated spatial or temporal localization within a neural network.

In doing so, this monograph underscores the significance of interdisciplinary integration in scientific inquiry, promoting a coherent and interconnected framework for understanding complex phenomena. By bridging the conceptual and methodological divides between disciplines, the goal is to enrich theoretical and applied science, paving the way for novel approaches to studying human cognition, behavior, and creativity through the lens of music psychology.

2.3 Psychoacoustics–Between Physics and Psychology

One of the primary areas of research in music psychology is music perception, which explores how individuals process and interpret musical stimuli. This field examines the emergent patterns and dynamic interactions that occur as the brain decodes complex auditory information, revealing the underlying mechanisms of perception. In this context, music psychology draws primarily on the results of psychophysics and psychoacoustics, with the integration of these two domains serving as a fundamental aspect of understanding the mechanisms underlying music perception.

Psychophysics investigates the relationship between physical stimuli and the sensations they evoke, employing quantitative methods such as threshold measurement and signal detection theory to analyze perceptual processes. Initially formalized by Fechner [12], psychophysics provided a systematic framework for linking the physical world to subjective experience, forming the foundation of experimental psychology. Contemporary psychophysical research extends to diverse applications, including sensory system modeling and data compression techniques, where auditory insensitivity to certain waveform differences is exploited for efficient encoding.

Psychoacoustics, a specialized branch of psychophysics, focuses on auditory perception, bridging objective acoustic properties with subjective auditory experiences. It examines the mechanisms by which the auditory system transforms physical sound waves into neural signals and perceptual phenomena. Human hearing operates within a frequency range of approximately 20 Hz to 20 kHz, with peak sensitivity between 1 and 5 kHz. Psychoacoustics plays a crucial role in technological advancements such as audio compression (audio coding formats, e.g. MP3 and AAC) and noise reduction by leveraging the nonlinear and selective characteristics of auditory processing.

The study of psychoacoustics often involves linking psychological perception to physical stimuli, using measurement scales that map perceptual attributes onto physical parameters. For instance, pitch perception is correlated with frequency, loudness with sound pressure level, and brightness with the spectral centroid. However, auditory perception is not merely a collection of isolated attributes; rather, it results from the interaction of multiple perceptual dimensions. Recent research highlights the interdependence of pitch, timbre, loudness, and rhythm in shaping auditory experience [13].

Theoretical and methodological advances in psychoacoustics emphasize the need for an integrated approach to sound perception. Periodicity and spectral harmonicity underpin pitch perception, while consonance and tonal organization reflect both auditory system function and the inherent structure of sound. Historical and cultural contexts further shape psychoacoustic phenomena, as seen in research on tone systems and scale formation. Additionally, binaural processing integrates these perceptual elements across multiple neural structures, from the inferior colliculus to cortical areas, demonstrating the complexity of auditory cognition.

Ultimately, the perception of musical sound arises from the dynamic interplay of temporal and spectral structures, reinforcing the necessity of holistic models in psychoacoustic research. Rather than treating pitch, timbre, loudness, and rhythm as independent parameters, their interrelation is central to understanding auditory perception. These elements collectively form the foundation of music perception, blending psychological and physiological principles to shape the diverse auditory experiences observed across cultures. The following sections focus on core psychoacoustic concepts relevant to the modeling of music perception, as explored in Chap. 5.

2.3.1 Pitch

Pitch is a fundamental perceptual attribute of sound, essential for music and speech, enabling the ordering of sounds along a continuum from low to high. Unlike a purely physical property, pitch is a psychological construct influenced primarily by the frequency content of a sound, with secondary contributions from factors such as sound pressure and waveform characteristics. Standards organizations (e.g., ASA 1960; ANSI 1994) formally define pitch as the perceptual attribute that allows the recognition of melodies, harmonies, and patterns in auditory stimuli. Pitch perception is inherently psychological, often described as "virtual pitch", reflecting its dependence on the spectral content of sound rather than a direct physical correlate.

The human auditory system perceives pitch across a frequency range of approximately 20 Hz to 4–5 kHz, facilitated by logarithmic compression that accommodates a broad spectrum of frequencies. Two key dimensions characterize pitch perception: pitch height, which correlates linearly with frequency, and pitch chroma, representing the cyclical nature of pitch classes (e.g., octaves). Shepard's pitch spiral exemplifies these components, with pitch height corresponding to the spiral's vertical axis and chroma represented by its circular path. Beyond ~4 kHz, pitch perception transitions into a sensation of "brightness", reflecting physiological constraints on auditory processing [14, 15].

Pitch perception is closely associated with auditory illusions and psychoacoustic phenomena. Shepard tones, for instance, demonstrate how pitch perception can be independent of absolute pitch height, creating the illusion of a continuously ascending or descending scale. Similarly, studies on categorical perception, tonal fusion, and the preference for stretched octaves challenge traditional periodicity-based models, underscoring the complexity of pitch perception beyond simple harmonic analysis.

Pitch sensation arises when environmental sounds exhibit repetitive waveform structures, with pitch corresponding to the waveform's repetition rate. The auditory system encodes pitch through intricate neural mechanisms, including phase-locking in auditory nerve fibers and the transformation of phase-locked activity into rate–place representations at higher auditory centers. The pitch of complex tones is processed through distinct mechanisms: resolved harmonics, which are spatially separated by the cochlea and represented tonotopically, and unresolved harmonics, which are processed temporally.

Pitch plays a crucial role in auditory scene analysis, aiding the segregation and grouping of sound sources based on fundamental frequency and harmonic relationships. These processes rely on both bottom-up mechanisms, such as cochlear responses, and top-down influences, including prior experience and acoustic context (see also Sect. 2.6.2). Neuroimaging studies suggest that pitch processing involves both temporal and frontal lobes, while computational models highlight the significance of synchrony-to-rate conversion in neural coding [16].

Despite considerable advancements, fundamental questions about pitch perception remain unresolved. Further progress requires interdisciplinary approaches, integrating psychophysics, neurophysiology, computational modeling, and neuroimaging. A deeper understanding of pitch perception holds broad implications for technological and clinical applications, including improved speech recognition systems, more effective auditory prosthetics such as cochlear implants and hearing aids, and enhanced models of auditory processing that refine our understanding of human hearing and music cognition.

2.3.2 *Timbre*

Timbre is a fundamental attribute of musical tones, defined by the spectral and temporal characteristics that distinguish sounds beyond pitch and loudness. It arises from the interaction of a sound's harmonic and inharmonic components, including the fundamental frequency, overtones, noise elements, and their temporal evolution. Among these, the attack behavior, referring to the initial spectral and amplitude changes within the first milliseconds of a tone, plays a crucial role in timbre perception. Without these early acoustic cues, identifying specific instruments becomes significantly more challenging. A defining aspect of timbre is the presence of formants, which are characteristic peaks and troughs in the overtone spectrum that remain largely stable across different pitches and are primarily determined by an instrument's construction.

Historically, timbre has been treated as secondary to pitch and rhythm, with traditional Western music theory emphasizing abstract musical structures over the sensory qualities of sound. Nineteenth-century theorists often regarded timbre as decorative rather than structural, favoring "neutral" timbres such as strings. Composers who foregrounded timbre, like Hector Berlioz, encountered skepticism. However, the advent of electronic and electroacoustic music in the twentieth century challenged these hierarchical assumptions, prompting a reassessment of timbre's role in musical structure and perception. Despite growing academic attention, timbre has often been relegated to specialized areas, such as contemporary or non-Western music. The Acoustical Society of America defines timbre as the residual attribute after pitch and loudness, reflecting both its complex nature and its resistance to straightforward categorization [17].

Timbre is an inherently multidimensional perceptual phenomenon, requiring interdisciplinary approaches for comprehensive understanding. It is often more intricate than pitch perception, with studies demonstrating that pitch changes tend to dominate perceptual judgments when pitch and timbre are altered simultaneously. This highlights their hierarchical yet interdependent relationship [15, 18]. Timbre perception involves both harmonic and inharmonic properties: harmonic sounds exhibit sinusoidal partials in harmonic relationships, whereas inharmonic sounds do not, with most natural sounds occupying a continuum between these extremes. Moreover, psychoacoustic phenomena such as virtual pitch illustrate how the brain infers a fundamental frequency even when it is weak or absent, emphasizing the complex cognitive mechanisms underlying timbre perception.

Timbre not only plays a central role in musical aesthetics but also shapes auditory scene analysis, influencing how listeners segregate and group sound sources based on spectral and temporal cues. The interplay between timbre and pitch is crucial for musical perception, affecting auditory stream segregation, instrument identification, and the perception of musical texture. Given its multifaceted nature, the study of timbre extends beyond psychoacoustics to encompass music production, room acoustics, and computational auditory modeling. A deeper understanding of timbre has implications for fields ranging from music cognition and audio engineering to hearing science, reinforcing its significance in both artistic and scientific contexts.

2.3.3 Loudness

Loudness is a perceptual correlate of sound intensity, but it cannot be directly derived from just-noticeable intensity variations. Instead, it must be quantified using methods such as magnitude estimation. The concept of loudness level, introduced by Barkhausen [19], is measured in psychoacoustic units like sone and phon, with equal-loudness contours used to map perceived intensity variations across frequencies. These contours illustrate the nonlinear nature of loudness perception, which follows a power-law relationship with sound intensity. Typically, a 10-dB increase in intensity results in a perceived doubling of loudness. However, this relationship steepens at lower intensities, where smaller increments in intensity lead to equivalent perceptual changes.

Masking effects significantly shape loudness perception. When a sound is partially masked by another sound, its perceived loudness is reduced, falling between its masked threshold and unmasked levels. This masking effect occurs in both spectral and temporal domains, depending on the presence and timing of the masker. Additionally, the directional dependence of sound perception plays a role, with differences in loudness perception between plane and diffuse sound fields necessitating adjustments in sound pressure level (SPL). The auditory system exhibits peak sensitivity in the frequency range of 2–5 kHz at higher intensities, which is reflected in characteristic dips in equal-loudness contours.

The A-weighted SPL scale provides an approximation of loudness for pure and narrow-band tones but falls short when applied to complex sounds. This limitation arises from factors such as bandwidth, frequency content, and duration, which contribute additional complexities to loudness perception. Partial masking, in which a sound's intensity perception lies between complete masking and its unmasked level, further complicates loudness measurement. This partial masking can occur due to simultaneous or temporally displaced maskers, highlighting both spectral and temporal influences on loudness perception.

Loudness perception is influenced by the interplay between auditory sensitivity, sound intensity, masking effects, and spatial considerations, making its quantification more intricate than a simple decibel measure [13]. Absolute thresholds of hearing (ATH) vary with frequency and age, shaping equal-loudness contours such as the Fletcher–Munson curves, which define human auditory sensitivity across different sound pressure levels. Although loudness is closely related to physical intensity, its perceptual characteristics exhibit nonlinearities driven by factors such as masking, spectral energy distribution, and temporal modulation. Standard loudness scales, such as phon and sone, may fail to fully capture the complexity of real-world auditory perception, particularly in the context of complex sounds such as music. For instance, the continued use of A-weighted SPL measurements for high-intensity music genres raises concerns about their adequacy in representing spectral energy distribution and their potential limitations in assessing auditory risk. The prevalence of pulse-like sound structures in modern music genres, including electronic and rock music, presents additional challenges in quantifying the auditory impact of these complex temporal and spectral interactions.

2.3.4 Rhythm

Rhythm, a fundamental temporal dimension of music, is perceived through divisive and additive structures. Divisive rhythms, characteristic of Western music, establish hierarchical beat organization, whereas additive rhythms construct cyclic patterns by combining long and short beats. Both types align with the short-term memory span of 3–5 s, facilitating perceptual coherence [15]. Neurophysiological research indicates that rhythmic processing engages motor and sensory cortical regions, with complex rhythms activating higher-order brain areas associated with pattern recognition. Emerging research also links rhythmic ability to language processing and identifies rhythm perception deficits in neurological disorders like Parkinson's disease [20]. Here, the therapeutic application of rhythm highlights its role in motor coordination, underscoring the intrinsic link between rhythm perception and movement. Despite advances in rhythm modeling, challenges remain in accounting for individual differences in beat perception and the variability of metric and grouping perception across listeners and tempi.

Rhythm perception encompasses figural and metric coding, where figural coding organizes rhythmic patterns into perceptual groups, and metric coding structures them into hierarchical frameworks of beat and meter [21]. While some rhythms naturally conform to a metrical grid, others depend on figural organization. Tempo, defined as the speed of musical progression, is closely tied to rhythm perception and is modeled through either interval-based frameworks, which encode durations as discrete memory representations, or entrainment models, which rely on neural oscillators adapting to temporal patterns.

Empirical studies indicate that perceptual and motor preferences for tempo emerge early and decelerate with age. Musicians tend to prefer slower tempi than nonmusicians, likely due to their training. Tempo discrimination studies reveal that human perception is most precise within an optimal range centered around 600 milliseconds, corresponding with findings on preferred tempo and absolute tempo memory.

Acoustic features such as intensity, duration, and frequency influence rhythmic grouping, while metrical structure enhances memory and reproduction, with metrically strong events being more salient. The role of melodic accents in metric coding remains unclear due to inconsistent findings, though the interaction between pitch and temporal cues is gaining research attention. Entrainment models, such as dynamic attending theory, provide insights into how rhythmic structure integrates with melodic elements.

Additionally, the temporal envelope of sounds plays a crucial role in rhythmic perception across speech and music [22]. While evenly spaced sound bursts with steep amplitude envelopes generate a uniform rhythm, gradual amplitude increases necessitate temporal adjustments for perceptual consistency. In speech, rhythm aligns with syllabic loudness peaks, whereas in music, rhythmic perception correlates with fluctuation strength, with dominant periodicities around 4 Hz. Furthermore, musical performance deviates from notated durations, as subjective rhythm emerges from perceptual rather than physical timing. Overall, rhythmic perception arises from complex interactions between amplitude modulation, temporal structure, and cognitive processing.

2.3.5 *Binaural Hearing and Sound Localization*

Binaural hearing enables the brain to determine the spatial origin of sound by processing interaural differences in time, intensity, and frequency. The auditory system relies on interaural time differences (ITDs) and interaural level differences (ILDs) as primary localization cues. ITDs, which arise from phase and group delays, are most effective for frequencies below 1 000 Hz, while ILDs become dominant above 1 500 Hz, increasing in magnitude with frequency. In the transitional frequency range between 1 000 Hz and 1 500 Hz, both cues contribute to localization accuracy. A critical factor in spatial hearing is the just-noticeable interaural delay, typically around 50 μs for frequencies below 1 500 Hz. This sensitivity supports binaural masking-level differences (BMLDs), in which variations in interaural phase or intensity enhance signal detection in noise [22].

The auditory system further refines spatial perception through spectral cues introduced by the outer ear. Acting as a direction-selective filter, the pinna imprints frequency-specific resonances based on the direction of the incoming sound. These spectral modifications facilitate precise localization, although perceptual errors such as front-back confusions can occur. Lateralization, the perceived shift of sound within the head when using headphones, is influenced by interaural time and level differences. Additionally, spatial hearing is shaped by psychoacoustic phenomena such as the Haas effect (law of the first wavefront), which describes the precedence of direct sound in localization perception [23]. Frequency-dependent localization also plays a crucial role in auditory perception and has significant implications for room acoustics, stereo reproduction, and spatial audio technologies [13].

Sound localization relies on precise neural computations occurring primarily in the superior olivary complex (SOC) and midbrain structures, where interaural timing and intensity differences are processed. Cortical and subcortical networks (see also Sects. 4.3.1.1 and 5.3.1) further contribute to selective auditory attention, enabling the "cocktail party effect", in which listeners can focus on a specific voice amid competing background noise. Bilateral connectivity enhances spatial hearing, while deficits in these pathways can impair sound localization and speech perception in complex acoustic environments.

2.4 Perspectives of Complex Systems

The concept of a complex system lacks a universally accepted definition. However, one practical approach to characterizing such systems lies in examining how many interrelated elements collectively give rise to emergent behaviors. These emergent phenomena cannot be solely attributed to the nonlinearity of individual components or the diversity and structure of their interactions. Instead, it is the interplay between the intrinsic properties of the elements and the overarching network of their interrelations that drives these phenomena. Key examples of emergent behaviors include spontaneous order, self-organization, and adaptive dynamics, which often defy straightforward reductionist explanations. Consequently, complex systems are often studied as either complex dynamics occurring on networks or the dynamics of complex networks themselves.

Complex systems are pervasive in natural and engineered domains, encompassing diverse examples such as ecosystems, social networks, biological systems, climate patterns, and even economic markets. These systems exhibit inherent nonlinearity, which is a defining feature that has captured the interest of researchers since the early twentieth century. This nonlinearity, combined with the intricate network of interactions, allows for the emergence of novel phenomena like synchronization, tipping points, and cascading failures.

Among the earliest and most intensively studied complex systems are nonlinear oscillators. Their significance became apparent in the context of technological advancements, such as the development of radio and radar systems during World War II, as well as laser technologies. These applications fueled the growth of chaos theory, which formalized the study of unpredictability and intricate behaviors in deterministic systems. By the mid-twentieth century, computational advances transformed the field, complementing traditional theoretical and experimental approaches with computational physics. This progress enabled the investigation of more sophisticated nonlinear models, paving the way for deeper insights into complex systems.

This section provides an introduction to the fundamental characteristics of oscillator models in nonlinear dynamics, outlining key concepts such as emergent behavior and the relationship between element-level dynamics and network properties. Readers interested in a deeper understanding are referred to foundational publications such as [24–27].

Complex systems serve as a unifying framework for studying phenomena in diverse disciplines, from physics and biology to sociology and engineering. At the heart of this framework are two central themes: the behavior of nonlinear systems and the role of networks as mediators of interactions. Nonlinear dynamics provides tools for understanding the often unpredictable and highly sensitive behaviors of individual system components, while network theory offers insights into the patterns and structures of their interconnections.

Many natural systems exhibit a blend of these characteristics. For instance, ecosystems are composed of nonlinear feedback loops (e.g., predator-prey dynamics) and network-like food webs. Similarly, in neuroscience, individual neurons display nonlinear activity while being embedded in a complex network of synaptic connections. These examples highlight the duality of complex systems as dynamics within networks and dynamics shaped by network topology.

The study of nonlinear dynamics has a rich history, beginning with the exploration of oscillatory systems in physics and extending into interdisciplinary domains. Early work focused on simple nonlinear oscillators, whose dynamics were described by differential equations. These studies laid the foundation for chaos theory, which investigates systems that are highly sensitive to initial conditions–a hallmark of nonlinearity.

The technological and scientific advances of the twentieth century catalyzed the study of nonlinear systems. During World War II, the development of radar systems brought about a deeper understanding of nonlinear feedback mechanisms. The subsequent invention of lasers highlighted the complex dynamics that arise in coupled systems. These breakthroughs emphasized the need for rigorous mathematical and computational tools to model and predict system behaviors.

The 1950s marked a turning point with the advent of digital computers, which allowed researchers to simulate and analyze nonlinear systems with unprecedented precision. This computational revolution expanded the scope of nonlinear dynamics, enabling the exploration of chaos, bifurcations, and higher-dimensional attractors. Today, these tools are integral to the study of complex systems, facilitating investigations into phenomena that defy analytical solutions.

This section introduces general concepts related to complex systems and their study through the lens of nonlinear dynamics and network theory. Section 2.4.1 delves into the mathematical tools of nonlinear dynamics, including chaos and bifurcation theory, which describe the transitions and stability of dynamical systems. In Sect. 2.4.2, an overview of network theory is provided, focusing on its role in capturing the structural and functional organization of complex systems. By integrating these perspectives, the aim is to offer a comprehensive foundation for understanding the intricate interplay between dynamics and structure that characterizes complex systems.

2.4.1 Differential Equations and Phase Space

Differential equations play a pivotal role in modeling dynamical systems across disciplines, not just in physics. As discussed in Sect. 1.3, they express the relationship between quantities and their rates of change. Since Newton's time, scientists have used differential equations to describe various natural phenomena, particularly deterministic time evolutions of physical states $\mathbf{x}$. These are expressed in general form as:

$$\dot{\mathbf{x}}(t) = \mathbf{f}\left(\mathbf{x}(t)\right), \tag{2.1}$$

where $\mathbf{x}$ represents an n-dimensional state vector, $\dot{\mathbf{x}}$ its time derivative, and $\mathbf{f}$ a function dictating the temporal evolution of the system's state. The advent of computational technologies has further elevated the significance of differential equations. As noted in Sect. 2.5, these technologies have made it feasible to solve complex equations numerically, a process termed *numerical integration*, which incrementally computes solutions using the differential quotient:

$$\dot{\mathbf{x}}(t) = \lim_{\Delta t \to 0} \frac{\mathbf{x}(t + \Delta t) - \mathbf{x}(t)}{\Delta t}. \tag{2.2}$$

By discretizing time into sufficiently small steps Δt, one iteratively maps the current state $\mathbf{x}(t)$ to the subsequent state $\mathbf{x}(t + \Delta t)$:

$$\mathbf{x}(t + \Delta t) = \mathbf{x}(t) + \mathbf{f}\left(\mathbf{x}(t)\right)\Delta t. \tag{2.3}$$

Numerical integration is particularly useful in chaos theory, where analytical solutions are often unattainable, and small variations in initial conditions can lead to vastly different outcomes. Such sensitivity is characteristic of nonlinear systems, where $\mathbf{f}$ depends nonlinearly on $\mathbf{x}$, e.g., through terms like $\mathbf{x}^k$ with powers $k > 1$.

The concept of phase space, developed in the nineteenth century, provides a geometric representation of all possible solutions of a differential equation. Each solution corresponds to a point in the n-dimensional space defined by the system's degrees of freedom. The temporal evolution of a solution traces a trajectory in this space. For example, consider the Newtonian equation of motion for a simple harmonic oscillator $\ddot{x} = -x$, which can be rewritten as a system of first-order differential equations with $\mathbf{x} = (x, \dot{x})^\top$. The phase space is then a two-dimensional plane, with x and $\dot{x}$ as axes, as illustrated in Fig. 2.1.

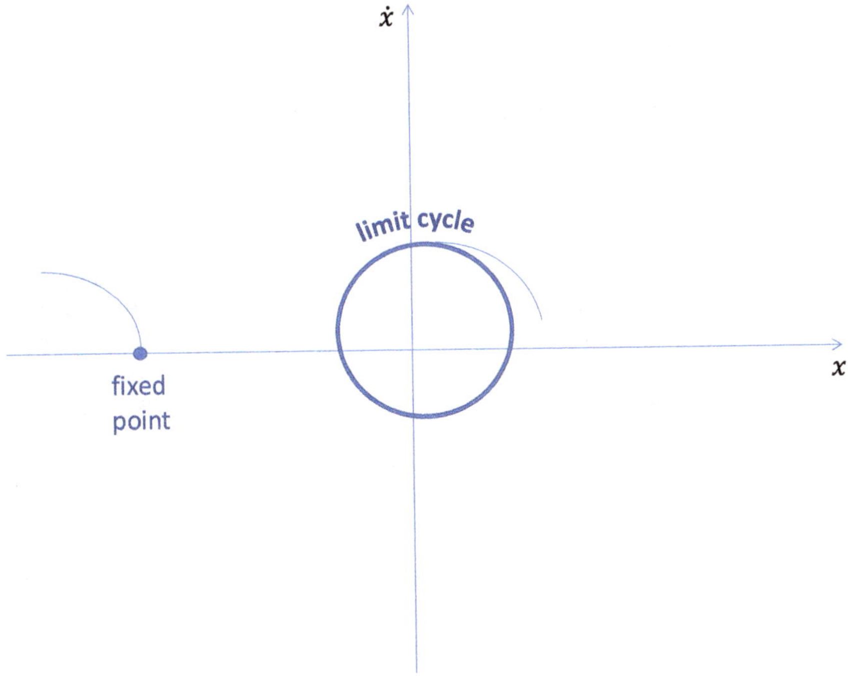

Fig. 2.1 Phase space as the figurative representation of dynamical systems: In the phase plane $(x, \dot{x})$ two invariant sets, a fixed point and a limit cycle, are shown. An exemplary trajectory is depicted in the neighborhood in case of a stable invariant set, respectively. Figure adapted from [28]

Even when analytical solutions are unavailable, certain key trajectories in phase space can provide critical insights. Fixed points, defined by:

$$\dot{\mathbf{x}}^*(t) = 0 = \mathbf{f}\left(\mathbf{x}^*\right), \tag{2.4}$$

are equilibrium solutions where the system remains stationary:

$$\dot{\mathbf{x}}^*(t) = 0 \quad \Rightarrow \quad \mathbf{x}^*(t) = \text{const.} \tag{2.5}$$

Another important type of trajectory is the limit cycle, a periodic solution forming a closed loop in phase space with period T_{LC}:

$$\mathbf{x}_{\text{LC}}(t) = \mathbf{x}_{\text{LC}}(t + T_{\text{LC}}). \tag{2.6}$$

Both fixed points and limit cycles, as shown in Fig. 2.1, are examples of *attractors*, which are subsets of phase space toward which trajectories converge. Attractors can be classified by dimensionality: fixed points are zero-dimensional, limit cycles are one-dimensional, and chaotic attractors, such as the Lorenz or Rössler attractors, are higher-dimensional. In contrast, *repellers* drive trajectories away.

The stability of trajectories, including attractors, can be analyzed using linear stability analysis. For a small perturbation $\boldsymbol{\delta x}(0) = \mathbf{x}(0) - \mathbf{x}^*$ around a fixed point, the time evolution of the perturbation is given by:

$$\dot{\boldsymbol{\delta x}}(t) = \boldsymbol{\delta x}(t)\mathbf{f}'(\mathbf{x}^*), \tag{2.7}$$

where $\mathbf{f}'$ is the Jacobian matrix. The eigenvalues of this matrix, particularly their real parts (Lyapunov exponents Λ), determine stability. A fixed point is stable if the largest Lyapunov exponent $\Lambda_{\max} < 0$ and unstable if $\Lambda_{\max} > 0$. The concept of Lyapunov exponents extends to periodic and chaotic trajectories [29]. For limit cycles the Floquet theory answers the question about stability. Limit cycles are periodic attractors.

Nonlinear systems often exhibit *bifurcations*, where qualitative changes in system behavior occur as a bifurcation parameter $\mathfrak{p}$ varies. Key bifurcations include:

Zero-eigenvalue bifurcation: This simplest local bifurcation occurs when a change of a system parameter $\mathfrak{p}$ causes a turn from positive to negative value of a single real eigenvalue. At the bifurcation point $\mathfrak{p}_0$ the eigenvalue is zero (center manifold). Examples for these bifurcations are the saddle-node, transcritical, super- and subcritical pitchfork bifurcation of fixed points.

Hopf bifurcation: This bifurcation occurs when a pair of complex-conjugate eigenvalues of a fixed point crosses the imaginary axis transversely. The fixed point changes from stable to unstable focus and a periodic attractor or limit cycle emerges at $\mathfrak{p}_0$ and its amplitude increases with $\sqrt{\mathfrak{p} - \mathfrak{p}_0}$. A distinction can be made between the supercritical and subcritical Hopf bifurcation.

Local bifurcation of limit cycles: These bifurcations transform periodic attractors. Examples for such bifurcations are limit-cycle bifurcation by condensation of paths (stable and unstable limit cycle around the same fixed point coalesce), period-doubling bifurcation of a limit cycle (limit cycle with period twice the period of an original cycle replaces the latter one) and secondary Hopf bifurcation of a limit cycle (limit cycle is replaced by a two-dimensional torus).

Global bifurcation of limit cycles: Global bifurcations involve global changes of the topological configuration, e.g., when "larger" invariant sets, such as limit cycles, collide with fixed points. Examples for such bifurcations are the omega explosion, homoclinic, heteroclinic, or saddle-node infinite-period (SNIPER) bifurcation.

Bifurcation of spatial patterns: Considering high-dimensional dynamical systems, e.g., nonuniform spatial or spatio-temporal dissipative structures, there exists a wide variety of more complex bifurcations; among them scenarios like the Turing instability.

Bifurcations are often visualized through bifurcation diagrams, which map the behavior of attractors as a function of the bifurcation parameter $\mathfrak{p}$. This monograph will detail examples of nonlinear oscillators and their bifurcation scenarios, highlighting their relevance across scientific fields. Generic models will be used to study the behavior of dynamical systems. The paradigmatic Kuramoto phase oscillator is a simple generic model and has been successfully applied in the modeling of adaptive networks and synchronization phenomena in a wide range of natural and technological systems. For oscillatory systems near the Hopf bifurcation, the Stuart-Landau oscillator acts as a representative model [30]. To represent nonlinear relaxation-type oscillators, which exhibit non-sinusoidal oscillations, the Van der Pol model will be employed. Notably, the Van der Pol oscillator was the first model in which synchronization was studied analytically [31, 32].

Additionally, neural systems are described by excitable models, which are classified based on their bifurcation properties into type-I and type-II excitability, both of which will be discussed in detail later. As an example of type-II excitability, the FitzHugh-Nagumo system [33, 34] will be explored in Chaps. 4 and 5.

2.4.2 Foundations of Network Theory

The previous section focused on differential equations and their dynamics, uncovering complex bifurcation behaviors, but was largely confined to single elements. In contrast, complex systems are characterized by interactions and relationships among multiple components. Therefore, this section shifts focus to the theoretical foundations of network theory, which provides a powerful framework for describing and analyzing these interactions.

At its core, network theory represents relationships between discrete entities, often referred to as nodes (or vertices) and links (or edges). Classical graph theory forms the foundation of this approach, where the objects are vertices and their connections are edges. The physical terminology of nodes and links is adopted here, as it better aligns with the broader applications of network theory in natural and engineered systems.

For a given network with N nodes, the structure and distribution of links between these nodes define the network topology. One standard way to represent this topology is through the $N \times N$ adjacency matrix $\mathbf{A}$:

$$\mathbf{A} = \{A_{ij}\} = \begin{pmatrix} A_{11} & \dots & A_{1N} \\ \vdots & \ddots & \vdots \\ A_{N1} & \dots & A_{NN} \end{pmatrix}, \tag{2.8}$$

where the elements of the matrix are given by:

$$A_{ij} = \begin{cases} 1, & \text{if there exists a link from node } j \text{ to node } i, \\ 0, & \text{if no link exists from node } j \text{ to node } i. \end{cases} \tag{2.9}$$

Network topology can either be *directed* or *undirected*. In directed networks, the existence of a link from node i to node j does not necessarily imply the presence of a reciprocal link from node j to node i. Conversely, undirected networks feature bidirectional links, meaning that $A_{ij} = A_{ji}$, and thus, the adjacency matrix is symmetric.

While the binary adjacency matrix defined above captures whether a link exists between nodes, many real-world networks require a representation of the strength of connections. This leads to the distinction between *unweighted* (binary) adjacency matrices and *weighted* adjacency matrices, where the value of A_{ij} reflects the strength of the link between nodes i and j. Weighted adjacency matrices are commonly referred to as coupling matrices in dynamical network studies. This monograph explores both weighted and unweighted types of adjacency matrices, emphasizing their applications to real-world systems.

The dynamics of complex systems can be modeled by extending the local dynamical equations of individual elements to include network interactions. Using the adjacency matrix, the state of each node $\mathbf{x}_i \in \mathbb{R}^n$ at time t evolves according to:

$$\dot{\mathbf{x}}_i = \underbrace{\mathbf{f}(\mathbf{x}_i)}_{\text{local dynamics}} \underbrace{+\sigma \sum_{j=1}^{N} A_{ij}\mathbf{H}[\mathbf{x}_j - \mathbf{x}_i],}_{\text{network interaction}} \tag{2.10}$$

where $\mathbf{f}(\mathbf{x}_i)$ represents the intrinsic dynamics of node i, $\mathbf{H}$ is the coupling scheme (an $n \times n$ matrix), σ is the coupling strength, and A_{ij} determines the network connectivity.

This framework captures both the local behaviors of individual nodes and their interactions through the network topology. The inclusion of the adjacency matrix allows for systematic exploration of how network structure influences the collective dynamics of the system. In Sect. 4.3.5, analytical tools grounded in the network-based approach, including the master stability equation, will be presented, especially in Sect. 4.3.5.4. The master stability equation, in particular, is employed to assess the stability of the synchronous state in a dynamical system composed of multiple identical subsystems which are coupled together.

In some cases, networks may feature multiple connections between the same pair of nodes, forming a multigraph. A classical example of this is Euler's bridge problem, depicted in Fig. 2.2, which cannot be fully represented using a simple adjacency matrix as defined above. To address such challenges, the concept of multilayer networks has been introduced. These frameworks allow for the representation of multiple types of links or connections within a single network, enabling more accurate modeling of complex systems that involve layered or multiplex interactions.

Network theory provides a versatile and robust toolkit for analyzing the interactions in complex systems. Its applications extend beyond the physical sciences to include fields such as biology, sociology, and engineering. By combining local dynamics with network interactions, researchers can study emergent behaviors, such

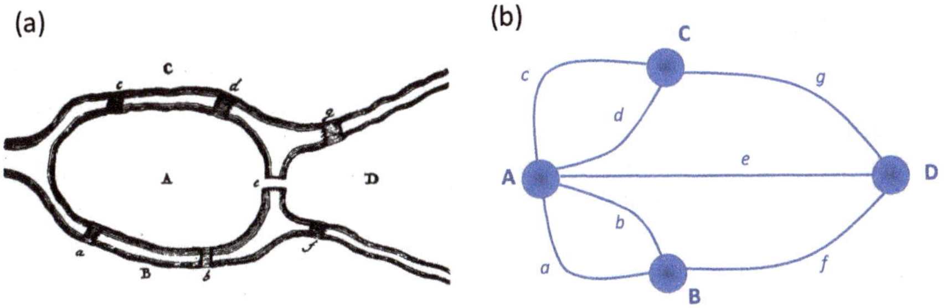

Fig. 2.2 The first graph in history: In 1736, Leonhard Euler solved a mathematical problem using graph theory, marking the inception of this field. He published a seminal paper titled *The Seven Bridges of Königsberg*, which introduced a mathematical framework for representing nodes A, B, C, D (depicted as circles) and links *a, b, c, d, e, f, g* (represented as lines), as illustrated in panel (**b**). The original problem, based on the geographical layout of Königsberg and its bridges, is shown in panel (**a**) [35]

as adaptation, synchronization, and pattern formation, which are hallmarks of complex systems. As such, the insights gained from network-based models are indispensable for understanding and addressing challenges in diverse real-world contexts, ranging from neuroscience to global communication systems.

2.5 Birth of Computational Neuroscience

Computational neuroscience, also referred to as theoretical or mathematical neuroscience [36], combines mathematics, computer science, and theoretical analysis to investigate the principles governing the development, structure, physiology, and cognitive functions of the nervous system [37, 38]. Through computational simulations, this field validates and solves mathematical models, often aligning with theoretical neuroscience, though the term "mathematical neuroscience" may highlight its quantitative nature.

Focusing on biologically plausible models of neurons and neural systems, computational neuroscience distinguishes itself from domains such as machine learning, artificial intelligence, and cybernetics, which frequently rely on biologically unrealistic frameworks. Despite this distinction, the field draws inspiration from these areas, fostering interdisciplinary innovation. Computational models operate across multiple spatial and temporal scales, from cellular and synaptic processes to higher-order psychological functions, enabling the formulation of testable hypotheses for biological and psychological experiments.

The term "computational neuroscience" was first introduced in 1985 during a conference dedicated to standardizing terminology within the emerging discipline [39]. Anchored in the convergence of neuroscience, mathematics, physics, and computer science, computational neuroscience exemplifies the broader movement toward interdisciplinary research. This integration has not only deepened the understanding of neural dynamics but has also established connections with related fields, including music psychology and the study of complex systems, further highlighting its transformative impact.

The origins of computational neuroscience trace back to efforts to formalize neural behavior using mathematical frameworks. Seminal work by Hodgkin and Huxley in the mid-twentieth century established a foundation for modeling neuronal activity through differential equations (see Sect. 4.3.3.1), marking the advent of a systematic connection between biological processes and theoretical constructs [40]. Subsequent developments expanded the scope to include network models and simulations of large-scale brain activity, enabling the study of phenomena such as oscillations, synchronization, and plasticity.

As the field matured, it incorporated principles from dynamical systems, network science, and statistical mechanics, fostering convergence with complex systems theory. This interdisciplinary evolution has positioned computational neuroscience as a transformative field, offering critical insights into the neural basis of perception, cognition, and behavior while bridging connections with other scientific domains.

Complex systems theory, which investigates the emergent behavior of interconnected components, shares conceptual and methodological parallels with computational neuroscience. Both fields emphasize the nonlinearity, adaptability, and self-organization inherent in their respective domains. The application of complex systems principles to neuroscience has illuminated patterns of synchronization and adaptation in neural networks, phenomena critical for understanding brain function [41, 42]. These concepts are equally significant in music psychology, where the perception of rhythm, melody, and harmony reflects the brain's capacity for dynamic organization and pattern recognition.

Music psychology, as an interdisciplinary domain, benefits from computational neuroscience's theoretical and empirical tools. By leveraging models of neural dynamics and connectivity, researchers can explore how music is processed, perceived, and produced by the brain. Synchronization and coherence in neural networks, for instance, are essential for understanding auditory perception and the neural underpinnings of musical engagement [43]. Moreover, the study of music often serves as a natural laboratory for examining complex systems, as musical structures inherently exhibit emergent properties and hierarchical organization.

The interplay between computational neuroscience, music psychology, and complex systems theory thus represents a fertile ground for advancing our understanding of both neural and cognitive processes. By integrating these domains, researchers can address fundamental questions about how the brain processes complex stimuli, such as music, while uncovering general principles of organization and adaptation in biological and artificial systems. This interdisciplinary perspective highlights the value of computational approaches in unraveling the intricate mechanisms underlying music perception and neural dynamics, bridging theoretical and applied research in neuroscience and psychology.

In the following, Sect. 2.5.1 examines the evolution of music neuroscience, tracing its development from early theoretical models to modern neuroimaging techniques that map neural networks involved in music processing. The section also addresses challenges in integrating findings across diverse methodologies. Section 2.5.2 explores the convergence of computational neuroscience and artificial intelligence (AI), highlighting advances in machine learning (ML), connectomics, and neural modeling. Lastly, Sect. 2.5.3 focuses on the computational modeling of tonality, emphasizing AI-driven frameworks such as neural networks, self-organizing maps, and Bayesian generative models, which bridge music theory, cognitive psychology, and machine learning (ML).

2.5.1 Neurodynamics of Music

The investigation of music and its neural correlates has evolved significantly, from early phrenological theories to sophisticated neuroimaging methodologies. Gall's identification of musical ability as a distinct mental faculty in 1791 set the stage for later explorations into the physiological and psychological underpinnings of musical cognition. By the late nineteenth century, neurologists such as Jackson and Knoblauch advanced this field by studying the neural substrates of music perception, with Knoblauch introducing the concept of amusia and proposing the existence of specialized cortical regions dedicated to music processing.

The twentieth century witnessed transformative developments in neurophysiological research. Early electroencephalography (EEG) studies in the 1920s facilitated investigations into musicogenic epilepsy and auditory processing, while the advent of positron emission tomography (PET) and functional magnetic resonance imaging (fMRI) in the 1980s allowed researchers to identify distinct neural activations associated with musical perception and motor functions. By the mid-1990s, fMRI became an essential tool for neuromusical research, providing unprecedented insights into how the brain processes music at both structural and functional levels. More recently, diffusion tensor imaging (DTI) has further refined our understanding of neural connectivity in music perception and cognition. Despite these advancements, the complexity of integrating findings across various methodologies and levels of analysis continues to pose challenges in fully elucidating the neural basis of music [43].

Musical perception is deeply intertwined with psychoacoustic, neurophysiological, and cognitive mechanisms. While tone perception is best understood within musical contexts, psychoacoustic studies using simplified stimuli have been instrumental in elucidating the relationships between acoustics, auditory physiology, and perception. Recent research integrating psychoacoustics, neurophysiology, and neuroimaging has provided deeper insights into how the brain encodes and represents pitch, timbre, and loudness, both for isolated tones and complex musical structures [44].

The cognitive processing of music involves the extraction of acoustic features, engaging neural pathways from the cochlea to the auditory cortex. These fundamental auditory processes support the formation of Gestalt structures, syntactic organization, and hierarchical tonal representations, all of which contribute to expectancy-based processing in music perception [45–52]. Structural violations within a musical sequence elicit distinct electrophysiological markers, such as early right anterior negativity (ERAN), which reflects deviations from anticipated harmonic patterns [53]. The emergence of emotional responses to music is closely linked to these expectancy mechanisms, engaging neural networks implicated in prediction, memory, and social cognition. Additionally, music perception and motor action are tightly coupled, with auditory predictions and motor planning relying on shared neural mechanisms involving forward modeling, efference copies, and sensory feedback integration.

Music neuroscience extends beyond perception and cognition to encompass broader cultural, clinical, and computational perspectives [54, 55]. Research underscores the interplay between biological predispositions and cultural influences in shaping musical perception, while computational models aim to quantify cultural variation in music cognition. Comparative studies have drawn parallels between human musical aesthetics and the complexity of birdsong, suggesting evolutionary underpinnings for music processing. Neuroimaging and network-based approaches have refined our understanding of auditory-motor and auditory-limbic interactions, highlighting the real-time dynamics of musical processing. Music engages extensive neural networks implicated in memory, executive function, attention, and emotion. While evidence suggests that musical experience enhances cognitive functions and emotional regulation, the causal mechanisms underlying these effects remain under debate. Neurochemical research has demonstrated that music modulates neurotransmitter and hormone levels, influencing mood and psychological well-being. Further, neuroaesthetics explores the subjective nature of musical preference and its underlying neural correlates, while investigations into the relationship between music and language reveal both overlapping and distinct neural mechanisms.

Musicianship exemplifies the profound effects of neuroplasticity, with practice-induced adaptations influencing both brain structure and function. Genetic studies suggest a heritable component to musical aptitude, while research on practice-related neuroplasticity has revealed both beneficial adaptations and maladaptive changes, such as musician's dystonia. Investigations into musical improvisation and auditory imagery have provided key insights into the neural basis of creativity and the internal representation of musical structures. Across the lifespan, music plays a crucial role in cognitive development, language acquisition, social bonding, and neurorehabilitation. Music-based interventions, particularly neurologic music therapy (NMT), have shown promise in addressing motor, speech, and cognitive disorders, though further research is required to optimize therapeutic outcomes. Studies of amusia and musical anhedonia offer additional perspectives on the neural mechanisms underlying music perception and affective engagement with music. As the field progresses, emerging research in connectivity modeling, genomics, and clinical applications is expected to further refine our understanding of music's impact on the brain, cognition, and health.

Music, as a high-level cognitive faculty, shares key similarities with language in its universality and learned nature across cultures. However, unlike language, music does not primarily reference external objects but rather consists of self-contained auditory structures. These ubiquitous yet internally coherent patterns suggest that music perception and behavior may be governed by fundamental principles of neural dynamics. Such principles could impose intrinsic constraints that facilitate the acquisition of musical knowledge, particularly during early development. Recent studies propose that core dimensions of musical experience–including pitch, tonality, and rhythm–can be directly attributed to the intrinsic oscillatory dynamics of the nervous system.

The neural resonance framework posits that music perception emerges not through explicit computational inference but through the brain's natural resonance with musical structures [56]. Exposure to music induces dynamic neural activity patterns governed by oscillatory principles, rhythmic bursting, and synchrony, linking physiological mechanisms with perceptual experiences. A central tenet of this model is nonlinear resonance, which provides an alternative to traditional linear approaches in auditory processing. Unlike Fourier-based models, which rely on statistical learning and spectral decomposition, nonlinear resonance accounts for cochlear resonance, cortical entrainment, and tonal hierarchy, suggesting that pitch perception and tonal stability arise from intrinsic neural properties rather than learned associations. This framework also extends to rhythm and meter perception, proposing that endogenous neural oscillations entrain to external auditory stimuli, thereby shaping beat perception and integrating auditory and motor functions.

This neurodynamic approach to music cognition stands in contrast to computational models of language, which often struggle to reconcile cognitive and neurobiological mechanisms. Whereas linguistic theories necessitate abstract intermediaries to bridge the gap between syntactic structures and neural circuits, music perception appears to be more directly governed by neurodynamic principles. The auditory system's inherent nonlinear properties align closely with dynamical systems theory, providing a robust physiological foundation for musical cognition. In this view, the qualitative experience of music emerges naturally from the interaction of sound with the brain's intrinsic oscillatory and resonant properties, offering a unified theoretical framework for understanding musical perception.

The relationship between language and music has been examined through the lenses of resource-sharing and modularity frameworks [57]. The resource-sharing model posits that while music and language maintain distinct representations, they rely on overlapping neural networks, particularly in syntax and motor-related processes. Conversely, the modularity framework highlights their partial independence, suggesting that specialized neural mechanisms differentiate speech and music processing. Interactions between cortical and subcortical structures mediate both shared and domain-specific processes, with auditory experience shaping subcortical plasticity. While both language and music serve communicative functions, their fundamental structures and processing mechanisms diverge–language conveys propositional meaning, whereas music is primarily defined by rhythmic, tonal, and affective dimensions. A network-based approach may offer a more comprehensive integration of these frameworks, advancing our understanding of the neural mechanisms that govern both domains.

2.5.2 Artificial Intelligence and Neuroscience

Computational neuroscience is a rapidly evolving discipline shaped by advancements in experimental methods and data collection. Moreover, computational neuroscience serves as a critical bridge between artificial intelligence (AI), modeling, and data analysis, evolving alongside advancements in experimental techniques that enable the collection of multi-modal data [58]. Initially focused on single neurons, the field has expanded to explore neural circuits' processing of sensory stimuli and motor outputs, delving into complex concepts such as low-dimensional manifolds and neural attractors [59]. Breakthroughs in connectomics and machine learning have significantly advanced our understanding of neural activity, connectivity, and behavior, yet challenges remain in translating molecular insights into learning models and addressing the complexity of learning rules [60]. Computational models are fundamental to unraveling the brain's complexity, offering valuable insights into synaptic plasticity and signaling networks. With the rapid expansion of neuroscience data, sophisticated data analysis tools have become indispensable for extracting meaningful insights [61].

Artificial intelligence (AI) is a multidisciplinary field of computer science dedicated to the automation of intelligent behavior and the development of machines capable of learning and decision-making. The concept of AI, difficult to define precisely due to the lack of a universal definition of intelligence itself, was formally introduced by John McCarthy in 1955. Broadly, AI represents the simulation of human cognitive functions, enabling machines to perceive their environment, process information, and take actions to achieve specific goals. AI systems achieve this through various methodologies, including pre-programmed rules or self-learning algorithms such as machine learning and deep learning.

The evolution of artificial intelligence (AI) has been closely linked to advancements in computing power and data availability. Early theoretical foundations, such as Alan Turing's pioneering work on machine self-awareness and the introduction of the Turing Test, laid the groundwork for the development of intelligent machines [62]. Recent studies show that large language models like GPT-4 and GPT-4.5 (Generative Pre-trained Transformer) can pass Turing Test variants, with GPT-4.5 even outperforming human baselines in some setups [63, 64]. However, success often depends on superficial cues, not true understanding. Critics argue that the Turing Test oversimplifies intelligence, and humans themselves can perform poorly in such evaluations [65]. This challenges its relevance as a measure of general intelligence. AI encompasses diverse subfields, including knowledge representation, reasoning, natural language processing, perception, and robotics. While general artificial intelligence capable of performing any human task remains a long-term goal, contemporary AI applications often focus on specific tasks such as autonomous vehicles, recommendation systems, virtual assistants, and pattern recognition in large datasets.

Machine learning (ML), a core subset of Artificial intelligence (AI), enables algorithms to learn from data without explicitly programmed solutions. ML algorithms build models from training data and generalize these models to make predictions or decisions about new, unseen data. Supervised learning, where models are trained on labeled datasets, and unsupervised learning, which identifies patterns and structures within unlabeled data, are prominent approaches. Reinforcement learning, in which agents learn optimal actions through feedback from rewards, represents another important paradigm. Neural networks, inspired by biological neurons, serve as a fundamental architecture for many ML models, particularly in the context of deep learning.

Deep learning, an advanced branch of machine learning (ML), utilizes multilayered artificial neural networks to process complex data patterns. These networks, often referred to as deep neural networks, excel in extracting hierarchical features from large datasets. Applications of deep learning include image and speech recognition, natural language processing, and autonomous systems. Unlike traditional ML, deep learning models can autonomously refine their internal parameters in response to external data, enabling highly accurate predictions and decision-making across diverse domains.

Artificial intelligence (AI) research has advanced rapidly, particularly with breakthroughs in deep learning and transformer architectures since 2017. This progress has driven widespread adoption but also raised concerns regarding ethical implications, unintended consequences, and the necessity of regulatory frameworks for safe deployment. As AI systems become increasingly sophisticated, interdisciplinary insights from psychology, neuroscience, economics, and sociology are essential for addressing the complexities of AI's societal integration. While AI enhances fields such as medical diagnostics, it functions as an assistive tool rather than a replacement for human expertise.

Persistent challenges related to reliability, ethics, regulation, and accessibility must be addressed, with key concerns including decision-making transparency, computational efficiency, responsible content moderation, legal uncertainties, data bias, and misinformation. Additionally, AI's role in creativity [66] and human-like interactions presents ethical dilemmas, necessitating a balance between innovation and regulation to align AI with societal values. Despite strong task performance, recent studies indicate that AI models often lack coherent internal representations, making them vulnerable in novel contexts [67]. More rigorous evaluation methods are required to ensure that AI systems genuinely capture the underlying structure of the domains they model.

2.5.3 Computational Modeling of Tonality

In 1957, the first computer-generated music, Guttman's "The Silver Scale", and the first computer-composed score, "The Illiac Suite", marked the beginnings of algorithmic composition, utilizing stochastic models and rule-based filtering. Unlike handcrafted methods, modern machine learning, particularly deep learning, learns directly from large datasets, enabling music generation across diverse genres without predefined rules. Deep learning's ability to process raw data and extract hierarchical features makes it robust and adaptable to complex tasks, outperforming traditional rule-based systems. Recent research has introduced a multi-criteria framework–objective, representation, architecture, challenge, and strategy–to evaluate approaches in deep learning-based music generation, providing insights for advancing this field [68].

Composing by means of artificial intelligence (AI) is closely related to the topic of modeling tonality. AI-driven composition leverages models of tonality to produce coherent music. Tonality refers to the structured organization of pitches and chords, where certain tones are perceived as more stable or central, typically around a tonic. This principle underlies much of Western art music from the seventeenth to nineteenth centuries and remains fundamental in popular music. The sense of expectancy and resolution in tonal music is thought to contribute to its expressive and emotional qualities. Moreover, tonal hierarchies are fundamental to music across cultures and historical periods, characterized by the prominence of certain tones that function as reference pitches. These tones are structurally emphasized, frequently repeated, and rhythmically accentuated, yet their specific hierarchical organization varies across musical traditions. This variability suggests that tonal hierarchies are not solely dictated by invariant acoustic properties, such as the harmonic overtone structure, but are instead shaped by cognitive processes that rely on general psychological principles found in other perceptual and cognitive domains.

A proposed framework for tonal hierarchies comprises three key propositions [69]: their psychological reality, their status as musical facts, and their abstraction through statistical learning. Empirical evidence supports the cognitive representation of tonal hierarchies, influencing perception, memory, and expectation based on musical exposure. Structurally, tonal hierarchies manifest in musical compositions, with statistical analyses confirming that higher-ranked tones occur more frequently and have longer durations. Listeners internalize these hierarchies by detecting statistical patterns, a process validated through experimental and computational studies. From a cognitive perspective, tonal hierarchies function as perceptual reference points, yet they are uniquely musical due to their context-dependent nature. Furthermore, sensitivity to tonal distributions facilitates the learning of musical structures. The investigation of tonality spans interdisciplinary approaches linking mathematical and computational modeling with human perception and cognition on the one side and music theory on the other side. The following discussion highlights key computational approaches to modeling tonality, emphasizing the role of artificial intelligence (AI) and machine learning (ML) in advancing this field.

Numerous computational models have been developed to explain tonal hierarchy and have been evaluated against psychological data. These models differ in their underlying assumptions regarding the processes involved in tonal hierarchy perception and the levels at which these processes operate. One class of models focuses on psychoacoustic mechanisms, such as Parncutt's model [70], which is based on the concept of pitch salience derived from Terhardt's pitch perception theory [71]. This approach posits the existence of virtual pitches—perceptual constructs that emerge through interactions between existing spectral components. The model effectively predicts tonal hierarchies observed in experimental contexts, with strong correlations to probe-tone profiles, suggesting that tonal hierarchy perception extends beyond simple tone distribution effects. An extension of this model, incorporating sensory memory decay, further improves its predictive capacity.

Neural network models provide additional insights into tonal processing, incorporating theories like Terhardt's subharmonic pitch extraction and Goldstein's harmonic analysis [72, 73]. These models use multi-layer architectures with an input layer, a hidden layer, and an output layer coding different chord types. Training involves adjusting connection weights based on input stimuli, with parameters controlling the rate and extent of learning. Another approach integrates both psychoacoustic and cognitive processes, as exemplified by Leman's model, which includes virtual pitch processing alongside a short-term memory component to account for temporal integration and decay [74]. When applied to tonal hierarchy experiments, this model produces results resembling probe-tone profiles, suggesting that immediate perceptual context significantly influences tonal perception. However, methodological limitations, such as the use of composite rather than individual harmonic contexts, highlight discrepancies between model predictions and empirical data, particularly regarding scale context representations.

Neural network models of tonality, inspired by biological neural systems, have been instrumental in simulating the perception of chords, tonal structures, and pitch [75, 76]. Many of these models employ self-organizing maps (SOMs), unsupervised machine learning systems based on Kohonen maps [77]. SOMs create low-dimensional representations of high-dimensional datasets, preserving their topological structure [78]. This process enables clustering and visualization of similar patterns, such as tonal relationships, in a two-dimensional grid. Training occurs through competitive learning rather than error correction, using weight adjustments that respond to input patterns. After training, the map can classify new inputs by identifying the closest match in the weight space. Another dynamic approach involves self-organizing neural networks, mapping input data onto neural network units and forming structured tonal representations over time. A three-layer interconnected system–comprising tones, chords, and keys–trained on harmonic sequences successfully replicates experimental results, capturing the evolving sense of key in chord sequences and reflecting tonal similarity judgments [79]. These findings suggest that the perception of tonal hierarchy may emerge from the internalization of statistical regularities in musical exposure.

Expanding on these concepts, adaptive resonance theory (ART) models the interaction of bottom-up and top-down processes in neural networks [80]. ART systems consist of connected neurons that adapt weights during training phases, leading to resonance states representing learned or recognized patterns [81]. This framework has been applied to various musical phenomena, including tonality, pitch, melody, and texture [82]. Additionally, several computational approaches explore different facets of tonal hierarchy, such as the Spiral Array approach [83, 84], behavioral modeling [85], connectionist methods [86], and algorithmic approaches grounded in cognitive science [87]. Further works on tonality are collated in [88, 89].

More recent advancements incorporate Bayesian generative frameworks for modeling pitch-class distributions (PCDs) in music, such as the Tonal Diffusion Model (TDM) [90]. Unlike conventional template-based models, TDM integrates geometric and algebraic structures from music theory with cognitive insights, modeling tone generation as a sequence of intervallic steps on the so-called Tonnetz, a graphical representation of tonal pitch space. Bayesian generative models, a class of machine learning algorithms, use Bayesian statistics to generate data and update beliefs based on prior knowledge and new observations. These models excel in handling uncertainty, are robust with limited data, and adapt continuously as new data is introduced. The Prime Factor-based Generalized Tonnetz (PFG Tonnetz) further extends the traditional Tonnetz using Prime Factor Representation, facilitating tonal organization in polyphonic music [91]. Additionally, a model for tonal tension profiling in chord progressions was introduced using the Tonal Interval Space (TIS), which quantifies instantaneous tonal tension based on dissonance, tonal distance, voice leading, and hierarchical tension, creating a continuous tension curve [92]. This approach provides a generalized framework for tonal analysis and aligns music theory with cognitive psychology, building upon Lerdahl's model [93, 94].

Collectively, these computational models highlight different psychological mechanisms underlying tonal hierarchy perception, including psychoacoustic interactions, memory-based processing, and learning from harmonic structures over time [15]. The integration of artificial intelligence (AI)-driven computational methods serves as both a key frontier and a dynamic bridge in the study and modeling of tonal perception and cognition within the framework of music theory.

2.6 Network Structures in the Auditory System

The auditory system functions as a highly specialized and dynamically interconnected network, enabling the perception, processing, and interpretation of sound. It comprises peripheral auditory structures responsible for capturing and conducting sound stimuli, the cochlea, where spectral decomposition and signal transduction occur, and the auditory nerve, which relays neural signals to the central auditory nervous system. Within this network, ascending and descending pathways interact to refine auditory perception, integrate sensory inputs, and facilitate adaptive responses to environmental auditory cues. These pathways exhibit complex feedback mechanisms that modulate auditory gain, filter relevant stimuli, and dynamically adjust auditory processing based on contextual demands.

Music perception and production rely on this intricate auditory network, operating within a nonlinear and multistable framework that encompasses brain dynamics, human interaction, and the interplay between individuals and musical instruments–whether acoustic, electronic, or digital. The neural processes underlying music-related functions extend beyond localized brain regions, engaging a distributed network that integrates auditory, motor, emotional, and cognitive domains. Self-organization, synchronization, and strong time dependence characterize these networks, which encompass not only brain-body dynamics and instrumental performance but also broader social, cultural, and historical contexts.

A comprehensive understanding of these complex musical phenomena requires models and analytical frameworks that account for the multidimensional nature of auditory network interactions. Traditional descriptive approaches are insufficient to capture the emergent properties of music perception and production, necessitating the integration of perceptual, ecological, sociological, ethnological, and historical perspectives. By embedding these dynamic processes within a network-based framework, predictive accuracy in auditory and musical phenomena can be improved, providing insights into the physiological and interactive mechanisms underlying sound processing.

Music perception and performance engage multisensory and motor networks, inducing neuroplastic changes that enhance cognitive and motor functions [55]. These processes influence both cortical and subcortical structures, integrating sensory inputs and engaging neuromodulatory systems such as dopamine and serotonin, which contribute to emotional and reward-related aspects of musical engagement. Especially, music performance as a dynamic interaction between musicians, music, and listeners, provides a valuable framework for studying network phenomena. These processes shape the cognitive, emotional, and social aspects of musical expression. In Sect. 3.1.5, interdisciplinary research at the intersection of music psychology and complex systems explores the interconnected stages of preparation, execution, and reception, offering insights into musical creativity and collective engagement.

Long-term musical training induces structural adaptations in both white and gray matter, particularly when initiated early, fostering enhanced motor coordination, emotional regulation, and cognitive flexibility. The therapeutic potential of music-based interventions, such as melodic intonation therapy and music-supported rehabilitation, has been demonstrated in facilitating neuroplastic recovery following neurological impairments, underscoring the clinical significance of auditory network research [55].

Despite significant progress in understanding the neural coding and network interactions within the auditory system, many aspects of descending auditory pathways remain unresolved. The role of these pathways in modulating auditory plasticity, learning, and selective attention requires further investigation, particularly regarding their influence on subcortical structures and their integration into broader cognitive and motor networks. Section 2.6.1 examines the structural organization of the auditory system. The auditory system converts sound into neural signals through the ear, with the cochlea breaking down frequencies. Feedback mechanisms refine auditory perception, supporting communication and environmental awareness. Section 2.6.2 demonstrates that the auditory system uses interconnected ascending and descending pathways to process and refine sound. Ascending pathways transmit sound from the ear to the brain, while descending pathways modulate this information. These loop structures enhance perception and attention and play a key role in hearing and auditory processing.

A network-based approach in music psychology shifts from traditional localization models to an integrated perspective, emphasizing dynamic neural interactions. Network neuroscience reveals that music perception engages distributed brain networks, integrating sensory, motor, cognitive, and emotional processes. Section

2.6.3 explores the potential of network neuroscience in music psychology and its applications. By leveraging network models, researchers can systematically examine how music influences brain connectivity, providing insights into both fundamental cognitive mechanisms and therapeutic interventions.

2.6.1 Auditory System

The auditory system processes sound through a highly sophisticated and interconnected network, overcoming significant computational challenges. All auditory input consists of variations in air pressure that move the eardrum in a unidimensional manner. These pressure changes, originating from multiple simultaneous sources, become entangled before reaching the ear, necessitating the brain's ability to separate and interpret them. Al Bregman's metaphor of observing water movement in narrow channels without direct access to the sea illustrates this complexity, highlighting how our perception must infer large-scale structure from limited, local information. Despite these challenges, the auditory system efficiently extracts information regarding sound sources, locations, and movements.

The auditory system comprises anatomically distinct yet functionally integrated components that enable the perception of sound. Peripheral auditory structures–including the outer, middle, and inner ear shown in Fig. 2.3–capture and conduct auditory stimuli. Within the cochlea, spectral decomposition and signal transduction convert mechanical energy into neural signals, which are then transmitted via the auditory nerve to the central auditory nervous system for higher-order processing. This intricate system supports essential auditory functions such as communication, environmental awareness, sound localization, and speech comprehension.

Sound enters through the external ear, where the pinna collects and directs sound waves into the external auditory canal. The pinna's anatomical features, including the helix, antihelix, tragus, antitragus, and concha, enhance directional hearing and sound amplification. Upon reaching the tympanic membrane, sound waves induce vibrations, converting airborne acoustic energy into mechanical motion. These vibrations propagate through the ossicular chain in the middle ear, consisting of the malleus, incus, and stapes. The stapes, the smallest bone in the human body, transmits vibrations to the oval window, converting mechanical energy into hydraulic waves within the cochlear fluids. The middle ear also contains the tensor tympani and stapedius muscles, which regulate ossicular motion to prevent excessive auditory stimulation and potential damage.

Within the cochlea, auditory vibrations induce perilymph movement in the scala vestibuli, propagating pressure waves toward the cochlear apex (helicotrema) before descending through the scala tympani. These fluid dynamics displace the basilar membrane, where sound is spectrally decomposed based on frequency. The cochlea's tonotopic organization ensures that high-frequency sounds are detected at the basal region, while low-frequency sounds are processed at the apical end. The organ of

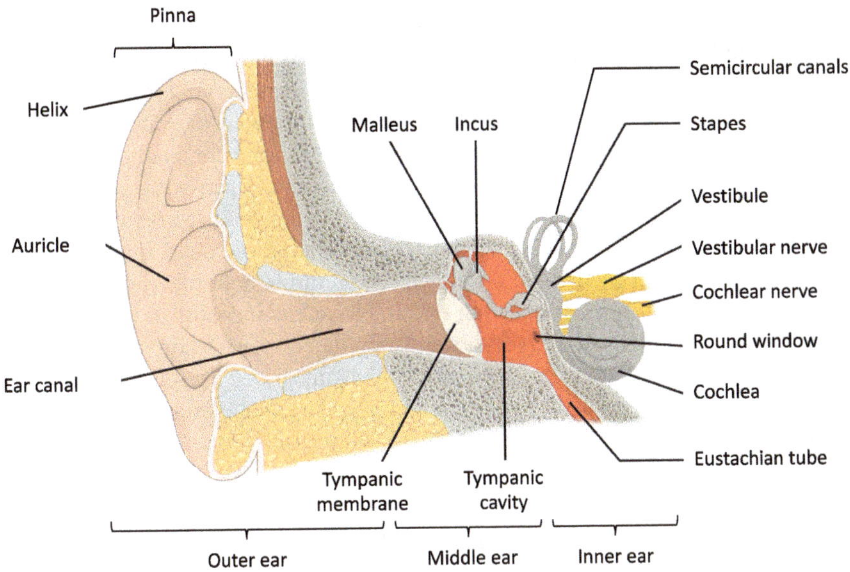

Fig. 2.3 Anatomy of the ear: The peripheral auditory system can be separated into three sections of outer, middle, and inner ear. The outer ear comprises the pinna with the helix, which functions as a sound-collecting structure, the external auditory canal, and the tympanic membrane, which transduces sound waves into mechanical vibrations. The middle ear contains the ossicular chain, consisting of the malleus, incus, and stapes, which amplify and transmit vibrations to the inner ear. The inner ear includes the cochlea, a spiral-shaped organ responsible for auditory signal processing, and the semicircular canals, which contribute to vestibular function by detecting head movements and maintaining balance. Figure adapted from OpenStax (CC BY 4.0)

Corti, situated atop the basilar membrane, contains mechanosensory hair cells responsible for converting mechanical energy into neural signals. Inner hair cells function as primary transducers, encoding auditory stimuli into action potentials, while outer hair cells amplify basilar membrane displacement through active motility, enhancing auditory sensitivity and frequency resolution. Stereocilia on hair cells deflect upon membrane movement, triggering mechanotransduction channels that allow potassium influx, ultimately leading to neurotransmitter release and excitation of the auditory nerve.

The auditory nerve, composed of fibers originating from the cochlear ganglia, conveys these neural signals to the brainstem's cochlear nuclei, marking the first site of auditory processing. Each auditory nerve fiber bifurcates upon entry, enabling parallel processing within the dorsal, anteroventral, and posteroventral cochlear nuclei. From here, projections extend to the superior olivary complex, where binaural integration facilitates spatial localization through interaural time and intensity differences (see Sect. 2.3). Ascending fibers then travel via the lateral lemniscus to the inferior colliculus, a midbrain structure responsible for integrating spatial and temporal auditory information. The medial geniculate nucleus of the thalamus further refines these signals before relaying them to the primary auditory cortex in the superior temporal gyrus, which is tonotopically organized to map frequency-specific auditory input.

The auditory nervous system follows a hierarchical processing pathway, beginning with vibratory motion transduction by hair cells in the cochlea and signal relay via the eighth cranial nerve to the cochlear nucleus. From there, auditory information is processed through subcortical nuclei, including the superior olivary complex, inferior colliculus, and medial geniculate body of the thalamus, before reaching the auditory cortex. Tonotopy, the systematic representation of sound frequency, is a fundamental organizing principle that is preserved throughout the ascending auditory pathway. Two primary processing streams have been identified: the lemniscal (primary) system, which delivers fast, frequency-tuned input to core cortical areas, and the non-lemniscal (auxiliary) system, which integrates broader auditory and multisensory information across widespread cortical and subcortical regions.

The auditory cortex exhibits a hierarchical and parallel processing organization, consisting of core, belt, and parabelt regions. The core areas, located on Heschl's gyrus, receive direct input from the medial geniculate body and exhibit sharp frequency tuning, which is essential for processing fine acoustic details such as musical pitch. The belt and parabelt regions, characterized by weaker tonotopy, contribute to higher-order auditory integration and multisensory processing. Functional imaging and lesion studies indicate that core auditory areas play a crucial role in precise frequency discrimination, particularly in the right hemisphere, where enhanced functional connectivity may contribute to pitch processing. Moreover, descending projections from the cortex to subcortical structures, which outnumber ascending fibers, suggest the presence of dynamic feedback mechanisms that refine auditory perception. These hierarchical and bidirectional pathways collectively enable the neural encoding of complex auditory features, fundamental for speech and music perception.

Higher-order auditory processing engages cortical and subcortical networks involved in language comprehension, speech production, and auditory memory. Wernicke's area, located in the superior temporal lobe, is essential for speech decoding, while Broca's area, situated in the frontal lobe, governs speech production. These regions are interconnected via the arcuate fasciculus, a key white matter tract that facilitates coherent language processing. Additionally, the frontal and parietal lobes contribute to the integration of auditory signals with cognitive and motor functions, refining auditory perception and response.

Through its complex organization and functional adaptability, the auditory system enables the detection, interpretation, and contextualization of sound. Frequency discrimination, spatial localization, and higher-order cognition collectively support essential human functions, including speech communication and environmental awareness.

2.6.2 Auditory Feedback Mechanisms and Loops

The auditory system exhibits a preserved tonotopic organization across its ascending and descending (afferent end efferent) pathways, with neurons tuned to specific frequencies at multiple levels, from cochlear hair cells to cortical processing centers. Frequency analysis is facilitated by phase-locking at lower frequencies and intensity coding via firing rate and population recruitment. The auditory cortex, particularly the superior temporal gyrus, integrates complex sound features such as speech and music, while also interacting with higher cognitive processes. Overall, the auditory system functions as a dynamic network of ascending [95] and descending [96] pathways that process and modulate auditory information. These pathways are described in the literature using a variety of equivalent terms, including ascending/descending, afferent/efferent, bottom-up/top-down, and ear-to-brain/brain-to-ear. While each pair emphasizes slightly different perspectives—whether anatomical, functional, or directional—they all refer to the same bidirectional flow of auditory information and modulation. In the following, we use these terms interchangeably, depending on context and emphasis.

The vestibulocochlear nerve consists of two components as shown in Fig. 2.3: the cochlear nerve, which mediates hearing, and the vestibular nerve, which regulates equilibrium. Cochlear nerve fibers originate from bipolar neurons in the spiral ganglion, extending peripheral processes to hair cells and central projections to the cochlear nucleus [97]. The auditory system is structured into ascending and descending pathways [98, 99], forming a bidirectional network (see Sect. 2.4.2) that facilitates complex auditory processing. Ascending pathways relay auditory signals from the cochlea to the brainstem nuclei, including the cochlear nucleus (CN), superior olivary complex (SOC), nucleus of lateral lemniscus (NLL), inferior colliculus (IC), and medial geniculate nucleus (MGN) of the thalamus, ultimately projecting to the auditory cortex (AC). Descending projections originate from the AC and extend to subcortical structures, including the CN, SOC, IC, and even the cochlea, forming intricate feedback loops that regulate auditory sensitivity and response. A schematic diagram of this auditory network is depicted in Fig. 2.4a, whereas the corresponding auditory neural pathways are shown in Fig. 2.4b.

Ascending pathways transmit sensory information from the periphery to the brain. Auditory signals travel through the vestibulocochlear nerve and first synapse in the cochlear nucleus of the superior medulla, where input is received exclusively from the ipsilateral cochlea. The superior olivary nucleus serves as the first binaural integration site, processing interaural time and intensity differences to localize sound sources. The inferior colliculus (IC) in the midbrain acts as a major integration center, receiving inputs from both ascending auditory projections and descending corticofugal pathways, as well as non-auditory sensory inputs. The medial geniculate nucleus of the thalamus further processes auditory information before relaying it to the auditory cortex, where higher-order perception and interpretation occur. This hierarchical, parallel-processing structure allows for robust spatial hearing, binaural integration, and sound discrimination.

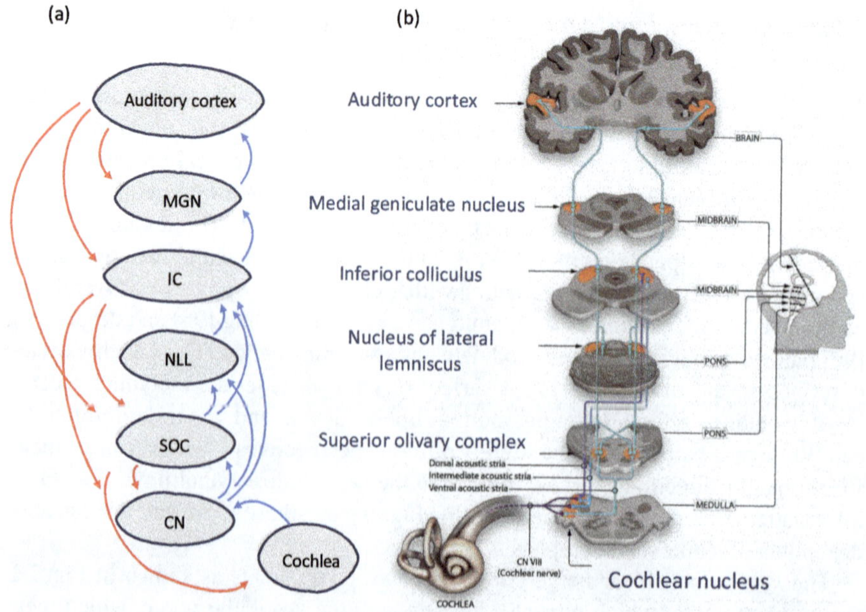

Fig. 2.4 Schematic diagram of the auditory system between the choclea and the auditory cortex: **a** The neural network consists of interconnected ascending (afferent) and descending (efferent) pathways, represented by blue and red arrows, respectively. These pathways relay auditory signals from the cochlea to the brainstem nuclei, including the cochlear nucleus (CN), superior olivary complex (SOC), nucleus of lateral lemniscus (NLL), inferior colliculus (IC), and medial geniculate nucleus (MGN) of the thalamus, ultimately projecting to the auditory cortex. A simplified model highlights key corticofugal and subcortical circuits within this system of feedback loops. Information for this panel taken from [100, 101]. **b** Schematic figure of the auditory neural pathway starting at the cochlear nucleus, then the superior olivary complex, then the inferior colliculus, and finally the medial geniculate nucleus. The information is decoded and integrated by each relay nucleus in the pathway and finally projected to the auditory cortex. Panel provided by Jonathan E. Peelle (CC BY 4.0)

Descending pathways play a crucial role in modulating auditory processing by dynamically controlling sensitivity to auditory stimuli. The efferent system regulates cochlear function and central auditory processing by modulating the synapses between medial olivocochlear (MOC) neurons and outer hair cells (OHCs). This system allows for adaptive regulation of cochlear gain, auditory attention, and signal-to-noise enhancement [100, 102, 103]. Emerging evidence highlights extensive corticofugal projections, which form a complex efferent network influencing auditory perception, plasticity, and cognitive functions such as selective attention. Descending pathways interact with neuromodulatory systems, including serotonergic, dopaminergic, and noradrenergic circuits, suggesting a broader role in auditory learning and behavioral adaptability. Despite advances in understanding these pathways, questions remain regarding their precise synaptic organization, neurotransmitter roles, and functional integration with intrinsic auditory circuits.

Descending circuits, including the corticofugal system, exert top-down modulation of auditory processing through bidirectional feedback loops. The olivocochlear bundle, originating from the superior olivary complex (SOC), regulates cochlear sensitivity via inhibitory mechanisms mediated by acetylcholine. Descending projections from the AC extend to subcortical structures, forming both localized feedback loops and extended multisynaptic chains. Feedback loops allow cortical and thalamic structures to regulate their own inputs, while descending chains provide hierarchical modulation of auditory processing at multiple levels. These circuits enhance auditory perception, facilitate selective attention, and contribute to experience-dependent plasticity.

Afferent and efferent pathways form interdependent networks within the auditory system, with efferent circuits being a key focus of contemporary research [104–107]. The auditory efferent system originates in the AC and extends to the medial geniculate nucleus (MGN), inferior colliculus (IC), cochlear nucleus (CN), and superior olivary complex (SOC), ultimately reaching the cochlea via olivocochlear fibers. This network operates through afferent-efferent feedback loops, enabling bidirectional communication between cortical and subcortical structures. Recent studies suggest that corticofugal influences on auditory processing occur through three parallel pathways connecting the AC to olivocochlear neurons [101].

The interaction between ascending and descending pathways forms an intricate system of neuronal feedback loops. Feedback mechanisms, such as cortico-thalamic circuits, provide insights into the cognitive roles of descending pathways, including selective attention and adaptive auditory processing. Descending chains serve as anatomical substrates for higher-order effects to influence cochlear function, contributing to intensity encoding, dynamic range adaptation, and auditory scene analysis. These mechanisms are crucial for auditory perception in noisy environments, where descending modulation enhances signal detection and suppresses irrelevant sounds.

A notable example of a feedback mechanism is observed in the cochlea: Descending pathways refine auditory perception by regulating cochlear responses, modulating outer hair cell (OHC) electromotility [108–110], and influencing synaptic plasticity. OHCs enhance sound sensitivity and frequency tuning through electromotility, a rapid length change in response to membrane potential variations, thereby generating mechanical force to modulate basilar membrane motion. Unlike inner hair cells, OHCs receive sparse afferent input but are regulated by the medial olivocochlear efferent system, which adjusts their activity to control cochlear sensitivity [111].

Another remarkable example of a feedback mechanism in the human auditory system is the residual neural control of pinna orientation. Although largely vestigial, minimal ear muscle activation persists. Evolutionary transitions to diurnal lifestyles led to reduced ear mobility and the loss of associated reflexes. However, the neural pathways remain adapted for more mobile pinnae, which may account for unexpected muscle activations during auditory attention [112]. Recent research [113] suggests that superior auricular muscle (SAM) activity could serve as an indicator of effortful listening, representing an attempt to modify pinna shape to enhance auditory processing.

The auditory system is characterized by extensive descending projections that influence processing at multiple levels. These pathways modulate thalamic, midbrain, and brainstem structures and exert functional control even at the cochlear level. The auditory cortico-collicular (CC) system, linking the AC to the auditory midbrain, mediates diverse functions, including predictive coding, attention regulation, and plasticity in response to peripheral damage [114]. Cortico-collicular (CC) neurons contribute to high-level cognitive processes, acoustico-motor behaviors, and experience-dependent auditory adaptation. Advances in circuit-mapping techniques have clarified the role of descending projections in these processes, demonstrating their critical involvement in auditory learning and perception.

In summary, the auditory system operates as an interconnected network of ascending and descending pathways, with feedback mechanisms and loops playing a fundamental role in refining sensory processing. These circuits enable dynamic adjustments to auditory perception in response to environmental and cognitive demands, supporting functions such as selective attention, auditory learning, and neuroplasticity. Research on the interactions between descending pathways and intrinsic auditory circuits enhances our understanding of how the brain integrates top-down control with sensory input. Network neuroscience (see next Sect. 2.6.3) provides a transformative framework for studying these interactions, offering novel insights into cognition, auditory perception, and clinical applications in hearing disorders.

2.6.3 *Network Neuroscience and Perception*

In many studies, perceptual mechanisms and the associated processes are attributed to specific brain regions, or attempts are made to assign functions to individual neural systems. This localization approach to perception may be influenced by the traditional view that sensory processing follows a unidirectional flow from peripheral receptors to cortical regions, which might be conceptually linked to the uniform motion of sound waves. This linear perspective is also reflected in the conventional representation of auditory pathways, such as the schematic diagram of the auditory system (see Fig. 2.4a). Typically, ascending (afferent) and descending (efferent) pathways define the route between the cochlea and the auditory cortex. From a physiological standpoint, such an approach can outline important connections relevant to specific perceptual mechanisms and functions.

Recent studies have examined the distinct roles of the ventral and dorsal auditory streams, which divide the human brain along the vertical axis and are both essential for music perception, albeit serving different functions [115]: The ventral stream is primarily responsible for the retention, integration, and long-term encoding of auditory information, supporting the perception of musical structures and the formation of predictive models based on past auditory experiences. It plays a crucial role in working memory, statistical learning, and expectation-based perception, facilitating the binding of sound elements and the generation of auditory representations. Disruptions in this system, as observed in cases of amusia, highlight its significance in auditory cognition.

The dorsal stream, in contrast, serves as the sensory-motor interface, enabling precise coordination between auditory and motor processes. This pathway facilitates the transformation of sound into motor actions, which is crucial for rhythm perception, synchronization in ensemble performance, and temporal precision. Moreover, it supports cognitive functions such as working memory and the manipulation of musical elements. The dorsal stream also contributes to abstract auditory transformations and musical creativity, allowing for tasks like transposition, covert rehearsal, and prediction.

Thus, while the ventral stream is more involved in auditory representation and predictive processing, the dorsal stream plays a key role in music production and the real-time, dynamic processing of musical information.

Another important aspect of auditory processing is hemispheric specialization, which plays a crucial role in cognitive functions, with distinct responsibilities assigned to the left and right hemispheres for speech and music [115]: The left hemisphere, traditionally associated with language processing, exhibits greater sensitivity to temporal resolution, making it particularly adept at detecting rapid acoustic changes, as commonly found in speech. The right hemisphere, in contrast, specializes in spectral resolution, which is essential for perceiving fine frequency variations important for melodic and harmonic processing in music. These differences are driven by local neural connectivity and broader network organization, with the left hemisphere characterized by more focused connections and the right hemisphere displaying more diffuse integration across both hemispheres.

Memory processes also play a vital role in music perception, requiring the perceptual system to generate an internal representation of musical structure. This involves encoding individual musical events (e.g., pitch, timbre, loudness, see also Sect. 2.3), grouping events into higher-order structures (e.g., chords, voices, and phrases), and recognizing relationships that allow the perception of larger musical forms and themes. Such representations are stored across various timescales, from short-lived echoic memory to long-term storage, including schematic knowledge of musical styles and episodic memories of musical experiences [116]. These stored representations create top-down expectations that shape ongoing perception and emotional responses to music. Moreover, early exposure to musical styles and explicit musical training can significantly modify perceptual abilities, further illustrating the dynamic nature of music cognition.

This complex perceptual interplay among neural pathways, cortical and subcortical regions, recognition, memory, emotions, and motor functions challenges the strict (spatial or temporal) localization of perceptual processes. Debates over resource-sharing versus modularity frameworks [57], as discussed in Sect. 2.5.1, highlight these challenges, with experimental evidence supporting both perspectives.

The approach advocated in this monograph is based on network theory (see in detail the Sects. 5.4.1 and 5.4.6), which becomes evident when reinterpreting Fig. 2.4a

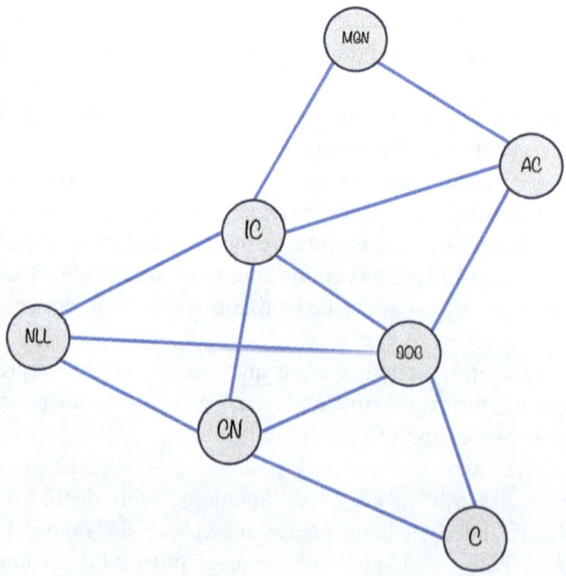

Fig. 2.5 Network of the auditory system between the choclea and the auditory cortex represented by the adjacency matrix **A** in Eq. 2.11: The neural network consists of connections (pathways) between the cochlea (C), cochlear nucleus (CN), superior olivary complex (SOC), nucleus of lateral lemniscus (NLL), inferior colliculus (IC), medial geniculate nucleus (MGN) of the thalamus, and auditory cortex (AC). Every nodes position is determined by a so called force atlas [118], i.e. (dis-)connected nodes are (repelled) attracted to one another. Same information as in Fig. 2.4a

in an alternative manner: Figure 2.5 presents the same information but explicitly visualizes the underlying network of nodes and links. The average clustering coefficient C–a measure in network science providing insight into the network's local cohesion and structure–for this network is approximately $C = 0.667$ [117]. This indicates a relatively high tendency for nodes to form interconnected clusters and may be characterized by a wide range of dynamic activities. The corresponding adjacency matrix **A** in Fig. 2.5 is given by

$$\mathbf{A} = \{A_{ij}\} = \begin{pmatrix} 0\,1\,1\,0\,1\,0\,0 \\ 1\,0\,0\,0\,0\,0\,0 \\ 0\,1\,0\,0\,1\,1\,0 \\ 0\,0\,1\,0\,0\,0\,0 \\ 0\,0\,1\,1\,0\,1\,1 \\ 0\,0\,1\,1\,1\,0\,0 \\ 0\,0\,0\,0\,0\,1\,0 \end{pmatrix}, \tag{2.11}$$

and–along with the high clustering coefficient–indicates that the auditory network exhibits characteristics of a mesh-like topology rather than a purely linear chain configuration.

As outlined in Sect. 2.4.2, network theory analyzes relationships between discrete entities represented as nodes (vertices) and connections (edges), with classical graph theory forming its foundation. This approach captures both the behavior of individual nodes and their interactions within the network topology. The adjacency matrix allows for a systematic examination of how structural organization influences collective dynamics. Analytical tools such as the master stability equation assess the stability of synchronized states in coupled subsystems. In networks with multiple connections between nodes (multigraphs), multilayer networks provide a more accurate representation of complex systems (see Chap. 3).

Applying network theory to the auditory system reveals it as an interconnected structure of ascending and descending pathways, with feedback mechanisms playing a crucial role in refining sensory processing. These circuits enable dynamic adjustments in auditory perception based on environmental and cognitive demands, supporting selective attention, auditory learning, and neuroplasticity. Unlike traditional localization models, network neuroscience in the framework of complex systems offers a transformative perspective on cognition and auditory perception by analyzing interdependent neural interactions.

The concept of network neuroscience is well established, with graph theory and network-based analyses significantly advancing the understanding of music perception [119]. These methods offer a statistically rigorous framework for investigating complex neural interactions. Early studies [120] applied network science to musical experiences, and subsequent research using electroencephalography (EEG), diffusion tensor imaging (DTI), diffusion spectrum imaging (DSI), and functional magnetic resonance imaging (fMRI) has demonstrated that both musical training and listening can modulate large-scale brain networks. Findings indicate that music shapes functional connectivity within the default mode network, enhances information transfer across cortical hubs, and engages regions associated with self-referential thought and emotional processing.

However, this monograph highlights that the full potential of network neuroscience still remains underexplored and underappreciated in music psychology. By drawing on examples such as the successful modeling of epileptic seizures in Sect. 4.5, innovative therapeutic approaches for Parkinson's disease in Sect. 4.1.1.3, the emergence of network physiology [121, 122] in Sect. 1.4.2, and its applications in predicting tumor progression and sepsis in Sect. 4.2, this work underscores the vast and transformative potential of complex systems and a network-based approach.

A crucial aspect of network theory is its emphasis on dynamics within and across networks, which manifest in dynamic activity patterns. In general, each system component (node) can attain multiple internal states that evolve over time. At any given moment, the system's state is represented by an activity pattern across the network, and its dynamics are described by the temporal evolution of this pattern. As illustrated in Fig. 2.5, network nodes are interconnected, and a node's activity often depends on its coupling with other nodes. These coupling effects, though sometimes small and difficult to measure, can significantly influence network activity. Additionally, the activation of one node can correlate with the inhibition of another. In Sect. 3.3, so-called pacemakers are introduced, which can drive specific activity patterns within a network. In Sect. 4.4, structural asymmetries between brain hemispheres are discussed, which contribute to distinct dynamic activity patterns and heightened activity in specific nodes. Thus, rather than focusing solely on individual network components, it is often more insightful to examine their interactions.

The integration of network neuroscience into music research has expanded the scope of experimental paradigms beyond traditional activation-based studies, providing a more comprehensive understanding of the brain's dynamic response to music. Changes in network topology induced by music listening have been linked to neuroplasticity, emotional regulation, and potential therapeutic applications, demonstrating the relevance of these methods for neurorehabilitation. By leveraging graph theory, researchers can assess local and global brain connectivity, systematically exploring music's influence on neural function.

This intricate network of auditory structures, pathways, and processing centers enables the detection, interpretation, and contextualization of sound. Through frequency discrimination, spatial localization, and higher-order cognition, the auditory system supports essential functions, from speech communication to environmental awareness. This monograph adopts a network-based modeling approach, emphasizing the balance of relations within complex auditory processing. The applications of this approach are diverse, particularly in music perception (see Chap. 5) and music performance research (see Sect. 3.1.5).

References

1. Gerson LP (2006) Aristotle and other platonists, 1st print., Cornell Paperbacks edn. Cornell University Press, Ithaca, NY U.A
2. Bowman WD (1998) Philosophical perspectives on music. Oxford University Press, New York. https://doi.org/10.1093/oso/9780195112086.001.0001
3. Aristotle (1957) Metaphysica, Vol 7, 17, 1041b. Oxford University Press. https://doi.org/10.1093/actrade/9780198145134.book.1
4. Boring EG (1950) A history of experimental psychology, 2nd edn. Appleton-Century-Crofts, East Norwalk, CT, US
5. Köhler W (1929) Gestalt psychology. Gestalt psychology, Liveright, Oxford, England
6. Meyer L (1956) Emotion and meaning in music. ACLS Humanities E-Book, University of Chicago Press. https://doi.org/10.7312/meye91434
7. Shannon CE (1948) A mathematical theory of communication. Bell Syst Tech J 27:379. https://doi.org/10.1002/j.1538-7305.1948.tb01338.x
8. Wiener N (1948) Cybernetics; or control and communication in the animal and the machine. MIT Press
9. von Bertalanffy L (2003) General system theory: foundations, development, applications. International library of systems theory and philosophy, G. Braziller
10. Lerdahl F, Jackendoff RS (1996) A generative theory of tonal music. The MIT Press. https://doi.org/10.7551/mitpress/12513.001.0001
11. Barabási AL (2002) Linked: the new science of networks. Perseus Books Group, New York
12. Fechner GT (1860) Elemente der Psychophysik. Breitkopf und Härtel, Leipzig. https://doi.org/10.3931/e-rara-10879
13. Bader R (2018) Springer handbook of systematic musicology, 1st edn. Springer Handbooks, Springer, Berlin Heidelberg, Berlin, Heidelberg. https://doi.org/10.1007/978-3-662-55004-5

14. Schneider A, von Ruschkowski A (2011) Systematic musicology: empirical and theoretical studies. Peter Lang Verlag, Berlin, Deutschland. https://doi.org/10.3726/978-3-653-01290-3
15. Bader R (2013) Nonlinearities and synchronization in musical acoustics and music psychology. Springer, Berlin. https://doi.org/10.1007/978-3-642-29014-4
16. Oxenham AJ (2019) Pitch: perception and neural coding. The MIT Press, Cambridge, MA, US, pp 3–32. https://doi.org/10.7551/mitpress/9780262039402.003.0001
17. Dolan E, Rehding A (2021) The Oxford handbook of timbre. Oxford handbooks. Oxford University Press. https://doi.org/10.1093/oxfordhb/9780198807835.001.0001
18. Iverson P, Krumhansl CL (1993) Isolating the dynamic attributes of musical timbre. J Acoust Soc Am 94(5):2595–2603. https://doi.org/10.1121/1.407371
19. Barkhausen H (1926) Ein neuer Schallmesser für die Praxis. Zeitschrift für Technische Physik 7:599–601
20. Grahn JA, Brett M (2009) Impairment of beat-based rhythm discrimination in parkinson's disease. Cortex 45(1):54–61. https://doi.org/10.1016/j.cortex.2008.01.005
21. McAuley JD (2010) Tempo and rhythm. Springer handbook of auditory research. Springer Science, New York, NY, US, pp 165–199. https://doi.org/10.1007/978-1-4419-6114-3_6
22. Fastl H, Zwicker E (2007) Psychoacoustics: facts and models. Springer series in information sciences. Springer Berlin Heidelberg. https://doi.org/10.1007/978-3-540-71048-3
23. Haas H (1949) Über den Einfluß eines Einfachechos auf die Hörsamkeit von Sprache. PhD thesis, Technische Hochschule Braunschweig
24. Strogatz SH (1994) Nonlinear dynamics and chaos, 1st edn. Perseus Books, Cambridge, MA
25. Schöll E (2001) Nonlinear spatio-temporal dynamics and chaos in semiconductors. Cambridge University Press, Cambridge
26. Ott E (2002) Chaos in dynamical systems. Cambridge University Press, Cambridge
27. Schöll E, Schuster HG (eds) (2008) Handbook of chaos control. Wiley-VCH, Weinheim
28. Sawicki J (2019) Delay controlled partial synchronization in complex networks. Springer Theses, Springer, Heidelberg. https://doi.org/10.1007/978-3-030-34076-6
29. Pikovsky A, Politi A (2016) Lyapunov exponents: a tool to explore complex dynamics. Cambridge University Press. https://doi.org/10.1017/CBO9781316416604
30. Kuramoto Y (1984) Chemical oscillations. Waves and turbulence, Springer, Berlin
31. van der Pol B (1920) A theory of the amplitude of free and forced triode vibrations. Rad Rev 1:701
32. van der Pol B (1926) On relaxation oscillations. Philos Mag 2:978–992
33. FitzHugh R (1961) Impulses and physiological states in theoretical models of nerve membrane. Biophys J 1:445–466. https://doi.org/10.1016/s0006-3495(61)86902-6
34. Nagumo J, Arimoto S, Yoshizawa S (1962) An active pulse transmission line simulating nerve axon. Proc IRE 50:2061–2070
35. Euler L (1741) Solutio problematis ad geometriam situs pertinentis. Commentarii academiae scientiarum Petropolitanae 8:128–140
36. Dayan P, Abbott LF (2005) Theoretical neuroscience: computational and mathematical modeling of neural systems. J Cogn Neurosci 15:154
37. Schwartz E (1993) Computational neuroscience. Bradford Books, MIT Press
38. Trappenberg TP (2002) Fundamentals of computational neuroscience. Oxford University Press UK. https://doi.org/10.1093/acprof:oso/9780195142781.001.0001
39. Schwartz EL (1990) Computational neuroscience. MIT Press, Cambridge, MA, USA
40. Hodgkin AL, Huxley AF (1952) A quantitative description of membrane current and its application to conduction and excitation in nerve. J Physiol 117:500–544. https://doi.org/10.1113/jphysiol.1952.sp004764
41. Pikovsky A, Rosenblum M, Kurths J (2001) Synchronization: a universal concept in nonlinear sciences, 1st edn. Cambridge University Press, Cambridge. https://doi.org/10.1017/CBO9780511755743
42. Strogatz SH (2001) Exploring complex networks. Nature 410:268–276. https://doi.org/10.1038/35065725

43. Thaut MH, Hodges DA (2019) The oxford handbook of music and the brain. Oxford University Press. https://doi.org/10.1093/oxfordhb/9780198804123.001.0001
44. Deutsch D (2013) The psychology of music. Academic Press series in cognition and perception, 3rd edn. Academic, Oxford (2013). https://doi.org/10.1016/B978-0-12-381460-9.00001-8
45. Gray CM, Singer W (1989) Stimulus-specific neuronal oscillations in orientation columns of cat visual cortex-1989. Proc Natl Acad Sci 86(5):1698–1702. https://doi.org/10.1073/pnas.86.5.1698
46. Tallon C, Bertrand O, Bouchet P, Pernier J (1995) Gamma-range activity evoked by coherent visual stimuli in humans. Eur J Neurosci 7(6):1285–1291. https://doi.org/10.1111/j.1460-9568.1995.tb01118.x
47. Keil A, Müller MM, Ray WJ, Gruber T, Elbert T (1999) Human gamma band activity and perception of a gestalt. J Neurosci 19(16):7152–7161. https://doi.org/10.1523/JNEUROSCI.19-16-07152.1999
48. Tallon-Baudry C, Bertrand O (1999) Oscillatory gamma activity in humans and its role in object representation. Trends Cogn. Sci. 3(4):151–162. https://doi.org/10.1016/s1364-6613(99)01299-1
49. Rodriguez E, George N, Lachaux JP, Martinerie J, Renault B, Varela FJ (1999) Perception's shadow: long-distance synchronization of human brain activity. Nature 397(6718):430–433. https://doi.org/10.1038/17120
50. Engel AK, Fries P, Singer W (2001) Dynamic predictions: oscillations and synchrony in top-down processing. Nat Rev Neurosci 2(10):704–716. https://doi.org/10.1038/35094565
51. Engel AK, Singer W (2001) Temporal binding and the neural correlates of sensory awareness. Trends Cogn Sci 5(1):16–25. https://doi.org/10.1016/s1364-6613(00)01568-0
52. Tan S, Pfordresher P, Harré R (2010) Psychology of music: from sound to significance, 2 edn. Taylor & Francis. https://doi.org/10.4324/9781315648026
53. Koelsch S (2012) Brain and music. Wiley-Blackwell. https://doi.org/10.1002/9781119242341
54. Altenmüller E, Finger S, Boller F (2015) Music, neurology, and neuroscience: historical connections and perspectives. Progress in brain research, vol 216, 1st edn. Elsevier, Amsterdam, Netherlands. https://doi.org/10.1016/bs.pbr.2014.11.019
55. Altenmüller E, Finger S, Boller F (2015) Music, neurology, and neuroscience: evolution, the musical brain, medical conditions, and therapies. Progress in brain research, vol 217, 1st edn. Elsevier, Amsterdam, Netherlands. https://doi.org/10.1016/bs.pbr.2014.11.020
56. Large EW (2010) Neurodynamics of music. Springer handbook of auditory research. Springer Science, New York, NY, US, pp 201–231. https://doi.org/10.1007/978-1-4419-6114-3_7
57. Rebuschat P, Rohmeier M, Hawkins JA, Cross I (2011) Language and music as cognitive systems. Oxford University Press. https://doi.org/10.1093/acprof:oso/9780199553426.001.0001
58. Verzelli P, Tchumatchenko T, Kotaleski JH (2024) Editorial overview: computational neuroscience as a bridge between artificial intelligence, modeling and data. Curr Opin Neurobiol 84:102835. https://doi.org/10.1016/j.conb.2023.102835
59. Ashwin P, Fadera M, Postlethwaite C (2024) Network attractors and nonlinear dynamics of neural computation. Curr Opin Neurobiol 84:102818. https://doi.org/10.1016/j.conb.2023.102818
60. Lansner A, Fiebig F, Herman P (2023) Fast hebbian plasticity and working memory. Curr Opin Neurobiol 83:102809. https://doi.org/10.1016/j.conb.2023.102809
61. Ramaswamy S (2024) Data-driven multiscale computational models of cortical and subcortical regions. Curr Opin Neurobiol 85:102842. https://doi.org/10.1016/j.conb.2024.102842
62. Turing AM (1950) Computing machinery and intelligence. Mind LIX(236):433–460. https://doi.org/10.1093/mind/LIX.236.433
63. Jones CR, Bergen BK (2024) People cannot distinguish GPT-4 from a human in a Turing test. arXiv 2405.08007. https://doi.org/10.48550/arXiv.2405.08007
64. Jones CR, Bergen BK (2025) Large Language Models pass the Turing Test. arXiv 2503.23674. https://doi.org/10.48550/arXiv.2503.23674

65. Giunti M (2025) ChatGPT-4 in the Turing Test: A Critical Analysis. arXiv 2503.06551. https://doi.org/10.48550/arXiv.2503.06551
66. Rohrmeier M (2022) On creativity, music's ai completeness, and four challenges for artificial musical creativity. Transactions of the international society for music information retrieval 5:50–66. https://doi.org/10.5334/tismir.104
67. Vafa K, Chen JY, Rambachan A, Kleinberg J, Mullainathan S (2024) Evaluating the world model implicit in a generative model. https://doi.org/10.48550/arXiv.2406.03689
68. Briot JP, Hadjeres G, Pachet FD (2019) Deep Learning techniques for music generation, 1st edn. Springer Publishing Company, Incorporated
69. Krumhansl CL, Cuddy LL (2010) A theory of tonal hierarchies in music. Springer handbook of auditory research., Springer Science + Business Media, New York, NY, US, pp. 51–87. https://doi.org/10.1007/978-1-4419-6114-3_3
70. Parncutt R (1988) Revision of terhardt's psychoacoustical model of the root (s) of a musical chord. Music Percept 6(1):65–93. https://doi.org/10.2307/40285416
71. Terhardt E (1979) Calculating virtual pitch. Hear Res 1(2):155–182. https://doi.org/10.1016/0378-5955(79)90025-X
72. Gjerdingen R (1990) Categorization of musical patterns by self-organizing neuronlike networks. Music Percept Interdiscip J 7:339–369. https://doi.org/10.2307/40285472
73. Todd PM, Loy G (2003) The representation of pitch in a neural net model of chord classification. https://doi.org/10.7551/mitpress/4804.003.0009
74. Leman M (2000) An auditory model of the role of short-term memory in probe-tone ratings. Music Percept 17(4):481–509. https://doi.org/10.2307/40285830
75. Leman M, Carreras F (1997) Schema and gestalt: testing the hypothesis of psychoneural isomorphism by computer simulation. In: Music, gestalt, and computing. Springer Berlin Heidelberg, pp 144–168. https://doi.org/10.1007/bfb0034112
76. Franklin J (2004) Recurrent neural networks and pitch representations for music tasks
77. Kohonen T, Schroeder MR, Huang TS (eds) (2001) Self-organizing maps. Springer-Verlag New York, Inc., Secaucus, NJ, USA. https://doi.org/10.1007/978-3-642-56927-2
78. Leman M (1995) Music and schema theory: cognitive foundations of systematic musicology. Springer Series in Information Sciences, Springer Berlin Heidelberg. https://doi.org/10.1007/978-3-642-85213-8
79. Tillmann B, Bharucha JJ, Bigand E (2000) Implicit learning of tonality: a self-organizing approach. Psychol Rev 107(4):885–913. https://doi.org/10.1037/0033-295X.107.4.885
80. Taylor I, Greenhough M (1994) Modelling pitch perception with adaptive resonance theory artificial neural networks. Connect Sci 6(2–3):135–154. https://doi.org/10.1080/09540099408915721
81. Brito da Silva LE, Elnabarawy I, Wunsch DC (2019) A survey of adaptive resonance theory neural network models for engineering applications. Neural Netw 120:167–203. https://doi.org/10.1016/j.neunet.2019.09.012
82. Gjerdingen RO (1991) Using connectionist models to explore complex musical patterns. In: Music and connectionism. The MIT Press. https://doi.org/10.7551/mitpress/4804.003.0013
83. Chew E (2014) Mathematical and computational modeling of tonality: theory and applications, vol 204. Springer. https://doi.org/10.1007/978-1-4614-9475-1
84. Herremans D, Chew E (2019) Towards emotion based music generation: a tonal tension model based on the spiral array. In: The annual meeting of the cognitive science society
85. Krumhansl CL (1979) The psychological representation of musical pitch in a tonal context. Cognit Psychol 11(3):346–374. https://doi.org/10.1016/0010-0285(79)90016-1
86. Laden B (1989) A model of tonality cognition which incorporates pitch and rhythm. University of Washington
87. Temperley D (1996) The perception of harmony and tonality : an algorithmic perspective. PhD thesis, Columbia University
88. Vos PG, Leman M (2000) Guest editorial: tonality induction. Music Percept 17(4):401–402. https://doi.org/10.2307/40285825

89. Milne A (2013) A computational model of the cognition of tonality. PhD thesis, The Open University. https://doi.org/10.13140/2.1.4958.3044
90. Lieck R, Moss F, Rohrmeier M (2020) The tonal diffusion model. In: Transactions of the international society for music information retrieval, vol 3, p 153. https://doi.org/10.5334/tismir.46
91. Shiramatsu S, Ozono T, Shintani T (2015) A computational model of tonality cognition based on prime factor representation of frequency ratios and its application. Zenodo. https://doi.org/10.5281/zenodo.851165
92. Navarro-Cáceres M, Caetano M, Bernardes G, Sánchez-Barba M, Merchán Sánchez-Jara J (2020) A computational model of tonal tension profile of chord progressions in the tonal interval space. Entropy 22(11). https://doi.org/10.3390/e22111291
93. Lerdahl F (1988) Tonal pitch space. Music Percept 5:315–350. https://doi.org/10.2307/40285402
94. Noll T (2005) Zwischen theoretischer strenge und milde: Anmerkungen zu fred lerdahls tonal pitch space. ZGMTH-Zeitschrift der Deutschen Gesellschaft für Musiktheorie. ISSN 1862-6742. https://doi.org/10.31751/208
95. Malmierca MS, Hackett TA (2010) Structural organization of the ascending auditory pathway. In: The oxford handbook of auditory science: the auditory brain. Oxford University Press. https://doi.org/10.1093/oxfordhb/9780199233281.013.0002
96. Schofield BR (2010) Structural organization of the descending auditory pathway. In: The oxford handbook of auditory science: the auditory brain. Oxford University Press. https://doi.org/10.1093/oxfordhb/9780199233281.013.0003
97. Palmer AR, Rees A (2010) The oxford handbook of auditory science: the auditory brain. Oxford University Press. https://doi.org/10.1093/oxfordhb/9780199233281.001.0001
98. Ryugo D, Fay R, Popper A (2010) Auditory and vestibular efferents. Springer handbook of auditory research. Springer New York. https://doi.org/10.1007/978-1-4419-7070-1
99. Kandler K (2019) The oxford handbook of the auditory brainstem. Oxford University Press. https://doi.org/10.1093/oxfordhb/9780190849061.001.0001
100. Schofield BR (2011) Central descending auditory pathways. Springer New York, New York, NY, pp 261–290. https://doi.org/10.1007/978-1-4419-7070-1_9
101. Terreros G, Delano PH (2015) Corticofugal modulation of peripheral auditory responses. Front Syst Neurosci 9. https://doi.org/10.3389/fnsys.2015.00134
102. Schofield BR, Beebe NL (2019) Descending auditory pathways and plasticity. In: The oxford handbook of the auditory brainstem. Oxford University Press. https://doi.org/10.1093/oxfordhb/9780190849061.013.17
103. Schofield BR, Beebe NL (2020) The efferent auditory system: central pathways that modulate peripheral input. In: Fritzsch B (ed) The senses: a comprehensive reference, 2nd edn. Elsevier, Oxford, pp 501–516. https://doi.org/10.1016/B978-0-12-809324-5.23826-1
104. Delano PH, Elgoyhen AB (2016) Editorial: auditory efferent system: ninsights from cortex to cochlea. Front Syst Neurosci 10:50. https://doi.org/10.3389/fnsys.2016.00050
105. Kraus N (2021) Descending control in the auditory system: a perspective. Front Neurosci 15. https://doi.org/10.3389/fnins.2021.769192
106. Wang Y, Sanghvi M, Gribizis A, Zhang Y, Song L, Morley B, Barson DG, Santos-Sacchi J, Navaratnam D, Crair M ()2021 Efferent feedback controls bilateral auditory spontaneous activity. Nat Commun 12(1):2449. https://doi.org/10.1038/s41467-021-22796-8
107. Zhao HB, Liu LM, Yu N, Zhu Y, Mei L, Chen J, Liang C (2022) Efferent neurons control hearing sensitivity and protect hearing from noise through the regulation of gap junctions between cochlear supporting cells. J Neurophys 127(1):313–327. https://doi.org/10.1152/jn.00468.2021
108. Brownell WE (2017) What is electromotility?-The history of its discovery and its relevance to acoustics. Acoust Today 13(1):20–27
109. Bidelman GM, Price CN, Shen D, Arnott SR, Alain C (2019) Afferent-efferent connectivity between auditory brainstem and cortex accounts for poorer speech-in-noise comprehension in older adults. Hear Res 382:107795. https://doi.org/10.1016/j.heares.2019.107795

110. Dewey JB, Altoè A, Shera CA, Applegate BE, Oghalai JS (2021) Cochlear outer hair cell electromotility enhances organ of corti motion on a cycle-by-cycle basis at high frequencies in vivo. Proc Natl Acad Sci 118(43):e2025206118. https://doi.org/10.1073/pnas.2025206118
111. Ashmore J (2019) Outer hair cells and electromotility. Cold Spring Harb Perspect Med 9(7). https://doi.org/10.1101/cshperspect.a033522
112. Strauss DJ, Corona-Strauss FI, Schroeer A, Flotho P, Hannemann R, Hackley SA (2020) Vestigial auriculomotor activity indicates the direction of auditory attention in humans. eLife 9:e54536. https://doi.org/10.7554/eLife.54536
113. Schroeer A, Corona-Strauss FI, Hannemann R, Hackley SA, Strauss DJ (2025) Electromyographic correlates of effortful listening in the vestigial auriculomotor system. Front Neurosci 18. https://doi.org/10.3389/fnins.2024.1462507
114. Lesicko AM, Geffen MN (2022) Diverse functions of the auditory cortico-collicular pathway. Hearing Res 425:108488. https://doi.org/10.1016/j.heares.2022.108488
115. Zatorre R (2024) From perception to pleasure: the neuroscience of music and why we love it. Oxford scholarship online, Oxford University Press. https://doi.org/10.1093/oso/9780197558287.001.0001
116. Pearce, MT (2023) Music Perception. In Oxford Research Encyclopedia of Psychology. https://doi.org/10.1093/acrefore/9780190236557.013.890
117. Watts DJ, Strogatz SH (1998) Collective dynamics of 'small-world' networks. Nature 393:440. https://doi.org/10.1038/30918
118. Jacomy M, Venturini T, Heymann S, Bastian M (2014) Forceatlas2, a continuous graph layout algorithm for handy network visualization designed for the gephi software. PLOS ONE 9(6):1–12. https://doi.org/10.1371/journal.pone.0098679
119. Wilkins RW (2019) Network neuroscience: an introduction to graph theory network-based techniques for music and brain imaging research. Oxford University Press, New York, NY, US, pp 122–144. https://doi.org/10.1093/oxfordhb/9780198804123.001.0001
120. Robin W, Hodges D, Laurienti P, Steen M, Burdette J (2012) Network science: a new method for investigating the complexity of musical experiences in the brain. Leonardo 45:282–283. https://doi.org/10.1162/LEON_a_00375
121. Ivanov PC (2021) The new field of network physiology: building the human physiolome. Front Netw Physiol 1:711778. https://doi.org/10.3389/fnetp.2021.711778
122. Schöll E, Sawicki J, Berner R, Ivanov PC (2022) Editorial: adaptive networks in functional modeling of physiological systems. Front Netw Physiol 2:996784. https://doi.org/10.3389/fnetp.2022.996784

Chapter 3
Adaptation and Synchronization in Complex Systems

This chapter provides a comprehensive overview of the pivotal roles of adaptation and synchronization in complex systems, emphasizing their mathematical underpinnings while also identifying areas where a complete theoretical framework remains to be established. Defining a complex system precisely remains challenging, yet one useful approach characterizes complex systems by their capacity to produce emergent collective behavior through the interactions of numerous interdependent components. Importantly, such collective behavior cannot be attributed solely to the nonlinearity of individual components or the heterogeneity of their interactions. Instead, it emerges from a nuanced interplay between the properties of individual elements and the network architecture, resulting in phenomena such as spontaneous order and self-organization.

The study of complex systems, therefore, encompasses both complex dynamics on networks–where intricate behaviors emerge from the activity within a fixed network structure–and dynamics on complex networks, where the structure itself may evolve in response to the dynamics [1]. This dual perspective underscores the adaptability of complex systems, enabling them to exhibit a wide range of behaviors depending on the configuration and interaction of their components.

Adaptation and synchronization serve as foundational mechanisms within this framework. Adaptation enables components within a system to modify their responses based on the states of other components, leading to resilience, flexibility, and the potential for self-organized patterns. Synchronization, on the other hand, fosters coherence across components, facilitating coordinated behavior critical for many natural and engineered systems. Both mechanisms are deeply interconnected with the topology and coupling strengths within the network, and their combined effects are essential for understanding emergent phenomena in complex systems.

Despite substantial progress, achieving a unified mathematical theory for adaptation and synchronization in complex systems remains a major scientific challenge. Various mathematical approaches, such as differential equations for describing

J. Sawicki, *Music Psychology—Balance of Relations*,
Current Research in Systematic Musicology 13,
https://doi.org/10.1007/978-3-031-97003-0_3

dynamic processes, graph theory for network structures, and stochastic models for handling uncertainties, have contributed valuable insights. However, a fully integrative theoretical model that encompasses the diverse dynamics observed across different types of complex systems is still elusive. This limitation highlights the ongoing need for interdisciplinary research to develop frameworks capable of capturing the multifaceted interactions underlying adaptation and synchronization in complex networks.

Complex systems can be viewed as dynamic entities where the network structure and individual element behaviors co-evolve, producing rich and often unpredictable emergent properties. This interplay between structure and dynamics represents a frontier in scientific research, with implications for fields ranging from biology and neuroscience to sociology and engineering. Thus, this Chapter aims to elucidate the current understanding of adaptation and synchronization within complex systems, while acknowledging the areas where further theoretical development is needed to fully comprehend the principles governing these phenomena.

Adaptation is a pervasive characteristic of both natural and artificial complex systems, generating substantial interest in modeling and understanding the diverse adaptive mechanisms present in real-world systems and developing control strategies grounded in these mechanisms.

Such adaptive control strategies are integral to complex systems science, as they offer insights into the underlying behavior of complex systems. Due to the intricate interactions, interdependencies, nonlinearities, and high-dimensional properties of these systems, their behavior is inherently challenging to model. Machine Learning methods are frequently employed for predictive tasks within complex systems; however, this approach is complicated by the need for training datasets that accurately reflect the systems' dynamic diversity. Consequently, datasets tend to be large, making Machine Learning particularly suitable for Big Data applications. Furthermore, complex systems research increasingly extends beyond systems with numerous interacting components to encompass interdisciplinary contributions, illustrated conceptually in Fig. 3.1.

Since the 1960s, partial frameworks for adaptive systems theory have been developed to address control and optimization challenges across deterministic and stochastic systems [2–6]. These include a systematic treatment of adaptation-learning interrelations [7] and the application of the speed gradient method [8], which has enabled adaptive control of network topologies [9]. These foundational advances continue to drive the evolution of adaptive complex systems science.

Synchronization in complex systems is a prominent field in nonlinear dynamics, with extensive applications across both natural and engineered systems. This phenomenon, first observed and described by Christiaan Huygens in 1665, is remarkably universal, manifesting in diverse contexts including clocks, chirping crickets, cardiac pacemakers, neuronal firing, and the coordinated applause of audiences. Such synchronization phenomena are foundational in multiple disciplines, including musicology, physics, chemistry, biology, neuroscience, engineering, and social sciences, and can be examined through a unified framework grounded in modern nonlinear dynamics.

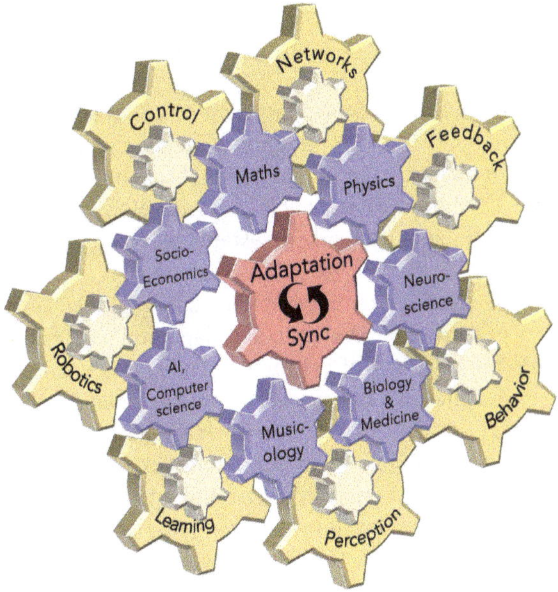

Fig. 3.1 Adaptation and synchronization ("Sync") across different scientific disciplines (blue) and applications (yellow) as well as their strong interlinking and interlocking, similar to a system of gears. The central rotating gear (red), which symbolizes the drive, shows the interrelationship of the two basic mechanisms, adaptation and synchronization: adaptive mechanisms can lead to synchronization, synchronization in turn can be the motor for adaptive processes

Mathematically, synchronization can be defined in various ways, with each approach offering unique insights into complex networks. The field has garnered significant scientific interest for its role in revealing the principles underlying synchronization phenomena in both natural and technological systems. Key factors influencing a system's capacity to synchronize include the local dynamics of its oscillators, the nature of their coupling, and the structural connectivity within the network. Understanding synchronization in oscillator networks has thus become central to elucidating the mechanisms driving collective behavior across a range of systems.

Synchronization phenomena span a spectrum from complete synchronization to intricate partial synchronization patterns, which include clusters and solitary states. These phenomena arise from the dynamic interplay between individual node dynamics and the network topology [10–12]. In recent years, research has increasingly focused on partial synchronization patterns, such as chimera states, which feature spatially coexisting domains of synchronized and desynchronized activity [13, 14]. These states are particularly intriguing as they reveal how coherent and incoherent dynamics can simultaneously exist within the same network, providing insight into complex organizational patterns in systems as diverse as ecological networks and social dynamics.

One of the most significant applications of synchronization studies is in neuroscience, where synchronization is crucial for normal brain function [15]. Neuronal synchronization underlies essential cognitive processes such as attention, perception, and memory consolidation. However, abnormal synchronization is also linked to pathological conditions, including seizures, where excessive synchrony can lead to debilitating disruptions in normal brain activity [16]. This dual role of synchronization–both as a facilitator of physiological function and as a contributor to disease–underscores the importance of understanding its mechanisms in neural networks.

Given its ubiquity, synchronization continues to be a focal point for research, with new findings broadening our comprehension of complex systems in fields ranging from biology to sociology. Through advances in this area, scientists aim to harness synchronization principles to develop innovative approaches for controlling and optimizing networked systems across a variety of scientific and engineering domains.

This chapter examines two fundamental mechanisms essential for the functionality of many complex systems: adaptation and synchronization. These mechanisms underpin a variety of complex phenomena, making them central to understanding system dynamics. Before delving into detailed analysis, we will present foundational concepts in Sect. 3.1 and introduce adaptation and synchronization, which are integral to the functionality of many complex systems. Subsequent sections extend these ideas to multilayered systems, beginning with Sect. 3.2, where we explore a two-layer multiplex network and investigate patterns of cluster synchronization. A method to control the dynamics within such two-layer networks involves the implementation of a pacemaker, as discussed in Sect. 3.3. This approach induces partial synchronization patterns, known as chimera states, which are effectively regulated by the pacemaker. In Sect. 3.4, an additional layer is introduced, with a focus on relay synchronization–a process by which two outer layers achieve synchronization via an intermediary layer. This concept is then generalized to accommodate topological heterogeneities, further enriching the synchronization landscape. Section 3.5 narrows in on relay synchronization within a three-layer network, specifically examining scenarios in which the middle (relay) layer consists of a single hub node.

3.1 Adaptation and Synchronization–Balance of Relations

This section investigates two foundational mechanisms integral to the functionality of many complex systems: adaptation and synchronization. These mechanisms underlie a range of complex behaviors, rendering them critical for a comprehensive understanding of system dynamics. We will demonstrate that both mechanisms are grounded in specific relational structures that can be systematically identified and mathematically characterized. From a formal perspective, numerous systems exhibit equilibrium points where these relations reach a state of quasi-balance. Identifying such equilibrium points is often essential for analyzing and predicting system behavior.

In Sect. 3.1.1, synchronization is rigorously defined and quantitatively analyzed through phase relations, which serve as a fundamental concept in applications such as sound generation in musical instruments. Equally significant are adaptive relations. We explore various definitions of adaptivity from the perspectives of nonlinear dynamics, control theory, and network science, demonstrating how adaptive systems provide insights into real-world networks of interacting units. Initially, a broad conceptual framework for adaptivity, focusing on the interplay between structure and function in dynamical network theory, is introduced in Sect. 3.1.2. Building on this foundation, adaptation is further examined as a slowly evolving feedback mechanism (Sect. 3.1.3). The discussion emphasizes the interplay between adaptivity and noise and the role of adaptive control mechanisms in facilitating critical transitions. Section 3.1.4 delves deeper into how adaptive mechanisms can drive synchronization from the perspective of complex networks, with a particular focus on the interplay between network dynamics and topology. Connections between different adaptive models are explored, showcasing adaptive network models as a robust approach to modeling real-world dynamical systems. In particular, the field of music performance relies on adaptation and synchronization to integrate cognitive, emotional, and social processes across preparation, execution, and reception. Section 3.1.5 illuminates how musicians and listeners create and experience music as a unified and dynamic interplay. Finally, Sect. 3.1.6 synthesizes the findings, consolidates the findings, emphasizing the conditions under which a balance of relations is achieved.

3.1.1 Synchronization–A Matter of Phase Relation

Synchronization is a fundamental emergent phenomenon observed across a diverse range of dynamical systems, encompassing processes such as neural signaling in the brain, the rhythmic beating of the heart, and the coordinated flashing of firefly light waves. It refers to the coordination of events or oscillations to enable a system to operate in unison. Despite its ubiquity, providing a comprehensive and universally applicable definition of synchronization that encompasses all manifestations in interacting dynamical systems remains a significant challenge.

A widely accepted approach conceptualizes synchronization as the adjustment of rhythms resulting from interaction between coupled systems [10, 17]. Within this framework, a particularly effective methodology for studying complex systems has been developed by focusing on the phase relationships between their components. Specifically, the concept of phase synchronization is introduced to describe the state in which a consistent relation between the phases of interacting systems is established, even if their amplitudes remain uncorrelated. This perspective provides valuable insights into synchronization phenomena by isolating phase dynamics as a key aspect of the interaction.

The fundamental mechanisms underlying synchronization become more accessible and comprehensible when the number of interacting objects in a system is restricted. For this reason, the foundational dynamics and synchronization scenarios can be effectively explored and understood using models with just two interacting elements. These simplified models enable a detailed examination of the transition to synchronized states and the conditions under which synchronization emerges, providing critical insights into the universal principles governing the phenomenon. Understanding synchronization in such minimal systems lays the groundwork for extending these principles to more complex networks, offering profound implications for fields as varied as neuroscience, cardiology, and the study of ecological systems.

In this section, we examine the synchronization dynamics of two coupled oscillators, using organ pipes as a representative example. Synchronization and reflective phenomena in organ pipes are well-documented for their potential to cause undesired weakening of the sound under specific conditions. Recent experimental studies highlight the highly complex and nonlinear nature of sound interactions in these systems. Despite this complexity, we demonstrate that a simplified model of two delay-coupled Van der Pol oscillators provides a remarkably effective representation of the observed dynamical behaviors.

In this model, the coupling is implemented as distance-dependent, which can be equivalently described as a time delay. Our analysis focuses on key dynamical features, including the synchronization frequency and the bifurcation scenarios that arise at the boundaries of Arnold tongues. These bifurcations delineate the transition between synchronized and desynchronized states, providing critical insights into the stability and robustness of synchronization.

We further validate the analytical findings by successfully comparing them with experimental data, underscoring the model's capability to capture essential aspects of the underlying dynamics. The work presented in this section builds including on the findings reported in my prior publications [1, 18–22], providing a theoretical and experimental foundation for understanding the synchronization phenomena in coupled oscillatory systems like organ pipes. This section is organized as follows to systematically explore the nonlinear dynamical behavior of coupled organ pipes through both theoretical modeling and experimental validation:

First, in Sect. 3.1.1.1, we present the foundational concept of nonlinear dynamics in the context of organ systems. Here, we introduce a simplified yet insightful model comprising two delay-coupled Van der Pol oscillators, which serve as an analog representation of coupled organ pipes. This model allows for an effective exploration of the complex interactions and synchronization phenomena inherent in such systems. Next, in Sect. 3.1.1.2, we employ two distinct analytical methods to gain deeper insights into the underlying synchronization mechanisms. These methods provide complementary perspectives on the dynamic interplay between the oscillators, elucidating key aspects of their behavior under varying coupling conditions.

The central focus of this section is detailed in Sect. 3.1.1.4, where we rigorously compare the predictions derived from the analytical approaches with empirical data obtained from acoustic experiments. This comparison not only validates the theoretical models but also enhances our understanding of the interplay between theory and experiment in capturing the complex dynamics of coupled acoustic systems. This structured approach enables a comprehensive examination of both the theoretical framework and the experimental observations, thereby providing robust insights into the nonlinear dynamics and synchronization.

3.1.1.1 Organ Pipes as Nonlinear Oscillators

The study of the physics of organ pipes represents an interdisciplinary domain, integrating principles from various scientific fields. This topic is particularly intriguing due to its incorporation of nonlinear dynamical system theory [23, 24], aeroacoustic modeling [25], and synchronization theory [10, 26]. At the core of these investigations lies the organ, often referred to as the "queen of instruments", renowned for its majestic sound and visual grandeur (see Fig. 3.2a). Although we examine the physics of the organ pipe as a specific example, self-organization and nonlinearities play crucial roles in sound production across musical instruments in general [27]. The pipe organ is a musical instrument that generates sound by channeling pressurized air ("wind") through pipes selected via a keyboard. Each pipe produces a single pitch, and pipes are grouped into sets called ranks, characterized by a consistent timbre, volume, and construction. Most organs feature multiple ranks with varying tonal qualities, which can be combined using controls known as stops.

Organ consoles house one or more keyboards (manuals) and a pedalboard, each controlling a division of stops (see Fig. 3.2b). Organs range in size from portable instruments with a few dozen pipes to monumental designs exceeding thousands of pipes and multiple manuals as the largest fully pneumatic romantic organ in the Berlin Cathedral. Organ pipes, constructed from wood or metal, produce sound when wind passes through them. Their pitch depends on pipe length, while timbre and volume depend on airflow and construction as depicted in Fig. 3.3. Ranks, consisting of pipes

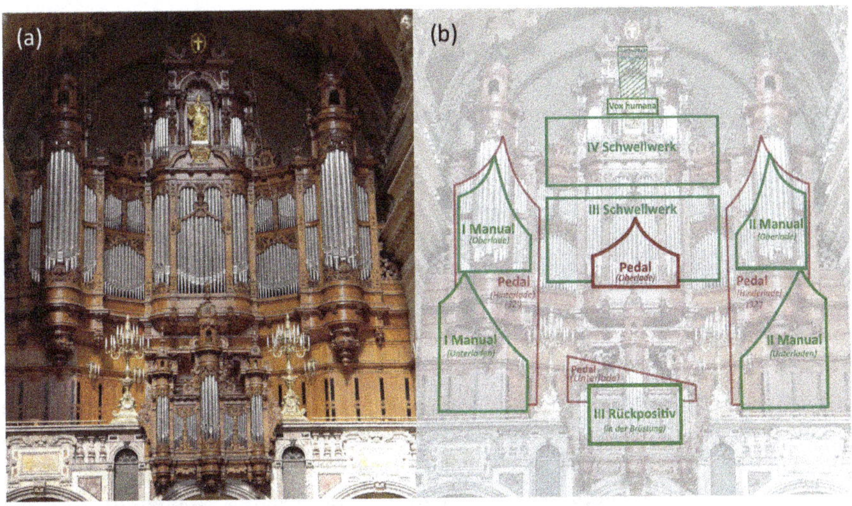

Fig. 3.2 The "queen of instruments": **a** The impressive organ of the Berlin Cathedral build by Wilhelm Sauer in 1905 with in total 7 269 pipes. The largest preserved organ in its original state–dating back to the 'Late Romantic' period–has a width of 14 m and height of about 20 m. Picture presented by Wolfgang Reich. Panel **b** shows the structure of the organ in its individual divisions each controlled by a separate keyboard

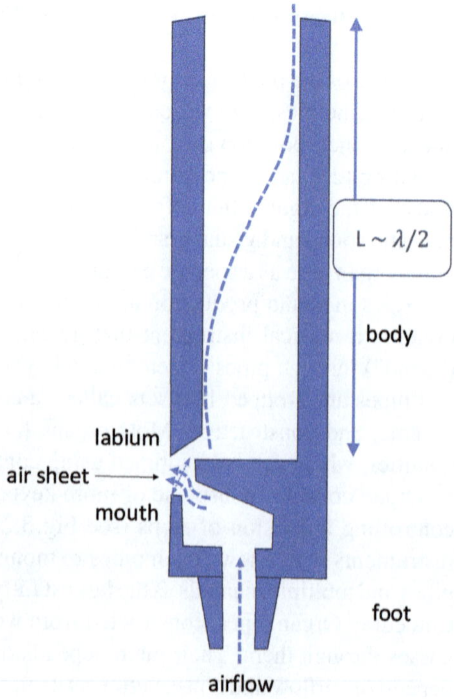

Fig. 3.3 Organ pipe as a self-sustained oscillator: The pressure waves driven by the supplied airflow trigger the "air sheet" which exits at the pipe mouth. The open organ pipe has a length of $L \sim \lambda/2$, where λ denotes the wavelength of the tone of the pipe

with similar timbre but varying pitches, are grouped into divisions, each controlled by a separate keyboard.

The wind system, comprising components to produce, store, and regulate air pressure, is vital for organ operation. Traditional wind pressures are measured in millimetre of water, ranging from 56 mm in Renaissance organs to over 1 300 mm for modern stops. Historically, bellows operated manually, but since the nineteenth century, mechanical blowers powered by water (similar to the *hydraulis*–the earliest (water) organs in the 3rd century BCE), steam, gasoline, or electricity have been used, enabling consistent wind supply and more accessible practice.

In this context, we explore a compelling nonlinear phenomenon: the synchronization of organ pipes positioned in close proximity. Both experimental and theoretical approaches have recently advanced the understanding of this effect [28–31]. The implications of synchronization in musical contexts vary; it may stabilize the pitch of certain organ pipes, which is musically advantageous, but it can also lead to undesirable sound attenuation. This attenuation, observed in the prospect pipes of an organ, manifests as an amplitude minimum resulting from destructive interference between adjacent pipes, particularly during the operation of the swell box [32–34].

A qualitative understanding of the nonlinear mechanisms governing organ pipes can be derived using the framework outlined in [29, 30]. A single organ pipe is effectively modeled as a self-sustained oscillator, where the oscillatory element is the air jet, or "air sheet", emerging from the pipe's mouth. The resonator, naturally, is the pipe body itself; within it, sound waves generated at the labium (the sharp edge of the pipe's upper opening) propagate longitudinally. These waves interact with the air sheet, inducing periodic oscillations. The energy required to sustain this oscillation is supplied by a pressure reservoir beneath the pipe, which maintains a near-constant flow rate. The schematic structure of a closed flue pipe is depicted in Fig. 3.3. Experimental and numerical studies by Abel et al. [29] suggest that the dynamics of an organ pipe can be accurately approximated using a Van der Pol oscillator model.

Building on this foundation, Fischer [30] investigated how nonlinearities in sound generation influence synchronization properties. Here, we extend the analysis to examine the effects of finite separation between two coupled organ pipes. This spatial separation introduces a delay into the coupling function, which we study by analyzing bifurcation scenarios. Specifically, we represent the coupled system using two delay-coupled Van der Pol oscillators, analogous to the experimental setup described by Bergweiler et al. [35].

To gain a deeper insight into the synchronization phenomena of two coupled organ pipes, we model the pipes by Van der Pol oscillators with delayed cross-coupling:

$$\ddot{x}_i + {\omega_i}^2 x_i - \mu\left[\dot{x}_i - \dot{f}(x_i) + \kappa(\tau)x_j(t-\tau)\right] = 0, \tag{3.1}$$

where $i, j = 1, 2$. These equations represent a harmonic oscillator with an intrinsic angular frequency ω_i, supplemented with linear and nonlinear damping of strength $\mu > 0$. The nonlinear damping can be described by the nonlinear function

$$f(x_i) = \frac{\gamma}{3} x_i^3, \tag{3.2}$$

where γ is the anisochronicity parameter and $\dot{f}(x_i) = \gamma x_i^2 \dot{x}_i$. The coupling delay is τ, and the delay-dependent coupling strength in Eq. (3.1) is $\kappa(\tau)$. For the clarity of the calculations we keep the coupling strength constant ($\kappa(\tau) = \kappa$), but all analytical results hold for general $\kappa(\tau)$. Since synchronization is crucially determined by the frequency difference of the two oscillators, we introduce the detuning parameter $\Delta \in \mathbb{R}$ by

$$\omega_1^2 = \omega_2^2 + \mu\Delta. \tag{3.3}$$

In Fig. 3.4, we display frequency locking as obtained by numerical simulation of Eq. (3.1) with symmetric initial conditions. The observed angular frequencies Ω are plotted versus the detuning Δ of two Van der Pol oscillators. A pronounced synchronization region and a sharp transition to synchronization is observed. Within the synchronization region, only for small $|\Delta|$ the in-phase synchronized solution is observed (lower branch), while for larger $|\Delta|$ the anti-phase synchronized solution (upper branch) is obtained. This solution arises even in the case of symmetric initial conditions (full circles). Note that for small $|\Delta|$, the anti-phase synchronized solution

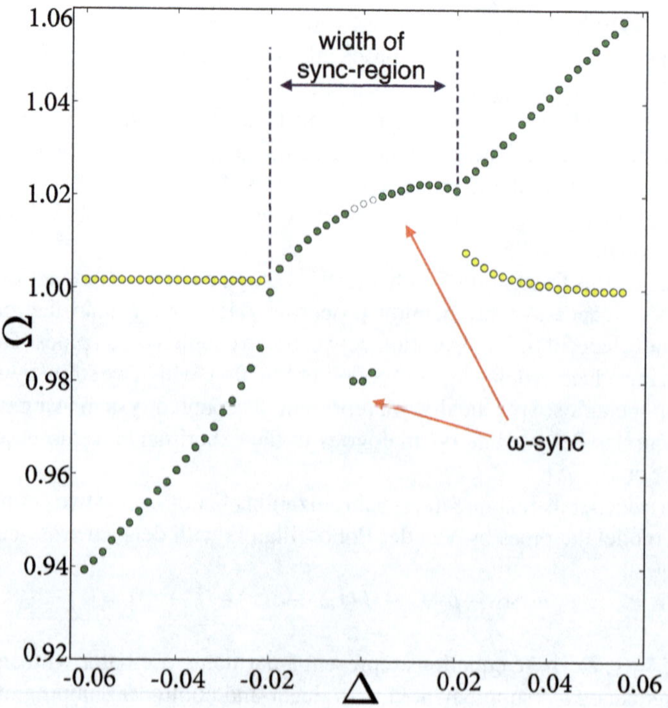

Fig. 3.4 Angular frequency Ω of oscillator x_1 (dark green circles), and oscillator x_2 (light yellow circles) versus the detuning Δ of the oscillators. Full (empty) circles correspond to symmetric (non-symmetric) initial conditions. Parameters: $\omega_2 = 1$, $\mu = 0.1$, $\gamma = 1$, $\kappa = 0.4$, $\tau = 0.1\pi$. Figure from [20]

is also observed for any non-symmetric initial conditions (empty circles). Our goal is to analyze the synchronization frequency, the width of the synchronization region, the phase difference in the synchronized state, its stability, and the bifurcation scenarios which occur at the boundaries.

3.1.1.2 Analytical Approaches and Arnold Tongue

The method of averaging (quasiharmonic reduction) describes weakly nonlinear oscillations in terms of slowly varying amplitude and phase.

For $\mu = 0$ the uncoupled system reduces to the harmonic oscillator $\ddot{x}_i + {\omega_i}^2 x_i = 0$ with solution

$$x_i = R_i \sin(\omega_i t + \phi_i), \tag{3.4}$$

with constant amplitude R_i and phase ϕ_i. For $0 < \mu \ll 1$ we look for a solution in the form Eq. (3.4) but assume that the amplitude $R_i \geq 0$ and the phase ϕ_i are time-dependent functions:

$$\begin{aligned} x_i &= R_i(t)\sin(\omega_i t + \phi_i(t)), \\ \dot{x}_i &= R_i(t)\omega_i\cos(\omega_i t + \phi_i(t)), \end{aligned} \tag{3.5}$$

where terms involving the slowly varying functions $\dot{R}_i$, $\dot{\Phi}_i$ are neglected. Without loss of generality, we choose $\omega_2 = 1$. For small μ we use the method of averaging, assuming that the product $\mu\tau$ is small, and Taylor expand $R_i(t-\tau)$ and $\phi_i(t-\tau)$ in the following way:

$$R_i(t-\tau) = R_i(t) - \tau\dot{R}_i(t) + \frac{\tau^2}{2}\ddot{R}_i(t) + \dots \quad . \tag{3.6}$$

We introduce the phase difference $\psi(t) = \phi_1(t) - \phi_2(t)$. Defining a new time scale $\tilde{t} = \frac{2t}{\mu}$, we find the equations which describe the system (3.1) on a slow time scale:

$$\dot{R}_{1/2}(\tilde{t}) = R_{1/2}(\tilde{t})\left(1 - \frac{\gamma R_{1/2}(\tilde{t})^2}{4}\right) \mp \kappa R_{2/1}(\tilde{t})\sin(\psi(\tilde{t}) + \tau), \tag{3.7}$$

$$\dot{\psi}(\tilde{t}) = -\Delta + \kappa\left[\frac{R_1(\tilde{t})}{R_2(\tilde{t})}\cos(\psi(\tilde{t}) - \tau) - \frac{R_2(\tilde{t})}{R_1(\tilde{t})}\cos(\psi(\tilde{t}) + \tau)\right]. \tag{3.8}$$

For the sake of simplicity, we omit the tilda $\sim$ in the following. The combined effect of the method of averaging and the truncation of the Taylor expansion in τ, is a reduction of the infinite-dimensional problem to a finite-dimensional one (valid only if the product $\mu\tau$ is small). This key step enables us to handle the original delay differential equation as a system of ordinary differential equations [36, 37]. We now have two dynamical equations (3.7) for the amplitudes R_1 and R_2, and one equation (3.8) for the phase difference $\psi(t)$, which is also called the *slow phase*. The latter equation is a generalized Adler equation [38] and contains the main features of synchronization.

The equilibria of the Adler equation correspond to the locking of phase and frequency, since the difference between the phases is constant. To investigate the stability and bifurcation scenario of such fixed points we take a closer look at the generalized Adler equation (3.8), written in general form:

$$\dot{\psi}(t) = -\Delta + \kappa q(\psi(t)), \tag{3.9}$$

where the averaged forcing term $q(\psi(t))$ is the 2π-periodic function

$$q(\psi(t)) = \frac{R_1(t)}{R_2(t)}\cos[\psi(t)-\tau] - \frac{R_2(t)}{R_1(t)}\cos[\psi(t)+\tau]\,. \tag{3.10}$$

The generalized Adler equation (3.8) is a valuable tool for the calculation of the Arnold tongue, which is one of the main characteristics of synchronization in nonlinear systems (see Fig. 3.9).

Furthermore, we gain information about the stability of the synchronization state: For $\psi = 0$ (or, equivalently, $\psi = 2\pi$) we have an unstable equilibrium, and for $\psi = \pi$ (anti-phase oscillation) a stable equilibrium, since $\dot{\psi} < 0$ for $\psi > 0$ and $\dot{\psi} > 0$ for $\psi < 0$. This is in accordance with experimental results [30], as discussed in Fig. 3.8a. The experimentally observed decrease of the amplitude at $\Delta = 0$ indicates an anti-phase oscillation [28]. The in-phase and anti-phase mode correspond to the enhancement or cancellation of sound in organ pipe experiments, respectively. As we recognize from Fig. 3.4 the center of the synchronization region plays a special role. This motivates a first investigation of the solutions and their stability for vanishing detuning $\Delta = 0$.

The describing function method (also called the method of harmonic balance) uses frequency domain techniques to investigate limit cycle behavior in nonlinear systems [1]. An approach with the describing function method determines the synchronization frequency ω_s, if the phase difference ψ is known. The method of averaging on the other hand yields the generalized Adler equation determining the dynamics of ψ.

In order to compare the numerical simulations with the results of the describing function method, we set $\omega_1 = \omega_2 = 1$, i.e., $\Delta = 0$. In this case the synchronization frequency can be simplified to

$$\omega_s^2 = 1 \pm \mu\kappa\cos(\omega_s\tau) \tag{3.11}$$

where $+$ and $-$ correspond to anti-phase and in-phase oscillations, respectively. In Fig. 3.5, we plot the synchronization frequency ω_s versus the time delay τ for $\Delta = 0$. The congruence between the numerical (from Eq. (3.1)) and analytical result for the in-phase mode (dark blue line, from Eq. (3.11)) is excellent. It is remarkable that the synchronization frequency ω_s is modulated around the single oscillator frequencies $\omega_i = 1$ in dependence upon the delay time. For small delay time, for instance, the in-phase oscillation frequency is lowered, while the anti-phase oscillation frequency (light green line) is increased. The stability of the two branches changes as τ is varied: At the extrema of the frequency curve in Fig. 3.5 we find bistability.

In Fig. 3.4, the upper frequency branch in the synchronization region stays in the stable anti-phase mode for non-zero detuning Δ (in congruence with experimental data [28]), whereas the lower branch, i.e., the stable in-phase mode, which is close to its instability point, is only observable in a small range of Δ.

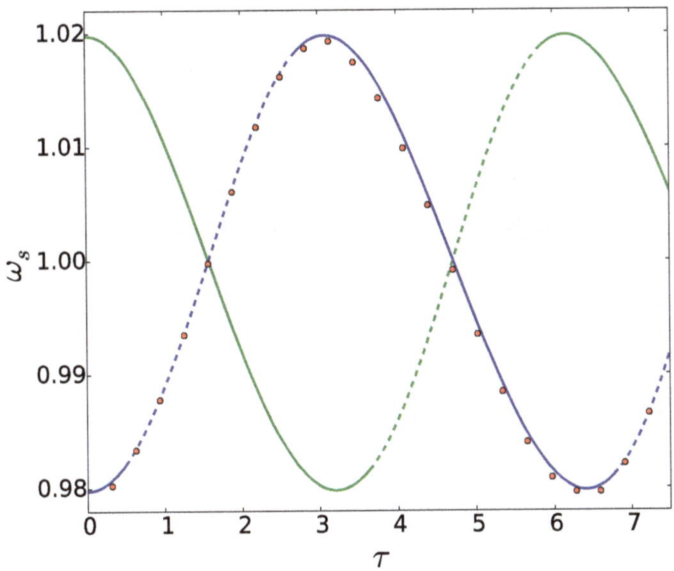

Fig. 3.5 Comparison of the analytic (line) and numeric (dots) results of the synchronization frequency ω_s versus the delay time τ for $\Delta = 0, \omega_1 = \omega_2 = 1, \mu = 0.1, \gamma = 1, \kappa = 0.4$. The analytic solution gives an in-phase (dark blue line) and an anti-phase mode (light green line), while the numeric solution (dots) with symmetric initial conditions only reproduces the in-phase mode. Solid line means stable solution, whereas dashed line stands for an unstable one. Figure from [20]

3.1.1.3 Dependence on Near- and Far-Field Coupling

In previous studies [30, 31], the synchronization of organ pipes has been analyzed using a model of two delay-coupled Van der Pol oscillators with a constant coupling coefficient. In this section, we extend this approach by introducing a time-delay-dependent coupling strength, $\kappa(\tau)$, replacing the constant coupling term. This refined model accounts for both near-field ($\sim\frac{1}{\tau^2}$) and far-field ($\sim\frac{1}{\tau}$) contributions. A time-dependent coupling strength is essential for accurately capturing the non-monotonic behavior of the Arnold tongue, particularly its pronounced widening at small delay times τ, as observed experimentally (see Fig. 3.7a for $d = 5$ cm).

The delay in coupling arises because the sound propagates over a finite distance d between the pipes, and its strength is distance-dependent due to attenuation, which follows the radiation characteristics of a spherical wave emitted from the pipe mouth. The coupling strength $\kappa(\tau)$ incorporates two distinct terms: a near-field component ($\propto \frac{1}{\tau^2}$) and a far-field component ($\propto \frac{1}{\tau}$), each characterized by positive coefficients $\kappa_1 = \kappa_n$ and $\kappa_2 = \kappa_f$, respectively:

$$\kappa(\tau) = \frac{\kappa_1}{\tau^2} + \frac{\kappa_2}{\tau}, \tag{3.12}$$

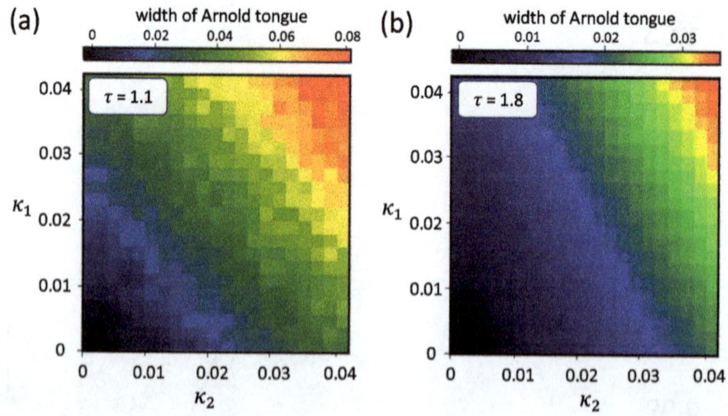

Fig. 3.6 Dependence on near- and far-field parameters: Shown is the dependence of the with of the synchronization region of the Arnold tongue against κ_1 (near-field coupling) and κ_2 (far-field coupling). The width of the synchronization region is color-coded and varies comparing $\tau = 1.1$ in (**a**) and $\tau = 1.8$ in (**b**). The other parameters are given by $\omega_1 = 1$, $\mu = 0.1$, $\gamma = 1$. Figure adapted from [20, 21]

where $\kappa_1 = \kappa_n$ and $\kappa_2 = \kappa_f$ are the near-field and far-field coupling parameters and τ is the time delay.

By substituting $\kappa(\tau)$ from Eq. (3.12) into Eq. (3.1), we analyze the boundaries of the Arnold tongue as a function of near- and far-field coupling contributions. The synchronization region's width varies with the delay time τ, as depicted in Fig. 3.6. In Fig. 3.6a, where $\tau = 1.1$, the Arnold tongue exhibits greater width, whereas increasing the delay time to $\tau = 1.8$ in Fig. 3.6b results in more sharply defined synchronization regions. A fundamental distinction between these two delay times lies in the asymmetry of the resulting representations. For the smaller delay time $\tau = 1.1$ in Fig. 3.6a, the matrix remains symmetric, meaning that interchanging the near-field ($\kappa_1 = \kappa_n$) and far-field ($\kappa_2 = \kappa_f$) coupling coefficients does not affect the width of the Arnold tongue.

In contrast, for the larger delay time $\tau = 1.8$ in Fig. 3.6b, the matrix exhibits asymmetry, implying that swapping the coupling parameters alters the width of the Arnold tongue. This observation suggests that the shape of the Arnold tongue can be precisely adjusted by modifying the relative contributions of near-field and far-field coupling. This indicates that the Arnold tongue shape can be fine-tuned by adjusting the relative contributions of near-field and far-field coupling. Notably, for small delay times, a significant widening of the Arnold tongue is observed, which is consistent with experimental findings (see Fig. 3.7b). This widening effect is attributed to the strong increase in coupling strength for small τ, as described by Eq. (3.12):

$$\lim_{\tau \to 0} \kappa(\tau) = \infty. \tag{3.13}$$

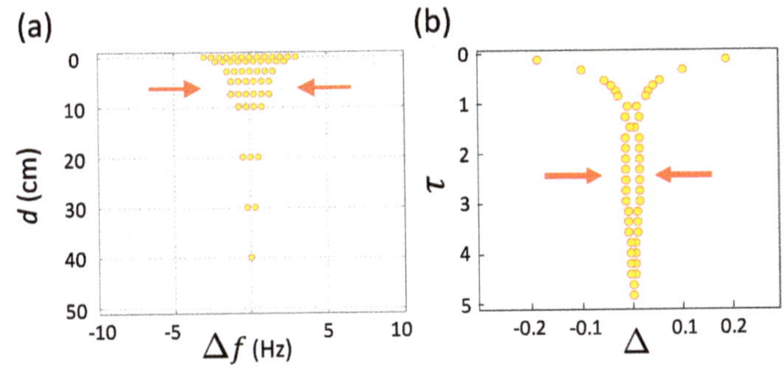

Fig. 3.7 Comparison of experiment and theory: Arnold tongue in the plane of coupling strength κ vs. detuning Δ: **a** experiment [30, 31], where the coupling strength is given by the distance d (cm) of the organ pipes, and **b** numerical result of Eq. (3.1) with $\omega_2 = 1$, $\mu = 0.1$, $\gamma = 1$, $\kappa_1 = \kappa_n = \kappa_2 = \kappa_f = 0.04$. The red arrows in (**a**) and (**b**) indicate the non-monotonic behavior of the Arnold tongue, respectively. Figure from [20, 21]

Another key consequence of the delay-dependent coupling strength is the destabilization of the in-phase solution for $\tau = 0$ [20]. Specifically, in the case where $\kappa_n = \kappa_f = \kappa$, the far-field coupling term $\kappa_2(0)$ satisfies:

$$\lim_{\tau \to 0} \kappa_2(\tau) = \frac{\tau^2}{(1+\tau)\sin\tau} = 0. \tag{3.14}$$

As a result, for $\tau = 0$, only the anti-phase mode remains stable. Introducing a delay-dependent coupling function $\kappa(\tau)$ thus leads to a distortion of the bifurcation curves, altering the stability properties of the synchronization modes.

3.1.1.4 Comparison with Acoustic Experiments

Comparing a complex experimental system to a simplified oscillator model is a challenging yet insightful task. On one side, we have an organ pipe characterized by a rich spectrum of overtones and intricate aeroacoustic behavior, while on the other, we utilize a basic Van der Pol oscillator model. Despite this disparity, the simplified model is capable of capturing complex dynamical scenarios, as demonstrated earlier. Furthermore, these scenarios qualitatively align well with experimental observations, as evidenced by the comparative analysis presented below. This agreement highlights the model's ability to provide a deeper understanding of the dynamical behavior exhibited by organ pipes. This supports the widely accepted perspective in dynamical systems theory that complex behavior can emerge from the interactions within simple systems.

For a visual comparison between experimental data and theoretical predictions, we present the synchronization region as a function of the detuning frequency of two oscillators in Fig. 3.8. Both experimental and theoretical plots display strikingly

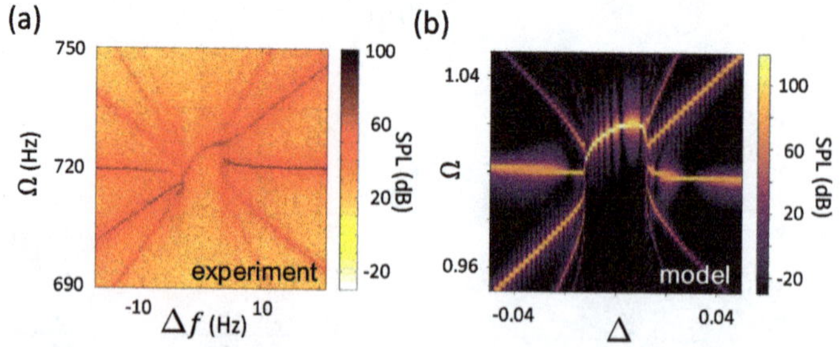

Fig. 3.8 Comparison of synchronization region in experiment and theory: **a** experimentally observed sound pressure level (SPL) in the plane of observed frequency ν versus frequency detuning Δf (in Hz) [30] and **b** angular frequency Ω versus dimensionless detuning Δ numerically calculated from Eq. (3.1) by a Fast Fourier Transform (FFT) as introduced in Sect. 1.3.5.6 for $\omega_2 = 1$, $\mu = 0.1$, $\gamma = 1$, $\kappa = 0.4$, $\tau = 1.1\pi$. Panel (**a**) from [20]

similar features. Notably, the dynamics in the transition regions at the boundaries of the locking interval are particularly well-captured, as is the concave curvature of the synchronization region itself (see also Fig. 3.7).

The methodology employed to determine frequency in Fig. 3.8 involves transforming time series data from the time domain to the frequency domain via the Fourier transform. Specifically, the Fast Fourier Transform (FFT) is utilized as an efficient computational implementation of the Fourier transform. Frequency identification is performed by locating the frequency bin with the highest amplitude within the transformed data. Accurate frequency determination requires a transform with high spectral resolution, which is achieved by analyzing a sufficiently large number of discrete time points (i.e., increasing the length of the sample). In accordance with the Nyquist-Shannon sampling theorem [39], the maximum resolvable frequency is limited to half of the sampling rate. The sampling rate, defined as the number of discrete samples taken per unit time, governs the ability to resolve frequencies and must be sufficiently high to capture the desired frequency components without aliasing.

In Fig. 3.9, we compare the experimentally observed and the analytically calculated Arnold tongue in the plane of the coupling strength κ and the detuning Δ. Experimentally, the coupling strength is determined by the distance of the two organ pipes. By analytical calculation we obtain an Arnold tongue with nonlinear, curved boundaries, see Fig. 3.9b. This is a noticeable result which occurs already for a small delay time τ and agrees well with experiments [30, 31], see Fig. 3.9a. The curvature of the boundaries may be further adjusted in our calculations by replacing the constant κ by a τ-dependent coupling strength $\kappa(\tau)$ [1].

To gain a deeper understanding of the bifurcation scenarios observed, we developed two complementary analytical approaches and successfully compared the results with experiments. Using the method of averaging, we derived a generalized Adler equation to describe the phase dynamics, enabling the stability analysis of equilibria corresponding to frequency locking between oscillators. However, the averaging method does not provide insight into the synchronization frequency within

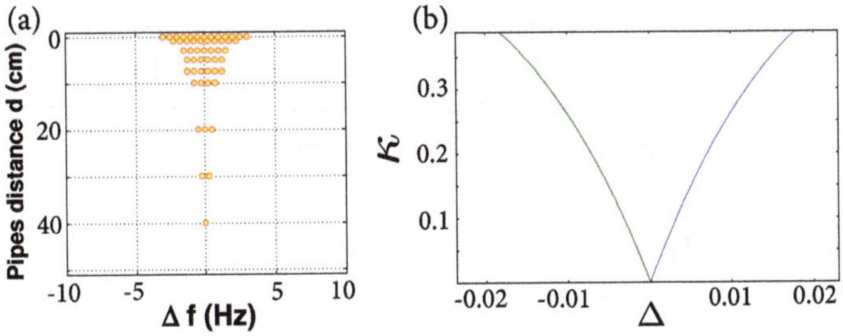

Fig. 3.9 Comparison of experiment and theory: Arnold tongue in the plane of coupling strength κ versus detuning Δ: **a** experiment [30], where the coupling strength is given by the distance d of the organ pipes, and **b** analytic result for $\omega_2 = 1$, $\mu = 0.1$, $\gamma = 1$, and delay time $\tau = 0.1\pi$. Figure from [20]

the locking region. This limitation is addressed by the describing function method, which allows us to determine the synchronization frequency and explain the curvature observed in the frequency-versus-detuning relationship. This behavior is consistent with numerical simulations and experimental observations (see Fig. 3.8). Together, these analytical methods offer a comprehensive picture of both the relative phase and the frequency dynamics. Overall, the agreement between the analytic results, numerical simulations, and experimental data is excellent. A detailed bifurcation analysis further confirms the presence of both in-phase and anti-phase synchronization states. Each synchronization state exhibits a distinct synchronization frequency, in perfect agreement with our analytic predictions.

3.1.2 Adaptive Relations in Dynamical Networks

Adaptivity is a broadly recognized concept that refers to the capacity or process by which a system adjusts or modifies its behavior in response to changing (often external) conditions. This ability is critical for maintaining functionality or optimizing performance under varying circumstances. When discussing adaptivity, it is implicit to differentiate between the 'conditions' (X) that drive change and the adaptation property (Y), which represents the system's responsive capacity or mechanisms. These two components are central to understanding and analyzing adaptivity in complex systems.

In the following discussion, an effort is made to formally define these two variables, (X) and (Y), and explore their interrelation, with particular focus on the theoretical framework of adaptive dynamical networks. Adaptive dynamical networks serve as a foundational model for examining how systems evolve structurally and functionally in response to external stimuli or internal states. Understanding this interplay is essential for characterizing the feedback mechanisms and emergent behaviors inherent in such networks.

This section integrates and extends concepts and findings drawn from previously established research, including elements from the publication [40], to provide a comprehensive perspective on the subject. The aim is to delineate a systematic framework for adaptivity that is both generalizable and applicable to the study of complex systems in various scientific domains.

- The structure Y is the adaptation matter, the part of the system responsible for the adaptation properties. In adaptive dynamical networks, this is usually understood as a network structure represented by connectivity and/or connection weights. By analogy with dynamical networks and neuroscience in general, we refer to this variable as *structure*.
- The function X represents the conditions that trigger the adaptation. In adaptive dynamical networks, this is usually the dynamic state of the network, i.e., the collective and individual dynamics of the nodes. This factor may also include stochastic or external perturbations. These variables usually change with time, i.e., $X(t)$ in the case of temporal adaptation. Following the terminology of the dynamical networks, we generally refer to this variable as *function*.

The non-adaptive systems correspond to a constant structure $Y = Y_0$ which is independent of the function $X(t)$. By assuming that X is governed by a system of differential equations, a general representation of a *non-adaptive* system is

$$\dot{X}(t) = f(X, Y), \tag{3.15}$$

$$\dot{Y} = 0. \tag{3.16}$$

We assume here the general case that the structure Y influences the function X. Systems (3.15)–(3.16) are often used for modeling neural networks with fixed connectivity Y. An example of a *non-adaptive* dynamical network is the coupled system

$$\dot{x}_i = f_i(x_i, t) + \sum_{j=1}^{N} \kappa_{ij} g_{ij}(x_i, x_j), \tag{3.17}$$

where $x_i(t)$ determines the state of node $i = 1, \ldots, N$ and κ_{ij} is the connection weight ($\kappa_{ij} = 0$ if there is no connection). The absence of network adaptivity is indicated by the fixed structure κ_{ij}. The function variable in this example is $X = (x_1, \ldots, x_N)$ while the structure variable is $Y = \left\{\kappa_{ij}\right\}_{i,j=1,\ldots,N}$, and it is constant. The class of non-adaptive networks is extremely useful for modeling many processes and phenomena in nature and technology [10, 12, 41].

When the structure depends on the function, we obtain an adaptive system

$$\dot{X}(t) = f(X, Y), \tag{3.18}$$

$$\dot{Y}(t) = g(X, Y), \tag{3.19}$$

with a mutual structure-function interaction [42]. An example of an adaptive dynamical network is

$$\dot{x}_i = f_i(x_i, t) + \sum_{j=1}^{N} \kappa_{ij} g(x_i, x_j), \tag{3.20}$$

$$\dot{\kappa}_{ij} = h(x_i, x_j, \kappa_{ij}), \tag{3.21}$$

where the rule (3.21) is responsible for the adaptation and the temporal changes of the structure Y. The rule (3.21) is the case when the connection weight between node i and node j depends only on the function of these nodes $x_i(t)$ and $x_j(t)$. Of course, this is not the only possible adaptation rule. Particular realizations of the adaptation rule (3.21) are neuronal systems with plasticity. Specifically, when the plasticity is long-term, i.e., the structural changes act on a slower timescale than the functional dynamics (neuronal spiking) [43–47], this leads to systems with multiple timescales. As a representative system, the paradigmatic adaptive network of phase oscillators

$$\dot{\phi}_i = \omega_i - \sum_{j=1}^{N} \kappa_{ij} \sin(\phi_i - \phi_j + \alpha), \tag{3.22}$$

$$\dot{\kappa}_{ij} = -\varepsilon \left(\kappa_{ij} + \sin(\phi_i - \phi_j + \beta)\right), \tag{3.23}$$

appears to be very useful to study various phenomena in adaptive networks, such as synchronization, frequency clustering, recurrent synchronization, adaptivity-induced resistance to noise, and others [46, 48–53]. Equations (3.22) and (3.23) are a special case of the more general Eqs. (3.24) and (3.25) in Sect. 3.1.4. All of these phenomena are also revealed in more realistic and complex models such as Hodgkin-Huxley neurons with spike-timing-dependent plasticity [46, 54]. Thus, paradigmatic models of the type (3.22) and (3.23) have demonstrated their effectiveness in studying and predicting novel phenomena characteristic for large classes of adaptive networks.

3.1.2.1 High Dimensionality and Multiscale Structure

The primary challenges in studying the aforementioned classes of adaptive dynamical networks are outlined as follows:

- High dimensionality. If the number of nodes in the network is N, the number of possible connections is N^2. Thus, the dimensionality of the model increases dramatically compared to dynamical networks with a fixed structure.
- If the adaptation is slow, i.e., $\varepsilon \ll 1$ in Eq. (3.23), the system becomes multi-scale with the slow manifold of dimension N^2. This additional multiscale structure provides opportunities for analysis [55], but for large networks, it goes far beyond the standard results employing geometric singular perturbation theory.

Despite recent advancements in the study of dynamical adaptive networks, several complex and unresolved challenges persist. Among these are issues related to mean-field theory, the application of adaptive network models to climate network analysis, and the understanding of adaptivity's role in machine learning. Additionally, the development of effective dimensionality reduction techniques remains a significant obstacle, particularly methods designed to address the intricacies of extremely high-dimensional slow manifolds.

In addition to the challenges posed by large networks, small adaptive networks exhibit a particularly complex bifurcation structure, distinct from that of their non-adaptive counterparts. The nontrivial nature of these bifurcations demands further investigation, including the identification of typical bifurcation scenarios in such systems. This task is akin to the study of phenomena such as Eckhaus instability or Busse balloons in partial differential equations (PDEs), which present critical yet unresolved areas of research in the context of adaptive dynamical systems.

3.1.3 Forms of Adaptation and the Relation to Feedback and Noise

Adaptation is frequently characterized as the slow evolution of network connectivity patterns driven by feedback from nodal dynamics, drawing parallels to synaptic plasticity in neuronal systems [56]. However, it is important to recognize that adaptation can also directly influence the intrinsic properties of nodal dynamics. Examples include frequency adaptation in clapping audiences or flashing fireflies [57] and scenarios where the availability of metabolic resources affects neuronal excitability [58, 59] or helps maintain neuronal systems near critical states [60].

Although these two forms of adaptation–affecting either coupling or nodal dynamics–can occur independently, they frequently act in unison, guiding the self-organization of the system [61, 62]. Systematic insights into adaptation's role have primarily focused on its impact on synchronization, such as the emergence of (partial) synchronization states [45, 50, 63–66], its influence on the order of synchronization transitions [57], and its role in the nucleation processes associated with these transitions [67].

Another active research area investigates adaptation as a versatile control mechanism, emphasizing its role in driving critical transitions [59, 60] and facilitating the onset of alternating or cyclic activity patterns [68–70]. This highlights adaptation's broad significance in shaping the dynamics of complex systems. This section incorporates parts of the publication [40].

3.1.3.1 Interplay of Adaptation and Noise

An important yet insufficiently understood problem involves the interaction between adaptation and noise, a phenomenon particularly relevant in neuroscience applications. While noise typically exerts a desynchronizing influence, studies have demonstrated that the interplay of adaptation and noise can lead to self-organized network activity, promoting the overall growth of synaptic strength [46]. This counterintuitive outcome arises from classical synaptic plasticity rules, such as spike-timing dependent plasticity [56], which favor synaptic potentiation when coupled neurons are approximately synchronized and maintain their relative firing order [71]. However, such noise-resilient synchronization has primarily been evidenced in coupled oscillator networks rather than systems involving excitable or mixed excitable-oscillatory populations. Investigating the latter cases is crucial for neuroscience applications, where local dynamics often feature excitability and heterogeneity [72–74].

In addition to influencing the mean coupling strength, adaptation-noise interactions also induce stochastic fluctuations, primarily studied at the microscopic scale. In motifs of coupled stochastic excitable units, this interaction can result in switching dynamics, characterized by slow stochastic transitions between coexisting metastable states. These transitions manifest in both nodal dynamics and effective coupling configurations, defined by the coupling strengths [75]. For instance, in a system of two identical excitable units, noise can induce two oscillatory modes with distinct firing orders between the units. When slow adaptation is present, the metastable states alternate, leading to a dynamic change in coupling configurations, where strong coupling in one direction alternates with strong depression in the opposite direction. In neuroscience, this effect corresponds to switching between two functional neuronal motifs with directed couplings on the same structural motif [76].

At the level of single excitable systems, slowly adapting feedback acting as a low-pass filter can interact with noise to induce a novel behavior termed stochastic bursting. This involves alternating episodes of quiescence and irregular spiking, observed in parameter regimes where an infinite timescale separation between nodal dynamics and adaptation supports bistability between noise-induced and noise-perturbed spiking [69]. Beyond introducing new behaviors, adaptation serves as a control mechanism, enhancing coherence resonance [69] and improving the efficiency of noise-induced spiking suppression within inverse stochastic resonance [70, 77].

3.1.3.2 Impact of Adaptation Rate

An often underappreciated aspect of adaptation in its influence on emergent dynamics is the role of the adaptation rate. Traditionally, adaptation rates are assumed to be sufficiently slow to allow the use of singular perturbation theory [55]. This framework enables a clear separation between the fast local dynamics of system units and the slow evolution of adaptation variables, simplifying the analysis. However, the adaptation rate itself has not been systematically explored, in part due to the absence of robust analytical methods tailored to this challenge.

Numerical studies in specific contexts suggest that intermediate adaptation rates can significantly alter the system's behavior compared to the predictions of singular perturbation theory [75]. Such deviations indicate that the classical separation of timescales may break down, leading to complex interactions between fast and slow dynamics that are not accounted for in conventional models. This intermediate regime, lying between the limits of fast and slow adaptation, remains poorly understood and represents a fertile area for future research.

Developing methods to analyze the impact of adaptation rate requires extending current theoretical frameworks. This includes exploring approaches that bridge timescales, such as generalized multiple timescale analysis, or employing computational techniques like machine learning to detect and classify patterns in high-dimensional, co-evolving systems. Further insights could also arise from systematically studying how varying adaptation rates affect key phenomena like synchronization, bistability, and critical transitions across different network models. Addressing this gap could not only deepen our understanding of adaptation in theoretical systems but also provide new tools for modeling real-world networks, such as those in neuroscience, power grids, and ecological systems, where adaptation rates are likely diverse and dynamically modulated.

3.1.3.3 Mathematical Approaches to Adaptation

From a broader perspective, the development of mathematical approaches to study adaptive networks poses significant challenges due to the need to reconcile diverse system behaviors, including criticality, feedback, multiscale dynamics, diversity, and noise. One promising tool in this context has been the extension of the master stability function framework [78], which effectively reduces synchronization problems by decoupling the dynamical and topological features of the system. This approach facilitates the classification of system states based on synchronization properties. For coupled phase or neural oscillators, it has revealed that adaptation can induce desynchronization transitions [50] and support the emergence of multi-frequency hierarchical cluster states and chimera-like partial synchronization states. However, the general question of how adaptation influences system multistability remains unresolved.

In specific examples, such as Kuramoto phase oscillators with asymmetric spike-timing-dependent plasticity-like rules, adaptation has been shown to induce multistability among synchronized, desynchronized, and partially synchronous states [45]. Similarly, adaptively coupled identical phase oscillators exhibit a high degree of multistability within multicluster states [63, 64]. Beyond synchronization, an important issue concerns the role of adaptation in generating cyclic activity patterns by controlling critical transitions in systems without adaptation. For instance, the onset of collective activity bursts in heterogeneous systems adaptively coupled to a shared resource pool [68] requires combining reduction techniques [79–81] with multiscale methods.

Despite progress, developing rigorous mathematical frameworks that integrate mean-field methods for layer dynamics with reduced models for adaptation remains an active area of research. Recent advances include hybrid approaches that combine Fokker-Planck formalisms with multiscale analysis to address the interaction between adaptation and noise [69]. A natural extension of these frameworks involves systems where the adaptation rate itself varies over time, requiring tools from nonequilibrium thermodynamics and information theory.

This generalization applies particularly to sensory adaptation [82, 83], where the optimization of information transmission occurs under constraints such as metabolic costs, dynamic range limitations, and intrinsic stochasticity [84]. From the viewpoint of nonequilibrium thermodynamics, sensory adaptation is a dissipative process governed by an energy-speed-accuracy tradeoff [82]. Here, the relationship between adaptation and irreversibility, quantified via entropy production, offers a novel angle for studying adaptive mechanisms [83]. These interdisciplinary connections highlight the rich potential for advancing both theoretical and applied understanding of adaptive systems.

3.1.4 Partial Synchronization Patterns in Adaptive Networks

This section examines the application of network models, as outlined in Sect. 3.1.2, across various domains and incorporates parts of the publication [40]. From the perspective of complex networks, the interaction between dynamics and network topology is a central focus. A hallmark of collective dynamics in networks of nonlinear oscillators is synchronization phenomena [10, 12], a topic with historical roots dating back to Christiaan Huygens in 1656. Among these, partial synchronization patterns have recently garnered significant research interest [85]. Notable examples include cluster or group synchronization, wherein all elements within each cluster are fully synchronized while exhibiting phase lags or frequency differences between clusters, along with numerous other manifestations.

A particularly intriguing instance of partial synchronization is *chimera states*, which are symmetry-breaking states characterized by the coexistence of coherent (synchronized) and incoherent (desynchronized) behavior. Recent reviews [1, 86, 87] highlight the growing attention these states have received. In dynamical networks, chimera states consist of spatially separated domains of synchronized and desynchronized dynamics, exemplifying spontaneous symmetry-breaking in systems of identical oscillators. Such states are observed in a variety of systems, including physical, chemical, biological, neuronal, ecological, technological, and socio-economic contexts.

Additional examples of partial synchronization include solitary states [88–90], where a single or a few elements exhibit behavior distinct from their neighboring elements or the background group, and hierarchical multifrequency clusters [49], which exhibit complex nested synchronization patterns. These phenomena have been studied in diverse dynamical systems, such as generalized Kuramoto-Sakaguchi models [88, 91–93], the Kuramoto model with inertia [89, 94, 95], the Stuart-Landau model [96], the FitzHugh-Nagumo model [97, 98], systems of excitable units [99], Lozi maps [100], and even in experimental setups involving coupled pendula [101]. Solitary states, in particular, are viewed as critical intermediaries in transitions between coherent and incoherent dynamics [89, 90, 102]. Hierarchical multifrequency clusters [103] also play a significant role in understanding the complex dynamics of synchronization in networked systems.

In adaptive networks, the coupling weights are not static but instead evolve dynamically through feedback mechanisms influenced by the system's dynamics. Consequently, both the local dynamics and the coupling weights undergo simultaneous co-evolutionary processes over time. Adaptive networks have been identified in various domains, including chemical systems [104], epidemic models [105], and biological and social systems [106]. A notable case of adaptively coupled phase oscillators has recently gained significant attention [49, 50, 50, 63, 64, 103, 107–113]. These models have demonstrated utility in predicting and describing phenomena observed in more realistic and detailed systems [47, 54, 114, 115], thereby serving as powerful frameworks for understanding complex dynamics in adaptive networked systems. It describes N adaptively coupled phase oscillators [49, 63] as a general case of Eqs. (3.22) and (3.23) in Sect. 3.1.2:

$$\dot{\phi}_i = \omega_i + \sum_{j=1}^{N} a_{ij}\kappa_{ij} f(\phi_i - \phi_j), \tag{3.24}$$

$$\dot{\kappa}_{ij} = -\epsilon \left(\kappa_{ij} + g(\phi_i - \phi_j)\right), \tag{3.25}$$

where $\phi_i \in [0, 2\pi)$ represents the phase of the ith oscillator ($i = 1, \ldots, N$), ω_i is its natural frequency, and κ_{ij} is the coupling weight of the connection from node j to i. Further, f and g are 2π-periodic functions where f is the coupling function and g is the adaptation rule, and $\epsilon \ll 1$ is the adaptation time constant. The connectivity between the oscillators is described by the entries $a_{ij} \in \{0, 1\}$ of the adjacency matrix A. In particular, for the Kuramoto phase oscillator [116], the coupling function is $f(\phi) = -\sin(\phi + \alpha)$ with phase lag parameter α, and synaptic neuronal plasticity may be described by $g(\phi) = -\cos(\phi + \beta)$ where the parameter β describes different adaptivity rules.

3.1.4.1 Adaptation and Inertia

This section aims to provide a unified perspective by illustrating that a diverse range of systems–spanning neuronal networks with synaptic plasticity, power grids, and physiological networks modeling tumor disease and sepsis–can be interpreted as adaptive oscillator networks. Within this framework, partial synchronization patterns across these systems can be analyzed consistently, enabling the transfer of methodologies and insights from one domain to another.

A prominent class of network models used to describe power grids consists of N coupled phase oscillators with inertia [117], commonly referred to as the swing equation. This model has been extensively employed in studies of synchronization in complex networks and serves as a fundamental paradigm for understanding the dynamics of modern power grids [95, 118–127]:

$$M\ddot{\phi}_i + \gamma\dot{\phi}_i = P_i + \sum_{j=1}^{N} a_{ij} h(\phi_i - \phi_j), \tag{3.26}$$

where M is the inertia coefficient, γ is the damping constant, P_i is the power of the ith oscillator (related to the natural frequency $\omega_i = P_i/\gamma$), h is the coupling function, and a_{ij} is the adjacency matrix as defined in Eq. (3.24).

It has been shown [128] that the class of phase oscillator models with inertia is a natural subclass of systems with adaptive coupling weights where the weights denote the power flows between the corresponding nodes. We first write Eq. (3.26) in the form

$$\dot{\phi}_i = \omega_i + \psi_i, \tag{3.27}$$

$$\dot{\psi}_i = -\frac{\gamma}{M}\left(\psi_i - \frac{1}{\gamma}\sum_{j=1}^{N} a_{ij} h(\phi_i - \phi_j)\right), \tag{3.28}$$

where ψ_i is the deviation of the instantaneous phase velocity from the natural frequency ω_i. We observe that this is a system of N phase oscillators (3.27) augmented by the adaptation (3.28) of the frequency deviation ψ_i. Similar systems with a direct frequency adaptation have been studied in [57, 129–131]. Note that the coupling between the phase oscillators is realized in the frequency adaptation which is different from the classical Kuramoto system [116]. In order to introduce coupling weights into system (3.27)–(3.28), we express the frequency deviation ψ_i as the sum $\psi_i = \sum_{j=1}^{N} a_{ij}\chi_{ij}$ of the dynamical power flows χ_{ij} from the nodes j that are coupled with node i. The power flows are governed by the equation $\dot{\chi}_{ij} = -\epsilon\left(\chi_{ij} + g(\phi_i - \phi_j)\right)$, where $g(\phi_i - \phi_j) \equiv -h(\phi_i - \phi_j)/\gamma$ are their stationary values [132] and $\epsilon = \gamma/M$. It is straightforward to check that ψ_i, defined in such a way, satisfies the dynamical equation (3.28).

As a result, the swing equation (3.27) and (3.28) can be written as the following system of adaptively coupled phase oscillators

$$\dot{\phi}_i = \omega_i + \sum_{j=1}^{N} a_{ij}\chi_{ij}, \tag{3.29}$$

$$\dot{\chi}_{ij} = -\epsilon\left(\chi_{ij} + g(\phi_i - \phi_j)\right). \tag{3.30}$$

The obtained system corresponds to (3.24) and (3.25) with coupling weights χ_{ij} and coupling function $f(\phi_i - \phi_j) \equiv 1$. The coupling weights form a pseudo coupling matrix χ describing the power flow between the nodes. Note that the base network topology a_{ij} of the phase oscillator system with inertia Eq. (3.26) is unaffected by the transformation.

3.1.4.2 Multifrequency Cluster and Splay States

In adaptive phase oscillator networks there exists a diversity of multifrequency cluster states [49, 64, 103], including chimera states [49] and solitary states [133]. In a multifrequency cluster state, all oscillators split into M groups (called clusters) each of which is characterized by a common cluster frequency Ω_μ. In particular, the temporal behavior of the ith oscillator of the μth cluster ($\mu = 1, \ldots, M$) is given by $\phi_i^\mu(t) = \Omega_\mu t + \rho_i^\mu + s_i^\mu(t)$ where $\rho_i^\mu \in [0, 2\pi)$ and $s_i^\mu(t)$ are bounded functions describing different types of phase clusters characterized by the phase relation within each cluster [63].

As an example, in Fig. 3.10a, c, we present a 4-cluster state of in-phase synchronous clusters on a globally coupled network. Hierarchical multicluster states are built out of single cluster states whose frequency scales approximately with the number N_μ of elements in the cluster. The coupling matrix displayed in Fig. 3.10e shows the characteristic block diagonal shape known for adaptive networks. In particular, the oscillators within each cluster are more strongly connected than the oscillators between different clusters.

A second example, which uses a splay state with $\phi_j = 2\pi kj/N$ and wavenumber $k \in \mathbb{N}$ as the building block for multiclusters, is shown in Fig. 3.10b, d, f. Splay states are characterized by the vanishing local order parameter $R_j = |\sum_{k=1}^{N} a_{jk} \exp(i\phi_k)| = 0$. Figure 3.10b, d, f presents a hierarchical mixed-type multicluster on a nonlocally coupled ring of phase oscillators. It consists of one large splay cluster with wavenumber $k = 2$ and a small in-phase cluster consisting of three solitary states.

In conclusion, the insights into partial synchronization in adaptively coupled phase oscillators can be extended to networks of phase oscillators with inertia. This generalization applies not only to simple homogeneous systems but also to heterogeneous, real-world networks, such as the German ultra-high voltage power grid [128]. Recent investigations into both oscillators with inertia and adaptively coupled oscillators have uncovered a wide array of shared dynamical phenomena, including solitary states [94, 95, 124, 133], multifrequency clusters [63, 64, 123, 135], chimera states [49, 110, 136], hysteresis, and non-smooth synchronization transitions [67, 108, 122, 137, 138].

These findings are relevant for understanding real-world systems of critical importance, such as power grids, neuronal networks with synaptic plasticity, and other adaptive networks. These systems exhibit partial synchronization patterns that are crucial for explaining the onset of instabilities. A particularly compelling application and future direction lies in the functional modeling of physiological two-layer networks, such as the immune system and parenchyma, adaptively coupled via cytokines [139,

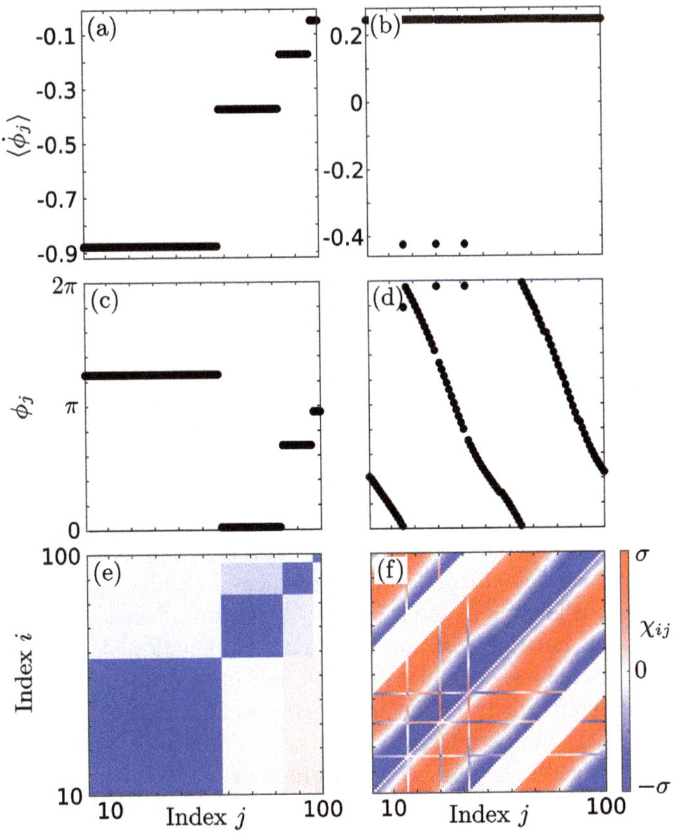

Fig. 3.10 Hierarchical multicluster states in networks of coupled phase oscillators with inertia. The panels (**a, b**), (**c, d**) and (**e, f**) show the temporally averaged phase velocities $\langle \dot{\phi}_j \rangle$, phase snapshots $\phi_j(t)$ and the pseudo coupling matrices $\chi_{ij}(t)$, respectively, at $t = 10000$. In **e** the oscillator indices are sorted in increasing order of their mean phase velocity. The states were found by numerical integration of (3.26) with identical oscillators $P_i = 0, h(\phi) = -\sigma\gamma \sin(\phi + \alpha)$, and uniform random initial conditions $\phi_i(0) \in (0, 2\pi)$, $\psi_i(0) \in (-0.5, 0.5)$. The parameter α is a phase-lag of the interaction [134]. Parameters: (**a, c, e**) globally coupled network, $M = 1$, $\gamma = 0.05$, $\sigma = 0.016$, $\alpha = 0.46\pi$; (**b, d, f**) nonlocally coupled ring network with coupling radius $P = 40$, $M = 1$, $\gamma = 0.3$, $\sigma = 0.033$, $\alpha = 0.8\pi$; $N = 100$. Figure taken from [40]

140] (see Sect. 4.2). This framework has potential applications in modeling tumor diseases and sepsis, where the immune layer serves as a reference point. In this context, the healthy state corresponds to complete frequency synchronization, while the pathological state manifests as a multifrequency cluster state.

3.1.5 Adaptation and Synchronization in Music Performance

The mathematical principles underlying adaptation and synchronization form an essential conceptual backdrop. We now shift our focus to the phenomenological level. In this section, we explore the intricate interplay between the musician–encompassing their physiological, psychological, and neuronal development–the music itself, and the listener within the domain of music performance. Adaptive mechanisms and synchronization phenomena are central to this interplay, offering insight into various facets of music performance, though the examples provided here are not exhaustive. The work presented in this section builds including on the findings reported in my publication [141].

Music performance studies is an emerging interdisciplinary field [142] within the framework of music psychology, combining research methods from multiple disciplines to investigate performance as both a process and an object of study. As a discipline, it is inherently broad, encompassing artistic and aesthetic performances such as concerts. The field employs a mix of methodologies, including practice-led and practice-based research, which integrate traditional approaches like literature reviews and oral histories with performance practices. Its multi- and interdisciplinary nature makes it challenging to define [143, 144], yet this diversity allows for a rich exploration of how musical ideas are realized and communicated to listeners.

Musical performance is integral to the music-making process, as the act of creating sound is deeply intertwined with the pleasure derived from music itself. Early manifestations of music and performance are rooted in human physiological rhythms, such as heartbeat and breathing, as well as in vocal expressions of emotions like grief, pleasure, and desire. These elements evolved into song and structured rhythm. Archaeological evidence, such as rock paintings and artifacts, demonstrates the role of musical performance in ritual, dance, work, and recreational activities.

Music performance provides a valuable context for studying the integration of cognitive and motor skills. Empirical research highlights factors contributing to the formation of interpretations and their transformation into motor actions within the performance context. Adaptation and synchronization are pivotal at every stage of the performance process: before, during, and after the act itself. These processes enable musicians to adjust to their own physiological states, coordinate with other performers, and interact with the audience. We illustrate these phenomena through three examples.

Section 3.1.5.1 gives an example for adaptation and synchronization in music education (before performance). Training musicians involves the synchronization of cognitive, emotional, and motor systems to prepare for performance. Practice sessions integrate rhythmic entrainment and dynamic attending mechanisms [145, 146], enabling musicians to anticipate and synchronize with complex temporal structures. These adaptations are essential for skill acquisition and effective performance.

During a performance (Sect. 3.1.5.2), musicians engage in self-organization and adapt to real-time auditory and visual cues. Emotions play a critical role, as synchronization among subsystems–such as autonomic responses, experiential feelings, and behavioral expressions–enhances the emotional depth and cohesion of the performance [147, 148]. Musicians synchronize not only with their internal states but also with fellow performers and the conductor, demonstrating the importance of adaptive coordination for collective expression.

After a performance (Sect. 3.1.5.3), adaptation and synchronization extend to the interaction between performers and the audience. Applause provides a striking example of rhythmic synchronization among listeners, reflecting shared emotional responses, interpersonal synchrony, and social cohesion [149, 150]. This phenomenon illustrates the broader impact of music performance as a dynamic interplay between creators and recipients of music.

Music performance, as an adaptive and synchronized activity, offers profound insights into the relationships between cognition, emotion, and social interaction. By examining the stages of preparation, execution, and reception, we can better understand the mechanisms underlying musical expression and its impact on both performers and listeners. This interdisciplinary exploration reinforces the importance of synchronization and adaptation in shaping the rich emotional and social dimensions of musical experiences.

3.1.5.1 Practicing and Flow

Flow, a concept introduced by Mihály Csíkszentmihályi in the 1970s, refers to a psychological state of total absorption and intrinsic motivation, where individuals experience heightened focus, diminished self-consciousness, and a distorted sense of time. This state arises during tasks that balance personal skills with challenges, fostering a seamless integration of action and awareness. Csíkszentmihályi first observed flow in creative professionals, such as painters, who became deeply engrossed in their work to the point of neglecting basic physical needs, yet often lost interest once the task was complete [151].

Flow has since been recognized across multiple domains, including sports, education, and the arts. In music psychology, for instance, the application of flow theory has elucidated the immersive states achieved during musical practice [152]. Despite its wide applicability, research on flow remains fragmented due to variability in conceptual and methodological approaches, which has impeded the formation of a unified theoretical framework [153].

Flow states are underpinned by dynamic interactions between cognitive, neurocognitive, and environmental factors. The hallmark of flow–peak performance–emerges from a balance between task difficulty and individual skill, facilitating transitions from explicit cognitive control to implicit, automatic processing [154]. Key theoretical models offer insights into the neural correlates of flow:

- Transient Hypofrontality Hypothesis posits a reduction in frontal lobe activity during flow, leading to decreased self-referential thought and enhanced focus on the task at hand [155].

- Synchronization Theory of Flow (STF) suggests that flow involves synchronized neural activity between attentional and reward networks. Specifically, the frontoparietal attention networks and dopamine-mediated reward pathways operate in a coordinated manner, optimizing neuronal efficiency and engagement [156].

The basal ganglia are central to skill automaticity, with dopamine playing a pivotal role in reinforcing task-related behaviors. During flow, synchronized activity in the basal ganglia, prefrontal cortex, and parietal regions facilitates a streamlined transition from conscious effort to automatic execution. This synchronization enhances task immersion, reduces cognitive effort, and promotes a sense of intrinsic reward.

Emerging interventions, such as transcranial direct current stimulation (tDCS), have shown potential in inducing flow states. By modulating neural excitability, tDCS can enhance skill acquisition and optimize performance, particularly in domains requiring sustained attention and rapid adaptation.

The Synchronization Theory of Flow (STF) provides a unifying framework to understand the flow experience. According to STF, the coordinated activation of frontoparietal attention networks and reward systems underpins the immersive and rewarding nature of flow. These networks exhibit inter-network synchronization, which enhances neuronal efficiency and fosters the seamless integration of action, awareness, and reward [156].

A remarkable example of the emergence of flow is also musical improvisation. Musical improvisation engages a distributed network of brain regions responsible for attention, motor control, sensory integration, and emotion, highlighting domain-general cognitive processes [157]. Expert improvisers transition from deliberate control to spontaneous creativity, characterized by increased activation in the medial prefrontal cortex while suppressing external distractions to achieve a flow state. This process closely parallels language production, where learned patterns are flexibly combined in real time. Investigating improvisation offers valuable insights into the neural mechanisms underlying creativity and how cognitive systems dynamically reorganize to facilitate spontaneous artistic expression.

Flow research continues to expand, revealing its potential as a coping mechanism for stress and anxiety and as a tool for optimizing performance. However, the diversity of findings underscores the need for standardized methodologies to further explore the role of synchronization and other neurocognitive mechanisms in flow states. Understanding these processes could inform interventions across domains, from education to mental health, ultimately advancing both theory and practice in this field.

3.1.5.2 Adaptation and Synchronization During Performance

Emotions can be understood as complex, patterned, and adaptive physiological responses to environmental and internal demands, playing a vital role in survival and well-being [158]. This perspective underscores the interplay between emotion and adaptation, positioning emotions as mechanisms that facilitate the organism's ability to respond effectively to challenges. Beginning with Darwin [159], researchers have emphasized that emotions involve coordinated changes across physiological and neuronal subcomponents to support adaptive responses [147, 160]. Synchronization among these subcomponents–autonomic (e.g., physiological changes), experiential (e.g., self-reported feelings), and behavioral (e.g., observable expressions)–is a defining feature of emotional processes [146]. Moreover, neuroimaging methods such as positron emission tomography (PET) and functional magnetic resonance imaging (fMRI) have provided new insights into the neural correlates of emotions [161, 162].

Music, as a uniquely potent stimulus, evokes strong emotional responses, often surpassing other stimuli in intensity. Emotional experiences with music arise from the dynamic interaction between the musical content, the listener, and the situational context. Notably, music's evolutionary origins as a social activity make it particularly relevant for exploring the synchronization and adaptation of emotional components in social and individual contexts.

Synchronization of subsystems is fundamental to emotional experiences [148], serving as a mechanism that integrates physiological, cognitive, and behavioral elements. This synchronization not only supports adaptive behavior and regulation but also underpins subjective emotional experiences, commonly referred to as "feelings" [145, 149]. The extent and mechanisms of synchronization remain subjects of ongoing research, but evidence suggests that some degree of coherence is essential for emotional processes [147, 163]. Consequently, multi-component approaches that examine the interaction of various subcomponents are increasingly favored.

In the context of music, synchronization plays a central role in shaping emotional responses. For instance, the synchronization of physiological and neural processes with rhythmic or harmonic structures in music enhances the emotional impact. Specific emotions, such as surprise or anxiety, may exhibit distinct patterns of synchronization, often involving greater cognitive engagement. Furthermore, the coherence among emotional components tends to increase with the intensity of the emotion, resembling feedback mechanisms in complex adaptive systems.

Research in this domain highlights the importance of identifying emotion-specific patterns of psychophysiological activity. Multivariate approaches, which monitor multiple variables simultaneously, have been instrumental in uncovering these patterns. By integrating insights from physiological, experiential, and neural domains, studies of music-induced emotions provide a comprehensive framework for understanding how synchronization and adaptation operate in emotional experiences.

For emotions to emerge during musical performances, it is essential that ensembles achieve a high degree of coordination and cohesion in their playing. Here, synchronization is fundamental to how musicians align their individual temporal structures to create a cohesive and unified ensemble. However, a degree of moderate asynchrony can enhance the perceptual clarity of individual musical lines, enabling listeners to discern distinct parts while remaining unaware of substantial temporal deviations [164]. As the size of the ensemble increases, achieving synchronization becomes progressively more challenging. For larger ensembles, the critical task of ensuring synchronized performance and interpretative unity is typically entrusted to the conductor.

The relationship between conductors and musical ensembles underscores the conductor's role in shaping the ensemble's sound, interpreting musical selections, and ensuring temporal synchronization among its members. Nevertheless, evidence indicates that synchronization is a shared responsibility rather than exclusively the conductor's [165]. Synchronization in musical ensembles is a dynamic and multifaceted process involving auditory, visual, cognitive, and physiological mechanisms [143]. The conductor plays an integral but not exclusive role, with ensemble members negotiating their timing through a combination of internal and external cues [166]. This intricate interaction reflects a complex network of shared responsibilities rather than a singular dependence on the conductor.

3.1.5.3 Interpersonal Synchrony

Research on interpersonal synchrony underscores its critical role in social interactions, including the alignment of physiological and motor responses among individuals exposed to shared stimuli. Recent investigations have demonstrated physiological and motor synchrony among audience members during music concerts, linking these phenomena to aesthetic experiences and individual personality traits.

Recent studies have also explored inter-brain synchronization, a phenomenon where brain waves align during social interactions such as collaborative music-making [167], singing in a choir [168], or joking [169], kissing [170], and playing chess [171]. Using electroencephalography (EEG) and related electrophysiological techniques, researchers have uncovered neural and physiological mechanisms underpinning these interactions. These findings advance our understanding of how interconnected neural networks form and synchronize, enhancing insights into cognitive abilities and social behaviors. This research highlights the importance of complex systems approaches in understanding inter-individual neural network interactions.

In particular, studies [172, 173] investigating live classical music concerts have revealed robust physiological synchrony among audience members, which is closely linked to shared aesthetic experiences and individual traits. Researchers measured heart rate, heart rate variability, skin conductance, and breathing during these performances, uncovering significant synchronization in physiological responses. These findings suggest that music induces embodied cognition, profoundly impacting both mind and body. This aligns with the 4E cognition model–embodied, embedded, enacted, and extended cognition–which examines the interplay between cognition, perception, and action while emphasizing the dynamic interaction of mind and body [174]. Synchrony in this context arises from continuous feedback loops between physiological states and environmental stimuli, underscoring the reciprocal relationship between mental and bodily processes.

Even in minimal-interaction settings such as concerts, synchrony–a hallmark of interpersonal interactions–was evident. Higher levels of synchrony were associated with reduced negative emotions, increased positive affect, and traits such as openness. Notably, emotional responses to music inversely correlated with synchrony, while analytical engagement enhanced it. Synchrony also varied by music type, with less familiar contemporary pieces eliciting the highest levels of synchronization [173].

Synchronization is also considered as a core mechanism in musical improvisation. Musical improvisation is a complex and dynamic form of social interaction that requires real-time coordination among performers, encompassing creativity, memory retrieval, and motor control. Synchronization–behavioral, neural, and social–is central to this process, facilitating the emergence of novel musical expressions and the maintenance of cohesive group performance. Studies spanning cognitive neuroscience, dynamical systems, and audience research provide insights into how synchronization functions across multiple levels during improvisation, fostering creativity and collaboration.

Improvisation relies on performers' ability to synchronize actions spontaneously, adapt to each other's input, and co-create musical expressions. Using cross-wavelet spectral analysis, researchers have uncovered multiscale movement dynamics between improvising musicians, revealing the intricate coordination of actions over time [175]. This process allows for self-organization within the group, where mutual constraints and adaptations among musicians result in novel outcomes. Such findings support the perspective that improvisation emerges from distributed, coupled interactions rather than individual cognitive efforts alone.

Neural synchronization plays a pivotal role in improvisation, as shown by mobile brain-body imaging (MoBI) and EEG hyperscanning studies. During live jazz performances, increased inter-brain synchrony, particularly in beta and gamma frequency bands (13–50 Hz), has been observed among musicians [176]. This synchrony is linked to shared intentionality, anticipation, and mutual adjustments to co-performers' actions. Notably, anticipatory cortical activity preceding synchronized events suggests that performers engage in predictive coordination, enhancing the fluency and creative flow of improvisation, including an appropriate timing [177].

Improvisation fosters synchronization not only among musicians but also with the audience. Research demonstrates that improvisational performances elicit movement synchronization between audience members and between the audience and performers, contributing to the collective experience of the music [178]. These embodied, non-verbal interactions underpin the perception of group flow and enrich the shared creative experience, highlighting the interconnectedness of performers and listeners in the improvisational process.

Live accompaniment has been shown to amplify synchronization during improvisation, maximizing motor responses, reward, and engagement [179]. The real-time interaction between live accompanists and improvisers promotes social cooperation and responsiveness, reinforcing shared intentions and enhancing the collaborative nature of music-making. These effects suggest that live accompaniment augments the intrinsic rewards of improvisation through increased synchronization and mutual feedback.

The central role of synchronization in musical improvisation offers insights into creativity, cognition, and social dynamics. Synchronization not only supports the generation of novel musical ideas but also fosters cognitive flexibility, cooperation, and adaptability. This understanding has practical applications in areas such as physical rehabilitation and social training, where coordinated actions and shared intentionality are critical for success. By emphasizing the emergent, decentralized nature of improvisation, studies highlight how synchronization mediates the interaction between individual and collective processes. This interplay extends the value of improvisation beyond the musical domain, offering lessons in cooperation, compromise, and flexibility for addressing complex problems in broader societal contexts [180].

Moreover, recent studies [181, 182] have examined the role of common input synchronization in collective audience responses, particularly focusing on heart rate synchronization during music listening. These investigations revealed that intra-subject heart rate synchronization–consistency in an individual's physiological response over time–was significantly higher than inter-subject synchronization across individuals, regardless of mood or music preference. Results indicate that heart rate synchronization depends on the reliability of physiological responses rather than external factors such as mood or motivation.

Together, these studies highlight music's profound ability to foster social bonds and emphasize physiological alignment as a key marker of collective experiences. Moreover, it has been shown that synchronization and self-organization are fundamental mechanisms underlying audience behavior during applause, e.g., at concerts [183, 184].

Synchronized clapping during applause exemplifies the powerful role of synchronization in social self-organization and complex system dynamics. This phenomenon begins with incoherent clapping that transitions into a synchronized phase, where individuals clap in unison at a slower, periodic rhythm. This synchronization is facilitated by a process of period doubling, wherein the clapping frequency shifts from the fast, dispersed rhythm of the initial phase to a slower and more uniform rhythm. The reduced dispersion in clapping frequencies is a critical condition for achieving synchronization, as predicted by the principles of globally coupled oscillators.

Detailed observations from Eastern European theater performances and controlled experiments confirm that synchronization reduces noise intensity and establishes periodic order, demonstrating the audience's ability to collectively self-organize. However, the conflicting desire to increase noise intensity often disrupts synchronization, as individuals shorten the clapping period, returning to an incoherent phase. These findings underscore synchronization's crucial role in structuring collective behavior, offering an important pathway to understanding synchronization in both social and natural systems [183].

Moreover, the role of synchronization in the social contagion of applause during academic presentations has been explored, focusing on how individual behaviors align with the actions of others in a crowd [184]. Using Bayesian model selection, the researchers demonstrated that individuals were more likely to start clapping as the number of others already clapping increased, reflecting a synchronization process driven by social influence rather than spatial proximity. The cessation of applause was also synchronized, with individuals' decisions to stop clapping being influenced by the broader group's behavior.

Interestingly, it has been found that synchronization, rather than just the desire to express appreciation, governed the timing and duration of applause, with periods of synchronized clapping interspersed with moments of spontaneous cessation and renewal. The findings suggest that synchronization is a key factor in coordinating collective behavior, as it regulates the flow of applause and ensures its periodic nature. The study's results align with models of social contagion, showing that external social cues–such as the actions of others–are crucial for maintaining synchronization in group dynamics. These insights into the synchronization process provide a novel perspective on how collective behaviors emerge and are sustained within social groups, highlighting the importance of synchronization in a range of social, psychological, and economic phenomena.

3.1.6 Summary

Synchronization is a fundamental emergent phenomenon observed in a broad range of dynamical systems. Specifically, the concept of phase synchronization has been introduced to describe the state in which a consistent relationship between the phases of interacting systems is established. Additionally, the notion of adaptation is employed across various disciplines–from nonlinear dynamics and socioeconomic systems to cognitive science and musicology–though its interpretation can vary between fields. In general, 'adaptability' refers to a system's ability to adjust its properties in response to dynamic, either external or intrinsic, changes. However, the specific details of adaptive mechanisms differ based on the context and community, including the identification of which system components can amend or adapt (i.e., adaptation rules) and what strategies facilitate the perception or sensing of such changes. Furthermore, the mathematical frameworks used to describe adaptive systems and mechanisms also vary widely across disciplines.

In this section, we have examined the synchronization of organ pipes using the tools of nonlinear dynamics, with particular attention given to the coupling delay that naturally arises from the finite spatial separation of the pipes. We utilize a simplified nonlinear oscillator model to represent the pipes, specifically two coupled Van der Pol oscillators interacting via a dissipative, direct, delayed coupling. To theoretically analyze the dynamics of this system, we explore various locking scenarios. Additionally, we perform numerical integration of the system and find that the results are in good agreement with existing experimental data. Based on these findings, we systematically vary the coupling parameters, including both the coupling strength and delay.

In general, synchronization may not only occur as complete or isochronous (zero-lag) synchronization, but also as partial synchronization of parts of the system [85]. Also, frequency synchronization (in which only the frequencies but not the phases are the same) should be distinguished from phase synchronization (in which the phases are also synchronized). Partial synchronization includes cluster or group synchronization (in which within each cluster all the elements are synchronized, but there is no synchrony between the clusters), and solitary states (in which isolated nodes form single-node clusters), and chimera states (spontaneously symmetry-breaking states with partially synchronized (spatially coherent) and partially desynchronized (spatially incoherent) dynamics [1, 87, 185–187]).

Furthermore, this section has discussed various approaches to defining adaptation within the contexts of nonlinear dynamics, control theory, and network science, highlighting how adaptive systems can be utilized to understand real-world systems of interacting units (i.e., networks). A common foundation across these fields is the concept of networks, in which the idea of adaptivity is well-established. Adaptive networks are applied in diverse areas, such as power grids, neural systems, and machine learning. Another shared feature across these disciplines is the link between adaptivity and feedback mechanisms, which are pervasive in both natural and engineered systems.

The similarities and differences across these fields present valuable opportunities for cross-disciplinary collaboration, particularly regarding adaptivity as a unifying mechanism. For instance, adaptive networks offer a potent modeling paradigm for realistic dynamical systems, with potential applications extending beyond current domains to fields like cognitive science, musicology, and active matter. Moreover, a promising avenue for future research involves drawing inspiration from adaptive mechanisms observed in natural systems to inform the design and control of artificial (intelligent, cooperative) systems.

The field of music performance represents a dynamical interplay among musicians, the music itself, and listeners. Moreover, music performance serves as a rich domain for examining adaptation and synchronization, revealing their central roles in shaping the cognitive, emotional, and social facets of musical expression. By analyzing the interconnected stages of preparation, execution, and reception, interdisciplinary research in the interface of music psychology and complex systems offers profound insights into the mechanisms driving musical creativity and collective engagement.

Concluding, it becomes clear that adaptation and synchronization are often intricately related and essential for the proper functioning of a wide variety of systems. These interconnections underscore the importance of further exploring their relation in the context of both theoretical and applied research.

3.2 Adaptation and Cluster Synchronization in Multilayer Networks

This section introduces a framework for generating and stabilizing diverse partial synchronization patterns, specifically phase clusters, in adaptive networks–structures prevalent across neuroscience, sociology, biology, engineering, and related fields. Through theoretical analysis and computational simulations, it is demonstrated that symmetry in multiplexed, multilayer networks can promote the formation of stable phase clusters, even in cases where such clusters are absent or unstable within a single network layer. Additionally, a method for analyzing the Laplacian matrices of multiplex networks is proposed to elucidate their spectral characteristics, enabling a reduction of the multilayer stability problem to the stability properties of single layers. Using the paradigmatic Kuramoto phase oscillator model–as a widely applicable model for synchrony across numerous natural and technological systems–as the basis for local dynamics, analytical insights into the stability of multilayer synchronization patterns are derived.

Complex networks are omnipresent in natural and technological contexts, with broad applications spanning physics, chemistry, biology, neuroscience, engineering, and socio-economic systems. A particular focus is adaptive networks, in which connections evolve over time. For example, synaptic links between neurons adjust based on the timing of neuronal spikes [47, 188–191], leading to dynamic reorganization of the network in response to its activity. Similar adaptive dynamics have been observed in chemical systems [104], where reaction rates vary based on system variables, and in activity-dependent plasticity seen in epidemic models [192] and social or biological systems [106]. Synchronization plays a key role in networks of coupled nonlinear oscillators [10, 12, 193–195], with notable synchronization patterns including cluster synchronization–where the network forms synchronized subgroups [196]–and partial synchronization states, such as chimera states that combine synchronized and desynchronized regions [13, 185, 186]. Adaptive networks have been shown to support these and other dynamic scenarios, including the emergence of coexisting frequency clusters and chimera-like states [49, 63, 64, 110, 111, 197–207]. Additionally, adaptations to network topology have been effective in controlling cluster synchronization, especially in networks with delay coupling [9].

Multilayer networks, a recent focus in network science, encompass interconnected systems with different types of links [208–211]. Social networks, for instance, are often structured into groups of individuals with distinct patterns of interaction [212–214], while communication, supply, and transportation networks–such as power grids, subway systems, and air traffic networks–offer additional applications [215]. In neuroscience, multilayer networks represent neuron groups in different brain areas, connected by either chemical or electrical synapses, or brain regions with modular connectivity [216–224]. Multiplex networks, a specific type of multilayer network, consist of layers with identical node sets linked pairwise across neighboring layers [102, 108, 225–242].

Despite considerable research interest in adaptive networks, the interplay among adaptively coupled multilayer or multiplex networks remains underexplored [110, 243, 244]. These adaptive networks naturally emerge in neuronal systems, such as interacting neuron populations governed by different plasticity mechanisms within each group [245, 246] or influenced by various forms of plasticity [247]. Beyond neuroscience, coexisting plasticity mechanisms are also being investigated in neuro-inspired devices to create self-learning circuits for artificial intelligence applications [248].

In Sect. 3.2.1, it is demonstrated that multiplexing adaptive networks can yield a rich diversity of novel dynamic patterns. Specifically, partial synchronization structures, such as phase clusters and more intricate cluster states that are inherently unstable in their monolayer counterparts, can be stabilized through multiplexing.

Furthermore, it is revealed that certain states–unattainable within a single-layer network for a given parameter set–can emerge uniquely in the multiplex configuration, as detailed in Sects. 3.2.2 and 3.2.3. Our objective is to deepen understanding of the interplay between network adaptivity and multiplexed topologies, as presented in Sect. 3.2.4, and to demonstrate the conditions under which adaptivity within multilayer networks enhances synchronization dynamics (see Sect. 3.2.5). This investigation highlights the delicate balance between adaptation and multiplexing–a characteristic of many complex networks beyond neuroscience [192, 249–251]. Our findings are summarized in Sect. 3.2.6. This section is related to our publication [252].

3.2.1 Multiplex Network with Adaptation

In this section, a detailed examination of the relationship between network topology and dynamic behavior is conducted. For the local dynamics within the system, the well-established Kuramoto phase oscillator model is applied–a fundamental, generic model widely recognized for effectively simulating synchronization phenomena across various natural and technological systems [12]. Accordingly, a general multiplex network is described consisting of L layers, each containing N identical phase oscillators that are coupled adaptively:

$$\dot{\phi}_i^\mu = \omega - \frac{1}{N}\sum_{j=1}^{N} \kappa_{ij}^\mu \sin(\phi_i^\mu - \phi_j^\mu + \alpha^{\mu\mu}) - \sum_{\nu=1,\nu\neq\mu}^{L} \sigma^{\mu\nu} \sin(\phi_i^\mu - \phi_i^\nu + \alpha^{\mu\nu}), \tag{3.31}$$

$$\dot{\kappa}_{ij}^\mu = -\epsilon\left(\kappa_{ij}^\mu + \sin(\phi_i^\mu - \phi_j^\mu + \beta^\mu)\right), \tag{3.32}$$

where $\phi_i^\mu \in [0, 2\pi)$ represents the phase of the ith oscillator ($i = 1, \ldots, N$) in the μth layer ($\mu = 1, \ldots, L$), and ω is the natural frequency. The interaction between the oscillators within each layer is determined adaptively by the intra-layer coupling weights $\kappa_{ij}^\mu \in [-1, 1]$, whereas between the layers the inter-layer coupling weights $\sigma^{\mu\nu} \geq 0$ are fixed. The parameters $\alpha^{\mu\nu}$ are the phase lags of the interaction [134]. The adaptation rate $0 < \epsilon \ll 1$ separates the time scales of the slow dynamics of the coupling weights and the fast dynamics of the oscillatory system. The phase lag parameter β^μ of the adaptation function $\sin(\phi_i^\mu - \phi_j^\mu + \beta^\mu)$, also called plasticity rule in the neuroscience terminology [197], describes different rules that may occur in neuronal networks. For instance, for $\beta^\mu = \underset{(+)}{-}\pi/2$, an (anti-) Hebbian-like rule [253–255] is obtained where the coupling κ_{ij} increases (decreases) between any two systems with close-by phases [256]. If $\beta = 0$, the link κ_{ij} will be strengthened if the ith oscillator is advancing the jth. Such a relationship is typical for spike-timing dependent plasticity in neuroscience [47, 190, 257, 258].

Let us note important properties of our model (3.31), which has been widely used as a paradigmatic model for adaptive networks [49, 63, 64, 110, 197–205] and generalizes the Kuramoto-Sakaguchi model with fixed coupling topology [116, 259–262]. First, ω can be set to zero without loss of generality due to the shift-symmetry of Eq. (3.31), i.e., considering the co-rotating frame $\phi \to \phi + \omega t$. Moreover, due to the existence of the attracting region $G \equiv \left\{\left(\phi_i^\mu, \kappa_{ij}^\mu\right) : \phi_i^\mu \in (0, 2\pi], |\kappa_{ij}^\mu| \leq 1,\ i, j = 1, \ldots, N, \quad \mu = 1, \ldots, L\right\}$, one can restrict the range of the coupling weights to the interval $-1 \leq \kappa_{ij} \leq 1$ [49]. Finally, based on the parameter symmetries of the model

$$(\boldsymbol{\alpha}, \boldsymbol{\beta}, \boldsymbol{\phi}, \boldsymbol{\kappa}) \mapsto (-\boldsymbol{\alpha}, \pi - \boldsymbol{\beta}, -\boldsymbol{\phi}, \boldsymbol{\kappa}), \tag{3.33}$$

$$(\alpha^{\mu\mu}, \beta^\mu, \phi_i^\mu, \kappa_{ij}^\mu) \mapsto (\alpha^{\mu\mu} + \pi, \beta^\mu + \pi, \phi_i^\mu, -\kappa_{ij}^\mu), \tag{3.34}$$

where $\boldsymbol{\alpha}, \boldsymbol{\beta}, \boldsymbol{\phi}, \boldsymbol{\kappa}$ abbreviate the whole set of variables and parameters, it is sufficient to analyze the system within the parameter region $\alpha^{11} \in [0, \pi/2), \alpha^{\mu\mu} \in [0, \pi)$ $(\mu \neq 1)$, $\alpha^{\mu\nu} \in [0, 2\pi)$ $(\mu \neq \nu)$ and $\beta^\mu \in [-\pi, \pi)$.

Before considering multiple layers, it is suggested that each solution of Eq. (3.31) for $L = 1, 2$ is called a monoplex or duplex state, respectively. Already for a single layer, Eq. (3.31) possesses a huge variety of dynamical (monoplex) states such as multiclusters with respect to frequency synchronization, chaotic attractors, and chimera-like states, which have been studied numerically and analytically [49, 63, 197–200]. In particular, it has been shown that starting from uniformly distributed random initial condition $\phi_i \in [0, 2\pi)$, $\kappa_{ij} \in [-1, 1]$ the system can reach different frequency multicluster states with hierarchical structure depending on the parameters α and β. The frequency multiclusters in turn consist of several one-clusters which determine the existence and stability of the former [64]. Therefore, these one-cluster states (with identical frequency, but different phase distributions) constitute the building blocks of adaptively coupled phase oscillators, and their generalization to the multiplex case will be in the focus of this section. The reason for this focus is that one-cluster states, which are analytically very well understood, are building blocks for more complex dynamical states. Chimera-like states as they were studied in [49, 110] exist close to the borders of these states, so the existence and stability of one-clusters may pave the way for observing those hybrid patterns.

In general, one-cluster states are given by equilibria relative to a co-rotating frame [63]

$$\begin{aligned} \phi_i^\mu &= \Omega t + a_i^\mu, \\ \kappa_{ij}^\mu &= -\sin(a_i^\mu - a_j^\mu + \beta^\mu), \end{aligned} \tag{3.35}$$

with collective frequency Ω and relative phases a_i^μ. Hence the second moment order parameter $R_2(\mathbf{a}^\mu) = \frac{1}{N}\left|\sum_{j=1}^N e^{\mathrm{i}2a_j^\mu}\right|$ with $\mathbf{a}^\mu \equiv (a_1^\mu, \ldots, a_N^\mu)^T$ can be used as a characteristic measure. In the case of monoplex systems ($L = 1$), three types of solutions exist (see Fig. 3.11) which are characterized by corresponding frequencies Ω as a function of (α^{11}, β^1) [63]:

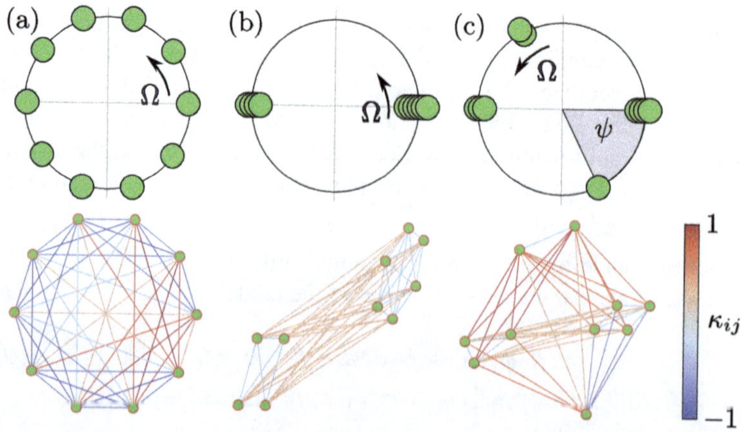

Fig. 3.11 Illustration of the three types of monoplex one-cluster states of Eq. (3.35) ($L = 1$) for an ensemble of 10 oscillators (green circles) with frequencies Ω (upper panels) and coupling structure with weights κ_{ij} (lower panels): One-cluster **a** of splay type ($R_2(\mathbf{a}) = 0$), **b** of antipodal type ($R_2(\mathbf{a}) = 1$), and **c** of double antipodal type with $Q = 7$. Parameters are given by $\alpha = 0.1\pi$, $\beta = 0.1\pi$. Figure taken from [252]

$$\Omega = \begin{cases} \frac{1}{2}\cos(\alpha^{11} - \beta^1) & \text{if } R_2(\mathbf{a}^1) = 0 \quad \text{(Splay)}, \\ \sin\alpha^{11}\sin\beta^1 & \text{if } R_2(\mathbf{a}^1) = 1 \quad \text{(Antipodal)}, \\ & \text{with } a_i^1 \in \{0, \pi\} \\ \frac{1}{2}\cos(\alpha^{11} - \beta^1) - & a_i^1 \in \{0, \pi, \psi_Q, \quad \text{(Double} \\ \frac{1}{2}R_2(\mathbf{a})\cos(\psi_Q) & \qquad \psi_Q + \pi\} \quad \text{antipodal)}, \end{cases} \tag{3.36}$$

with ψ_Q being the unique solution (modulo 2π) of

$$(1 - q)\sin(\psi_Q - \alpha^{11} - \beta^1) = q\sin(\psi_Q + \alpha^{11} + \beta^1), \tag{3.37}$$

where $q = Q/N$ and $Q \in \{1, \ldots, N-1\}$ denotes the number of relative phases $a_i^1 \in \{0, \pi\}$. Here, splay states are defined in a more general sense by $R_2(\mathbf{a}^1) = 0$, which includes the states $a_i^1 = 2\pi i/N$ usually referred to as splay state [263].

Let us now consider these one-cluster states in multiplex structures. Therefore, the notion of *lifted* one-cluster states is introduced, where in each layer the state $(\phi_i^\mu(t), \kappa_{ij}^\mu(t))$ is a monoplex one-cluster, i.e., the phases a_i^μ of the oscillators are of splay, antipodal, or double antipodal type which solves Eq. (3.37). It can be shown that in duplex systems ($L = 2$) the phase difference of oscillators between the layers $\Delta a \equiv a_i^1 - a_i^2$ takes only two values and solves

$$\Delta\Omega = \sigma^{12}\sin(\Delta a + \alpha^{12}) + \sigma^{21}\sin(\Delta a - \alpha^{12}), \tag{3.38}$$

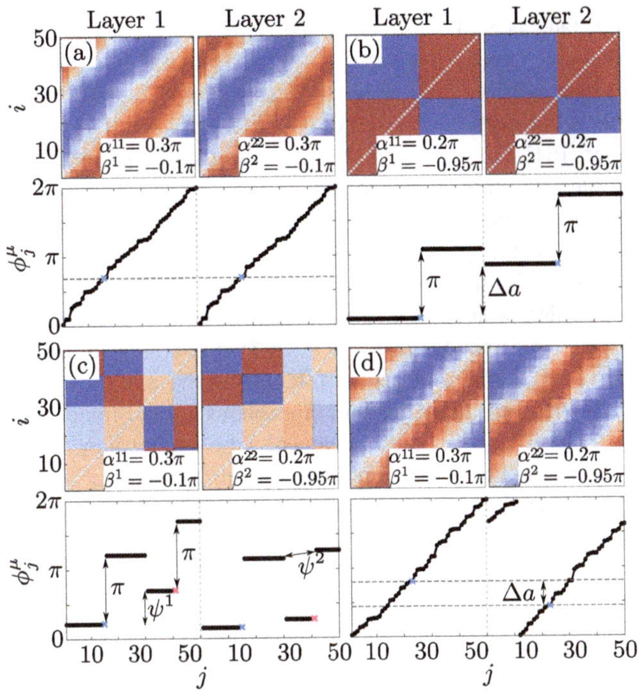

Fig. 3.12 Different duplex states of Eq. (3.35) ($L = 2$) for an ensemble of 50 oscillators in each layer with color-coded coupling weights κ_{ij}^{μ} (upper panels, color code as in Fig. 3.11), phases ϕ_j^{μ} (lower panels): Duplex one-cluster states **a** of lifted splay type ($R_2(\mathbf{a}^{\mu}) = 0$) for $\alpha^{12/21} = 0.3\pi$, $\sigma^{12/21} = 0.07$; **b** of lifted antipodal type ($R_2(\mathbf{a}^{\mu}) = 1$) for $\alpha^{12} = 0.3\pi$, $\alpha^{21} = 0.75\pi$, $\sigma^{12/21} = 0.62$; **c** of double antipodal type (not a lifted state) for $\alpha^{12/21} = 0.05\pi$, $\sigma^{12/21} = 0.28$; **d** of lifted splay type for $\alpha^{12} = 0.3\pi$, $\alpha^{21} = 0.4\pi$, $\sigma^{12/21} = 0.8$, and $\epsilon = 0.01$. In the lower panels phase differences between the two layers are indicated by $\Delta a \equiv a_i^1 - a_i^2$, and between the two new antipodal states (c) by ψ^1, ψ^2. Figure taken from [252]

where $\Delta\Omega \equiv \Omega(\alpha^{11}, \beta^1) - \Omega(\alpha^{22}, \beta^2)$ is given above for the three different one-cluster states (splay, antipodal, double antipodal). Moreover, the condition holds: $(\sigma^{12})^2 + (\sigma^{21})^2 + 2\sigma^{12}\sigma^{21}\cos(\alpha^{12} + \alpha^{21}) \geq \Delta\Omega^2$. This condition is a relation between all parameters of the system which has to be fulfilled for the existence of duplex relative equilibria. Note that for any given inter-layer coupling $\sigma^{12} \neq 0$ and $\alpha^{12} + \alpha^{21} \neq \pm\pi/2$ or $\pm 3\pi/2$ there exists a minimum coupling weight $\sigma^{21} < \infty$ such that the lifted one-clusters exist. In case of unidirectional coupling, i.e., $\sigma^{12} = 0$, the condition gives the minimum weight $\sigma^{21} \geq \Delta\Omega$.

Figure 3.12 displays lifted states of splay (a), antipodal (b), and splay type (d). The phase distributions in both layers are the same but shifted by the constant value Δa in agreement with Eq. (3.38). In contrast to the lifted states, Fig. 3.12c shows another possible one-cluster for the duplex network. Due to the interaction of the two layers, a phase distribution can be found that is of double antipodal type in each layer, but it is not a lifted state, as neither ψ^1 nor ψ^2 solve Eq. (3.37) for $Q = 30$. This means that these states are born by the duplex set-up. Moreover, in contrast to the other examples the phase distribution between the layers does not agree, $\psi^1 \neq \psi^2$. For the monoplex case, it has been shown that double antipodal states are unstable for any set of parameters [64]. Hence, finding stable double antipodal states which interact through the duplex structure is unexpected.

For more insight into the birth of phase-locked states by multiplexing, Fig. 3.13 displays the emergence of double antipodal states in a parameter regime where they do not exist in single-layer networks. They are characterized by the second moment order parameter R_2. It is remarkable that the new double antipodal state can be found for a wide range of the inter-layer coupling strength larger than a certain critical value σ_c, and is clearly different from those of the monoplex. Moreover, these states are even robust for inhomogeneous natural frequencies as will be shown in Sect. 3.2.3. Below the critical value σ_c, the double antipodal states are no longer stable, and more complex temporal dynamics occurs which causes temporal changes in R_2. This leads to non-vanishing temporal variance indicated by the error bars in Fig. 3.13.

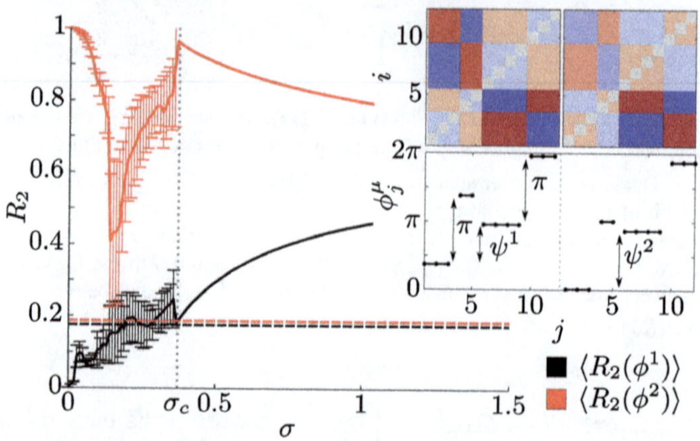

Fig. 3.13 Birth of double antipodal state in a duplex network ($N = 12$) for a wide range of inter-layer coupling strength $\sigma = \sigma^{12} = \sigma^{21}$. The solid lines are the temporal averages for the second moment order parameter R_2 of the individual layers (layer 1: black, layer 2: red). The error bars for $\sigma < \sigma_c$ denote the standard deviation of the temporal evolution of R_2. The dashed horizontal lines represent the unique values of R_2 for the double antipodal state in a monoplex network. The plot was obtained by adiabatic continuation of a duplex double antipodal state (see inset) in both directions starting from $\sigma = 0.5$. Parameters are given by $\alpha^{11/22} = 0.3\pi$, $\alpha^{12/21} = 0.05$, $\beta^1 = 0.1\pi$, $\beta^2 = -0.95\pi$, and $\epsilon = 0.01$. Figure taken from [252]

3.2.2 Existence and Stability of Cluster Synchronization

Suppose we have two one-cluster states where each is of either splay, antipodal, or double antipodal type which form a duplex one-cluster (see Eq. (3.35)) and $\phi_i^\mu = \Omega(\alpha^{\mu\mu}, \beta^\mu)t + \chi^\mu t + a_i^\mu$ ($\mu = 1, 2$), where $\Omega(\alpha^{\mu\mu}, \beta^\mu)$ is given by

$$\Omega = \begin{cases} \cos(\alpha^{\mu\mu} - \beta^\mu)/2 & \text{if } R_2(\mathbf{a}^\mu) = 0 \quad \text{(Splay)}, \\ \sin\alpha^{\mu\mu}\sin\beta^\mu & \text{if } R_2(\mathbf{a}^\mu) = 1 \quad \text{(Antipodal)}, \\ & \text{with } a_i^\mu \in \{0, \pi\} \\ \cos(\alpha^{\mu\mu} - \beta^\mu)/2 - & a_i^\mu \in \{0, \pi, \psi_Q, \quad \text{(Double} \\ \frac{1}{2}R_2(\mathbf{a})\cos(\psi_Q) & \qquad \psi_Q + \pi\} \quad \text{antipodal)}, \end{cases} \tag{3.39}$$

and the coupling weights are given by $\kappa_{ij}^\mu = -\sin(a_i^\mu - a_j^\mu + \beta^\mu)$. We verify by direct insertion that ϕ_i^μ and κ_{ij}^μ solve Eq. (3.32). For the given ansatz, Eq. (3.31) reads

$$\begin{aligned} \Omega(\alpha^{11}, \beta^1) + \chi^1 = \frac{1}{2}\cos(\alpha^{11} - \beta^1) - \frac{1}{2}\Re\left(e^{-\mathrm{i}(2a_i^1 + \alpha^{11} + \beta^1)} Z_2(\mathbf{a}^1)\right) \\ - \sigma^{12}\sin(\Delta\Omega t + \Delta\chi t + a_i^1 - a_i^2 + \alpha^{12}), \end{aligned} \tag{3.40}$$

and

$$\begin{aligned} \Omega(\alpha^{22}, \beta_2) + \chi^2 = \frac{1}{2}\cos(\alpha^{22} - \beta^2) - \frac{1}{2}\Re\left(e^{-\mathrm{i}(2a_i^2 + \alpha^{22} + \beta^2)} Z_2(\mathbf{a}^2)\right) \\ + \sigma^{21}\sin(\Delta\Omega t + \Delta\chi t + a_i^1 - a_i^2 - \alpha^{21}). \end{aligned} \tag{3.41}$$

where $\Delta\Omega = \Omega(\alpha^{11}, \beta^1) - \Omega(\alpha^{22}, \beta^2)$ and $\Delta\chi = \chi^1 - \chi^2$, respectively. In order to obtain a phase-locked state, the time-dependent part within the sin-function of the two Eqs. (3.40)–(3.41) must vanish. Thus, $\phi = (\phi^1, \phi^2)$ is a duplex equilibrium if

$$\Delta\Omega + \Delta\chi = 0. \tag{3.42}$$

Since the values a_i^μ are chosen such that in each layer we already have a one-cluster of either splay, antipodal, or double antipodal type, Eqs. (3.40) and (3.41) can be rewritten as

$$\chi^1 = -\sigma^{12}\sin(\Delta\Omega t + \Delta\chi t + a_i^1 - a_i^2 + \alpha^{12}), \tag{3.43}$$

$$\chi^2 = \sigma^{21}\sin(\Delta\Omega t + \Delta\chi t + a_i^1 - a_i^2 - \alpha^{21}). \tag{3.44}$$

Thus the values of χ^μ are introduced to account for the changes of the one-cluster frequencies of the monoplex system due to the inter-layer coupling. Hence, $\Delta\Omega + \Delta\chi = 0$ is equivalent to

$$\Delta\Omega = \sigma^{12}\sin(a_i^1 - a_i^2 + \alpha^{12}) + \sigma^{21}\sin(a_i^1 - a_i^2 - \alpha^{21}) \tag{3.45}$$

for all $i = 1, \ldots, N$. Note that $\Delta\Omega$ is not necessarily zero even if the phase-lag parameters for both layers agree. They can still differ in the type of one-cluster state. The former equation can be written as

$$\frac{\Delta\Omega}{C} = \sin(a_i^1 - a_i^2 + \nu) \tag{3.46}$$

with

$$\begin{aligned}\sin(\nu) &= \frac{1}{C}\left(\sigma^{12}\sin(\alpha^{12}) - \sigma^{21}\sin(\alpha^{21})\right),\\ \cos(\nu) &= \frac{1}{C}\left(\sigma^{12}\cos(\alpha^{12}) + \sigma^{21}\cos(\alpha^{21})\right),\end{aligned} \tag{3.47}$$

where

$$C = \sqrt{(\sigma^{12})^2 + (\sigma^{21})^2 + 2\sigma^{12}\sigma^{21}\cos(\alpha^{12} + \alpha^{21})}. \tag{3.48}$$

Whenever $(\sigma^{12})^2 + (\sigma^{21})^2 + 2\sigma^{12}\sigma^{21}\cos(\alpha^{12} + \alpha^{21}) \geq 0$ and

$$(\sigma^{12})^2 + (\sigma^{21})^2 + 2\sigma^{12}\sigma^{21}\cos(\alpha^{12} + \alpha^{21}) \geq \Delta\Omega^2, \tag{3.49}$$

Equation (3.46) has the two solutions $a_i^1 - a_i^2 = \arcsin(\Delta\Omega/C) - \nu$ and $a_i^1 - a_i^2 = \pi - \arcsin(\Delta\Omega/C) - \nu$. Considering the inverse function $\arcsin : [-1, 1] \to [-\pi/2, \pi/2]$ applied to Eq. (3.47) determines ν to be either ν' or $\pi - \nu'$, where $\nu' := \arcsin(\sin(\nu))$ and $\sin(\nu)$ as given in Eq. (3.47). The second equation for $\cos(\nu)$ then fixes ν to take one of the values.

The condition in Eq. (3.49) is a relation between all parameters of the system which has to be fulfilled for the existence of duplex relative equilibria. Note that for any given inter-layer coupling $\sigma^{12} \neq 0$ and $\alpha^{12} + \alpha^{21} \neq \pm\pi/2$ or $\pm 3\pi/2$ there exists a minimum coupling weight $\sigma^{21} < \infty$ such that the lifted one-clusters exist. In case of unidirectional coupling, i.e., $\sigma^{12} = 0$, the condition gives the minimum weight $\sigma^{21} \geq \Delta\Omega$. Due to the fact that Eq. (3.46) has to hold for all $i = 1, \ldots, N$, ϕ_1 and ϕ_2 are not allowed to be of different one-cluster type. To see this, assume that ϕ_1 and ϕ_2 are of different monoplex equilibrium type and form a duplex equilibrium state. We know from the former consideration that there are two possible values which the difference $\Delta a := a_{i,1} - a_{i,2}$ can take. By definition of the types of monoplex equilibria, there are more than two different phase differences $a_{i,1} - a_{i,2}$ which contradicts the assumption.

So far, the existence of lifted one-clusters has been studied, and several examples of these states have been discussed. In the following, the dynamics in a neighborhood of these states are shown to be liftable, i.e., their local stability is investigated. The linearization of Eq. (3.31) around the one-cluster states described by Eq. (3.35) is exemplified for antipodal states but can be generalized to the other states as well:

$$\dot{\delta\phi}_i^\mu = \frac{1}{N}\sum_{j=1}^{N}\big[\sin(\Delta a + \beta^\mu)\cos(\Delta a + \alpha^{\mu\mu})\Delta_{ij}^{\mu\mu}\delta\phi -$$

$$\sin(\Delta a + \alpha^{\mu\mu})\delta\kappa_{ij}^\mu\big] - \sum_{\nu=1}^{M}\sigma^{\mu\nu}\cos(\Delta a + \alpha^{\mu\nu})\Delta_{ij}^{\mu\nu}\delta\phi, \tag{3.50}$$

$$\dot{\delta\kappa}_{ij}^\mu = -\epsilon\left(\delta\kappa_{ij}^\mu + \cos(\Delta a + \beta^\mu)\Delta_{ij}^{\mu\mu}\delta\phi\right)$$

where $\Delta_{ij}^{\mu\nu}\delta\phi \equiv \delta\phi_i^\mu - \delta\phi_j^\nu$.

In duplex networks, the coupling structure is given by a 2×2 block matrix M with the $N \times N$ unity matrix $\mathbb{I}_N$:

$$M = \begin{pmatrix} A & m \cdot \mathbb{I}_N \\ n \cdot \mathbb{I}_N & B \end{pmatrix}. \tag{3.51}$$

If A and B are diagonalizable $N \times N$ matrices which commute ($m, n \in \mathbb{R}, n \neq 0$), the following relation for the characteristic polynomial can be proven using Schur's decomposition [264, 265]:

$$\mu^2 - ((d_A)_i + (d_B)_i)\,\mu + (d_A)_i(d_B)_i - mn = 0 \tag{3.52}$$

where $(d_A)_i$ and $(d_B)_i$ are the diagonal elements of the corresponding diagonal matrices of A and B, respectively. Note that Eq. (3.52) not only simplifies the calculation for the eigenvalues in the case of a duplex structure, moreover, it is a general result on linear dynamical systems on duplex networks. Therefore, this result is important for the investigation of stability and symmetry in multiplex networks.

In the case of a duplex antipodal one-cluster state Eq. (3.31) with $a_i^1 \in \{0, \pi\}$ and $a_i^2 = a_i^1 - \Delta a$, Eq. (3.50) can be brought to the form of Eq. (3.51) and possesses the following set of Lyapunov exponents $\mathcal{S} = \{-\epsilon, \left(\lambda_{i,1}, \lambda_{i,2}, \lambda_{i,3}, \lambda_{i,4}\right)_{i=1,\dots,N}\}$ where $\lambda_{i,1,\dots,4}$ are the solutions of polynomials containing the eigenvalues of the monoplex system.

Thus, the stability analysis of the duplex system is reduced to that of the monoplex case. We are able to analyze the stabilizing and destabilizing features of a duplex network numerically and analytically. To illustrate the effect of multiplexing, the interaction between two clusters of antipodal type is presented in Fig. 3.14. The stability of these states is determined by integrating Eq. (3.31) numerically starting with a slightly perturbed lifted antipodal state. The states are stable if the numerical trajectory is approaching the lifted antipodal state. Otherwise, the state is considered as unstable. The black contour lines in Fig. 3.14 show the borders of the stability regions in dependence of the coupling strength σ^{21}, as calculated from the Lyapunov exponents in Eq. (3.86). The borders are in remarkable agreement with the numerical results.

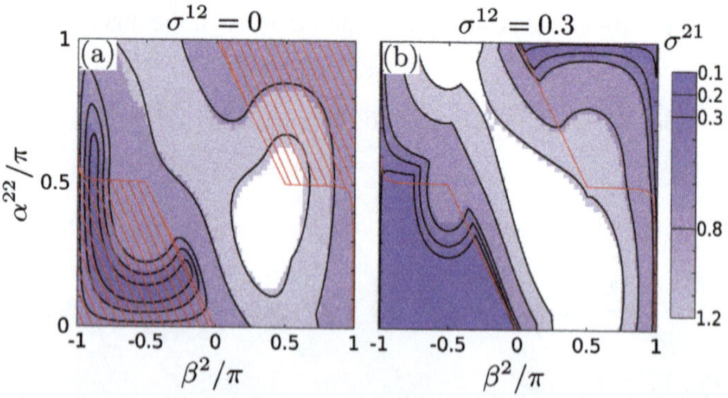

Fig. 3.14 Regions of stability (blue) and instability (white) of the lifted antipodal state in the (α^{22}, β^2) parameter plane for different values of interlayer coupling (indicated by different blue shading) σ^{21}, where regions of stronger coupling σ^{21} (lighter blue) include such of weaker σ^{21} (darker blue). Stability regions for single-layer antipodal clusters are indicated by red hatched areas. The inter-layer coupling is considered as **a** unidirectional ($\sigma^{12} = 0$) and **b** bidirectional ($\sigma^{12} = \sigma^{21}$). Parameters are given by $\alpha^{11} = 0.2\pi$, $\beta^1 = -0.8\pi$, $\alpha^{12} = 0$, $\alpha^{21} = 0.3\pi$, and $\epsilon = 0.01$. Figure taken from [252]

In Fig. 3.14, the parameters for the first layer α^{11}, β^1 are chosen such that the antipodal state is stable without inter-layer coupling. The stability of the duplex antipodal states is displayed in the (α^{22}, β^2) parameter plane for several values of the inter-layer coupling σ^{21} (the stability regions for smaller values of σ^{21} are always contained in regions of larger ones). To compare the effects of the duplex network with the mono-layer case, the stability regions for monoplex antipodals states are displayed as red hatched areas. They are markedly different. In Fig. 3.14a, the two layers are connected unidirectionally ($\sigma^{12} = 0$). It can be seen that with increasing inter-layer coupling weight σ^{21} the region of stability for the lifted antipodal state also grows. Already for small values of the inter-layer couplings σ^{21}, a stabilizing effect of the duplex network can be noticed. For $\sigma = 0.1$ there exist already regions for which the duplex antipodal state is stable but the corresponding monoplex state would not be stable. The opposite effect is found as well where the duplex network destabilizes a lifted state. Figure 3.14b shows the results for two layers with bidirectional coupling. Here, the duplex structure can have stabilizing and destabilizing effects. Further, for the bidirectional coupling, a growth of the stability region with increasing σ^{21} is also observed, similar to the unidirectional case. However, the regions of stability grow at different rates in dependence on σ^{21} and non-monotonically with respect to the parameters α^{22}, β^2. Comparing the size of the stability region for both cases, one can see that for small values of σ^{21} the region for bidirectional coupling is larger. In turn, for higher inter-layer coupling, the regions for the unidirectional case are larger.

3.2.3 Robustness of Cluster Synchronization

So far, a system of identical oscillators has been investigated. There the existence of particular phase cluster states of double-antipodal type is demonstrated. In order to show that these states are also present in a system of heterogeneous phase oscillators, we modify Eqs. (3.31) and (3.32) as follows:

$$\frac{d\phi_i^\mu}{dt} = \omega_i^\mu - \frac{1}{N}\sum_{j=1}^{N} \kappa_{ij}^\mu \sin(\phi_i^\mu - \phi_j^\mu + \alpha^{\mu\mu}) \\ - \sum_{\nu=1,\nu\neq\mu}^{L} \sigma^{\mu\nu} \sin(\phi_i^\mu - \phi_i^\nu + \alpha^{\mu\nu}), \tag{3.53}$$

$$\frac{d\kappa_{ij}^\mu}{dt} = -\epsilon\left(\kappa_{ij}^\mu + \sin(\phi_i^\mu - \phi_j^\nu + \beta^\mu)\right). \tag{3.54}$$

For the numerical analysis of Eqs. (3.53)–(3.54), randomly uniformly distributed natural frequencies on the interval $[-\Delta\omega, \Delta\omega]$ are considered. To check for the robustness of the phase cluster presented in the inset of Fig. 3.13, the following steps are performed: a random realization of a uniform distribution is fixed. For any inter-layer coupling strength $\sigma = \sigma^{\mu\nu}$, the final state from the simulation with $\omega_i^\mu = \omega_i^\nu = 0$ in Fig. 3.13 is taken as the initial condition. An adiabatic continuation of the state is then performed by running the simulation for $t = 5\,000$ and increasing the width of the distribution $\Delta\omega$ with a stepsize of 0.01. This is done until $\Delta\omega$ reaches 0.5. The continuation is performed for each value of σ and for 10 different realizations of the uniform distribution. Afterwards, the final state is first checked to see if it still has the same form as the one presented in Fig. 3.13. This is done by calculating the second moment order parameter for both states in each layer individually, determining the difference between the order parameters for both layers, and setting the upper limit to 0.01. States with a difference smaller than this limit are considered to possess the same form. Secondly, it is checked whether the final state is still a phase-locked state, meaning all oscillators are frequency-synchronized. The range and the boundaries up to which the final state is still a duplex one-cluster state with or without the form from Fig. 3.13 are presented in Fig. 3.15. For the boundaries and the range the mean value over the 10 realizations is determined and the error bars indicate the standard deviation.

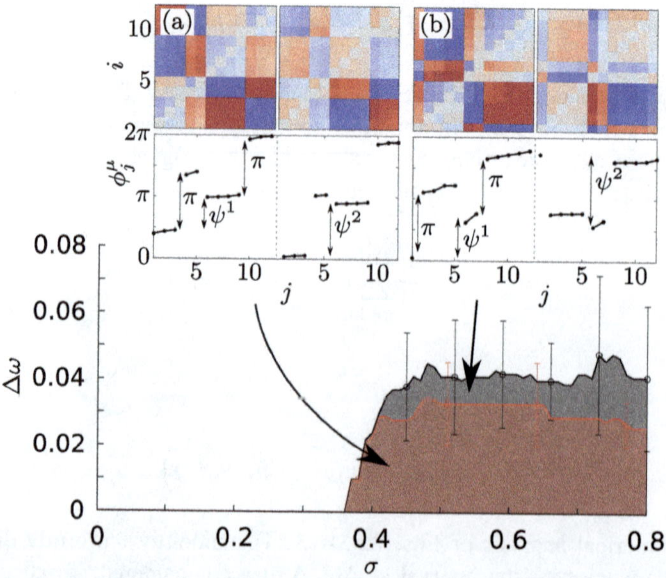

Fig. 3.15 The figure shows the range $\Delta\omega$ vs where duplex one-cluster states in general (gray) and of the form presented in Fig. 3.13 (red) can be found. For this the Eqs. (3.53), (3.54) are integrated numerically for 10 different random uniform distributions of the natural frequencies in the interval $[-\Delta\omega, \Delta\omega]$. The results are obtained by adiabatic continuation starting with the phase clusters found for $\Delta\omega = 0$ (see Fig. 3.13). Duplex one-cluster states of double antipodal type with **a** $\sigma = 0.5$, $\Delta = 0.02$ and **b** $\sigma = 0.5$, $\Delta = 0.07$ are shown as insets. Parameters are given by $\alpha^{11/22} = 0.3\pi$, $\alpha^{12/21} = 0.05$, $\beta^1 = 0.1\pi$, $\beta^2 = -0.95\pi$, $\epsilon = 0.01$, and $N = 12$. Figure taken from [252]

It is clearly visible that duplex one-cluster states of double-antipodal type are still present for a considerable range of heterogeneity $\Delta\omega$ of the natural frequencies. In the inset of Fig. 3.15a, a duplex one-cluster double-antipodal state of the same form is presented as shown in Fig. 3.13. The phases are distorted slightly due to the frequency distribution but the double-antipodal configuration is still clearly visible. The inset Fig. 3.15b shows another one-cluster state of double-antipodal type. However, due to the frequency mismatch the phase distribution becomes different compared with the one presented in Fig. 3.13.

A similar result, as shown in Fig. 3.15, is obtained when a Gaussian distribution of frequencies is considered instead of a uniform distribution, as seen in Fig. 3.16.

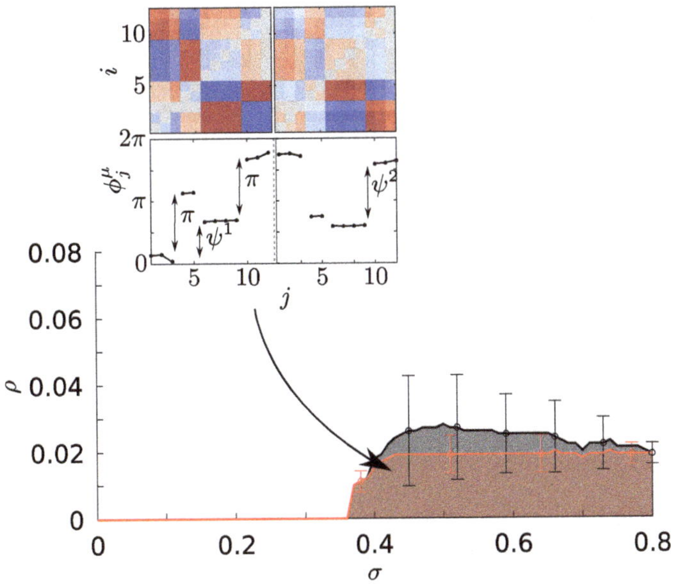

Fig. 3.16 The figure shows the standard deviation ρ where duplex one-cluster states in general (gray) and of the form presented in Fig. 3.13 (red) can be found. For this the Eqs. (3.53) and (3.54) are integrated numerically for 10 different random normal distributions of the natural frequencies with standard deviation ρ and zero mean. The results are obtained by adiabatic continuation starting with the phase clusters found for $\rho = 0$ (see Fig. 3.13). A duplex one-cluster states of double antipodal type with $\sigma = 0.5$, $\Delta = 0.02$ is shown as insets. Parameters are given by $\alpha^{11/22} = 0.3\pi$, $\alpha^{12/21} = 0.05$, $\beta^1 = 0.1\pi$, $\beta^2 = -0.95\pi$, $\epsilon = 0.01$, and $N = 12$. Figure taken from [252]

3.2.4 Multiplex Networks and their Decomposition

This section provides important tools and theorems for finding the spectrum of multiplex networks.

Theorem 1 *Let $\mathcal{R}$ be a commutative subring of $\mathbb{C}^{N\times N}$ and let $M \in \mathcal{R}^{L\times L}$. Then,*

$$\det_{\mathbb{C}} M = \det_{\mathbb{C}} (\det_{\mathcal{R}} M) . \tag{3.55}$$

The proof can be found in [266]. This rather abstract result allows for a very nice decomposition for pairwise commuting matrices and yields a useful tool to study the local dynamics in multiplex systems.

Proposition 2 *Let $M \in \mathbb{C}^{N\times N}$ be a unitary diagonalizable matrix with $M = U D_M U^H$ where U, U^H and D_M are a unitary, its adjoint and a diagonal matrix, respectively. Let further $\mathcal{D}_M$ be the set of simultaneously diagonalizable matrices to M, i.e., the set of all matrices which commute pairwise and with M. Then,*

$$\det\begin{pmatrix} A_{11} & \cdots & A_{1L} \\ \vdots & \ddots & \vdots \\ A_{L1} & \cdots & A_{LL} \end{pmatrix} = \det\left(\sum_{\sigma \in S_L}\left[sgn(\sigma)\prod_{\mu=1}^{L} D_{A_{\mu,\sigma(\mu)}}\right]\right) \tag{3.56}$$

where $A_{\mu\nu} \in \mathcal{D}_M$ for $\mu, \nu = 1, \ldots, L$ and S_L is the set of all permutations of the numbers $1, \ldots, L$.

Proof Consider any $A, B \in \mathcal{D}_M$, then they are simultaneously diagonalizable with M and hence $A = D_A U^H$ and $B = U D_B U^H$ with the same U. Thus, all $A_{\mu\nu}$ can be diagonalized with the same U. Since U is unitary,*i.e.* $(\det U)^2 = 1$, we find

$$\det\begin{pmatrix} A_{11} & \cdots & A_{1L} \\ \vdots & \ddots & \vdots \\ A_{L1} & \cdots & A_{LL} \end{pmatrix} = \det\begin{pmatrix} D_{A_{11}} & \cdots & D_{A_{1L}} \\ \vdots & \ddots & \vdots \\ D_{A_{L1}} & \cdots & D_{A_{LL}} \end{pmatrix} \tag{3.57}$$

by applying the block diagonal matrices $\text{diag}(U, \cdots, U)$ and $\text{diag}(U^H, \cdots, U^H)$ from the left and right, respectively. The set of diagonal matrices with usual matrix multiplication and addition form a commutative subring of $\mathbb{C}^{N\times N}$. Applying Theorem 1 and using the well-known determinant representation of Leibniz, Eq. (3.56) follows. □

Remark 3 The set $\mathcal{D}_M$ consists of all matrices which commute with M and all the other elements of $\mathcal{D}_M$. In particular, the identity matrix $\mathbb{I}_N \in \mathcal{D}_M$ for any $M \in \mathbb{C}^{N\times N}$.

In the following, the last result is applied to a duplex and triplex system, connecting the local dynamics on the one-layer network to the multiplex case. The consideration is specified by defining two special multiplex systems.

Definition 4 Suppose $A, B, C \in \mathbb{C}^{N\times N}$ and $m_{ij} \in \mathbb{C}$ $(i, j = 1, \ldots, 3)$. Then, the $2N \times 2N$ block matrix

$$M^{(2)} = \begin{pmatrix} A & m_{12}\mathbb{I} \\ m_{21}\mathbb{I} & B \end{pmatrix} \tag{3.58}$$

and the $3N \times 3N$ block matrix

$$M^{(3)} = \begin{pmatrix} A & m_{12}\mathbb{I} & m_{13}\mathbb{I} \\ m_{21}\mathbb{I} & B & m_{23}\mathbb{I} \\ m_{31}\mathbb{I} & m_{32}\mathbb{I} & C \end{pmatrix} \tag{3.59}$$

are called (complex) duplex and triplex network, respectively.

Suppose we know how to diagonalize the individual layer topologies. The next result shows how the eigenvalues of the individual layers are connected to eigenvalues of the multiplex system. This will be done for the duplex and triplex network. Two different proofs are provided for the following statement. The first approach makes use of Schur's decomposition [264, 265] which will be used, later on, in order to derive the characteristic equations. In particular, any $m \times m$ matrix $M = \begin{pmatrix} A & B \\ C & D \end{pmatrix}$ in the 2×2 block form can be written as

$$M = \begin{pmatrix} \mathbb{I}_p & BD^{-1} \\ 0 & \mathbb{I}_q \end{pmatrix} \begin{pmatrix} A - BD^{-1}C & 0 \\ 0 & D \end{pmatrix} \begin{pmatrix} \mathbb{I}_p & 0 \\ D^{-1}C & \mathbb{I}_q \end{pmatrix}. \tag{3.60}$$

With this, a simplified form of the determinant of a 2×2 block matrix $M = \begin{pmatrix} A & B \\ C & D \end{pmatrix}$ is derived, namely

$$\det(M) = \det(A - BD^{-1}C) \cdot \det(D), \tag{3.61}$$

An extension of the first approach to any number of layers in the network can be found by induction but is very technical, see [267]. The second approach uses Proposition 2 which allows for a straightforward extension to any number of layers in a multiplex network.

Proposition 5 *Suppose $A, B, C \in \mathbb{C}^{N \times N}$, they commute pairwise, and are diagonalizable with diagonal matrices D_A, D_B, D_C and unitary matrix U. Then, the eigenvalues μ for the multiplex networks $M^{(2)}$ and $M^{(3)}$ can be found by solving the N quadratic*

$$\mu^2 - ((d_A)_i + (d_B)_i)\,\mu + (d_A)_i(d_B)_i - m_{12}m_{21} = 0 \tag{3.62}$$

and cubic polynomial equations

$$\mu^3 + a_{2,i}\mu^2 + a_{1,i}\mu + a_{0,i} = 0, \tag{3.63}$$

respectively, with

$$a_{2,i} = -\left((d_A)_i + (d_B)_i + (d_C)_i\right) \tag{3.64}$$

$$\begin{aligned} a_{1,i} &= (d_A)_i(d_B)_i + (d_A)_i(d_C)_i + (d_B)_i(d_D)_i \\ &\quad - m_{12}m_{21} - m_{13}m_{31} - m_{23}m_{32} \end{aligned} \tag{3.65}$$

$$\begin{aligned} a_{0,i} &= m_{12}m_{21}(d_C)_i + m_{13}m_{31}(d_B)_i + m_{23}m_{32}(d_A)_i \\ &\quad - (d_A)_i(d_B)_i(d_C)_i - m_{12}m_{23}m_{31} - m_{13}m_{32}m_{21} \end{aligned} \tag{3.66}$$

and $(d_A)_i$, $(d_B)_i$, and $(d_C)_i$ the respective diagonal elements of D_A, D_B, and D_C.

Proof Since A, B, C are diagonalizable and commute, Proposition 2 can be applied to both matrices $M^{(2)}$, $M^{(3)}$. For the matrix $M^{(2)}$, another proof will be provided using Schur's decomposition.

The determinant is an antisymmetric multi-linear form. Thus, it can be written as

$$\det\left(M^{(2)} - \mu\mathbb{I}_{2N}\right) = \det\begin{pmatrix} A - \mu\mathbb{I}_N & m_{12}\cdot\mathbb{I}_N \\ m_{21}\cdot\mathbb{I}_N & B - \mu\mathbb{I}_N m \end{pmatrix} \tag{3.67}$$

$$= (-1)^N \det\begin{pmatrix} m_{12}\cdot\mathbb{I}_N & A - \mu\mathbb{I}_N \\ B - \mu\mathbb{I}_N & m_{21}\cdot\mathbb{I}_N. \end{pmatrix} \tag{3.68}$$

By assumption A and B are both diagonalizable with respect to the unitary transformation matrix U, and so are $A - \mu\mathbb{I}$ and $B - \mu\mathbb{I}$. This allows us to write

$$\det\begin{pmatrix} m_{12}\cdot\mathbb{I}_N & A - \mu\mathbb{I}_N \\ B - \mu\mathbb{I}_N & m_{21}\cdot\mathbb{I}_N. \end{pmatrix} = \det\begin{pmatrix} m_{12}\mathbb{I}_N & D_A - \mu\mathbb{I}_N \\ D_B - \mu\mathbb{I}_N & m_{21}\mathbb{I}_N \end{pmatrix} \tag{3.69}$$

and we apply the block diagonal matrices $\mathrm{diag}(U, \cdots, U)$ and $\mathrm{diag}(U^H, \cdots, U^H)$ from the left and right, respectively. Now, using Schur's decomposition in Eq. (3.60) the determinant can written as

$$\begin{aligned} \det&\begin{pmatrix} m_{12}\mathbb{I}_N & D_A - \mu\mathbb{I}_N \\ D_B - \mu\mathbb{I}_N & m_{21}\mathbb{I}_N \end{pmatrix} \\ &= n^N \det\left(m - \frac{1}{n}\left(D_A - \mu\mathbb{I}_N\right)\left(D_B - \mu\mathbb{I}_N\right)\right) \\ &= \det\left(m_{12}m_{21}\mathbb{I}_N - \left(D_A - \mu\mathbb{I}_N\right)\left(D_B - \mu\mathbb{I}_N\right)\right). \end{aligned} \tag{3.70}$$

The last expression together with $\det\left(M^{(2)} - \mu\mathbb{I}_{2N}\right) = 0$ yields the N quadratic Eq. (3.62).

Using that $(A - \mu\mathbb{I})$, $(B - \mu\mathbb{I})$, $(C - \mu\mathbb{I})$ commute pairwise, Proposition 2 can be applied. We find

$$\begin{aligned} &\det\left(M^{(3)} - \mu\mathbb{I}_{3N}\right) = \\ &\quad \det\left((D_A - \mu\mathbb{I}_N)\left[(D_B - \mu\mathbb{I}_N)(D_C - \mu\mathbb{I}_N) - m_{23}m_{32}\mathbb{I}_N\right]\right. \\ &\qquad - m_{21}\left[m_{12}(D_C - \mu\mathbb{I}_N) - m_{13}m_{32}\mathbb{I}_N\right] \\ &\qquad\quad \left. + m_{31}\left[m_{12}m_{23}\mathbb{I}_N - m_{13}(D_B - \mu\mathbb{I}_N)\right]\right) \end{aligned} \tag{3.71}$$

The last expression together with $\det\left(M^{(3)} - \mu\mathbb{I}_{3N}\right) = 0$ yields the N cubic Eq. (3.63). □

Let us briefly discuss some special cases for both the duplex and triplex network. Consider a duplex network with master and slave layer, *i.e.*, either $m_{12} = 0$ or $m_{21} = 0$. Then, the quadratic Eq. (3.62) yield

$$\left(\mu - (d_A)_i\right)\left(\mu - (d_B)_i\right) = 0. \tag{3.72}$$

As shown in Proposition 2, the eigenvalues for special triplex networks can be found by solving cubic equations. For the solution even closed forms exist. Despite this, the explicit form of the solutions is rather tedious, in general. However, if we consider $A = B = C$ and a ring-like inter-layer connection between the networks, *i.e.*, $m_{12} =$

$m_{23} = m_{31} = 0$, then Eq. (3.63) has the following solutions for all $j = 1, \ldots, N$

$$\mu_1 = -(d_A)_j + (m_{13}m_{32}m_{21})^{\frac{1}{3}}, \tag{3.73}$$

$$\mu_2 = -(d_A)_j + \frac{1}{2}\mathrm{i}(\mathrm{i} + \sqrt{3})\,(m_{13}m_{32}m_{21})^{\frac{1}{3}}, \tag{3.74}$$

$$\mu_3 = -(d_A)_j - \frac{1}{2}(\mathrm{i} + \sqrt{3})\,(m_{13}m_{32}m_{21})^{\frac{1}{3}}, \tag{3.75}$$

where i denotes the imaginary unit. In analogy to Eq. (3.72), a decoupling for the eigenvalues can be found. Consider three pairwise commuting matrices A, B, C, and the structure between the layers is a directed chain, i.e., $m_{12} = m_{13} = m_{31} = m_{23} = 0$, then

$$(\mu - (d_A)_i)\,(\mu - (d_B)_i)\,(\mu - (d_C)_i) = 0. \tag{3.76}$$

In the following section, the theory is applied to determine the stability of phase clusters in adaptive networks.

3.2.5 Regions of Synchronization in Adaptive Multiplex Networks

For an arbitrary duplex equilibrium of the form $\phi_i^\mu = \Omega t + a_i^\mu$ with

$$\begin{aligned} a_k^1 &= (0, \frac{2\pi}{N}k, \ldots, (N-1)\frac{2\pi}{N}k)^T, \\ a_k^2 &= \mathbf{a}_k^1 - \Delta a, \end{aligned} \tag{3.77}$$

we start with the linearized Eq. (3.50). This can also be written in the block matrix form

$$\begin{pmatrix} \delta\dot{\phi}^1 \\ \delta\dot{\phi}^2 \\ \delta\dot{\kappa}^1 \\ \delta\dot{\kappa}^2 \end{pmatrix} = \begin{pmatrix} A_1 & m_1\mathbb{I}_N & B_1 & 0 \\ m_2\mathbb{I}_N & A_2 & 0 & B_2 \\ C_1 & 0 & -\epsilon\mathbb{I}_{N^2} & 0 \\ 0 & C_2 & 0 & -\epsilon\mathbb{I}_{N^2} \end{pmatrix} \begin{pmatrix} \delta\phi^1 \\ \delta\phi^2 \\ \delta\kappa^1 \\ \delta\kappa^2 \end{pmatrix} \tag{3.78}$$

with $(\delta\phi^\mu, \delta\kappa^\mu)^T = \left(\delta\phi_1^\mu, \ldots, \delta\phi_N^\mu, \delta\kappa_{11}^\mu, \ldots, \delta\kappa_{1N}^\mu,\ \delta\kappa_{21}^\mu, \ldots, \delta\kappa_{NN}^\mu\right)^T$, the matrices A^μ, B^μ, and C^μ follow from Eq. (3.51), and $m_1, m_2 \in \mathbb{R}$. With the help of Schur's decomposition the characteristic equation for the linearized system takes the form

$$(\lambda + \epsilon)^{2(N^2-N)} = 0$$

$$\det \begin{pmatrix} (\lambda \mathbb{I}_N - A_1)(\lambda + \epsilon) - B_1 C_1 & -(\lambda + \epsilon) m_1 \mathbb{I}_N \\ -(\lambda + \epsilon) m_2 \mathbb{I}_N & (\lambda \mathbb{I}_N - A_2)(\lambda + \epsilon) - B_2 C_2 \end{pmatrix} = 0. \quad (3.79)$$

The second equation has the block matrix form which is required from Proposition 5. All blocks can be diagonalized and commute since they all possess a cyclic structure; compare Lemma 4.1 in [63]. Thus, Proposition 5 can be applied, and it is used to diagonalize the matrix in Eq. (3.79). For the diagonalized matrix, the following equations for the diagonal elements μ_i are found:

$$(\lambda + \epsilon)^2 m_1 m_2 - (p_i^1(\lambda; \alpha^{11}, \beta^1, \alpha^{12}, \sigma^{12}) - \mu_i)(p_i^2(\lambda; \alpha^{22}, \beta^2, \alpha^{21}, \sigma^{21}) - \mu_i) = 0 \quad (3.80)$$

where $i = 1, \ldots, N$, $p_i^{\mu}(\lambda; \alpha^{\mu\mu}, \beta^{\mu})$ is a second order polynomial in λ which depends continuously on α and β as well as functionally on the type of the one-cluster state. For every $i \in \{1, \ldots, N\}$, these equations will give us two eigenvalues $\mu_{i,1}$ and $\mu_{i,2}$ for the matrix in Eq. (3.79) depending on λ and the system parameters. Thus, Eq. (3.80) can be written as

$$\left(\mu_i - \mu_{i,1}(\lambda; \boldsymbol{\alpha}, \boldsymbol{\beta}, \boldsymbol{\sigma})\right)\left(\mu_i - \mu_{i,2}(\lambda; \boldsymbol{\alpha}, \boldsymbol{\beta}, \boldsymbol{\sigma})\right) = 0 \quad (3.81)$$

where $\boldsymbol{\alpha}, \boldsymbol{\beta}, \boldsymbol{\sigma}$ represent all system parameter chosen for Eqs. (3.31)–(3.32). In order to find the eigenvalue λ of the linearized Eq. (3.79) one of the eigenvalues μ has to vanish. This means that λ must be found such that Eq. (3.80) equals

$$\mu_i \left(\mu_i - \mu_{i,2}(\lambda; \boldsymbol{\alpha}, \boldsymbol{\beta}, \boldsymbol{\sigma})\right) = 0 \quad (3.82)$$

which is equivalent to finding λ such that the following quartic equation is solved

$$p_i^1(\lambda; \alpha^{11}, \beta^1, \alpha^{12}, \sigma^{12}) p_i^2(\lambda; \alpha^{22}, \beta^2, \alpha^{21}, \sigma^{21}) - (\lambda + \epsilon)^2 m_1 m_2 = 0. \quad (3.83)$$

Note that here the diagonal elements of A_1 are slightly different from those in Prop. 4.2 of [63] but they do not affect the result, i.e., the diagonal element equals $\rho_i(\alpha^{11}, \beta^1) - m_1(\alpha_{12})$. The same holds true for A_2. Thus, with the two possible eigenvalues $\rho_{i,1,2}(\alpha^{\mu\mu}, \beta^{\mu})$ for the monoplex system from Corr. 4.3 of [63] one finds the following quartic equation which give the Lyapunov exponents for the lifted duplex one-cluster

$$\begin{aligned} &\left[\left(\lambda - \rho_{i,1}(\alpha^{11}, \beta^1)\right) \cdot \left(\lambda - \rho_{i,2}(\alpha^{11}, \beta^1)\right) + m_1(\lambda + \epsilon)\right] \times \\ &\left[\left(\lambda - \rho_{i,1}(\alpha^{22}, \beta^2)\right)\left(\lambda - \rho_{i,2}(\alpha^{22}, \beta^2)\right) + m_2(\lambda + \epsilon)\right] - (\lambda + \epsilon)^2 m_1 m_2 = 0. \end{aligned} \quad (3.84)$$

In the case of a duplex antipodal one-cluster state given by Eq. (3.35) with $a_i^1 \in \{0, \pi\}$ and $a_i^2 = a_i^1 - \Delta a$, Eq. (3.50) possesses the following set of Lyapunov exponents

$$\mathcal{S}_{\text{Duplex}} = \{-\epsilon, \left(\lambda_{i,1}, \lambda_{i,2}, \lambda_{i,3}, \lambda_{i,4}\right)_{i=1,\dots,N}\} \tag{3.85}$$

where $\lambda_{i,1,\dots,4}$ solve the following N quartic equations

$$\begin{aligned}(\lambda + \epsilon)^2 m_1 m_2 - \left[\left(\lambda - \rho_{i,1}^1\right) \cdot \left(\lambda - \rho_{i,2}^1\right) + m_1(\lambda + \epsilon)\right] \times \\ \left[\left(\lambda - \rho_{i,1}^2\right)\left(\lambda - \rho_{i,2}^2\right) + m_2(\lambda + \epsilon)\right] = 0,\end{aligned} \tag{3.86}$$

with $m_1 \equiv \sigma^{12}\cos(\Delta a + \alpha^{12})$, $m_2 \equiv \sigma^{21}\cos(\Delta a - \alpha^{21})$ and the eigenvalues $\rho_{i,1,2}^{\mu} \equiv \rho_{i,1,2}(\alpha^{\mu\mu}, \beta^{\mu})$ for the monoplex system

$$\mathcal{S}_{\text{Monoplex}} = \left\{(0)_1, (-\epsilon)_{(N-1)N+1}, (\rho_1)_{N-1}, (\rho_2)_{N-1}\right\}, \tag{3.87}$$

where ρ_1 and ρ_2 solve $\rho^2 + \left(\epsilon - \cos(\alpha^{11})\sin(\beta^1)\right)\rho - \epsilon\sin(\alpha^{11} + \beta^1) = 0$. Here, the multiplicities for each eigenvalue are given as lower case labels. The proof and the results for other clusters can be found in [63, 64].

Using the Lyapunov exponents of the duplex antipodal clusters, the stability for duplex antipodal states can be found. The results are presented in Fig. 3.14.

3.2.6 Summary

In conclusion, a framework has been introduced to induce a variety of partial synchronization patterns, specifically phase clusters, within adaptively coupled networks of phase oscillators. While adaptive networks have recently gained significant attention across neuroscience, social sciences, biology, engineering, and other fields, and multilayer networks represent a key paradigm in modeling real-world complex systems, the interaction between multilayer structures and adaptivity remains underexplored. This work addresses this gap through rigorous theoretical analysis and computational simulations. It has been shown that symmetry in multiplexed multilayer networks can stabilize distinct phase cluster states, such as splay, antipodal, and double-antipodal configurations–patterns that may otherwise be unstable or nonexistent in single-layer networks.

Additionally, a novel method for analyzing the Laplacian matrices of duplex networks has been developed, offering insights into the spectral properties of these systems. This approach is readily extendable to networks with more than two layers. The multiplex decomposition method has broad applicability across physical, biological, socio-economic, and technological contexts, including plasticity dynamics in neurobiology, linear diffusive systems [268, 269], and generalizations of the master stability framework [78, 270] for adaptive networks. This multiplex decomposition has been applied to obtain analytic stability criteria for lifted states in multilayer systems. For local dynamics, the Kuramoto phase oscillator model was used, enriched with adaptive link strengths modulated by a phase-lag parameter. This setup allows for the modeling of various adaptation rules, including Hebbian, spike-timing-dependent plasticity, and anti-Hebbian forms of plasticity.

3.3 Pacemaker Control of Partial Synchronization in Multilayer Networks

Coupled networks of nonlinear oscillators facilitate the emergence of complex, partially synchronized spatiotemporal patterns, such as chimera states. In these states, synchronized (coherent) and desynchronized (incoherent) regions coexist within the same network. These regions emerge spontaneously, with no a priori knowledge of where the synchronized domains will form. Consequently, the ability to control the spatial distribution of coherent and incoherent domains within chimera states is of significant current research interest.

In this section, we explore mechanisms for controlling chimera patterns in multiplex networks composed of FitzHugh-Nagumo neurons. Specifically, we aim to demonstrate that remote control of chimera states is feasible by exploiting the layered structure of these multiplex networks. To achieve this, we introduce a pacemaker oscillator into the network. This pacemaker operates independently, receiving no input from the network but sending output to its neighboring oscillators. By positioning pacemakers in one or both network layers, we intentionally break the spatial symmetry within the layer(s), which in turn allows us to control the location of the incoherent domain.

We present evidence that remote control of chimera states is achievable under both unidirectional and bidirectional coupling conditions between the layers. Additionally, we discuss the limitations of these control strategies, particularly when extending the mechanisms from single-layer to multilayer network configurations.

3.3.1 Partial Synchronization and Chimera States

Chimera states represent a fascinating phenomenon of partial synchronization in networks of oscillators. First identified by Kuramoto and Battogtokh in 2002, these states are characterized by the coexistence of coherent (synchronized) and incoherent (desynchronized) regions within spatially symmetric oscillator networks [185]. Two years later, Abrams and Strogatz coined the term "chimera states" to describe these peculiar dynamical patterns [186]. Over the past two decades, chimera states have garnered significant attention within the nonlinear dynamics community [87,

271–287], attracting researchers from diverse fields including neuroscience, chemistry, and engineering. This growing interest has led to the identification of conceptual parallels between chimera states and various real-world dynamical systems [220, 275, 288–302]. Furthermore, the phenomenon has been experimentally verified in laboratory settings, with numerous studies successfully observing chimera states in physical systems [303–308].

Recently, a growing body of research has focused on the study of chimera states in multilayer networks [1, 102, 110, 225, 229, 230, 232, 234, 235, 309–316]. Multilayer networks serve as critical modeling frameworks for complex systems, including transportation systems, social networks, and the brain [208, 317]. Among these, neuronal networks represent a particularly promising area for the application of multilayer modeling [187]. For example, chimera states have been observed in two- and three-layer networks of Hindmarsh-Rose neurons [225].

In [309, 310], Majhi and colleagues investigated a two-layer network of Hindmarsh-Rose neurons, discovering that chimera states could emerge through multiplexing when neurons in one layer remained uncoupled. Further research has explored synchronization between chimera states in multilayer networks. In [229], the authors reported the presence of generalized synchronization between chimera states in a two-layer network of phase oscillators. Subsequently, [313] pioneered the investigation of chimera states in networks with layers of different sizes. In this work, a mean-field interlayer coupling was modeled, demonstrating that the phases of the order parameters in the two layers synchronized.

In [232], time delays were shown to control relay synchronization between chimera states in a three-layer multiplex network consisting of ring networks of FitzHugh-Nagumo oscillators. Additionally, the synchronization of chimera states in multiplex networks of phase oscillators with adaptive couplings within each layer was explored in [110], revealing further insights into the dynamics of these complex systems.

In finite-size single-layer networks, chimera states exhibit several instabilities. Firstly, these states are transient, eventually collapsing into a stable synchronous state, although they may persist for extended periods before doing so [272, 292]. Secondly, the coherent and incoherent groups that characterize chimera states drift across the network, causing the spatial configuration of the chimera to change over time [271]. These instabilities are particularly pronounced in smaller networks, where the initial positioning of the coherent and incoherent groups is highly sensitive to initial conditions.

Controlling such instabilities has been a focus of various studies on single-layer networks, involving different types of oscillators [318–325]. However, the control of chimera states in multilayer networks remains relatively unexplored. In 2019, Omelchenko and colleagues demonstrated that the "tweezers" control mechanism, originally introduced for single-layer networks [323], could also be applied to control chimera states in multiplex networks of Van der Pol oscillators [233]. Additionally, [315] showed that it is possible to clone a chimera state from one layer to another in a multiplex network, even when the coupling between layers is active only briefly. Furthermore, in a two-layer network of oscillatory FitzHugh-Nagumo units, a control strategy based on weak multiplexing was developed, enabling the induction or suppression of chimera states [102].

It is natural to explore whether alternative mechanisms for controlling chimera states can be effective in multilayer networks. In this section, we propose a method that leverages the interaction between a pacemaker oscillator [325] and the multiplex network structure. The primary advantages of this approach are its simplicity of implementation and its non-reliance on feedback from the system. By modifying only the connectivity structure while keeping the oscillators unchanged, we achieve control over the position of the incoherent region in one or both layers. The efficiency of this control depends on the specific configuration and the strength of interlayer coupling.

This section builds upon the results presented in [326] and is structured as follows: In Sect. 3.3.2, we describe the model and the numerical integration methods, introducing metrics for evaluating control efficiency and interlayer synchronization. The results, presented in Sect. 3.2.3, are divided into two parts, corresponding to the cases of unidirectional and bidirectional coupling. Finally, we conclude with a brief discussion in Sect. 3.2.4.

3.3.2 Pacemaker in a Neuronal Network

We study the dynamics of a two-layer network of FitzHugh-Nagumo oscillators. Each unit is characterized by two variables u and v. The layers are formed by N oscillators arranged in a ring topology. Following [274], inside each layer $l = 1, 2$, the oscillators are coupled nonlocally with range R and strength σ_l. The dynamics of the multiplex network is governed by the following set of differential equations for u_{lk}, v_{lk} and $k = 1, ..., N$ [102]:

$$
\begin{aligned}
\epsilon \frac{du_{1k}}{dt} &= u_{1k} - \frac{u_{1k}^3}{3} - v_{1k} + \frac{\sigma_1}{2R} \sum_{j=1}^{N} G_1(k, j)\{B_{uu}(u_{1j} - u_{1k}) + B_{uv}(v_{1j} - v_{1k})\} \\
&\quad + \sigma_{2\to1}(u_{2k} - u_{1k}), \\
\frac{dv_{1k}}{dt} &= u_{1k} + a + \frac{\sigma_1}{2R} \sum_{j=1}^{N} G_1(k, j)\{B_{vu}(u_{1j} - u_{1k}) + B_{vv}(v_{1j} - v_{1k})\}, \\
\epsilon \frac{du_{2k}}{dt} &= u_{2k} - \frac{u_{2k}^3}{3} - v_{2k} + \frac{\sigma_2}{2R} \sum_{j=1}^{N} G_2(k, j)\{B_{uu}(u_{2j} - u_{2k}) + B_{uv}(v_{2j} - v_{2k})\} \\
&\quad + \sigma_{1\to2}(u_{1k} - u_{2k}), \\
\frac{dv_{2k}}{dt} &= u_{2k} + a + \frac{\sigma_2}{2R} \sum_{j=1}^{N} G_2(k, j)\{B_{vu}(u_{2j} - u_{2k}) + B_{vv}(v_{2j} - v_{2k})\}. \quad (3.88)
\end{aligned}
$$

The bifurcation parameter a is set to 0.5 [274], so that each individual FitzHugh-Nagumo unit is in the oscillatory regime. The value of the time separation parameter ϵ is fixed at 0.05. The intralayer coupling scheme is constructed using a rotation matrix $\mathbf{B}$ (equal for both layers) and a coupling matrix $\mathbf{G}_l$ [274]:

$$\mathbf{B} = \begin{pmatrix} B_{uu} & B_{uv} \\ B_{vu} & B_{vv} \end{pmatrix} = \begin{pmatrix} \cos\phi & \sin\phi \\ -\sin\phi & \cos\phi \end{pmatrix}, \tag{3.89}$$

$$G_l(k,j) = \begin{cases} 1 & \text{if there is a link } j \to k \text{ in layer } l \\ 0 & \text{otherwise,} \end{cases} \tag{3.90}$$

where $j, k = 1, ..., N$ and $l = 1, 2$. For nonlocal coupling with a rectangular coupling kernel of range R, the coupling matrix has the structure shown in Fig. 3.17a ($N = 50$, $R = 18$). It is known that for ϕ slightly smaller than $\frac{\pi}{2}$ the uncoupled layers show chimera states, so we set $\phi = \frac{\pi}{2} - 0.1$ throughout all simulations [274]. Regarding the interlayer coupling scheme, Eq. (3.88) models a bidirectional multiplex configuration where each oscillator's u_{1k} variable in layer 1 is coupled to the corresponding u_{2k} variable in layer 2, and vice versa [102].

The interlayer coupling strength is a scalar parameter $\sigma_{2\to1}$ from layer 2 to layer 1 and $\sigma_{1\to2}$ in the other direction. We also consider unidirectional coupling, i.e., a driver-response configuration which corresponds to $\sigma_{1\to2} > 0$ and $\sigma_{2\to1} = 0$. In Fig. 3.18 we give a schematic representation of all the network configurations used in this study, without and with a control mechanism.

We integrate the system using a Runge-Kutta algorithm and exclude transients from the measurements by taking an evaluation interval I in which the first 10^4 steps are discarded. Following [274], for each layer, initial conditions are taken to be uniformly distributed on a circle of radius 2.

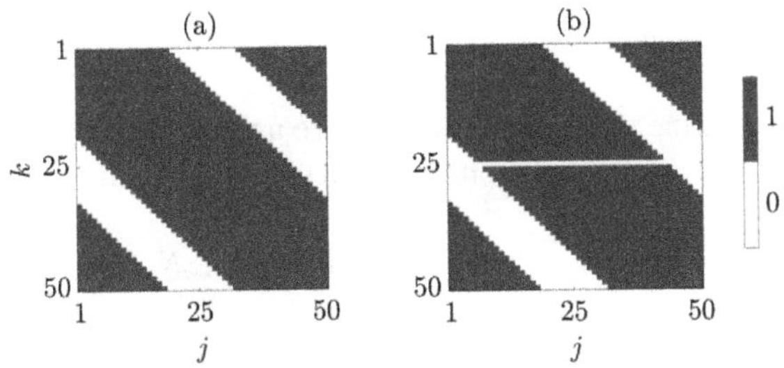

Fig. 3.17 Matrix representation of nonlocal coupling: **a** The connectivity matrix of a ring of nonlocally coupled oscillators. The network has $N = 50$ nodes and a rectangular coupling kernel with $R = 18$. **b** The modification of the connectivity matrix which corresponds to the presence of a pacemaker in position $p = 25$. All incoming connections of oscillator 25 are cut, which translates into the 25th row being set to 0. Figure from [326]

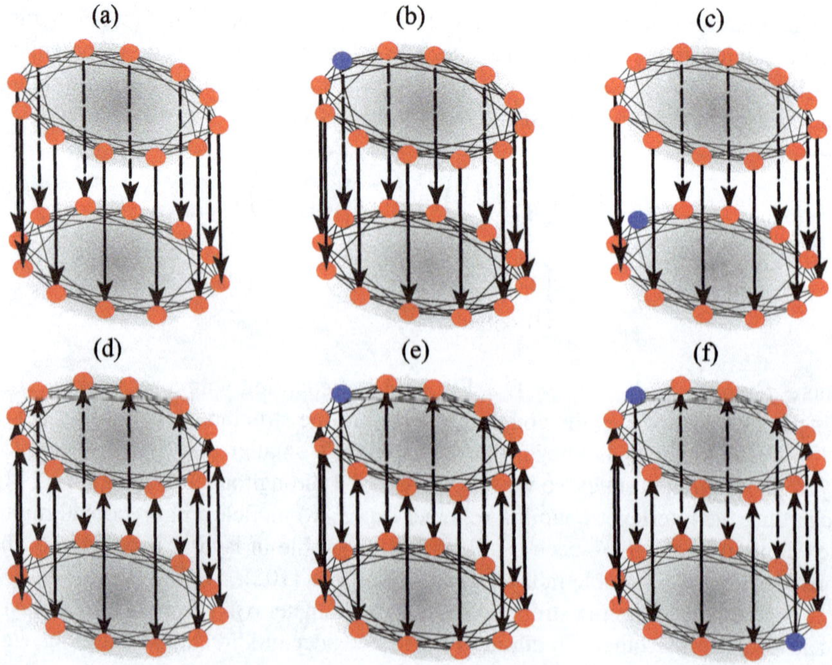

Fig. 3.18 Different combinations of multiplexing and control: All networks contain two rings of nonlocally coupled FitzHugh-Nagumo oscillators. The ring networks are highlighted in gray and the oscillator acting as a pacemaker in blue. The first row shows networks with unidirectional coupling, for which $\sigma_{2\to1} = 0$ and $\sigma_{1\to2} > 0$. **a** No control. **b** Pacemaker in layer 1, position 25. **c** Pacemaker in layer 2, position 25. In the second row we show networks with bidirectional coupling between the layers and $\sigma_{1\to2} = \sigma_{2\to1}$. **d** No control. **e** Pacemaker in layer 1, position 25. **f** Pacemaker in layer 1, position 25 and in layer 2, position 50. Figure from [326]

3.3.2.1 Control Impact

In this section we define some quantities that are used to analyze chimera state in a single-layer network. A geometric phase is calculated for each oscillator [274]:

$$\theta_{lk} = \arctan \frac{v_{lk}}{u_{lk}} \quad \text{for } k = 1, ..., N. \tag{3.91}$$

This quantity is used to compute local order parameters

$$Z_{lk} = \left| \frac{1}{2\delta + 1} \sum_{|m-k|\leq\delta} e^{i\theta_{lm}} \right| \tag{3.92}$$

and mean phase velocities, which serve as tools to display and detect chimeras:

$$\omega_{lk} = \frac{d\theta_{lk}}{dt} \quad \text{for } k = 1, ..., N \text{ and } l = 1, 2, \tag{3.93}$$

where i is the imaginary unit, δ determines the number of neighbors of one oscillator used to calculate its local order parameter, and $\langle\cdot\rangle_I$ is the time average over the evaluation interval I. As control mechanism, we use a pacemaker oscillator [325] in one or both layers. A pacemaker is an oscillator which is not receiving any input from the rest of the network but is sending output like all the others. Having a pacemaker in one of the layers corresponds to changing that layers' connectivity matrix $\mathbf{G}_l$ defined in Eq. (3.90). For example, a pacemaker in position $p_l = 25$ of layer l corresponds to setting to zero the 25th row of the matrix Gl defined in Eq. (3.90). This configuration is shown in Fig. 3.17b for $N = 50$.

When we have a pacemaker in position p_l of layer l we also multiply by 0 the interlayer coupling term $\sigma_{1\to2}(u_{1_{pl}} - u_{2_{pl}})$ and/or $\sigma_{2\to1}(u_{2_{pl}} - u_{1_{pl}})$, depending if the pacemaker is in layer 1, in layer 2 or in both layers. We want to study how the position of the chimera's incoherent group in the two layers of the network in Eq. (3.88) is affected by the interplay of multiplexing and control with a pacemaker. To do so, following the algorithm in Appendix B of [325], we calculate the position $c_l(t)$ of the center of the chimera's incoherent group. The center position $c_l(t)$ takes integer and half-integer values c in the interval $[0.5, N]$. We then define a binary function χ_l that codifies the evolution of the center position in space and time for each layer:

$$\chi_l(c, t) = \begin{cases} 1 & \text{if } c = c_l(t) \\ 0 & \text{otherwise.} \end{cases} \tag{3.94}$$

We define $\gamma_l(c) = \langle\chi_l(c, t)\rangle_I$, which determines the fraction of times for which the center of the incoherent group is in any possible position c. We then define the control impact $\Gamma_l(p_l, \Delta)$ for a pacemaker in position p_l as

$$\Gamma_l(p_l, \Delta) = \sum_{c=p_l-\Delta}^{p_l+\Delta} \gamma_l(c). \tag{3.95}$$

This control impact measures how many times the center of the chimera's incoherent group lies in a neighborhood of width $2\Delta + 1$ of the pacemaker position p_l during the interval I. In Fig. 3.19a we show a chimera state in a single-layer network of FitzHugh-Nagumo oscillators without any control. In this case the corresponding distribution $\gamma(c)$ is almost flat in Fig. 3.19b. If we introduce a pacemaker in position $p = 25$ and we solve Eq. (3.88) starting from the same initial condition, then we obtain the chimera state in Fig. 3.19c, which has the incoherent group centered around the pacemaker position. Now $\gamma(c)$ has a peak around position 25 in Fig. 3.19d.

The area highlighted in green in Figs. 3.19b and d corresponds to the control impact Γ_l. It follows from the definition in Eq. (3.95) that the control impact takes values between 0 and 1. In an uncontrolled network, there is no preferred position for the center of the incoherent group. Therefore, when the evaluation interval I becomes

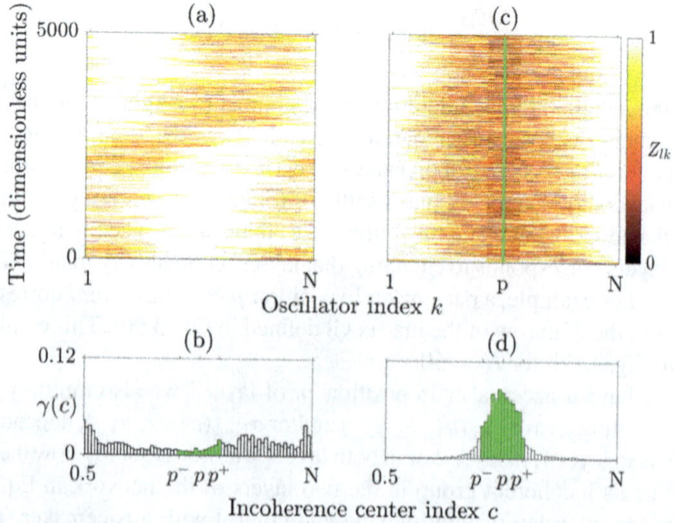

Fig. 3.19 A pacemaker controls the position of a chimera state in a single-layer network on FitzHugh-Nagumo oscillators: **a, c** Space-time plots of Z_{lk} without control (left column) and with a pacemaker in position $p = 25$ (right column). **b, d** The distribution γ of the position of the center of incoherence. The green area corresponds to the control impact $\Gamma(25, 5)$ where $p = 25$ and $\Delta = 5$. In panels **b** and **d** we define $p^- = p - \Delta$ and $p^+ = p + \Delta$. Network parameters: $N = 50$, $R = 18$, $\sigma = 0.2$, $\phi = \frac{\pi}{2} - 0.1$, $a = 0.5$, $\epsilon = 0.05$. Figure from [326]

arbitrarily large, the expected value of Γ_l is $\frac{4\Delta+1}{2N}$. This means that the center of the chimera's incoherent group is equally likely to occupy all positions $c \in [0.5, N]$ while it drifts along the network. However, here we show results for finite simulation times, therefore we cannot see a constant value of Γ even when there is no control. This is because there is no preferred initial position for the center of the incoherent group and the drifting is not fast enough to allow the center of the incoherent group to spend the same amount of time in all available positions.

3.3.2.2 Quantification of Synchronization

To assess the interplay between chimera states across the two layers we here introduce two quantities that measure the alignment of the incoherent groups of the chimera states and the interlayer synchronization of the two dynamics. First, we measure the degree to which the chimera states in the two layers align their position. To do this we calculate the difference between the position of the centers of the incoherent groups $c1$, $c2$ in the two layers on a circumference of length N:

$$D_{12} = \langle \min\{|c_1(t) - c_2(t)|, N - |c_1(t) - c_2(t)|\} \rangle_I. \tag{3.96}$$

A value of $D_{12} = 0$ denotes alignment of the incoherent groups, while if $D_{12} = N/2$ the center of the two incoherent groups are in antipodal positions. To quantify synchronization between the two layers we use the measures introduced in [232]. For $k = 1, ..., N$, the local interlayer synchronization error is defined as

$$E_{12}(k) = \langle \| x_{1k}(t) - x_{2k}(t) \| \rangle_I, \tag{3.97}$$

where $x_{lk} = \begin{bmatrix} u_{lk} \\ v_{lk} \end{bmatrix}$ for $l = 1, 2$. By taking also the spatial average in Eq. (3.97), one obtains the global interlayer synchronization error [232]:

$$E_{12} = \frac{1}{N} \sum_{k=1}^{N} E_{12}(k). \tag{3.98}$$

3.3.3 Control of Chimera States

In this section, we present results about control of chimera states obtained for unidirectional coupling and bidirectional coupling between the layers. The results are shown in Figs. 3.20 through 3.25, and detailed information about the parameters used in the simulations is given in the caption of Fig. 3.20. The results are presented with the same configuration for different combinations of multiplex coupling schemes and positions of the pacemaker. The figures are composed of two rows of three panels each, the first two columns show single-layer quantities, while the last one shows multilayer quantities. In Figs. 3.20 through 3.25 panels (a), (b), (c), and (d) show the control impact Γ values and a snapshot of the low coherence group [325] center position c for layer 1 and layer 2, respectively. For Γ, yellow means that the positions of chimera states are controlled. In this case the position of the center of the low coherence group takes values that are close to the pacemaker position. If there is no control, we observe a speckled pattern for c. In panel (e) we show the alignment of the low coherence group centers D_{12} and in panel (f) the global synchronization error E_{12}. We vary the interlayer coupling strength $\sigma_{1\to 2}$ between a minimum of 0 (corresponding to isolated layers) and a maximum of 0.05. Furthermore, we keep $\sigma_1 = 0.2$, and we vary σ_2 in the interval $[0.1, 0.2]$.

3.3.3.1 Unidirectional Coupling

First, we consider the case of unidirectional interlayer coupling from layer 1 to layer 2. We study three possible configurations: without any control, with a pacemaker in layer 1 in position $p_1 = 25$, and finally with a pacemaker in layer 2 in position $p_2 = 25$ while layer 1 is uncontrolled. Figure 4a and b confirm that in the absence of a control mechanism there cannot be any preferred position for the center of the

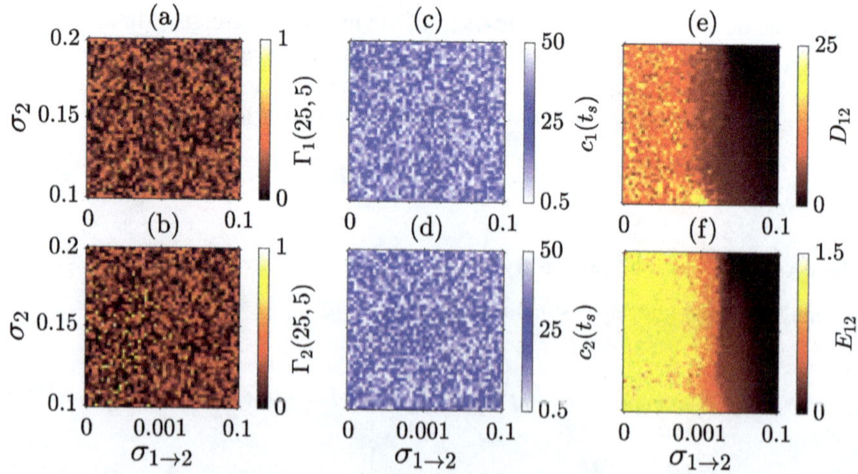

Fig. 3.20 Effect of unidirectional coupling on chimera states in a two-layer network of FitzHugh-Nagumo oscillators in the parameter space of the interlayer coupling $\sigma_{1\to2}$ and intralayer coupling σ_2 without pacemaker (see Fig. 3.18a). The interlayer coupling $\sigma_{1\to2}$ varies from 10^{-5} to 10^{-1} and is sampled on a logarithmic scale, while $\sigma_{2\to1}$ is kept equal to 0. We then added a column of results for $\sigma_{1\to2} = 0$, corresponding to isolated layers. The intralayer coupling $\sigma 2$ varies linearly from 0.1 to 0.2. Other parameters are $N = 50$, $R = 18$, $\sigma_1 = 0.2$, $\phi = \frac{\pi}{2} - 0.1$, $a = 0.5$, $\epsilon = 0.05$. Panels **a** and **b** show control impact values $\Gamma_1(25, 5)$ and $\Gamma_2(25, 5)$, while panels **c** and **d** are snapshots of the position $c_l(t_s)$ of the center of incoherence in layer $l = 1, 2$, respectively, at time $t_s = 500\,000$ time steps. **e** and **f** Synchronization measures: **e** alignment D_{12} and (f) global synchronization error E_{12}. Figure from [326]

incoherent group. This is reflected in values of $\Gamma_1(25, 5)$ and $\Gamma_2(25, 5)$ that are close to the expected value of $\frac{4\Delta+1}{2N}$. This means that in each realization the drifting causes the chimera's center to occupy all positions almost uniformly. This is confirmed by the snapshot of the centers of incoherence shown in Figs. 3.20c and d. At a certain moment in time, the center of incoherence is in different positions in different realizations. The incoherent groups of the chimera states align for strong interlayer coupling–the dark area in Fig. 3.20e–and, in general, the two layers synchronize their dynamics, as is shown in Fig. 3.20f, where the global synchronization error E_{12} reaches values close to zero.

The scenario described so far in the uncontrolled case is quite intuitive, while it is less obvious what happens when there is also a pacemaker present in the network. The results for the second configuration, obtained for a network with a pacemaker in position $p_1 = 25$ of the driving layer (layer 1), are shown in Fig. 3.21. Because of the unidirectional coupling, layer 1 is not receiving any input from layer 2, therefore it behaves as an isolated ring network. The almost constant value of $\Gamma_1(25, 5)$ close to 1 shows that the pacemaker can control the position of a chimera state in an isolated layer of FitzHugh-Nagumo oscillators as shown in Fig. 3.21a. These results generalize our observations for phase oscillators [325] to FitzHugh-Nagumo oscillators. The question now is whether the pacemaker in layer 1 can remotely control the

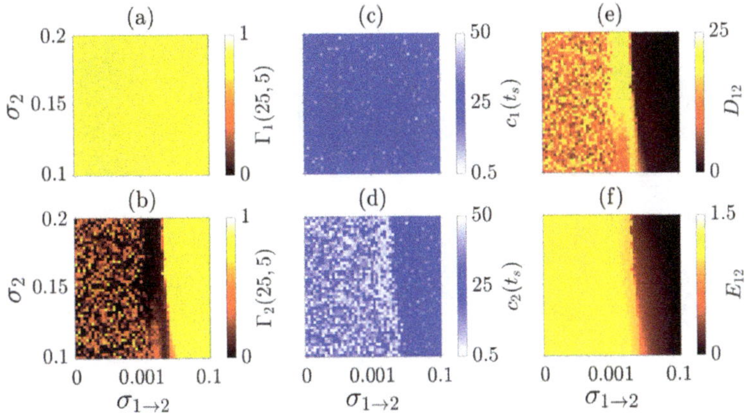

Fig. 3.21 With unidirectional coupling between the layers, it is possible to remotely control chimera states in layer 2 using a pacemaker only in layer 1. Same network configuration and parameters as in Fig. 3.20, but here there is a pacemaker in layer 1 in position $p_1 = 25$ (see Fig. 3.18b). Figure from [326]

position of the chimera state in layer 2 via the coupling. When the two chimeras in the two layers are aligned as depicted in Figs. 3.21e and f, we see that the answer is affirmative (yellow region in Fig. 3.21b). It is worth noticing that before the control becomes effective in layer 2 there is an intermediate region of parameters in which we observe that the center of the incoherent group in layer 2 tends to be diametrically opposed to the pacemaker position as indicated by the yellow region in Fig. 3.21e. Therefore the distance between the centers of the two low coherence groups reaches its maximum. The pacemaker has a repulsive action on the center of incoherence in layer 2 before the remote control starts to be effective.

Another interesting question is which one is stronger, the pacemaker or the driving? To address this problem, we move to the third configuration. In this case we still have unidirectional coupling from layer 1 to layer 2 and a pacemaker only in position $p_2 = 25$ of layer 2. As a consequence, there is no preferred position for the chimera states in layer 1, as we can see from the values of $\Gamma_1(25, 5)$ in Fig. 3.22a. In fact, Fig. 3.22a is qualitatively similar to Fig. 3.20a, since slight differences are due only to different realizations of the network in layer 1. As the chimera states align for high values of the interlayer coupling $\sigma_{1\to2}$ in Figs. 3.22b and e, the driving effect of layer 1 eventually wins over the controlling effect of the pacemaker in layer 2 in Fig. 3.22b. Nevertheless, it is interesting that there is a wide region of parameter space in which it is possible to control the position of the chimera in layer 2 despite the driving by layer 1.

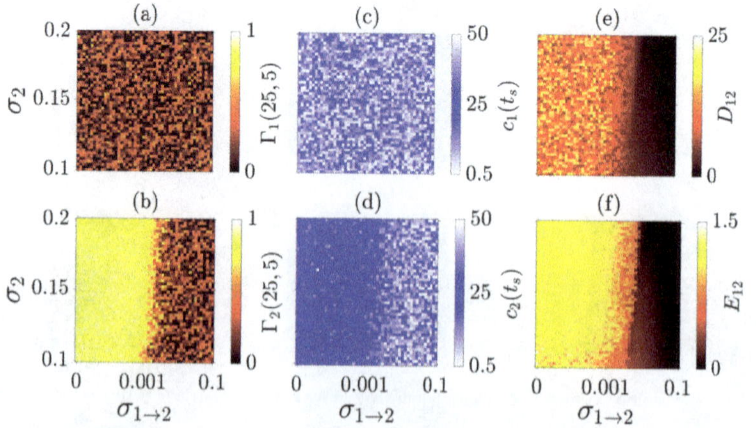

Fig. 3.22 The driving effect becomes stronger than the pacemaker effect for increasing interlayer coupling. Same network configuration and parameters as in Fig. 3.20, but here there is a pacemaker in layer 2 in position $p_2 = 25$ (see Fig. 3.18c). Figure from [326]

3.3.3.2 Bidirectional Coupling

We now consider bidirectional interlayer coupling and three possible configurations: without any control, with a pacemaker in layer 1 in position $p_1 = 25$, and finally with two conflicting pacemakers, one in layer 1 in position $p_1 = 25$ and one in layer 2 in position $p_2 = 50$. We vary the parameter σ_2 like in the unidirectional case and the parameter $\sigma_{1\leftrightarrow2}$ like $\sigma_{1\to2}$ in the unidirectional case.

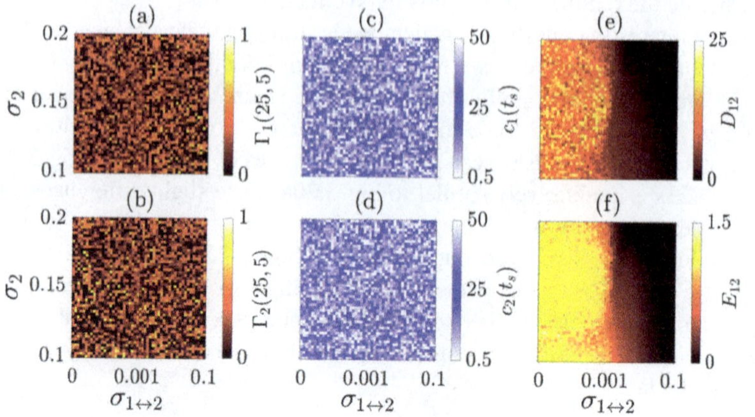

Fig. 3.23 Results for a duplex network of FitzHugh-Nagumo oscillators with bidirectional coupling. All network parameters are the same as in Fig. 3.20, except we have $\sigma_{2\to1} = \sigma_{1\to2} \equiv \sigma_{1\leftrightarrow2}$ (see Fig. 3.18d). Figure from [326]

In the first configuration (Fig. 3.23), we observe a similarity with the results obtained in the case of unidirectional coupling (Fig. 3.20). In this case, again there cannot be any preferred position for the chimera states in the two layers (see Fig. 3.23a and b). This is reflected also in the snapshots of the center position $c_{1,2}(t_s)$in Fig. 3.23c and d. The incoherent groups become aligned through a monotonic process (Fig. 3.23e), and in the same way the dynamics of the two layers become synchronized for increasing interlayer coupling (Fig. 3.23f).

The second configuration, in which a pacemaker is present in layer 1 at position $p_1 = 25$, leads to results that are different from the unidirectional case. In Fig. 3.24a we see that the pacemaker is able to control the chimera's position in layer 1 only up to a certain value of interlayer coupling $\sigma_{1\leftrightarrow 2}$ (yellow region). Above this value, the coupling between the layers takes over and the chimera states are aligned (Fig. 3.24e) but still do not have a preferred position. The main difference with the unidirectional case is that the remote control of the chimera state in layer 2 via the pacemaker in layer 1 and the coupling is possible only in a small region of the parameter space, indicated by the yellow stripe in Fig. 3.24b.

In the third and last configuration we have conflicting pacemakers trying to control the chimera states in the two layers, one in layer 1 in position $p_1 = 25$ and one in layer 2 in position $p_2 = 50$. In this case we see that the control works in both layers in a certain region of the parameter space, corresponding to the yellow areas in Figs. 3.25a and b. Note that there is another small region for which the control in layer 1 continues to work. A similar effect, even more pronounced, is observed with a single pacemaker in layer 1 (Fig. 3.24a) and is correlated with antipodal alignment in Fig. 3.24e.

An analogous effect can also be observed in the transition to remote control with unidirectional coupling and a pacemaker in layer 1 as shown in Fig. 3.21e.

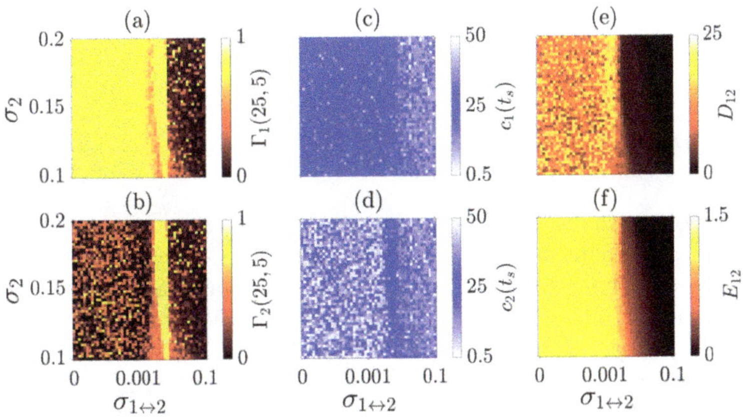

Fig. 3.24 Bidirectional coupling makes remote control via a pacemaker more difficult. Same network configuration and parameters as in Fig. 3.23, but here a pacemaker is present in layer 1, position $p_1 = 25$ (see Fig. 3.18e). Figure from [326]

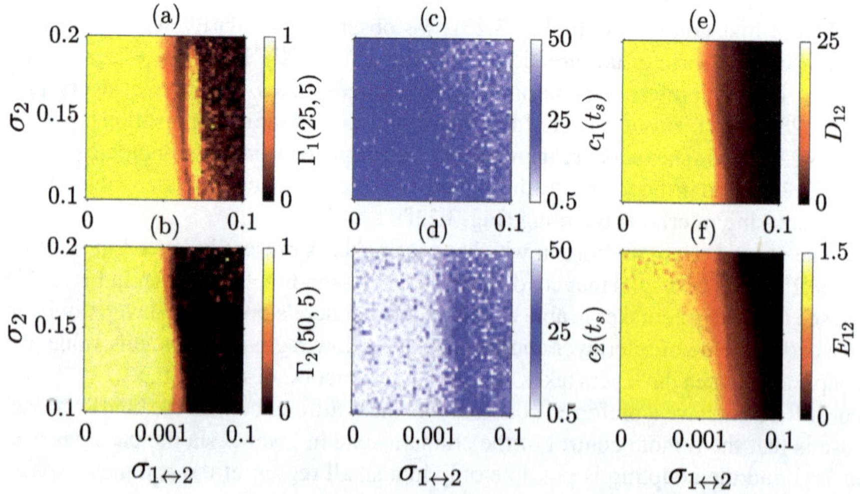

Fig. 3.25 Conflicting pacemakers and bidirectional coupling cause the chimera states to compromise on their position. Same network configuration and parameters as in Fig. 3.23, but here in each layer there is a pacemaker, in position $p_1 = 25$ and $p_2 = 50$ [see Fig. 3.18f]. Figure from [326]

For stronger interlayer coupling, the control effect ceases to prevail, but the chimeras become aligned, as is to be expected (see Fig. 3.25e). Interestingly, the two centers are aligned, but their positions do not coincide with either of the pacemakers' positions, as we can deduce from the low values of both coupling impact measures $\Gamma_1(25, 5)$ and $\Gamma_2(50, 5)$ in the rightmost part of Fig. 3.25a and b. Looking at the center snapshots in Fig. 3.25c and d, we see that the center is positioned close to 12.5 or 37.5, which are halfway from the pacemakers' positions. It is worth noticing that the synchronization region in Fig. 3.25f becomes smaller when two or more pacemakers are present in the network.

3.3.4 Summary

In summary, we find that in a duplex network with a driver-response configuration, characterized by unidirectional coupling between layers, there is a complex interaction between pacemaker control of chimera states and interlayer synchronization. This interaction can be exploited to design networks where chimera states appear in both layers at specific positions. Notably, the ability to remotely control chimera states in layer 2 via a pacemaker located in layer 1 is crucial in scenarios where direct access to parts of the network is limited. However, bidirectional coupling complicates this remote control, making it less effective. In cases where two pacemakers exert conflicting influences, at low interlayer coupling strengths, each pacemaker attracts the incoherent groups to its own location. As coupling increases, the incoherent groups in both layers synchronize in a position midway between the two pacemakers.

Interestingly, in prior studies, pacemakers were primarily used to induce full synchronization [327, 328]. It is notable that the same tool, when combined with nonlocal coupling, can be used for an entirely different objective: controlling chimera states. Transferring control methods from single-layer to multilayer networks is nontrivial due to the diverse configurations that can arise. Our results demonstrate that there are extensive regions of parameter space where the control mechanism proposed in [325] effectively controls chimera states in one or both layers. This section extends the findings of [325] in two key ways: from phase to FitzHugh-Nagumo oscillators, and from single-layer to two-layer multiplex networks.

An intriguing aspect for future investigation is the counterintuitive behavior observed in Fig. 3.21, where control becomes effective through a non-monotonic process. Initially, layer 2 resists being controlled by layer 1, with the center of incoherence positioning itself as far from the pacemaker as possible before control is eventually established.

The control mechanism we propose has several advantages for potential experimental applications. These include its straightforward implementation and the fact that it does not require feedback from the system or modifications to the intrinsic dynamics of the oscillators. While our model is simple, it effectively serves the purpose of exploring control strategies for partial synchronization patterns, particularly in contexts that may be linked to brain dynamics [16, 297, 298, 301]. As such, our findings could provide a foundation for developing methods to control chimera states in both experimental and real-world settings.

3.4 Relay Synchronization in Complex Networks

Relay synchronization in complex networks refers to the phenomenon where synchronization between distant regions of a network is mediated by an intermediary, or relay, node. In multilayer networks, synchronization between layers that are not directly connected can be achieved through signal transmission via intermediate relay layers. This section investigates relay synchronization in the context of partial synchronization patterns, such as chimera states, within three-layer networks composed of interacting FitzHugh-Nagumo oscillators. Our findings demonstrate that relay synchronization is robust even in the presence of small-world type topological irregularities within the network layers. Furthermore, we show that introducing randomness in the connectivity patterns of either the remote network layers or the relay layer increases the range of interlayer coupling strengths over which relay synchronization is sustained.

The study of synchronization in networks of coupled oscillatory units is a rapidly developing field with broad applications in both natural and technological systems [10, 208]. Gaining a deeper understanding of the complex mechanisms that give rise to the coexistence of synchrony and asynchrony is vital for advancing our knowledge of many biological [329] and engineered systems [330, 331]. A particularly challenging phenomenon in this area is remote synchronization, where distant components of a complex network synchronize despite lacking direct connections. In multilayer networks, synchronization between remote layers can occur through interactions with intermediate relay layers, a scenario commonly observed in multiplex networks [240]. These multiplex networks are a subclass of multilayer networks in which each layer contains an identical set of nodes, with one-to-one interlayer connections between corresponding nodes. Depending on the nature of the network nodes and intralayer topologies, complex partial synchronization patterns, such as clusters, chimera states, or solitary states, can emerge within the layers. Recently, relay synchronization of chimera states–characterized by spatial coexistence of coherent and incoherent regions–has attracted attention due to its wide-ranging applications [232]. In this section, we investigate the robustness of relay synchronization of chimera states in three-layer multiplex networks of FitzHugh-Nagumo oscillators, a well-established model in neuroscience. Our results indicate that relay synchronization is sensitive to variations in the connectivity structure of both the remote layers and the relay layer. The following section is related to my publication [316].

3.4.1 Relay Synchronization in Multilayer Networks

The analysis of complex networks is essential for understanding a wide range of real-world systems, including social networks [212], economics [332], ecology [333], finance [334], transportation systems [215, 335], and neural activity in the brain [217, 218, 220, 297]. Many of these systems can be effectively modeled as multilayer networks [211, 336], where elements are structured into distinct layers with different types of intra- and interlayer interactions. A specific type of multilayer network is the multiplex network, in which each layer contains the same set of nodes, with one-to-one connections between corresponding nodes in adjacent layers. Such network structures are particularly relevant to both social and technological systems [12], including dynamical processes where layers represent either different states of a system over time [337] or distinct types of interactions. Multilayer frameworks facilitate full [270] or partial [252] synchronization across layers, giving rise to various synchronization phenomena, such as explosive synchronization [108].

Relay (or remote) synchronization is a compelling phenomenon in multiplex networks, where non-adjacent layers achieve synchronization through an intermediate relay layer. This behavior has been observed in individual systems, such as lasers [330] and electronic circuits [331], as well as in networks of distant node pairs [338–340]. Relay synchronization can enhance overall synchrony in complex networks with heterogeneous units [340], and recent studies have demonstrated its occurrence in remote pairs of layers [232, 240, 286, 314, 341], including networks with intricate topologies such as Erdős-Rényi and scale-free structures [240]. This phenomenon has practical applications, such as improving distant coordination for encryption key distribution and secure communication [341], and plays a significant role in human brain networks, where relay structures like the thalamus [342] and hippocampus [329] facilitate communication between different brain regions. The brain's architecture exhibits small-world topologies at both macro and micro scales [343, 344]. In this section, we expand on our previous work on relay synchronization in nonlocally coupled ring networks [232], applying it to more complex small-world-like topologies. This extended investigation of relay synchronization mechanisms may provide valuable insights into certain brain functions.

Reference [340] has emphasized the significance of structural and dynamical symmetries in networks for the emergence of distant synchronization in relay synchronization phenomena. While several studies have explored the dynamical aspects of remote synchronization at the level of individual nodes [240, 338, 345], the relay synchronization of partial synchronization patterns, such as chimera states, remains underexplored. Chimera states are characterized by the coexistence of synchronized (coherent) and desynchronized (incoherent) spatial regions and have been identified in various networks of coupled oscillators [13, 14, 185, 186, 232, 282, 284–286, 346–348]. Similar phenomena occur in the brain, where partially synchronized patterns resembling chimera states have been observed in processes such as unihemispheric sleep [220, 349–351], perceptual organization [352], sensory behavior [353, 354], and epileptic seizures [297, 355, 356]. These chimera-like patterns have also been numerically demonstrated in networks of coupled neurons with nonlocal connectivity [273, 274, 281, 296, 357].

This section examines the impact of layer topologies on relay synchronization dynamics in a three-layer network of FitzHugh-Nagumo oscillators [358]. In Sect. 3.4.2, each layer is structured as a nonlocally coupled ring, which supports the formation of chimera states within individual layers [274, 281]. In this configuration, two remote layers achieve synchronization via an intermediate relay layer. Additionally, a distinct "double chimera" regime is identified, where only the coherent regions of the chimera states synchronize across the remote layers [232]. To assess the robustness of relay synchronization, perturbations are introduced to the network topologies within the layers in Sect. 3.4.3. First, using the Watts-Strogatz algorithm [359], links are randomly removed from the remote layers and replaced with random shortcuts. Second, topological inhomogeneities are introduced within the relay layer, while regular ring topologies are maintained in the remote layers. Interestingly, it is found that these topological modifications expand the range of interlayer coupling strengths over which relay synchronization is sustained.

3.4.2 Triplex Neuronal Network

A multiplex network consisting of three layers is considered, schematically shown in Fig. 3.26. Each layer, labeled by $i = 1, 2, 3$, contains N nodes which form a nonlocally coupled ring where each node is connected with its R nearest neighbors to both sides. Throughout the section, $N = 500$ and $R = 170$ are fixed. Between the

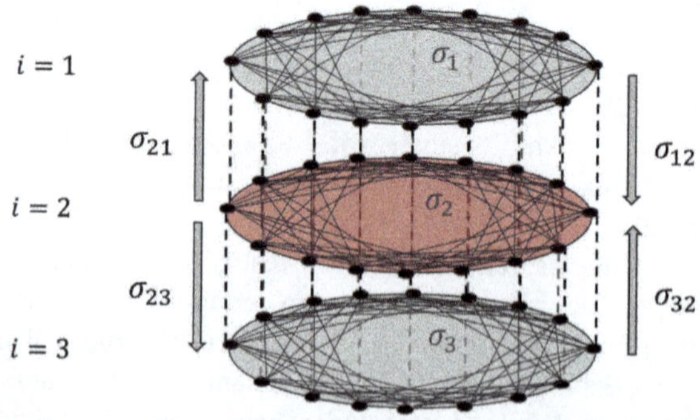

Fig. 3.26 Illustration of a triplex network with nonlocally coupled ring topology with intralayer coupling strength σ_i and interlayer coupling strength σ_{ij} $(i, j = 1, 2, 3)$. The remote layers $(i = 1, 3)$, depicted in grey are connected through the relay layer $(i = 2)$, marked in red. Figure from [316]

layers, the nodes are bidirectionally connected with their corresponding counterpart from the neighboring layers. The dynamical variable $\mathbf{x}_k^i = (u_k^i, v_k^i)^T$ of each node is governed by

$$\dot{\mathbf{x}}_k^i(t) = \mathbf{F}(\mathbf{x}_k^i(t)) + \frac{\sigma_i}{R}\sum_{l=1}^{N} a_{kl}^i \mathbf{H}[\mathbf{x}_k^i(t) - \mathbf{x}_l^i(t)]$$
$$+\sum_{j=1}^{3} \sigma_{ij}\mathbf{H}[\mathbf{x}_k^j(t) - \mathbf{x}_k^i(t)], \qquad (3.99)$$

where $k = 1, ..., N$ numbers the nodes, $i = 1, 2, 3$ labels the layer, and $a_{kl}^i \in [0, 1]$ are the elements of the adjacency matrix $\mathbf{A}^i$, determining the fundamental topology of layer i, σ_i is the intralayer coupling strength, and σ_{ij} is the interlayer coupling strength. The local dynamics of each oscillator is governed by the FitzHugh-Nagumo (FHN) system:

$$\mathbf{F}(\mathbf{x}) = \begin{pmatrix} \frac{1}{\epsilon}(u - \frac{u^3}{3} - v) \\ u + a \end{pmatrix} \qquad (3.100)$$

where u is the activator (membrane potential), and v is the inhibitor (recovery variable comprising all inhibitory processes). The threshold parameter $a = 0.5$ and the time-scale separation parameter $\epsilon = 0.01$ are fixed throughout the section, ensuring oscillatory dynamics of the individual units. The interlayer coupling scheme for the triplex network is defined as follows:

$$\boldsymbol{\sigma} = \begin{pmatrix} 0 & \sigma_{12} & 0 \\ \sigma_{21} & 0 & \sigma_{23} \\ 0 & \sigma_{32} & 0 \end{pmatrix}. \tag{3.101}$$

By choosing $\sigma_{12} = \sigma_{32}$ and $\sigma_{21} = \sigma_{23} = \frac{\sigma_{12}}{2}$ the interlayer coupling is bidirectional and has a constant row sum. Throughout this section the interaction between the neurons is realized through a rotational coupling matrix in order to include also cross-coupling between activator and inhibitor.

$$\mathbf{H} = \begin{pmatrix} \epsilon^{-1}\cos\phi & \epsilon^{-1}\sin\phi \\ -\sin\phi & \cos\phi \end{pmatrix} \tag{3.102}$$

Fixing the coupling phase to $\phi = \frac{\pi}{2} - 0.1$ allows for the observation of chimera states within the layers [274].

To analyze and distinguish different forms of synchronization between the layers in our networks, the following measures are employed: The global interlayer synchronization error E^{ij} introduced in Sect. 3.3.2.2 quantifies the synchronization between layers i and j:

$$E^{ij} = \lim_{T\to\infty} \frac{1}{NT} \int_0^T \sum_{k=1}^{N} \left\| \mathbf{x}_k^i(t) - \mathbf{x}_k^j(t) \right\| \mathrm{d}t, \tag{3.103}$$

where $\|\cdot\|$ is the Euclidean norm, and $i, j = 1, 2, 3$ denote to the layer number[232]. The global interlayer synchronization error is equal to zero when all corresponding oscillators in two layers are synchronized and perform identical dynamics. Based on this measure, one can distinguish the following types of synchronization:

1. **Full interlayer synchronization:** All layers are completely synchronized ($E^{12} = E^{13} = 0$).
2. **Relay interlayer synchronization:** Remote layers are completely synchronized, and not synchronized to the relay layer ($E^{13} = 0$ and $E^{12} \neq 0$).
3. **Interlayer desynchronization:** Different patterns in each layer ($E^{12} \neq E^{13} \neq 0$)

While the global interlayer synchronization error gives us information about the complete synchrony of the layers, it is not able to distinguish possible regimes of partial synchronization between the layers. To obtain more detailed information, the local interlayer synchronization error E_k^{ij} for corresponding pairs of nodes k from two layers is analyzed:

$$E_k^{ij} = \lim_{T\to\infty} \frac{1}{T} \int_0^T \left\| \mathbf{x}_k^i(t) - \mathbf{x}_k^j(t) \right\| \mathrm{d}t. \tag{3.104}$$

This measure uncovers complex regimes where only some of the oscillator pairs are synchronized, thus vanishing there.

Nonlocally coupled ring topologies within the layers allow for the observation of chimera states within the layers [274]. An important feature of these patterns is the difference of mean phase velocities for synchronized and desynchronized groups of oscillators [185, 186]. Usually, oscillators in synchronized domains are phase-locked and have identical frequencies, while oscillators from the incoherent domain speed up (or slow down) and the mean frequency profile assumes an arc-like shape. The mean phase velocity for each oscillator is calculated as

$$\omega_k = \frac{2\pi M_k}{\Delta T}, \; k = 1, 2, ...N, \tag{3.105}$$

where M_k is the number of complete oscillations of the kth node during the average time ΔT.

3.4.3 Effect of Topology upon Relay Synchronization

Before introducing topological inhomogeneities in the network, the case of the unperturbed network is addressed, where full and partial relay synchronization is observed [232]. In contrast to [232] the interlayer coupling delay is neglected. The main point of relay synchronization is that the two remote layers are not directly connected, but via the relay, which breaks the symmetry of the interlayer couplings between all three layers and assigns a special topological role to the relay as a mediator. As a result, diverse dynamical patterns may arise, including relay synchronization between the outer layers as well as full synchronization between all three layers, or desynchronization, depending upon the interlayer coupling strength. Figure 3.27 displays averaged results of five numerical simulations for different sets of random initial conditions for the three-layer network Eq. (3.99). Figure 3.27a shows the local synchronization errors between the remote layers and the relay layer (upper panel, orange), and between the two remote layers (middle panel, blue), with increasing interlayer coupling strength. Figure 3.27b condenses those results by depicting the global synchronization errors. One can identify three characteristic regimes. Note that the horizontal axis starts with $\sigma_{ij} > 0$, as the decoupled dynamics of the triplex network at very small σ_{ij} are not of interest.

In the first case, for small interlayer coupling strength $\sigma_{ij} \equiv \sigma_{12} = \sigma_{32} \geq 0.05$, relay synchronization with zero error E^{13} between the remote layers 1 and 3 is observed, along with a nonzero error E^{12} between the remote and relay layers. In the second regime, with further increase of σ_{ij}, all three layers are desynchronized. Finally, for large interlayer coupling strength, in the third regime, the whole network is completely synchronized ($E^{12} = E^{13} = 0$). The inset demonstrates an example of relay synchronization for $\sigma_{ij} = 0.075$ (region shaded orange in panel (b)): upper panels depict snapshots of chimera states in three layers, and bottom panels show the corresponding mean phase velocity profiles ω_k^i together with local synchronization errors E_k^{ij}.

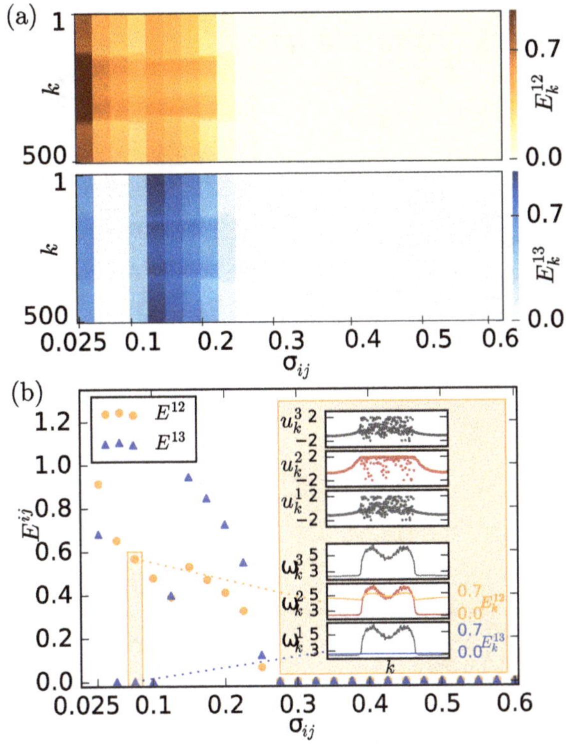

Fig. 3.27 **a** Local and **b** global synchronization error with identical layers versus interlayer coupling strength $\sigma_{ij} \equiv \sigma_{12} = \sigma_{32}$. For each parameter value, averaging is performed over 5 sets of random initial conditions, with different realizations used for the three layers. One can distinguish three regions of dynamics with increasing $\sigma_{ij} \geq 0.05$: (i) relay synchronization, (ii) desynchronization and (iii) full synchronization. The insets in (**b**) show snapshots u_k^i and mean phase velocity profiles ω_k^i together with local synchronization errors E_k^{12} (orange) and E_k^{13} (blue) for a coupling strength of $\sigma_{ij} = 0.075$. The relay layer is marked in red and the remote layers in grey. Parameters are $\epsilon = 0.05$, $a = 0.5$ $R = 170$, $\phi = \frac{\pi}{2} - 0.1$, $N = 500$, $\sigma_i = 0.2$ for layers $i = 1, 2, 3$. The simulation time is $t_{max} = 2000$ with time steps of $\Delta t = 0.05$. Figure from [316]

In the following, the connectivities inside the network layers are changed step by step towards a random topology following the Watts-Strogatz algorithm [359]. For this, the parameter p is introduced, which determines the probability for each link of the regular ring to be replaced by a random link, thereby serving as a measure for the degree of randomness in the network. First, the influence of the parameter p on the network properties is investigated.

With increasing probability of random shortcuts, the clustering coefficients C and average shortest path length L within the inhomogeneous layer are analyzed. Figure 3.28a demonstrates that increasing the randomness p results in a decreasing clustering coefficient, corresponding to the formation of low- and high-degree nodes (Fig. 3.28c), which leads to growing topological inhomogeneity within one layer.

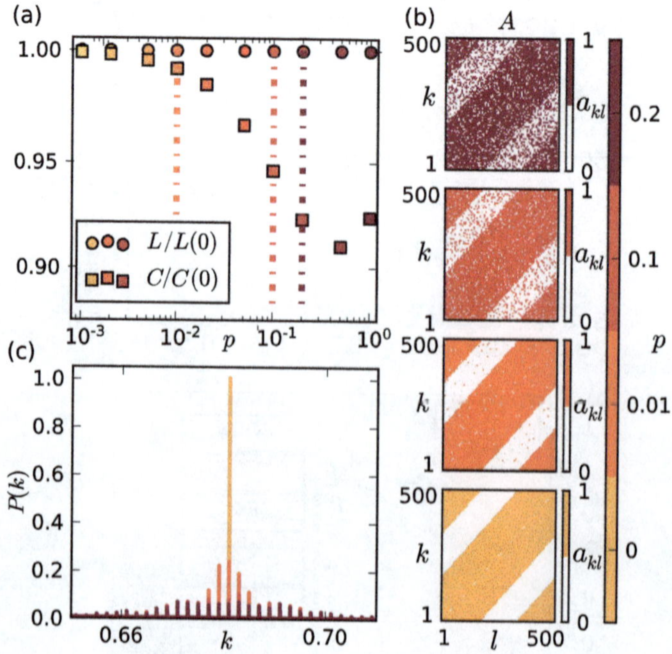

Fig. 3.28 **a** Normalized shortest path length $L(p)/L(0)$ (circles) and normalized clustering coefficient $C(p)/C(0)$ (squares) averaged over 5 random realizations of the Watts-Strogatz connectivity for increasing rewiring probability p (color coded, see color scale on the right), generated with the Watts-Strogatz algorithm [359]. **b** $N \times N$-adjacency matrices a_{kl} for different p ($p = 0, 0.01, 0.1, 0.2$, see color scale on the right). **c** Histograms showing the normalized degree distribution $P(k)$ of a Watts-Strogatz graph with $N = 500$, $R = 170$ and increasing p ($p = 0, 0.01, 0.1, 0.2$, see color scale on the right). Other parameters as in Fig. 3.27. Figure from [316]

Due to the small ensemble size used for averaging, artifacts may arise, such as the data point at $p = 1$, which does not continue to decrease. At the same time, the shortest path length remains almost constant, as shown in Fig. 3.28a. Due to the large nonlocal coupling range R the shortest path length is already quite small for the regular ring ($p = 0$) and is not changed much by the introduction of shortcuts. This is in contrast to classical small-world networks which start rewiring from a sparse nonlocally coupled ring network (small coupling range), and hence exhibit a pronounced transition from large C and L (sparse ring) to small C and L (random network) via an intermediate small-world regime with large C and small L.

The adjacency matrices of a layer are shown in Fig. 3.28b for different p. For $p = 0$ the matrix is symmetric and contains spatially separated homogeneous regions of coupling (colored) and no coupling (white), representing the regular nonlocally coupled ring. Increasing p leads to the creation of short-cuts in the network which connect distant spatial domains. In order to examine the influence of p upon the relay synchronization scenario, the focus is on the global synchronization error versus

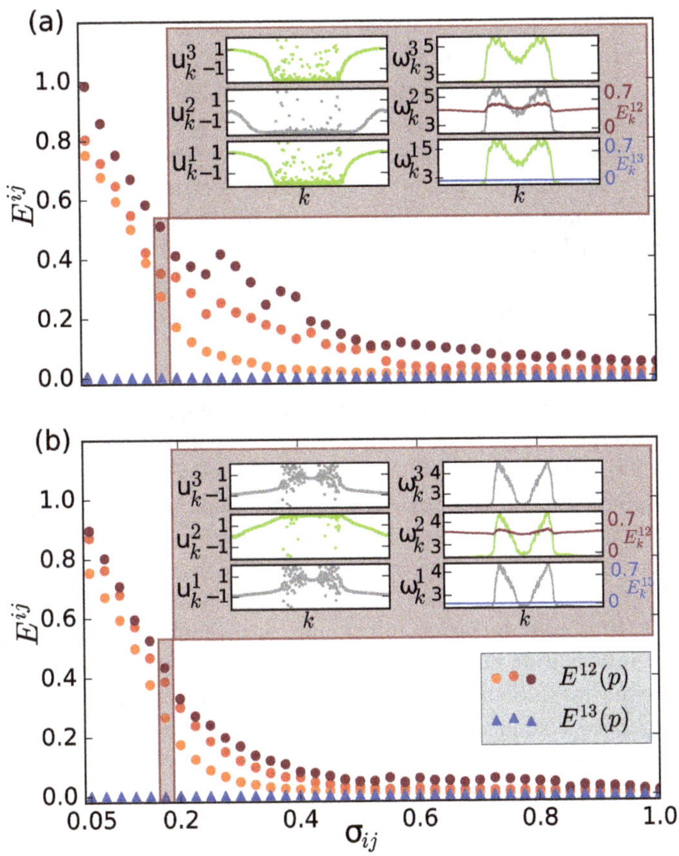

Fig. 3.29 Global synchronization error E^{ij} versus coupling strength $\sigma_{ij} \equiv \sigma_{12} = \sigma_{32}$, where $E^{13} = 0$ is shown by blue triangles, and E^{12} is marked by colored circles for 3 different values of p ($p = 0.01, 0.1, 0.2$, color-coded as in Fig. 3.28). For each value of p, averaging is performed over 5 network realizations with different random sets of initial conditions. **a** Topology of the remote layers is modified, **b** topology of the relay layer is changed. The insets show snapshots u_k^i and mean phase velocity profiles ω_k^i with local synchronization errors E_k^{12} (red), E_k^{13} (blue) for all layers for $\sigma_{ij} = 0.175$ and $p = 0.2$. The manipulated layers are marked green. Other parameters as in Fig. 3.27. Figure from [316]

the interlayer coupling strength (Fig. 3.29). Note that the horizontal axis starts with $\sigma_{ij} > 0$, as the decoupled dynamics of the triplex network at very small σ_{ij} are not of interest. In the following, the topology of the layers is modified with the rewiring probability p in two different ways: The topology of the remote layers is either changed (by applying the same random realization to both layers), or the relay layer is modified, while maintaining the regular nonlocally coupled ring structure in the relay layer or remote layers, respectively.

Our simulations with random initial conditions show that besides the familiar scenario (relay interlayer synchronization - desynchronization - full interlayer synchronization) observed for the unperturbed three-layer network, frequently a different scenario occurs, where the system changes its dynamics directly from relay to full synchronization without being desynchronized in between. This new transition is shown in Fig. 3.29. For both cases of inhomogeneity increasing p leads to a growing global synchronization error E^{12}, such that the threshold value of interlayer coupling σ_{ij}^*, up to which the system remains relay-synchronized, before becoming fully synchronized, shifts substantially towards higher values. Snapshots and mean phase velocity profiles with local synchronization errors are shown in the inset, depicting the relay synchronization of chimera states for an interlayer coupling of $\sigma_{ij} = 0.175$ and $p = 0.2$. Here, the emergence of a two-headed chimera state is further observed, i.e., two incoherent domains, in the case of an inhomogeneous relay layer. In case of inhomogeneous remote layers the two-headed chimera state is not observed. As it is known from Ref. [274], two-headed chimera states emerge due to a subtle interplay between coupling strength and nonlocality in the coupling structure. Remarkably, the relay layer is more sensitive to the introduction of nonlocality in the sense of Watts and Strogatz. Note that with increasing randomness $p > 0.2$ the intervals of σ_{ij} where chimeras exist successively shrink, but relay synchronization is still found.

Figure 3.30 illustrates the results of our numerical simulations for three non-zero values of the link rewiring probability p. It compares the critical interlayer coupling strength σ_{ij}^* at the transition from relay synchronization to complete synchronization, and for comparison also gives the threshold of relay synchronization for $p = 0$. It is found that networks with inhomogeneous remote layers have higher thresholds,

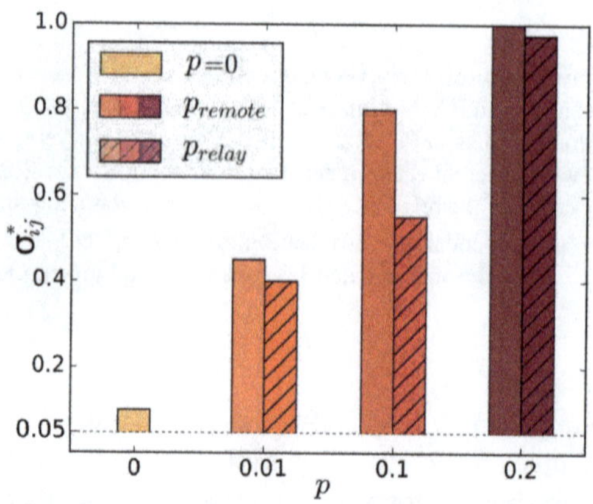

Fig. 3.30 Threshold value σ_{ij}^* up to which the system remains relay-synchronized depending on the interlayer coupling strength σ_{ij} and the link rewiring probability $p = 0, 0.01, 0.1, 0.2$. The values $\sigma_{ij}^*(p)$ are calculated from the data of Fig. 3.29 where the system is assumed to be fully synchronized if $E^{12} \leq 0.02$. The left and right bars refer to inhomogeneous remote (plain) or relay layer (hatched), respectively. Figure from [316]

while the transition to complete synchronization for a network with an inhomogeneous relay layer occurs at slightly lower coupling strengths. Our analysis shows that introducing randomness in the topology of either the remote or the relay layer has an advantageous effect upon the robustness of the regime of relay synchronization in the three-layer network. With increasing rewiring probability, the threshold values of the interlayer coupling strength, up to which the relay-synchronized state is preserved, increase significantly.

3.4.4 Summary

In conclusion, we have investigated complex synchronization dynamics in three-layer networks of FitzHugh-Nagumo oscillators with topological inhomogeneities generated using a small-world algorithm. In a triplex network with identical, homogeneous layers, varying the interlayer coupling strength results in different synchronization regimes, including relay synchronization, desynchronization, and complete synchronization. To assess the robustness of these regimes, we introduced topological inhomogeneities by randomly rewiring regular links in either the remote layers or the relay layer and replacing them with random shortcuts with a probability p.

In addition to the established scenarios for regular nonlocally coupled ring networks, we identified a novel phenomenon: a direct transition from relay synchronization to complete synchronization, bypassing the desynchronized state. This finding suggests that introducing inhomogeneities generally enhances synchronization. These results are consistent with previous studies on single-layer networks [360], which show that regular ring networks exhibit low synchronizability that significantly improves with the addition of random connections [361] or link rewiring [362]. Our findings extend these observations to multilayer systems, demonstrating that heterogeneous network topologies influence the transitions between relay and full synchronization [363], and generalizing them to include partial synchronization patterns such as chimera states.

We show that random inhomogeneities in network topology positively influence relay synchronization by broadening the parameter space over which it occurs, with this effect becoming more pronounced as the rewiring probability p increases. A comparison of topological inhomogeneities in the remote layers versus the relay layer reveals that inhomogeneity in the remote layers more effectively promotes relay synchronization, extending its range over a wider span of interlayer coupling strengths.

By maintaining a relatively large coupling range R within the initial network layers, chimera states are observed, each containing a single incoherent domain in each layer. In more diluted network topologies, characterized by nonlocally coupled rings with a smaller coupling range, chimera-like patterns featuring a greater number of alternating coherent and incoherent domains would emerge. However, even with a large R, the introduction of inhomogeneities in the relay layer can induce chimeras with two incoherent domains across all three layers.

3.5 Relay Synchronization via a Hub

In this section, we explore various relay synchronization scenarios in a three-layer network, where the intermediate (relay) layer is represented by a single node, functioning as a hub. The two remote layers consist of nonlocally coupled rings of FitzHugh-Nagumo oscillators, modeling neuronal dynamics. Each node in the remote layers is connected to the hub. The influence of the hub and its role in sustaining chimera states is analyzed with respect to the interlayer coupling strength and time delay between layers. We identify tongue-like regions in the parameter space that exhibit double chimeras–chimera states in the remote layers where the coherent regions are synchronized–as well as salt-and-pepper states. Remarkably, at very low intralayer coupling strengths, where chimera states do not naturally form in individual layers, the presence of the hub can induce their formation. Additionally, we investigate the impact of link dilution between the remote layers and the hub on the network dynamics. The most significant effect of dilution occurs when connections to the coherent regions of the chimeras are removed.

Relay synchronization plays a critical role in neuronal dynamics, particularly in the human brain, where certain regions are connected via smaller intermediary structures, or hubs. Physiological evidence suggests that various synchronization phenomena are closely related to the underlying network structure, significantly influencing brain activity. In such relay networks, the speed of communication is affected by the spatial distance between regions, introducing time delays for information transmission to remote areas. Both the time delay and the number and strength of connections between different brain regions are key factors in determining the emergence of synchronization patterns. These aspects are the primary focus of this section, which builds on my previous work [364].

3.5.1 *Synchronization Scenarios in Multilayer Networks*

Many fascinating phenomena in nature can be described using the framework of nonlinear dynamics. The study of such dynamical systems is a broad and well-established field, encompassing various aspects such as synchronization, chaos control, and methods for reconstructing attractors and dynamical systems from experimental time series [365–367]. Synchronization, in particular, is pervasive in biological systems and is of significant interest in networks of oscillators [10, 12]. Recent research has focused on partial synchronization patterns [1, 14, 86, 86, 87, 348], such as chimera states, which are characterized by the coexistence of coherent and incoherent oscillations within spatially distinct domains [185, 186]. A striking feature of these states is that they emerge in systems of identical oscillators arranged in symmetric ring topologies with symmetric interactions, coexisting with a completely synchronized state. In the past decade, there has been growing interest in phase and amplitude chimeras, both in time-continuous systems [13, 368–375] and time-discrete maps [97, 100, 234, 280, 311, 314, 376–388]. These patterns are not limited to phase oscillators but are found in various systems, including neural dynamics [16, 187, 220, 274, 281, 297, 389].

A crucial area of network science is multilayer networks, which have unveiled new perspectives by describing systems that are interconnected via different types of links. These systems include intra-layer interactions, as well as inter-layer interactions between nodes across layers [208–211]. Examples of such structures include neuronal and social networks, where individuals are linked by different patterns of contacts or interactions [212]. Other applications encompass communication, supply, and transport networks, such as power grids, subway systems, and air traffic networks [215]. Notably, multilayer networks can generate and stabilize diverse partial synchronization patterns in adaptive networks [112, 390], where connectivity evolves over time [103].

One particularly intriguing phenomenon in multiplex networks is relay, or remote, synchronization. In this scenario, layers that are not directly connected synchronize through an intermediate relay layer [240]. Remote synchronization, in which distant nodes synchronize despite their separation on the network graph, has been shown to depend on the symmetry of the network [331, 338, 339, 341, 391, 392]. In neuroscience, several instances have been identified where specific brain regions act as functional relays, influencing signal propagation and overall brain function [393, 394]. For example, thalamic relay cells serve as both the primary conduits of sensory information from the periphery to the cortex and as interactive hubs for cortical communication [342, 395–397]. These relay structures facilitate visual processing [398], and rapid coordination of spatially segregated cortical computations, which are critical for cognitive flexibility and control, as well as its disruption in disease states [399]. Parahippocampal regions similarly act as relay stations, modulating impulse traffic between the neocortex and hippocampus, with significant implications for neural activity propagation [400]. Furthermore, the hippocampus plays a key role in cortico-cortical theta synchronization [329, 330], with signal transmission between cortical and subcortical brain regions being fundamental to many brain functions [401]. Notably, partial relay synchronization has been observed in mice experiments [329], where only a portion of the hippocampal relay synchronizes with the cortical regions, which themselves exhibit partial zero-lag synchronization.

Going towards more realistic models, time-delay plays a significant role in the modeling of the dynamics of complex networks [1]. In brain networks, the communication speed is affected by the distance between regions and therefore a stimulation applied to one region needs time to reach a different region. Thus, the introduction of time delay between the layers in a network is not only a factor towards more realistic modeling, but entering into the system delay time can also work as a powerful parameter to control dynamical patterns.

A simple realization of a system which exhibits relay synchronization is a triplex network where a relay layer in the middle acts as a transmitter between the two outer layers. Recently the notion of relay synchronization has been extended from completely synchronized states to partial synchronization patterns in the individual layers of a three-layer multiplex network. It has been shown that the three-layer structure of the network allows for (partial) synchronization of chimera states in the outer layers via the relay layer [232, 286, 314, 316, 347]. In this section, we consider the special configuration where the relay layer is reduced to a single node, i.e., a hub. The dynamics on each node is given by the paradigmatic FitzHugh-Nagumo model of neuronal dynamics. In Sect. 3.5.2 we introduce the model. In Sect. 3.5.3 we investigate how the partial synchronization of chimera states depends upon the inter-layer coupling delay and strength. In Sect. 3.5.4 we show that chimera states which do not exist in single layers can be induced by the hub. In Sect. 3.5.5 we analyze the influence of network topology, specifically we study how a dilution of the inter-layer connections between the outer layers and the hub plays a role for the chimera states in the outer layers. Section 3.5.6 gives a conclusion.

3.5.2 *Three-Layer Networks with a Hub*

We consider a three-layer network, in which the outer layers consist of a ring of N identical FitzHugh-Nagumo (FHN) oscillators [358, 402, 403] with non-local (intra-layer) coupling of coupling range R^m and the middle layer consists of an individual node, i.e., a hub. An illustration of the considered model is presented in Fig. 3.31. Layers $m = 1$ and 3 (orange) are coupled through the intermediate layer 2 (hub), so that the middle layer acts as a relay between the two outer layers, but there is no inter-layer coupling between layers 1 and 3. The dynamical equations for the outer layers ($m = 1, 3$) are given by

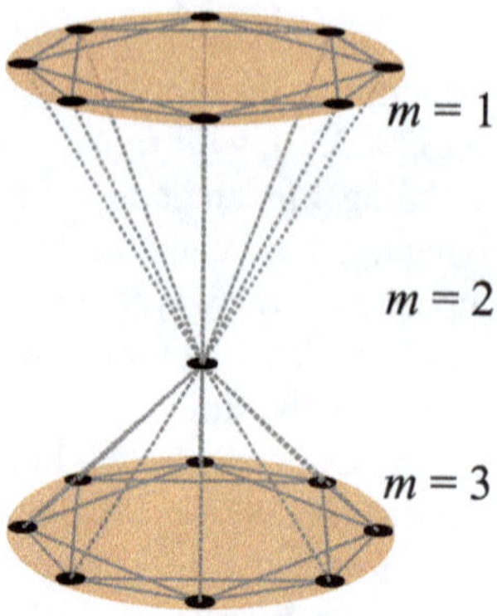

Fig. 3.31 Illustration of the three-layer network with a hub. All layers $m = 1, 2, 3$ consist of identical FitzHugh-Nagumo oscillators, where each of the outer layers $m = 1, 3$ includes 500 nodes arranged in a non-local ring topology and the relay layer $m = 2$ consists of a single node. The intra-layer coupling within the outer layers is instantaneous and is represented by solid lines, while the inter-layer coupling between the outer layers and the hub is time-delayed and is depicted by dotted lines. Figure taken from [364]

$$\dot{\mathbf{x}}_i^m(t) = \mathbf{F}(\mathbf{x}_i^m(t)) + \frac{\sigma^m}{2R^m} \sum_{j=i-R^m}^{i+R^m} \mathbf{H}[\mathbf{x}_j^m(t) - \mathbf{x}_i^m(t)] + \\ + \sum_{l=1}^{3} \sigma^{ml} \mathbf{G}[\mathbf{x}^2(t-\tau) - \mathbf{x}_i^m(t)], \tag{3.106}$$

and for the hub ($m = 2$) which acts as an active relay by

$$\dot{\mathbf{x}}^2(t) = \mathbf{F}(\mathbf{x}^2(t)) + \sum_{l=1}^{3} \sum_{i=1}^{N} \frac{\sigma^{2l}}{N} \mathbf{G}[\mathbf{x}_i^l(t-\tau) - \mathbf{x}^2(t)], \tag{3.107}$$

where $\mathbf{x}_i^m = (u, v)^T \in \mathbb{R}^2$, $m \in \{1, 2, 3\}$, $i \in \{1, ..., N\}$ with all indices modulo N, denotes the set of activator (u) and inhibitor (v) variables, $\mathbf{x}_i^2 \equiv \mathbf{x}^2$ for the hub, and the dynamics of each individual oscillator is governed by

$$\mathbf{F}(\mathbf{x}) = \begin{pmatrix} \varepsilon^{-1}(u - \frac{u^3}{3} - v) \\ u + a \end{pmatrix}, \tag{3.108}$$

where $\varepsilon > 0$ describes the time scale separation between fast activator and slow inhibitor, fixed at $\varepsilon = 0.05$ throughout this section. Depending on the threshold parameter a, the single FHN elements exhibit either oscillatory ($|a| < 1$) or excitable ($|a| > 1$) behavior. Here we choose the oscillatory regime ($a = 0.5$). The parameter σ^m denotes the intra-layer coupling strength (σ_{intra}), while σ^{ml} is the inter-layer coupling strength (σ_{inter}). We use time delay τ only in the inter-layer coupling, since in real-world systems the transfer of information between two different layers is often slower than within one layer. In order to ensure constant row sum we choose the inter-layer coupling as $\sigma_{\text{inter}} = \sigma^{12} = \sigma^{32}$, $\sigma^{21} = \sigma^{23} = \sigma^{32}/2$ and $\sigma^{11} = \sigma^{13} = \sigma^{22} = \sigma^{31} = \sigma^{33} = 0$. The inter-layer interaction is only through the activator variables, i.e., the coupling matrix $\mathbf{G}$, whereas the intra-layer interaction is realized through rotational coupling with coupling matrix $\mathbf{H}$

$$\mathbf{G} = \begin{pmatrix} \varepsilon^{-1} & 0 \\ 0 & 0 \end{pmatrix}, \quad \mathbf{H} = \begin{pmatrix} \varepsilon^{-1}\cos\phi & \varepsilon^{-1}\sin\phi \\ -\sin\phi & \cos\phi \end{pmatrix} \tag{3.109}$$

and coupling phase $\phi = \frac{\pi}{2} - 0.1$. The latter coupling scheme, which consists predominantly of activator-inhibitor cross-coupling, is similar to a phase-lag of approximately $\pi/2$ in the Kuramoto phase oscillator model and has been chosen such that chimera states are most likely to occur [274]. Chimera states exhibit typical arc-shaped mean phase velocity profiles.

Note that the hub receives input from N nodes in each outer layer, while each node of the outer layers receives inter-layer input only from the hub. In order to balance the inputs, and secure that the input which the hub receives does not outweigh the local dynamics by orders of magnitude, we introduce the normalization factor $\frac{1}{N}$ in the coupling term in Eq. (3.107), which scales the coupling term such that it becomes of the same order as the local hub dynamics; thus we have an active relay. If we drop the normalization factor $\frac{1}{N}$, which we will do only in Sect. 3.5.3, the local hub dynamics does not play a role any more, and the hub becomes effectively a passive relay.

To detect partial relay synchronization patterns between the outer layers, we compute the local synchronization error E_i^{13}:

$$E_i^{13} = \lim_{T\to\infty} \frac{1}{T} \int_0^T \left\| \mathbf{x}_i^1(t) - \mathbf{x}_i^3(t) \right\| dt. \tag{3.110}$$

This measure calculates the difference in variables $\mathbf{x}$ between two corresponding nodes with index i in the outer layers, averaged over time T [232]. Since we calculate the errors only between pairs of nodes, this measure is well suited to detect partial relay synchronization patterns, where only parts of the outer layers are synchronized with each other.

Introducing a time delay τ in the coupling term often leads to travelling patterns as shown, e.g., in Ref. [346]. To avoid any numerical artifacts caused by those, we will make use of the method of detrending introduced in Ref. [232]. By detrending the simulated data afterwards we can avoid this problem: After each time-step in the numerical simulation we re-index the nodes i in such a way that $i' = (i + c)|_N$, where | stands for the absolute value and c is given by the center of the largest domain of the ring where for all i's of that domain $\left\|\mathbf{x}_i^m(t) - \mathbf{x}_{i+1}^m(t)\right\| < \theta$. As the threshold parameter θ we choose $\theta = 0.25$.

3.5.3 Partial Synchronization of Chimera States

To analyze the spatiotemporal dynamics of our three-layer network, we run numerical simulations for initial conditions randomly distributed on the circle $u^2 + v^2 = 4$ with varying inter-layer time delay τ and inter-layer coupling strength σ_{inter}. In this Section, we are considering the passive relay, i.e., we drop the normalization factor $\frac{1}{N}$ in the coupling term in Eq. (3.107). Since we are particularly interested in synchronization patterns between chimera states in the outer layers, we set the parameters controlling intra-layer coupling and topology to values which are known to induce chimeras in isolated layers [274]. Chimera states are characterized by the spontaneous emergence of coexisting coherent and incoherent domains on completely symmetrical networks. Especially intriguing is their resemblance with some patterns observed in brain dynamics like epileptic seizures [16].

The results of our numerical simulations are summarized in Fig. 3.32, which shows the parameter plane of inter-layer coupling strength and inter-layer time delay with the domains of different observed partial relay synchronization patterns, organized in evenly spaced tongue-like regions. At least two more tongues exist for larger time delays but are not mapped out in detail. The cross-hatched orange and hatched red

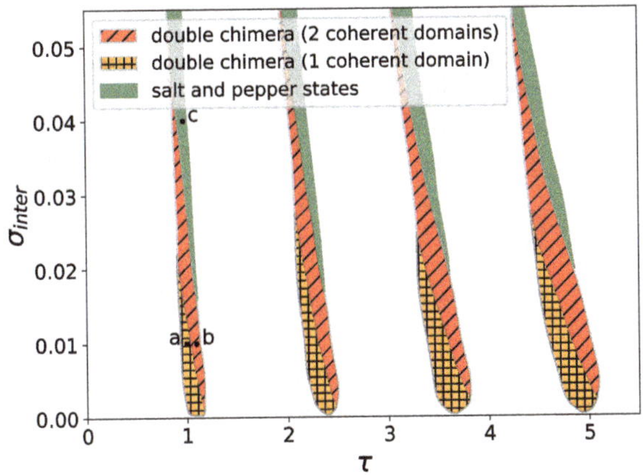

Fig. 3.32 Partial synchronization tongues in the parameter plane of the inter-layer coupling strength $\sigma_{\text{inter}} = \sigma^{12}$ and time delay τ. Partial synchronization patterns including chimera states (cross-hatched orange), double chimera states (hatched red), and salt and pepper states (green) emerge in tongue-like regions. All results have been obtained with Eqs. (3.106) and (3.107), where Eq. (3.107) has been used without the normalization factor $\frac{1}{N}$ in the coupling term (passive relay). The parameters are $N = 500$, $R^m = 170$, $\sigma_{\text{intra}} = 0.2$, $\epsilon = 0.05$, $a = 0.5$. Figure taken from [364]

domains are so-called double chimeras [232]. The name refers to the twofold occurrence of the typical chimera-like feature of the coexistence of coherence and decoherence: On the one hand within each outer layer, where we observe regular chimera states with a single or multiple coherent domain, and on the other hand in the partial inter-layer synchronization of chimera states (only the coherent cores of both outer layers are synchronized). The green areas in the synchronization tongues denote *salt and pepper states*, which are characterized by strong variations on very short length scales, so that the dynamical patterns have arbitrarily short wavelengths [285, 380, 404, 405]. In Fig. 3.32, our focus is on the overall structure of the resonance tongues in a wide range of coupling strengths and coupling delays, and only one simulation has been run for each point in the parameter plane. The transition of chimeras with one coherent domain to those with several coherent domains is common in many systems and has been analyzed in more detail in Refs. [274, 406].

The different partial synchronization patterns are shown in more detail in Fig. 3.33 for three sets of parameters marked by black dots and labeled a,b,c in Fig. 3.32: In the cross-hatched orange region we observe double chimeras comprising regular chimera states with one coherent domain in both outer layers (Fig. 3.33a), while the hatched red region exhibits double chimeras with two coherent domains in both outer layers (Fig. 3.33b). This is clearly shown by the snapshots (left column) and the mean phase velocity profiles (right column). The local synchronization error (right column, orange) indicates partial relay synchronization of the coherent domains. Note that the hub (red dot) synchronizes with the coherent domain. Ultimately we can also observe partially synchronized salt and pepper states in the outer layers for higher inter-layer coupling strengths, marked by the green areas

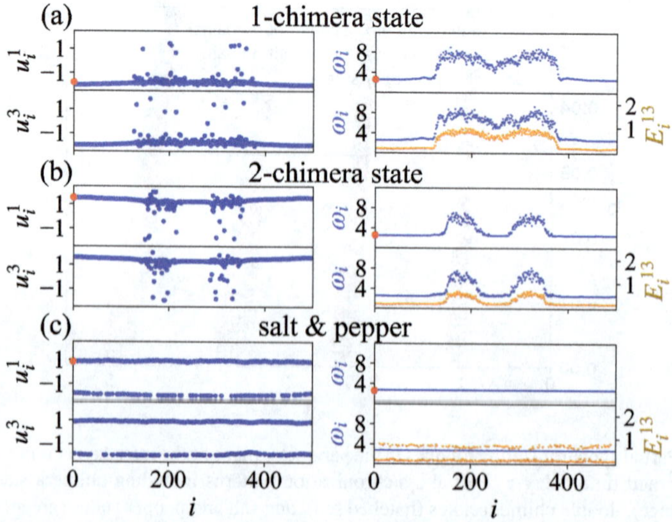

Fig. 3.33 Dynamics of the outer layers for different values of the inter-layer coupling strength $\sigma_{\text{inter}} = \sigma^{12}$ and time delay τ, marked in Fig. 3.32: **a** double chimera state with 1 coherent domain for $\sigma_{\text{inter}} = 0.01$, $\tau = 1.0$ **b** double chimera state with 2 coherent domains for $\sigma_{\text{inter}} = 0.01$, $\tau = 1.1$ **c** salt and pepper state for $\sigma_{\text{inter}} = 0.04$, $\tau = 0.98$. The left column shows snapshots of variables u_i^m for the outer layers $m = 1, 3$ (dark blue), whereas the right column shows the corresponding mean phase velocity profiles ω_i (dark blue) for each layer and local inter-layer synchronization error E_i^{13} (light orange), averaged over a time window $t_a = 7\,000$. The snapshot of u^2 and the mean phase velocity of the hub are marked by red dots in the upper panels. In panels (**a**) and (**b**), the coherent parts of the chimera states in the outer layers are synchronized, whereas the incoherent parts are desynchronized (double chimera states). Other parameters as in Fig. 3.32. Figure taken from [364]

(Fig. 3.33c). In salt and pepper states all oscillators move with the same mean phase velocity. In contrast to the completely synchronized state a large number of oscillators split from the main coherent cluster and move with a constant phase lag (on a slightly perturbed limit cycle) [285, 286].

We can observe a resonance effect with respect to the time delay τ: The partial relay synchronization tongues appear close to full or half integer multiples of the intrinsic period of the uncoupled system $T = 2.3$ (see Fig. 3.32). Furthermore we can see that the tongues are tilted to the left with increasing inter-layer coupling strength σ_{inter}. In Ref. [232] analytical calculations were performed for a triplex network. Note that in the present section the relay-layer of this triplex network is reduced to a single node, but the results carry over. yielding a decreasing period T for increasing inter-layer coupling strength, which is consistent with the observed tilt of the tongues in Fig. 3.32.

In between the tongues of Fig. 3.32 and for the special case of no time delay ($\tau = 0$) we observe desynchronization or complete synchronization or traveling waves where all nodes move with the same mean phase velocity but with a constant phase lag between them. For high inter-layer coupling strengths the tongues are only shown up to $\sigma_{\text{inter}} = 0.055$. Salt and pepper states cease to exist at approximately $\sigma_{\text{inter}} = 0.06$, and only completely synchronized states are observed. The simulations with the lowest inter-layer coupling strength for which partial relay synchronization still occurs are done with $\sigma_{\text{inter}} = 0.0002$. Below that value only desynchronization has been observed. The transition from desynchronization to chimera states is in the focus of the next section.

3.5.4 Induction of Chimera States via a Hub

We will now examine whether the hub can induce chimera states through inter-layer coupling in the outer layers if the single individual outer layers exhibit no chimera states. Accordingly, we investigate if the interplay of intra-layer coupling, inter-layer coupling, and time delay can result in chimera states which are not observed in the isolated outer layers. The regime of chimera states has been explored in the parameter plane of the intra-layer coupling strength σ_{intra} and the coupling radius r in Ref. [274]. With no connection to the hub ($\sigma_{\text{inter}} = 0$), the value $\sigma_{\text{intra}} = 0.00055$ is determined for which there are no chimera states in the outer layers after a simulation time of $t_s = 10\,000$. For larger coupling strength $\sigma_{\text{intra}} > 0.00055$, the dynamics of the isolated outer layers shows chimera states. Figure 3.34(a) shows the dynamics of the system at an intra-layer coupling value of $\sigma_{\text{intra}} = 0.00055$ without inter-layer coupling to the hub.

Throughout the rest of this section we will use the normalization factor $\frac{1}{N}$ in the coupling term in Eq. (3.107), such that the hub becomes an active relay, in contrast to the passive relay which we have considered in Figs. 3.32 and 3.33. This is appropriate since we are now considering very small values of the intra-layer and inter-layer coupling strengths, and there should be some balance between the local dynamics and the coupling terms.

Figure 3.34 shows the transition from a desynchronized state, when the hub is disconnected from the outer layers (Fig. 3.34a), to a completely synchronized state, when the inter-layer coupling strength is increased (Fig. 3.34c). If we choose appropriate values for the inter-layer coupling strength and time delay, chimera states can be induced in the outer layers by the hub (Fig. 3.34b). Furthermore, we investigate the hub dynamics. In the case where chimera states appear in both outer layers, snapshots and space-time plots reveal that the hub is not phase-synchronized with the coherent cores of the chimera states in the outer layers. On the contrary, the space-time plots (not shown here) demonstrate that the hub dynamics has a phase lag to that of the coherent core of the chimera states.

Figure 3.35 summarizes the results in the parameter plane of inter-layer coupling strength σ_{inter} and time delay τ, where 20 simulations with random initial conditions are performed for each set of parameters. The figure indicates that without coupling of the hub to the outer layers ($m = 1, 3$), no chimera states occur (light yellow). For large σ_{inter} both layers are fully synchronized through the hub (turquoise). By varying σ_{inter} and τ, chimera states can be induced for some parameter combinations of σ_{inter}

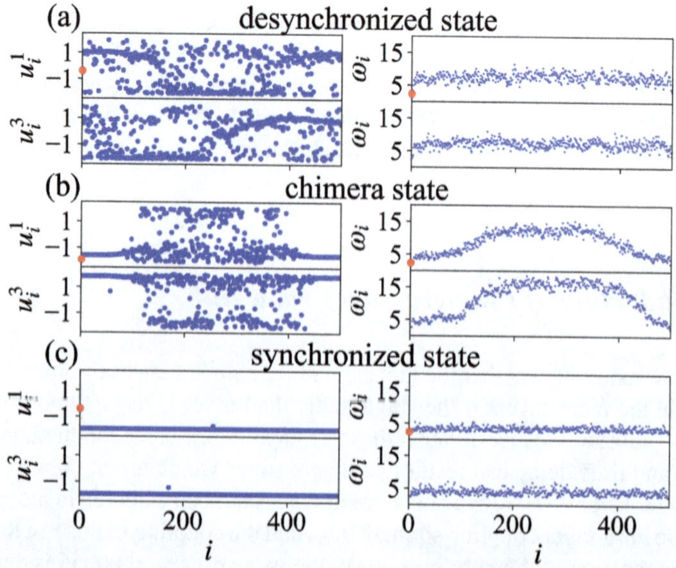

Fig. 3.34 Chimera states induced via a hub: dynamics for different values of weak inter-layer coupling strength σ_{inter} and time delay τ, marked by black circles (a,b,c) in Fig. 3.35. **a** desynchronized state for $\sigma_{\text{inter}} = 0$, $\tau = 0$ **b** chimera state in both layers for $\sigma_{\text{inter}} = 0.00006$, $\tau = 1.08$ **c** completely synchronized state for $\sigma_{\text{inter}} = 0.00015$, $\tau = 1.14$. Left panels: Snapshots of the activator variable u_i^m of the outer layers $m = 1$ and $m = 3$. The snapshots have been taken for the last time step of the simulation. Right panels: Mean phase velocity profiles ω_i of both outer layers. The snapshot of u^2 and the mean phase velocity of the hub are marked by red dots in the upper panels. The mean phase velocity has been averaged over the time window $t_a = t_s - t_{\text{trans}} = 8\,000$ with the total simulation time $t_s = 10\,000$ and the transient time $t_{\text{trans}} = 2\,000$. $\sigma_{\text{intra}} = 0.00055$, other parameters as in Fig. 3.32, but with the normalization factor $\frac{1}{N}$ in the coupling term (active relay). Figure taken from [364]

and τ. We have chosen $\tau \simeq 1$ because the first partial relay synchronization tongue in Fig. 3.32 is located at this value. If the simulations are repeated several times using different random initial conditions, different patterns may be found because of the high multistability of the system. When a chimera state emerges in one outer layer, in most cases no corresponding chimera state can be observed in the other outer layer. In some cases, however, the chimera states occur in both layers, and then we represent the number of such simulations by the red color intensity. Note that the chimera states which appear in both layers do not exhibit full or partial relay synchronization. This can be attributed to the low intra-layer coupling strength at with chimera states in the outer layers are just beginning to emerge.

In conclusion, we have shown that coupling the initially isolated and desynchronized outer layers via the hub can induce chimera states in these layers. We have, however, investigated a parameter regime of σ_{inter} and τ where chimera states are just beginning to emerge. Therefore the mean phase velocity profiles are not as pronounced as for higher values of intra-layer coupling strength.

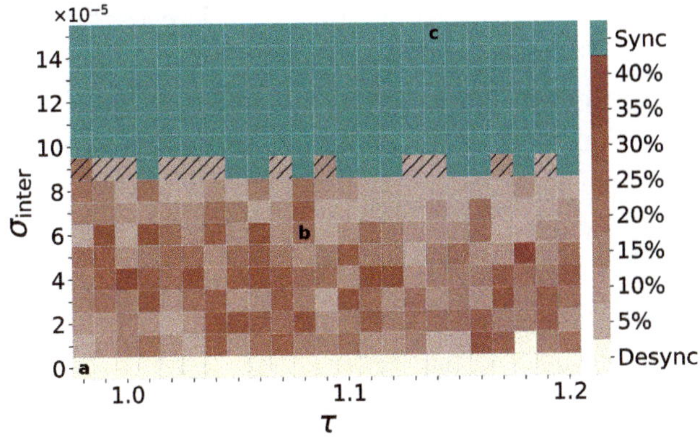

Fig. 3.35 Chimera states in the parameter plane of inter-layer coupling strength σ_{inter} and time delay τ. The color of each tile indicates the state for specific values for σ_{inter} and τ for 20 simulations with random initial conditions, respectively. Light yellow tiles represent desynchronized states, while turquoise tiles stand for states which are almost completely synchronized. All red tiles indicate chimera states which emerge in both layers. The intensity of the red color indicates the ratio of simulations in which chimera states emerge. In the hatched region synchronization can be found additionally. In the other cases desynchronization can be observed in at least one layer. The intra-layer coupling strength within both outer layers is set to $\sigma_{\text{intra}} = 0.00055$ for all simulations. The total simulation time is set to $t_s = 10\,000$ for all simulations. The dynamics of the three marked dots are shown in Fig. 3.34a, b, c. Other parameters as in Fig. 3.34. Figure taken from [364]

3.5.5 *Effect of Dilution of Links*

In the following, we investigate to what extent a dilution of the inter-layer connections between the outer layers and the hub plays a role for the chimera states in the outer layers. In particular, we address the question of whether the chimera states will persist if the links either to their coherent domain or to their incoherent domain are removed. Consequently, the inter-layer links between the hub and the outer layers are removed in different percentages of the links to coherent or incoherent nodes. The connections are cut in a controlled regular manner. For example, for a dilution $d = 50\,\%$ of the links with the coherent core, each second connection to the coherent core is removed. The effects of dilution of network links upon synchronization has also been studied in power grid models [122, 123].

We start the dilution at time $t = 15\,000$, where t corresponds to the time at which we have determined the dynamics of the chimera states occurring in both outer layers in Fig. 3.35. To study the dilution effects upon the system, the simulation is run until $t_s = 25\,000$. As a measure of dilution, we introduce the percentage d of removed links. It represents the percentage of cut connections between the hub ($m = 2$) and the coherent or incoherent core of the chimera states in the layers $m = 1, 3$. The effects of dilution for $d = 25\,\%$, $50\,\%$ and $75\,\%$ are demonstrated in Figs. 3.36 and 3.37.

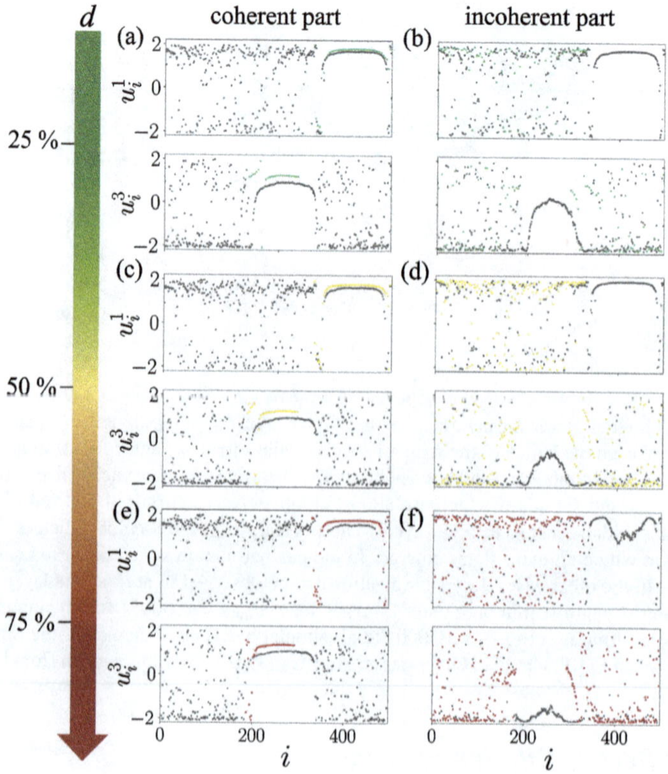

Fig. 3.36 Effect of dilution on chimera states in the outer layers: snapshots for three values of the dilution percentage **a**, **b** $d = 25\,\%$, **c**, **d** $d = 50\,\%$, **e**, **f** $d = 75\,\%$. Left column: d links of the coherent domain are cut, right column: d links of the incoherent domain are cut. The color-coded dots correspond to the u-value of the nodes in the outer layers to which the link from the hub was cut, whereas gray dots are still connected to the hub. Green dots: $d = 25\,\%$ links cut (**a**) of the coherent domain (72 links) or (**b**) of the incoherent domain (180 links). Yellow dots: $d = 50\,\%$ links cut (confirm that in the absence) of the coherent domain (142 links) or (**d**) of the incoherent domain (360 links). Red dots: $d = 75\,\%$ links cut (**e**) of the coherent domain (210 links) or (**f**) of the incoherent domain (538 links). Parameters: $\tau = 1.08$, $\sigma_{\text{inter}} = 0.00006$, $\sigma_{\text{intra}} = 0.00055$, simulation time $t_s = 25\,000$. Other parameters as in Fig. 3.34. Figure taken from [364]

A dilution of the inter-layer links to the incoherent core of the chimera states at different percentages does not cause a visible deterioration or even destruction of the chimera states. The corresponding snapshots in Fig. 3.36b, d, f are indicating the emergence of chimera states. The mean phase velocity profiles confirm this phenomenon in Fig. 3.37b, d, f, where all profiles show the typical arc-shape with a coherent and an incoherent part of a chimera state regardless of dilution to the incoherent region.

The dilution of links to the coherent core of the chimera states leads to splitting-off some nodes from the original coherent domain in layer 3. In Fig. 3.36a, c, e, the

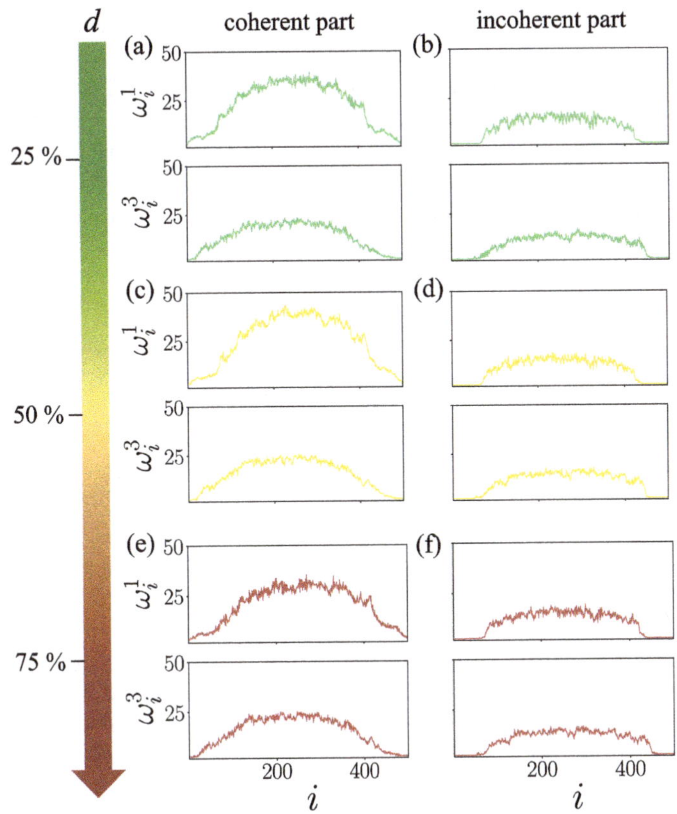

Fig. 3.37 Effect of dilution on chimera states in the outer layers: mean phase velocity profiles averaged over $t_a = 10\,000$ for the same parameters as in Fig. 3.36. Figure taken from [364]

nodes of the coherent domain to which the connection has been broken oscillate with a phase different from that of the original coherent core nodes. The splitting of the coherent core yields a modified detrended mean phase velocity profile exhibiting higher values in case of dilution, as shown in Fig. 3.37a, c, e. The increased values in the incoherent region can be attributed to the increased number of split-off nodes resulting from the dilution. On the other hand, the formation of another coherent range of u-values smears out the coherent region in the mean phase velocity profile which can no longer be clearly distinguished. The fact that a significant difference exists between the two coherent cores implies that the mean phase velocities are increased, and the constant regions vanish. For $d = 50\,\%$ dilution, this behavior is more pronounced in the first layer than in the third layer, which can be attributed to the different initial conditions underlying the two chimera states. The increase of the mean phase velocity profile with detrending in the upper panel of Fig. 3.37c can be traced back to the division of the coherent cores' velocities.

We can therefore conclude that the dilution of links to the coherent part of the chimera states has the most significant effect at $d = 50\,\%$ since here the size of the newly formed coherent core of each layer is the largest. This assumption has been confirmed by simulations in which the dilution of the inter-layer links has been applied at an earlier stage of the simulation when chimera states have not yet formed. Our investigation shows that at this stage diluting inter-layer links to the coherent core has a much more pronounced negative effect upon the formation of chimera states than dilution of inter-layer links to the incoherent core.

3.5.6 Summary

The focus of this section is the effect of inter-layer coupling delay and dilution upon chimera states in three-layer networks with a hub. While previous works have analyzed triplex networks of three identical layers [232, 286, 347], here we have chosen the (middle) relay layer as a hub, i.e., a single node. This is typical of the brain architecture, where distant brain areas communicate via a functional relay (e.g., thalamus or hippocampus) consisting of a significantly smaller number of neurons. First, we have elaborated the role of the hub and its importance for the existence of chimera states. We have mapped out the regimes of partial relay synchronization of chimera states between two remote layers in the parameter space of inter-layer coupling strength and time delay, and found tongue-like regions in the parameter plane exhibiting double chimeras, i.e., chimera states in the remote layers whose coherent cores are synchronized with each other, either with one or with two coherent cores, and salt-and-pepper states consisting of two intertwined groups of oscillators, all moving with the same phase velocity.

Secondly, we have considered a very low intra-layer coupling strength such that no chimera states occur in the outer layers when the inter-layer coupling to the hub is switched off. For certain combinations of τ and σ_{inter} we have found chimera states induced by the hub. In more detail, chimera states occur in one remote layer only as well as in both layers, for different random initial conditions, because of the high multistability of the system.

In the third part, we have investigated the influence of dilution of links to the hub upon the dynamics. Starting with chimera states induced by the hub, we have cut the inter-layer links either to the coherent domain or to the incoherent domain of the chimera states at a time when the chimera states have already emerged. The greatest effect of dilution has been observed when $d = 50\,\%$ of the links to the coherent domain were removed. In contrast, no significant change in the chimera states could be found when links to the incoherent domain were removed.

These results on partial relay synchronization scenarios induced by a passive or active hub might be useful for a deeper understanding of dynamical patterns and functions occurring in the brain.

References

1. Sawicki J (2019) Delay controlled partial synchronization in complex networks. Springer Theses, Springer, Heidelberg. https://doi.org/10.1007/978-3-030-34076-6
2. Yakubovich V (1968) Theory of adaptive systems. Sov Phys Doklady 13(9):852–855
3. Yakubovich V (1968) Adaptive systems with multistep goal conditions. Sov Phys Doklady 13(11):1096–1099
4. Fomin V, Fradkov A, Yakubovich V (1981) Adaptive control of dynamical systems. Nauka
5. Annaswamy AM, Fradkov AL (2021) A historical perspective of adaptive control and learning. Ann Rev Control 52:18–41. https://doi.org/10.1016/j.arcontrol.2021.10.014
6. Fradkov AL, Shepeljavyi AI (2022) The history of cybernetics and artificial intelligence: a view from Saint Petersburg. Cybern Phys 11(4), 253–263. https://doi.org/10.35470/2226-4116-2022-11-3-253-263
7. Tsypkin YZ (1971) Adaptation and learning in automatic systems. In: Mathematics in science and engineering. Academic Press
8. Fradkov A (2007) Cybernetical physics: from control of chaos to quantum control. Springer, Berlin Heidelberg. https://doi.org/10.1007/978-3-540-46277-4
9. Lehnert J, Hövel P, Selivanov AA, Fradkov AL, Schöll E (2014) Controlling cluster synchronization by adapting the topology. Phys Rev E 90(4):042914. https://doi.org/10.1103/physreve.90.042914
10. Pikovsky A, Rosenblum M, Kurths J (2001) Synchronization: a universal concept in nonlinear sciences, 1st edn. Cambridge University Press, Cambridge. https://doi.org/10.1017/CBO9780511755743
11. Boccaletti S, Hwang DU, Chavez M, Amann A, Kurths J, Pecora LM (2006) Synchronization in dynamical networks: evolution along commutative graphs. Phys Rev E 74:016102. https://doi.org/10.1103/physreve.74.016102
12. Boccaletti S, Pisarchik AN, del Genio CI, Amann A (2018) Synchronization: From Coupled Systems to Complex Networks. Cambridge University Press, Cambridge. https://doi.org/10.1017/9781316334870
13. Panaggio MJ, Abrams DM (2015) Chimera states: Coexistence of coherence and incoherence in networks of coupled oscillators. Nonlinearity 28:R67. https://doi.org/10.1088/0951-7715/28/3/r67
14. Schöll E (2016) Synchronization patterns and chimera states in complex networks: interplay of topology and dynamics. Eur Phys J Spec Top 225:891–919. https://doi.org/10.1140/epjst/e2016-02646-3
15. Singer W (2018) Neuronal oscillations: unavoidable and useful? Eur J Neurosci 48(7):2389–2398. https://doi.org/10.1111/ejn.13796
16. Gerster M, Berner R, Sawicki J, Zakharova A, Skoch A, Hlinka J, Lehnertz K, Schöll E (2020) FitzHugh-Nagumo oscillators on complex networks mimic epileptic-seizure-related synchronization phenomena. Chaos 30:123130. https://doi.org/10.1063/5.0021420
17. Sawicki J (2021) Modelling synchronisation in the brain and in nature. Res Featur. https://doi.org/10.26904/rf-136-1423398575
18. Sawicki J (2015) Synchronization of organ pipes. Technische Universität Berlin, MSc
19. Sawicki J, Abel M, Schöll E (2015) Synchronization in coupled organ pipes. In: Proceedings of the 7th international conference on physics and control (PhysCon 2015). IPACS Electronic Library (2015), Istanbul, Turkey
20. Sawicki J, Abel M, Schöll E (2018) Synchronization of organ pipes. Eur Phys J B 91(2):24. https://doi.org/10.1140/epjb/e2017-80485-8
21. Sawicki J (2025) Importance of near- and far-field coupling for the synchronization of organ pipes. Hausorgel Arbeitskreis of the Gesellschaft der Orgelfreunde e. V. arXiv:2504.00556
22. Sawicki J (2025) Synchronisation zweier Labialpfeifen. Hausorgel Arbeitskreis der Gesellschaft der Orgelfreunde e. V. arXiv:2504.02896
23. Fabre B, Hirschberg A (2000) Physical modeling of flue instruments: a review of lumped models. Acta Acust 86:599

24. Flunkert V, Fischer I, Schöll E (2013) Dynamics, control and information in delay-coupled systems. Theme Issue Phil Trans R Soc A 371:20120465. https://doi.org/10.1098/rsta.2012.0465
25. Howe MS (2003) Theory of vortex sound, vol 33. Cambridge University Press. https://doi.org/10.1017/CBO9780511546600
26. Stein S, Luther S, Parlitz U (2017) Impact of viscoelastic coupling on the synchronization of symmetric and asymmetric self-sustained oscillators. New J Phys 19(6):063040. https://doi.org/10.1088/1367-2630/aa6d4a
27. Bader R (2013) Nonlinearities and synchronization in musical acoustics and music psychology. Springer, Berlin. https://doi.org/10.1007/978-3-642-29014-4
28. Abel M, Bergweiler S, Gerhard-Multhaupt R (2006) Synchronization of organ pipes: experimental observations and modeling. J Acoust Soc Am 119(4):2467–2475. https://doi.org/10.1121/1.2170441
29. Abel M, Ahnert K, Bergweiler S (2009) Synchronization of sound sources. Phys Rev Lett 103:114301. https://doi.org/10.1103/physrevlett.103.114301
30. Fischer J (2014) Nichtlineare Kopplungsmechanismen akustischer Oszillatoren am Beispiel der Synchronisation von Orgelpfeifen. PhD thesis, Universität Potsdam
31. Fischer JL, Bader R, Abel M (2016) Aeroacoustical coupling and synchronization of organ pipes. J Acoust Soc Am 140:2344. https://doi.org/10.1121/1.4966517
32. Rayleigh JWS (1882) On the pitch of organ-pipes. Philos Mag 13:340
33. Fletcher NH (1978) Mode locking in nonlinearly excited inharmonic musical oscillators. J Acoust Soc Am 64:1566. https://doi.org/10.1121/1.382057
34. Stanzial D, Bonsi D, Gonzales D (2001) Nonlinear modelling of the Mitnahme-effekt in coupled organ pipes. Int Symp Muic Acoust 108:333
35. Bergweiler S (2006) Körperoszillation und Schallabstrahlung akustischer Wellenleiter unter Berücksichtigung von Wandungseinflüssen und Kopplungseffekten: Verändern Metalllegierung und Wandungsprofil des Rohrresonators den Klang der labialen Orgelpfeife? PhD thesis, Universität Potsdam
36. Wirkus S, Rand RH (2002) The dynamics of two coupled van der pol oscillators with delay coupling. Nonlinear Dyn 30(3):205. https://doi.org/10.1023/A:1015021621953
37. Semenov V, Feoktistov A, Vadivasova T, Schöll E, Zakharova A (2015) Time-delayed feedback control of coherence resonance near subcritical Hopf bifurcation: theory versus experiment. Chaos 25:033111. https://doi.org/10.1063/1.4915066
38. Adler R (1973) A study of locking phenomena in oscillators. Proc IEEE 61(10):1380–1385. https://doi.org/10.1109/PROC.1973.9030
39. Shannon C (1949) Communication in the presence of noise. Proc IRE 37(1):10–21. https://doi.org/10.1109/JRPROC.1949.232969
40. Sawicki J, Berner R, Loos SAM, Anvari M, Bader R, Barfuss W, Botta N, Brede N, Franović I, Gauthier DJ, Goldt S, Hajizadeh A, Hövel P, Karin O, Lorenz-Spreen P, Miehl C, Mölter J, Olmi S, Schöll E, Seif A, Tass PA, Volpe G, Yanchuk S, Kurths J (2023) Perspectives on adaptive dynamical systems. Chaos Interdiscip J Nonlinear Sci 33(7):071501. https://doi.org/10.1063/5.0147231
41. Yanchuk S, Roque AC, Macau EEN, Kurths J (2021) Dynamical phenomena in complex networks: fundamentals and applications. Eur Phys J Special Top 230(14–15):2711–2716. https://doi.org/10.1140/epjs/s11734-021-00282-y
42. Cabral J, Jirsa V, Popovych OV, Torcini A, Yanchuk S (2022) Editorial: from structure to function in neuronal networks: effects of adaptation, time-delays, and noise. Front Syst Neurosci 16:871165. https://doi.org/10.3389/fnsys.2022.871165
43. Abbott LF, Nelson SB (2000) Synaptic plasticity: taming the beast. Nat Neurosci 3:1178–1183. https://doi.org/10.1038/81453
44. Dan Y, Poo MM (2004) Spike timing-dependent plasticity of neural circuits. Neuron 44(1):23–30. https://doi.org/10.1016/j.neuron.2004.09.007
45. Maistrenko YL, Lysyansky B, Hauptmann C, Burylko O, Tass PA (2007) Multistability in the kuramoto model with synaptic plasticity. Phys Rev E 75:066207. https://doi.org/10.1103/PhysRevE.75.066207

46. Popovych OV, Yanchuk S, Tass PA (2013) Self-organized noise resistance of oscillatory neural networks with spike timing-dependent plasticity. Sci Rep 3:2926. https://doi.org/10.1038/srep02926
47. Lücken L, Popovych OV, Tass PA, Yanchuk S (2016) Noise-enhanced coupling between two oscillators with long-term plasticity. Phys Rev E 93(3):032210. https://doi.org/10.1103/physreve.93.032210
48. Aoki T, Aoyagi T (2009) Co-evolution of phases and connection strengths in a network of phase oscillators. Phys Rev Lett 102(3):034101. https://doi.org/10.1103/PhysRevLett.102.034101
49. Kasatkin DV, Yanchuk S, Schöll E, Nekorkin VI (2017) Self-organized emergence of multi-layer structure and chimera states in dynamical networks with adaptive couplings. Phys Rev E 96(6):062211. https://doi.org/10.1103/physreve.96.062211
50. Berner R, Vock S, Schöll E, Yanchuk S (2021) Desynchronization transitions in adaptive networks. Phys Rev Lett 126(2):028301. https://doi.org/10.1103/physrevlett.126.028301
51. Berner R, Mehrmann V, Schöll E, Yanchuk S (2021) The multiplex decomposition: an analytic framework for multilayer dynamical networks. SIAM J Appl Dyn Syst 20(4):1752–1772. https://doi.org/10.1137/21M1406180
52. Berner R, Yanchuk S (July2021) Synchronization in networks with heterogeneous adaptation rules and applications to distance-dependent synaptic plasticity. Front Appl Math Stat 7:714978. https://doi.org/10.3389/fams.2021.714978
53. Thiele M, Berner R, Tass PA, Schöll E, Yanchuk S (2023) Asymmetric adaptivity induces recurrent synchronization in complex networks. Chaos 33:023123. https://doi.org/10.1063/5.0123821
54. Röhr V, Berner R, Lameu EL, Popovych OV, Yanchuk S (2019) Frequency cluster formation and slow oscillations in neural populations with plasticity. PLoS ONE 14(11):e0225094. https://doi.org/10.1371/journal.pone.0225094
55. Kuehn C (2015) Multiple time scale dynamics. Springer International Publishing, Switzerland
56. Caporale N, Dan Y (2008) Spike timing-dependent plasticity: a hebbian learning rule. Ann. Rev. Neurosci. 31:25. https://doi.org/10.1146/annurev.neuro.31.060407.125639
57. Taylor D, Ott E, Restrepo JG (2010) Spontaneous synchronization of coupled oscillator systems with frequency adaptation. Phys Rev E 81(4):046214. https://doi.org/10.1103/physreve.81.046214
58. Fardet T, Levina A (2020) Simple models including energy and spike constraints reproduce complex activity patterns and metabolic disruptions. PLoS Comput Biol 16:e1008503. https://doi.org/10.1371/journal.pcbi.1008503
59. Bonvento G, Bolaños JP (2021) Astrocyte-neuron metabolic cooperation shapes brain activity. Cell Metab 33:1546. https://doi.org/10.1016/j.cmet.2021.07.006
60. Roberts JA, Iyer KK, Vanhatalo S, Breakspear M (2014) Critical role for resource constraints in neural models. Front Syst Neurosci 8:154. https://doi.org/10.3389/fnsys.2014.00154
61. Virkar YS, Shew WL, Restrepo JG, Ott E (2016) Feedback control stabilization of critical dynamics via resource transport on multilayer networks: How glia enable learning dynamics in the brain. Phys Rev E 94:042310. https://doi.org/10.1103/PhysRevE.94.042310
62. Levina A, Herrmann JM, Geisel T (2007) Dynamical synapses causing self-organized criticality in neural networks. Nat Phys 3:857. https://doi.org/10.1038/nphys758
63. Berner R, Schöll E, Yanchuk S (2019) Multiclusters in networks of adaptively coupled phase oscillators. SIAM J Appl Dyn Syst 18(4):2227–2266. https://doi.org/10.1137/18m1210150
64. Berner R, Fialkowski J, Kasatkin DV, Nekorkin VI, Yanchuk S, Schöll E (2019) Hierarchical frequency clusters in adaptive networks of phase oscillators. Chaos 29(10):103134. https://doi.org/10.1063/1.5097835
65. Kroma-Wiley KA, Mucha PJ, Bassett DS (2021) Synchronization of coupled kuramoto oscillators under resource constraints. Phys Rev E 104:014211. https://doi.org/10.1103/PhysRevE.104.014211
66. Thamizharasan S, Chandrasekar VK, Senthilvelan M, Berner R, Schöll E, Senthilkumar DV (2021) Exotic states induced by coevolving connection weights and phases in complex networks. Phys Rev E 105:034312. https://doi.org/10.1103/PhysRevE.105.034312

67. Fialkowski J, Yanchuk S, Sokolov IM, Schöll E, Gottwald GA, Berner R (2022) Heterogeneous nucleation in finite size adaptive dynamical networks. Phys Rev Lett 130:067402. https://doi.org/10.1103/physrevlett.130.067402
68. Franović I, Eydam SR, Yanchuk S, Berner R (2022) Collective activity bursting in a population of excitable units adaptively coupled to a pool of resources. Front Netw Physiol 2:841829. https://doi.org/10.3389/fnetp.2022.841829
69. Franović I, Yanchuk S, Eydam SR, Bačić I, Wolfrum M (2020) Dynamics of a stochastic excitable system with slowly adapting feedback. Chaos 30:083109. https://doi.org/10.1063/1.5145176
70. Bačić I, Klinshov V, Nekorkin VI, Perc M, Franović I (2018) Inverse stochastic resonance in a system of excitable active rotators with adaptive coupling. EPL 124:40004. https://doi.org/10.1209/0295-5075/124/40004
71. Knoblauch A, Hauser F, Gewaltig MO, Körner E, Palm G (2012) Does spike-timing-dependent synaptic plasticity couple or decouple neurons firing in synchrony? Front Comput Neurosci 6:55. https://doi.org/10.3389/fncom.2012.00055
72. Pazo D, Montbrió E (2006) Universal behavior in populations composed of excitable and self-oscillatory elements. Phys Rev E 73:055202(R). https://doi.org/10.1103/PhysRevE.73.055202
73. Lafuerza LF, Colet P, Toral R (2010) Nonuniversal results induced by diversity distribution in coupled excitable systems. Phys Rev Lett 105:084101. https://doi.org/10.1103/PhysRevLett.105.084101
74. Klinshov V, Franović I (2019) Two scenarios for the onset and suppression of collective oscillations in heterogeneous populations of active rotators. Phys Rev E 100:062211. https://doi.org/10.1103/PhysRevE.100.062211
75. Bačić I, Yanchuk S, Wolfrum M, Franović I (2018) Noise-induced switching in two adaptively coupled excitable systems. Eur Phys J Spec Top 227:1077. https://doi.org/10.1140/epjst/e2018-800084-6
76. Sporns O, Kötter R (2004) Motifs in brain networks. PLoS Biol 2:e369. https://doi.org/10.1371/journal.pbio.0020369
77. Bačić I, Franović I (2020) Two paradigmatic scenarios for inverse stochastic resonance. Chaos 30:033123. https://doi.org/10.1063/1.5139628
78. Pecora LM, Carroll TL (1998) Master stability functions for synchronized coupled systems. Phys Rev Lett 80(10):2109–2112. https://doi.org/10.1103/physrevlett.80.2109
79. Ott E, Antonsen TM (2008) Low dimensional behavior of large systems of globally coupled oscillators. Chaos 18:037113. https://doi.org/10.1063/1.2930766
80. Ott E, Antonsen TM (2009) Long time evolution of phase oscillator systems. Chaos 19:023117. https://doi.org/10.1063/1.3136851
81. Bick C, Goodfellow M, Laing CR, Martens EA (2020) Understanding the dynamics of biological and neural oscillator networks through exact mean-field reductions: a review. J Math Neurosci 10:9. https://doi.org/10.1186/s13408-020-00086-9
82. Lan G, Sartori P, Neumann S, Sourjik V, Tu Y (2012) The energy-speed-accuracy trade-off in sensory adaptation. Nat Phys 8:422. https://doi.org/10.1038/nphys2276
83. Conti D, Mora T (2022) Nonequilibrium dynamics of adaptation in sensory systems. Phys Rev E 106:054404. https://doi.org/10.1103/PhysRevE.106.054404
84. Tkačik G, Bialek W (2016) Information processing in living systems. Ann Rev Condens Matter Phys 7:89. https://doi.org/10.1146/annurev-conmatphys-031214-014803
85. Schöll E (2021) Partial synchronization patterns in brain networks. Europhys Lett 136:18001. https://doi.org/10.1209/0295-5075/ac3b97
86. Schöll E (2020) Chimeras in physics and biology: synchronization and desynchronization of rhythms. Nova Acta Leopold 425:67–95. https://doi.org/10.26164/leopoldina_10_00275
87. Zakharova A (2020) Chimera patterns in networks: interplay between dynamics, structure, noise, and delay. In: Understanding complex systems, Springer, Cham. https://doi.org/10.1007/978-3-030-21714-3

88. Maistrenko Y, Penkovsky B, Rosenblum M (2014) Solitary state at the edge of synchrony in ensembles with attractive and repulsive interactions. Phys Rev E 89:060901. https://doi.org/10.1103/physreve.89.060901
89. Jaros P, Maistrenko Y, Kapitaniak T (2015) Chimera states on the route from coherence to rotating waves. Phys Rev E 91:022907. https://doi.org/10.1103/physreve.91.022907
90. Semenov V, Zakharova A, Maistrenko Y, Schöll E (2016) Delayed-feedback chimera states: forced multiclusters and stochastic resonance. Europhys Lett 115:10005. https://doi.org/10.1209/0295-5075/115/10005
91. Wu H, Dhamala M (2018) Dynamics of kuramoto oscillators with time-delayed positive and negative couplings. Phys Rev E 98(3):032221. https://doi.org/10.1103/physreve.98.032221
92. Teichmann E, Rosenblum M (2019) Solitary states and partial synchrony in oscillatory ensembles with attractive and repulsive interactions. Chaos 29(9):093124. https://doi.org/10.1063/1.5118843
93. Chen B, Engelbrecht JR, Mirollo RE (2019) Dynamics of the Kuramoto-Sakaguchi oscillator network with asymmetric order parameter. Chaos 29(1):013126. https://doi.org/10.1063/1.5055758
94. Jaros P, Brezetsky S, Levchenko R, Dudkowski D, Kapitaniak T, Maistrenko Y (2018) Solitary states for coupled oscillators with inertia. Chaos 28:011103. https://doi.org/10.1063/1.5004757
95. Taher H, Olmi S, Schöll E (2019) Enhancing power grid synchronization and stability through time delayed feedback control. Phys Rev E 100(6):062306. https://doi.org/10.1103/physreve.100.062306
96. Sathiyadevi K, Chandrasekar VK, Senthilkumar DV, Lakshmanan M (2019) Long-range interaction induced collective dynamical behaviors. J Phys A 52(18):184001. https://doi.org/10.1088/1751-8121/ab111a
97. Rybalova E, Anishchenko VS, Strelkova GI, Zakharova A (2019) Solitary states and solitary state chimera in neural networks. Chaos 29:071106. https://doi.org/10.1063/1.5098477
98. Schülen L, Ghosh S, Kachhvah AD, Zakharova A, Jalan S (2019) Delay engineered solitary states in complex networks. Chaos Solitons Fractals 128:290. https://doi.org/10.1016/j.chaos.2019.05.013
99. Zaks MA, Tomov P (2016) Onset of time dependence in ensembles of excitable elements with global repulsive coupling. Phys Rev E 93(2):020201. https://doi.org/10.1103/physreve.93.020201
100. Rybalova E, Semenova N, Strelkova G, Anishchenko V (2017) Transition from complete synchronization to spatio-temporal chaos in coupled chaotic systems with nonhyperbolic and hyperbolic attractors. Eur Phys J Spec Top 226:1857. https://doi.org/10.1140/epjst/e2017-70023-1
101. Kapitaniak T, Kuzma P, Wojewoda J, Czolczynski K, Maistrenko Y (2014) Imperfect chimera states for coupled pendula. Sci Rep 4:6379. https://doi.org/10.1038/srep06379
102. Mikhaylenko M, Ramlow L, Jalan S, Zakharova A (2019) Weak multiplexing in neural networks: switching between chimera and solitary states. Chaos 29:023122. https://doi.org/10.1063/1.5057418
103. Berner R (2021) Patterns of synchrony in complex networks of adaptively coupled oscillators. Springer Theses, Springer, Cham. https://doi.org/10.1007/978-3-030-74938-5
104. Jain S, Krishna S (2001) A model for the emergence of cooperation, interdependence, and structure in evolving networks. Proc Natl Acad Sci 98(2):543–547. https://doi.org/10.1073/pnas.98.2.543
105. Gross, T., Dommar D'Lima, C.J., Blasius, B.: Epidemic dynamics on an adaptive network. Phys Rev Lett 96(208701). https://doi.org/10.1103/PhysRevLett.96.208701
106. Gross T, Blasius B (2008) Adaptive coevolutionary networks: a review. J R Soc Interface 5(20):259–271. https://doi.org/10.1098/rsif.2007.1229
107. Gutiérrez R, Amann A, Assenza S, Gómez-Gardeñes J, Latora V, Boccaletti S (2011) Emerging meso- and macroscales from synchronization of adaptive networks. Phys Rev Lett 107(23):234103. https://doi.org/10.1103/PhysRevLett.107.234103

108. Zhang X, Boccaletti S, Guan S, Liu Z (2015) Explosive synchronization in adaptive and multilayer networks. Phys. Rev. Lett. 114(3):038701. https://doi.org/10.1103/physrevlett.114.038701
109. Madadi Asl M, Valizadeh A, Tass PA (2018) Dendritic and axonal propagation delays may shape neuronal networks with plastic synapses. Front Physiol 9:1849. https://doi.org/10.3389/fphys.2018.01849
110. Kasatkin DV, Nekorkin VI (2018) Synchronization of chimera states in a multiplex system of phase oscillators with adaptive couplings. Chaos 28:093115. https://doi.org/10.1063/1.5031681
111. Kasatkin DV, Nekorkin VI (2018) The effect of topology on organization of synchronous behavior in dynamical networks with adaptive couplings. Eur Phys J Spec Top 227:1051. https://doi.org/10.1140/epjst/e2018-70034-9
112. Berner R, Sawicki J, Schöll E (2020) Birth and stabilization of phase clusters by multiplexing of adaptive networks. Phys Rev Lett 124(8):088301. https://doi.org/10.1103/physrevlett.124.088301
113. Feketa P, Schaum A, Meurer T (2020) Synchronization and multi-cluster capabilities of oscillatory networks with adaptive coupling. IEEE Trans Autom Control 66(7):3084–3096. https://doi.org/10.1109/tac.2020.3012528
114. Popovych OV, Xenakis MN, Tass PA (2015) The spacing principle for unlearning abnormal neuronal synchrony. PLoS ONE 10(2):e0117205. https://doi.org/10.1371/journal.pone.0117205
115. Chakravartula S, Indic P, Sundaram B, Killingback T (2017) Emergence of local synchronization in neuronal networks with adaptive couplings. PLoS ONE 12(6):e0178975. https://doi.org/10.1371/journal.pone.0178975
116. Kuramoto Y (1984) Chemical oscillations. Waves and turbulence, Springer, Berlin. https://doi.org/10.1007/978-3-642-69689-3
117. Filatrella G, Nielsen AH, Pedersen NF (2008) Analysis of a power grid using a Kuramoto-like model. Eur Phys J B 61(4):485–491. https://doi.org/10.1140/epjb/e2008-00275-9
118. Dörfler F, Bullo F (2012) Synchronization and transient stability in power networks and nonuniform kuramoto oscillators. SIAM J Control Optim 50(3):1616–1642. https://doi.org/10.1137/110851584
119. Rohden M, Sorge A, Timme M, Witthaut D (2012) Self-organized synchronization in decentralized power grids. Phys Rev Lett 109:064101. https://doi.org/10.1103/physrevlett.109.064101
120. Motter AE, Myers SA, Anghel M, Nishikawa T (2013) Spontaneous synchrony in power-grid networks. Nat Phys 9:191–197. https://doi.org/10.1038/nphys2535
121. Rodrigues FA, Peron TKDM, Ji P, Kurths J (2016) The Kuramoto model in complex networks. Phys Rep 610:1–98. https://doi.org/10.1016/j.physrep.2015.10.008
122. Tumash L, Olmi S, Schöll E (2018) Effect of disorder and noise in shaping the dynamics of power grids. Europhys Lett 123:20001. https://doi.org/10.1209/0295-5075/123/20001
123. Tumash L, Olmi S, Schöll E (2019) Stability and control of power grids with diluted network topology. Chaos 29:123105. https://doi.org/10.1063/1.5111686
124. Hellmann F, Schultz P, Jaros P, Levchenko R, Kapitaniak T, Kurths J, Maistrenko Y (2020) Network-induced multistability through lossy coupling and exotic solitary states. Nat Commun 11:592. https://doi.org/10.1038/s41467-020-14417-7
125. Kuehn C, Throm S (2019) Power network dynamics on graphons. SIAM J Appl Math 79(4):1271–1292. https://doi.org/10.1137/18m1200002
126. Totz CH, Olmi S, Schöll E (2020) Control of synchronization in two-layer power grids. Phys Rev E 102:022311. https://doi.org/10.1103/PhysRevE.102.022311
127. Zhang X, Ma C, Timme M (2020) Vulnerability in dynamically driven oscillatory networks and power grids. Chaos 30(6):063111. https://doi.org/10.1063/1.5122963
128. Berner R, Yanchuk S, Schöll E (2021) What adaptive neuronal networks teach us about power grids. Phys Rev E 103:042315. https://doi.org/10.1103/physreve.103.042315

129. Acebrón JA, Spigler R (1998) Adaptive frequency model for phase-frequency synchronization in large populations of globally coupled nonlinear oscillators. Phys Rev Lett 81(2229):2229. https://doi.org/10.1103/physrevlett.81.2229
130. Acebrón JA, Bonilla LL, Pérez Vicente CJ, Ritort F, Spigler R (2005) The Kuramoto model: a simple paradigm for synchronization phenomena. Rev Mod Phys 77(1):137–185. https://doi.org/10.1103/revmodphys.77.137
131. Skardal PS, Taylor D, Restrepo JG (2013) Complex macroscopic behavior in systems of phase oscillators with adaptive coupling. Phys. D 267:27–35. https://doi.org/10.1016/j.physd.2013.01.012
132. Schäfer B, Witthaut D, Timme M, Latora V (2018) Dynamically induced cascading failures in power grids. Nat Commun 9:1975. https://doi.org/10.1038/s41467-018-04287-5
133. Berner R, Polanska A, Schöll E, Yanchuk S (2020) Solitary states in adaptive nonlocal oscillator networks. Eur Phys J Spec Top 229(12–13):2183–2203. https://doi.org/10.1140/epjst/e2020-900253-0
134. Sakaguchi H, Kuramoto Y (1986) A soluble active rotater model showing phase transitions via mutual entertainment. Prog Theor Phys 76(3):576–581. https://doi.org/10.1143/PTP.76.576
135. Belykh IV, Brister BN, Belykh VN (2016) Bistability of patterns of synchrony in Kuramoto oscillators with inertia. Chaos 26:094822. https://doi.org/10.1063/1.4961435
136. Olmi S (2015) Chimera states in coupled Kuramoto oscillators with inertia. Chaos 25:123125. https://doi.org/10.1063/1.4938734
137. Olmi S, Navas A, Boccaletti S, Torcini A (2014) Hysteretic transitions in the Kuramoto model with inertia. Phys Rev E 90(4):042905. https://doi.org/10.1103/physreve.90.042905
138. Barré J, Métivier D (2016) Bifurcations and singularities for coupled oscillators with inertia and frustration. Phys Rev Lett 117(21):214102. https://doi.org/10.1103/physrevlett.117.214102
139. Sawicki J, Berner R, Löser T, Schöll E (2022) Modelling tumor disease and sepsis by networks of adaptively coupled phase oscillators. Front Netw Physiol 1:730385. https://doi.org/10.3389/fnetp.2021.730385
140. Berner R, Sawicki J, Thiele M, Löser T, Schöll E (2022) Critical parameters in dynamic network modeling of sepsis. Front. Netw Physiol 2:904480. https://doi.org/10.3389/fnetp.2022.904480
141. Sawicki J (2025) Adaptation and synchronization-basic mechanisms in music performance. Creative performance workshop. arXiv:2504.03958
142. Palmer C (1997) Music performance. Annu Rev Psychol 48:115–138. https://doi.org/10.1146/annurev.psych.48.1.115
143. McPherson GE (2022) The oxford handbook of music performance. Oxford handbooks series, vol 1. Oxford University Press. https://doi.org/10.1093/oxfordhb/9780190638748.001.0001
144. McPherson GE (2022) The oxford handbook of music performance, vol 2. Oxford University Press (2022). https://doi.org/10.1093/oxfordhb/9780190058869.001.0001
145. Juslin PN, Scherer KR (2005) Vocal expression of affect. Series in affective science. Oxford University Press, New York, NY, US, pp 65–135. https://doi.org/10.1093/acprof:oso/9780198529392.003.0003
146. Scherer KR, Zentner MR (2001) Emotional effects of music: production rules. Series in affective science. Oxford University Press, New York, NY, US, pp 361–392. https://doi.org/10.1093/acprof:oso/9780198529392.003.0011
147. Mauss IB, Levenson RW, McCarter L, Wilhelm FH, Gross JJ (2005) The tie that binds? Coherence among emotion experience, behavior, and physiology. Emotion 5(2):175–190. https://doi.org/10.1037/1528-3542.5.2.175
148. Grewe O, Nagel F, Kopiez R, Altenmüller E (2007) Emotions over time: synchronicity and development of subjective, physiological, and facial affective reactions to music. Emotion 7(4):774–788. https://doi.org/10.1037/1528-3542.7.4.774
149. Ekman P (1992) An argument for basic emotions. Cogn Emot 6(3–4):169–200. https://doi.org/10.1080/02699939208411068

150. Levenson RW (2003) Blood, sweat, and fears: the autonomic architecture of emotion. Ann N Y Acad Sci 1000:348–366. https://doi.org/10.1196/annals.1280.016
151. Nakamura J, Csikszentmihalyi M (2021) The concept of flow. In: Snyder C, Lopez S, Edwards L, Marques S (eds) Handbook of positive psychology, 3 edn. Oxford University Press. https://doi.org/10.1093/oso/9780195135336.003.0007
152. MacDonald R, Byrne C, Carlton L (2006) Creativity and flow in musical composition: an empirical investigation. Psychol Music 34(3):292–306. https://doi.org/10.1177/0305735606064838
153. Harris DJ, Vine SJ, Wilson MR (2017) Neurocognitive mechanisms of the flow state. Prog Brain Res 234:221–243. https://doi.org/10.1016/bs.pbr.2017.06.012
154. Gold J, Ciorciari J (2020) A review on the role of the neuroscience of flow states in the modern world. Behav Sci (Basel) 10(9). https://doi.org/10.3390/bs10090137
155. Dietrich A (2006) Transient hypofrontality as a mechanism for the psychological effects of exercise. Psychiatry Research 145(1):79–83. https://doi.org/10.1016/j.psychres.2005.07.033
156. Weber R, Fisher JT (2020) Advancing the synchronization theory of flow experiences. ICA Handbook Series, Routledge, pp 157–176. https://doi.org/10.4324/9781351235587-15
157. Erkkinen MG, Berkowitz AL (2018) Brain research in music improvisation. In: Thaut MH, Hodges DA (eds) The oxford handbook of music and the brain. Oxford University Press. https://doi.org/10.1093/oxfordhb/9780198804123.013.8
158. Lazarus RS (1991) Emotion and adaptation. Oxford University Press, New York, NY, US. https://doi.org/10.1093/acprof:oso/9780195069447.001.0001
159. Darwin C (1872) The expression of the emotions in man and animals. John Murray, London, England. https://doi.org/10.1037/10001-000
160. Scherer KR, Schorr A, Johnstone T (eds) (2001) Appraisal processes in emotion: theory, methods, research. Oxford University Press, New York, NY, US. https://doi.org/10.1093/0195105838.001.0001
161. Blood AJ, Zatorre RJ (2001) Intensely pleasurable responses to music correlate with activity in brain regions implicated in reward and emotion. Proc Natl Acad Sci USA 98(20):11818–11823. https://doi.org/10.1073/pnas.191355898
162. Koelsch S (2014) Brain correlates of music-evoked emotions. Nat Rev Neurosci 15:170. https://doi.org/10.1038/nrn3666
163. Juslin PN (2010) Handbook of music and emotion: theory, research, applications. Oxford University Press. https://doi.org/10.1093/acprof:oso/9780199230143.001.0001
164. Rasch RA (2001) Timing and synchronization in ensemble performance. In: Generative processes in music: the psychology of performance, improvisation, and composition. Oxford University Press. https://doi.org/10.1093/acprof:oso/9780198508465.003.0004
165. D'Ausilio A, Badino L, Li Y, Tokay S, Craighero L, Canto R, Aloimonos Y, Fadiga L (2012) Leadership in orchestra emerges from the causal relationships of movement kinematics. PLOS ONE 7(5):1–6. https://doi.org/10.1371/journal.pone.0035757
166. Ono K, Nakamura A, Maess B (2015) Keeping an eye on the conductor: neural correlates of visuo-motor synchronization and musical experience. Front Hum Neurosci 9. https://doi.org/10.3389/fnhum.2015.00154
167. Müller V, Perdikis D, Mende MA, Lindenberger U (2021) Interacting brains coming in sync through their minds: an interbrain neurofeedback study. Ann New York Acad Sci 1500(1):48–68. https://doi.org/10.1111/nyas.14605
168. Delius JAM, Müller V (2023) Interpersonal synchrony when singing in a choir. Front Psychol 13. https://doi.org/10.3389/fpsyg.2022.1087517
169. Tschacher W, Haken H (2023) A complexity science account of humor. Entropy 25(2). https://doi.org/10.3390/e25020341
170. Hughes SM, Harrison MA, Gallup GG Jr (2007) Sex differences in romantic kissing among college students: an evolutionary perspective. Evol Psychol 5(3):147470490700500320. https://doi.org/10.1177/147470490700500320
171. Liu T, Saito G, Lin C, Saito H (2017) Inter-brain network underlying turn-based cooperation and competition: a hyperscanning study using near-infrared spectroscopy. Sci Rep 7(1):8684. https://doi.org/10.1038/s41598-017-09226-w

172. Tschacher W, Greenwood S, Ramakrishnan S, Tröndle M, Wald-Fuhrmann M, Seibert C, Weining C, Meier D (2023) Audience synchronies in live concerts illustrate the embodiment of music experience. Sci Rep 13(1):14843. https://doi.org/10.1038/s41598-023-41960-2
173. Tschacher W, Greenwood S, Weining C, Wald-Fuhrmann M, Ramakrishnan C, Seibert C, Tröndle M (2024) Physiological audience synchrony in classical concerts linked with listeners'experiences and attitudes. Sci Rep 14(1):16412. https://doi.org/10.1038/s41598-024-67455-2
174. Newen A, De Bruin L, Gallagher S (2018) The oxford handbook of 4E cognition. Oxford University Press. https://doi.org/10.1093/oxfordhb/9780198735410.001.0001
175. Walton AE, Richardson MJ, Langland-Hassan P, Chemero A (2015) Improvisation and the self-organization of multiple musical bodies. Front Psychol 6. https://doi.org/10.3389/fpsyg.2015.00313
176. Ramírez-Moreno MA, Cruz-Garza JG, Acharya A, Chatufale G, Witt W, Gelok D, Reza G, Contreras-Vidal J (2023) Brain-to-brain communication during musical improvisation: a performance case study. F1000Res 11:989. https://doi.org/10.12688/f1000research.123515.3
177. Nelias C, Sturm EM, Albrecht T, Hagmayer Y, Geisel T (2022) Downbeat delays are a key component of swing in jazz. Commun Phys 5(1):237. https://doi.org/10.1038/s42005-022-00995-z
178. Nozawa T, Sas M, Dolan D, Rajpal H, Rosas F, Timmermann C, Mediano P, Honda K, Amano S, Miyake Y, Jensen H (2023) Swinging, fast and slow: multiscale synchronisation dynamics reveals the impact of an improvisatory approach to performance on music experience. PsyArXiv. https://doi.org/10.31234/osf.io/cqxya
179. Palumbo A, Groves K, Munoz-Vidal EL, Turry A, Codio R, Raghavan P, Schambra H, Voelbel GT, Ripollés P (2024) Improvisation and live accompaniment increase motor response and reward during a music playing task. Sci Rep 14(1):13112. https://doi.org/10.1038/s41598-024-62794-6
180. Borgo D (2006) Sync or swarm: musical improvisation and the complex dynamics of group creativity. Springer Berlin Heidelberg, Berlin, Heidelberg, pp 1–24. https://doi.org/10.1007/11780274_1
181. Nomura R, Liang Y, Okada T (2015) Interactions among collective spectators facilitate eyeblink synchronization. PloS one 10(10):e0140774. https://doi.org/10.1371/journal.pone.0140774
182. Nomura R (2024) Reliability for music-induced heart rate synchronization. Sci Rep 14(1):12200. https://doi.org/10.1038/s41598-024-62994-0
183. Néda Z, Ravasz E, Brechet Y, Vicsek T, Barabási AL (2000) Self-organizing processes: The sound of many hands clapping. Nature (London) 403:849–850. https://doi.org/10.1038/35002564
184. Mann RP, Faria J, Sumpter DJT, Krause J (2013) The dynamics of audience applause. J R Soc Interface 10(85):20130466. https://doi.org/10.1098/rsif.2013.0466
185. Kuramoto Y, Battogtokh D (2002) Coexistence of coherence and incoherence in nonlocally coupled phase oscillators. Nonlinear Phen Complex Syst 5(4):380–385. https://doi.org/10.1142/S0218127402006024
186. Abrams DM, Strogatz SH (2004) Chimera states for coupled oscillators. Phys Rev Lett 93(17):174102. https://doi.org/10.1103/physrevlett.93.174102
187. Majhi S, Bera BK, Ghosh D, Perc M (2019) Chimera states in neuronal networks: a review. Phys Life Rev 28:100–121. https://doi.org/10.1016/j.plrev.2018.10.002
188. Markram H, Lübke J, Frotscher M, Sakmann B (1997) Regulation of synaptic efficacy by coincidence of postsynaptic APs and EPSPs. Science 275(5297):213–215. https://doi.org/10.1126/science.275.5297.213
189. Abbott LF, Nelson S (2000) Synaptic plasticity: taming the beast. Nat Neurosci 3(11):1178–1183. https://doi.org/10.1038/81453
190. Caporale N, Dan Y (2008) Spike timing-dependent plasticity: a Hebbian learning rule. Annu Rev Neurosci 31(1):25–46. https://doi.org/10.1146/annurev.neuro.31.060407.125639

191. Meisel C, Gross T (2009) Adaptive self-organization in a realistic neural network model. Phys Rev E 80(6):061917. https://doi.org/10.1103/physreve.80.061917
192. Gross T, D'Lima CJD, Blasius B (2006) Epidemic dynamics on an adaptive network. Phys Rev Lett 96(20):208701. https://doi.org/10.1103/physrevlett.96.208701
193. Strogatz SH (2001) Exploring complex networks. Nature 410:268–276. https://doi.org/10.1038/35065725
194. Albert R, Barabási AL (2002) Statistical mechanics of complex networks. Rev Mod Phys 74:47–97. https://doi.org/10.1103/revmodphys.74.47
195. Newman MEJ (2003) The structure and function of complex networks. SIAM Rev 45(2):167–256. https://doi.org/10.1137/s0036144503
196. Dahms T, Lehnert J, Schöll E (2012) Cluster and group synchronization in delay-coupled networks. Phys Rev E 86(1):016202. https://doi.org/10.1103/physreve.86.016202
197. Aoki T, Aoyagi T (2009) Co-evolution of phases and connection strengths in a network of phase oscillators. Phys Rev Lett 102:034101. https://doi.org/10.1103/physrevlett.102.034101
198. Aoki T, Aoyagi T (2011) Self-organized network of phase oscillators coupled by activity-dependent interactions. Phys Rev E 84:066109. https://doi.org/10.1103/physreve.84.066109
199. Nekorkin VI, Kasatkin DV (2016) Dynamics of a network of phase oscillators with plastic couplings. AIP Conf Proc 1738(1):210010. https://doi.org/10.1063/1.4951993
200. Kasatkin DV, Nekorkin VI (2016) Dynamics of the phase oscillators with plastic couplings. Radiophys Quantum Electron 58(11):877–891. https://doi.org/10.1007/s11141-016-9662-1
201. Gushchin A, Mallada E, Tang A (2015) Synchronization of phase-coupled oscillators with plastic coupling strength. In: Information theory and applications workshop ITA 2015, San Diego, CA, USA. IEEE, pp 291–300. https://doi.org/10.1109/ita.2015.7309003
202. Picallo CB, Riecke H (2011) Adaptive oscillator networks with conserved overall coupling: sequential firing and near-synchronized states. Phys Rev E 83(3):036206. https://doi.org/10.1103/physreve.83.036206
203. Timms L, English LQ (2014) Synchronization in phase-coupled Kuramoto oscillator networks with axonal delay and synaptic plasticity. Phys Rev E 89(3):032906. https://doi.org/10.1103/physreve.89.032906
204. Ren Q, Zhao J (2007) Adaptive coupling and enhanced synchronization in coupled phase oscillators. Phys Rev E 76(1):016207. https://doi.org/10.1103/physreve.76.016207
205. Avalos-Gaytán V, Almendral JA, Leyva I, Battiston F, Nicosia V, Latora V, Boccaletti S (2018) Emergent explosive synchronization in adaptive complex networks. Phys Rev E 97(4):042301. https://doi.org/10.1103/physreve.97.042301
206. Papadopoulos L, Kim JZ, Kurths J, Bassett DS (2017) Development of structural correlations and synchronization from adaptive rewiring in networks of kuramoto oscillators. Chaos 27(7):073115. https://doi.org/10.1063/1.4994819
207. Kasatkin DV, Klinshov VV, Nekorkin VI (2019) Itinerant chimeras in an adaptive network of pulse-coupled oscillators. Phys Rev E 99:022203. https://doi.org/10.1103/PhysRevE.99.022203
208. Boccaletti S, Bianconi G, Criado R, del Genio CI, Gómez-Gardeñes J, Romance M, Sendiña Nadal I, Wang Z, Zanin M (2014) The structure and dynamics of multilayer networks. Phys Rep 544(1):1–122. https://doi.org/10.1016/j.physrep.2014.07.001
209. De Domenico M, Solé-Ribalta A, Cozzo E, Kivelä M, Moreno Y, Porter MA, Gómez S, Arenas A (2013) Mathematical formulation of multilayer networks. Phys Rev X 3(4):041022. https://doi.org/10.1103/physrevx.3.041022
210. De Domenico M, Nicosia V, Arenas A, Latora V (2015) Structural reducibility of multilayer networks. Nat Commun 6:6864. https://doi.org/10.1038/ncomms7864
211. Kivelä M, Arenas A, Barthélemy M, Gleeson JP, Moreno Y, Porter MA (2014) Multilayer networks. J Complex Netw 2(3):203–271. https://doi.org/10.1093/comnet/cnu016
212. Girvan M, Newman MEJ (2002) Community structure in social and biological networks. Proc Natl Acad Sci USA 99:7821. https://doi.org/10.1073/pnas.122653799
213. Amato R, Díaz-Guilera A, Kleineberg KK (2017) Interplay between social influence and competitive strategical games in multiplex networks. Sci Rep 7:7087. https://doi.org/10.1038/s41598-017-07332-x

214. Amato R, Kouvaris NE, San Miguel M, Díaz-Guilera A (2017) Opinion competition dynamics on multiplex networks. New J Phys 19:123019. https://doi.org/10.1088/1367-2630/aa936a
215. Cardillo A, Zanin M, Gòmez Gardeñes J, Romance M, Garcia del Amo A, Boccaletti S (2013) Modeling the multi-layer nature of the European air transport network: resilience and passengers re-scheduling under random failures. Eur Phys J ST 215:23–33. https://doi.org/10.1140/epjst/e2013-01712-8
216. Meunier D, Lambiotte R, Bullmore ET (2010) Modular and hierarchically modular organization of brain networks. Front Neurosci 4:200. https://doi.org/10.3389/fnins.2010.00200
217. Bentley B, Branicky R, Barnes CL, Chew YL, Yemini E, Bullmore ET, Vétes PE, Schafer WR (2016) The multilayer connectome of Caenorhabditis Elegans. PLOS Comput Biol 12(12):e1005283. https://doi.org/10.1371/journal.pcbi.1005283
218. Battiston F, Nicosia V, Chavez M, Latora V (2017) Multilayer motif analysis of brain networks. Chaos 27:047404. https://doi.org/10.1063/1.4979282
219. Vaiana M, Muldoon SF (2020) Multilayer brain networks. J Nonlinear Sci 30(5):2147–2169. https://doi.org/10.1007/s00332-017-9436-8
220. Ramlow L, Sawicki J, Zakharova A, Hlinka J, Claussen JC, Schöll E (2019) Partial synchronization in empirical brain networks as a model for unihemispheric sleep. Europhys Lett 126:50007. https://doi.org/10.1209/0295-5075/126/50007
221. Ashourvan A, Telesford QK, Verstynen T, Vettel JM, Bassett DS (2019) Multi-scale detection of hierarchical community architecture in structural and functional brain networks. PLoS ONE 14(5):e0215520. https://doi.org/10.1371/journal.pone.0215520
222. Zhou C, Zemanová L, Zamora G, Hilgetag CC, Kurths J (2006) Hierarchical organization unveiled by functional connectivity in complex brain networks. Phys Rev Lett 97:238103. https://doi.org/10.1103/physrevlett.97.238103
223. Zhou C, Zemanová L, Zamora-López G, Hilgetag CC, Kurths J (2007) Structure-function relationship in complex brain networks expressed by hierarchical synchronization. New J Phys 9:178. https://doi.org/10.1088/1367-2630/9/17/178
224. Wang R, Lin P, Liu M, Wu Y, Zhou T, Zhou C (2019) Hierarchical connectome modes and critical state jointly maximize human brain functional diversity. Phys Rev Lett 123(3):038301. https://doi.org/10.1103/physrevlett.123.038301
225. Maksimenko VA, Makarov VV, Bera BK, Ghosh D, Dana SK, Goremyko MV, Frolov NS, Koronovskii AA, Hramov AE (2016) Excitation and suppression of chimera states by multiplexing. Phys Rev E 94:052205. https://doi.org/10.1103/physreve.94.052205
226. Jalan S, Singh A (2016) Cluster synchronization in multiplex networks. Europhys Lett 113(3):30002. https://doi.org/10.1209/0295-5075/113/30002
227. Ghosh S, Kumar A, Zakharova A, Jalan S (2016) Birth and death of chimera: interplay of delay and multiplexing. Europhys Lett 115:60005. https://doi.org/10.1209/0295-5075/115/60005
228. Leyva I, Sevilla-Escoboza R, Sendiña-Nadal I, Gutiérrez R, Buldú JM, Boccaletti S (2017) Inter-layer synchronization in non-identical multi-layer networks. Sci Rep 7:45475. https://doi.org/10.1038/srep45475
229. Andrzejak RG, Ruzzene G, Malvestio I (2017) Generalized synchronization between chimera states. Chaos 27:053114. https://doi.org/10.1063/1.4983841
230. Ghosh S, Zakharova A, Jalan S (2018) Non-identical multiplexing promotes chimera states. Chaos Solitons Fractals 106:56–60. https://doi.org/10.1016/j.chaos.2017.11.010
231. Semenova N, Zakharova A (2018) Weak multiplexing induces coherence resonance. Chaos 28(5):051104. https://doi.org/10.1063/1.5037584
232. Sawicki J, Omelchenko I, Zakharova A, Schöll E (2018) Delay controls chimera relay synchronization in multiplex networks. Phys Rev E 98:062224. https://doi.org/10.1103/physreve.98.062224
233. Omelchenko I, Hülser T, Zakharova A, Schöll E (2019) Control of chimera states in multilayer networks. Front Appl Math Stat 4:67. https://doi.org/10.3389/fams.2018.00067
234. Rybalova E, Vadivasova T, Strelkova G, Anishchenko V, Zakharova A (2019) Forced synchronization of a multilayer heterogeneous network of chaotic maps in the chimera state mode. Chaos 29:033134. https://doi.org/10.1063/1.5090184

235. Nikitin D, Omelchenko I, Zakharova A, Avetyan M, Fradkov AL, Schöll E (2019) Complex partial synchronization patterns in networks of delay-coupled neurons. Phil Trans R Soc A 377:20180128. https://doi.org/10.1098/rsta.2018.0128
236. Blaha KA, Huang K, Della Rossa F, Pecora LM, Hossein-Zadeh M, Sorrentino F (2019) Cluster synchronization in multilayer networks: a fully analog experiment with lc oscillators with physically dissimilar coupling. Phys Rev Lett 122(1):014101. https://doi.org/10.1103/physrevlett.122.014101
237. Sevilla-Escoboza R, Sendiña-Nadal I, Leyva I, Gutiérrez R, Buldú JM, Boccaletti S (2016) Inter-layer synchronization in multiplex networks of identical layers. Chaos 26:065304. https://doi.org/10.1063/1.4952967
238. Requejo RJ, Díaz-Guilera A (2016) Replicator dynamics with diffusion on multiplex networks. Phys Rev E 94:022301. https://doi.org/10.1103/physreve.94.022301
239. Pitsik E, Makarov V, Kirsanov D, Frolov NS, Goremyko M, Li X, Wang Z, Hramov AE, Boccaletti S (2018) Inter-layer competition in adaptive multiplex network. New J. Phys. 20:075004. https://doi.org/10.1088/1367-2630/aad00d
240. Leyva I, Sendiña-Nadal I, Sevilla-Escoboza R, Vera-Avila VP, Chholak P, Boccaletti S (2018) Relay synchronization in multiplex networks. Sci Rep 8(1):8629. https://doi.org/10.1038/s41598-018-26945-w
241. Jalan S, Kumar A, Leyva I (2019) Explosive synchronization in frequency displaced multiplex networks. Chaos 29(4):041102. https://doi.org/10.1063/1.5092226
242. Frolov NS, Maksimenko VA, Makarov VV, Kirsanov D, Hramov AE, Kurths J (2018) Macroscopic chimeralike behavior in a multiplex network. Phys Rev E 98(2):022320. https://doi.org/10.1103/physreve.98.022320
243. Maslennikov OV, Nekorkin VI (2018) Hierarchical transitions in multiplex adaptive networks of oscillatory units. Chaos 28:121101. https://doi.org/10.1063/1.5066330
244. Goremyko MV, Kirsanov D, Nedaivozovb VO, Makarov VV, Hramov AE (2017) Pattern formation in adaptive multiplex network in application to analysis of the complex structure of neuronal network of the brain. In: Dynamics and fluctuations in biomedical photonics XIV. Proceedings of SPIE, vol 10063. SPIE, pp 100631C–1. https://doi.org/10.1117/12.2249842
245. Citri A, Malenka RC (2008) Synaptic plasticity: multiple forms, functions, and mechanisms. Neuropsychopharmacology 33(1):18–41. https://doi.org/10.1038/sj.npp.1301559
246. Edelmann E, Cepeda-Prado E, Leßmann V (2017) Coexistence of multiple types of synaptic plasticity in individual hippocampal ca1 pyramidal neurons. Front Synaptic Neurosci 9:7. https://doi.org/10.3389/fnsyn.2017.00007
247. Zenke F, Agnes EJ, Gerstner W (2015) Diverse synaptic plasticity mechanisms orchestrated to form and retrieve memories in spiking neural networks. Nat Commun 6(6922). https://doi.org/10.1038/ncomms7922
248. John RA, Liu F, Chien NA, Kulkarni MR, Zhu C, Fu QD, Basu A, Liu Z, Mathews N (2018) Synergistic gating of electro-iono-photoactive 2d chalcogenide neuristors: Coexistence of hebbian and homeostatic synaptic metaplasticity. Adv Mater 30(25):1800220. https://doi.org/10.1002/adma.201800220
249. Shai S, Dobson S (2013) Coupled adaptive complex networks. Phys Rev E 87(4):042812. https://doi.org/10.1103/physreve.87.042812
250. Wardil L, Hauert C (2014) Origin and structure of dynamic cooperative networks. Sci Rep 4(5725). https://doi.org/10.1038/srep05725
251. Klimek P, Diakonova M, Eguiluz VM, San Miguel M, Thurner S (2016) Dynamical origins of the community structure of an online multi-layer society. New J Phys 18(083045). https://doi.org/10.1088/1367-2630/18/8/083045
252. Berner R, Sawicki J, Schöll E (2020) Birth and stabilization of phase clusters by multiplexing of adaptive networks. Phys Rev Lett 124(8):088301. https://doi.org/10.1103/physrevlett.124.088301
253. Hoppensteadt FC, Izhikevich EM (1996) Synaptic organizations and dynamical properties of weakly connected neural oscillators ii. learning phase information. Biol Cybern 75(2):129–135. https://doi.org/10.1007/s004220050280

254. Seliger P, Young SC, Tsimring LS (2002) Plasticity and learning in a network of coupled phase oscillators. Phys Rev E 65(4):041906. https://doi.org/10.1103/physreve.65.041906
255. Aoki T (2015) Self-organization of a recurrent network under ongoing synaptic plasticity. Neural Netw 62:11–19. https://doi.org/10.1016/j.neunet.2014.05.024
256. Hebb D (1949) The organization of behavior: a neuropsychological theory, new, edition. Wiley, New York
257. Maistrenko Y, Lysyansky B, Hauptmann C, Burylko O, Tass PA (2007) Multistability in the kuramoto model with synaptic plasticity. Phys Rev E 75(6):066207. https://doi.org/10.1103/physreve.75.066207
258. Popovych OV, Yanchuk S, Tass PA (2013) Self-organized noise resistance of oscillatory neural networks with spike timing-dependent plasticity. Sci Rep 3:2926. https://doi.org/10.1038/srep02926
259. Strogatz SH (2000) From kuramoto to crawford: exploring the onset of synchronization in populations of coupled oscillators. Phys D 143:1–20. https://doi.org/10.1016/S0167-2789(00)00094-4
260. Acebrón JA, Bonilla LL, Vicente CJP, Ritort F, Spigler R (2005) The kuramoto model: A simple paradigm for synchronization phenomena. Rev Mod Phys 77(1):137. https://doi.org/10.1103/RevModPhys.77.137
261. Pikovsky A, Rosenblum M (2008) Partially integrable dynamics of hierarchical populations of coupled oscillators. Phys Rev Lett 101(26):264103. https://doi.org/10.1103/physrevlett.101.264103
262. Omel'chenko OE, Wolfrum M (2012) Nonuniversal transitions to synchrony in the sakaguchi-kuramoto model. Phys Rev Lett 109:164101. https://doi.org/10.1103/physrevlett.109.164101
263. Choe CU, Dahms T, Hövel P, Schöll E (2010) Controlling synchrony by delay coupling in networks: from in-phase to splay and cluster states. Phys Rev E 81(2):025205(R). https://doi.org/10.1103/physreve.81.025205
264. Boyd S, Vandenberghe L (2004) Convex optimization. Cambridge University Press. https://doi.org/10.1017/CBO9780511804441
265. Liesen J, Mehrmann V (2015) Linear algebra. Springer, Cham. https://doi.org/10.1007/978-3-319-24346-7
266. Silvester JR (2000) Determinants of block matrices. Math Gaz 84(501):460–467. https://doi.org/10.2307/3620776
267. Powell PD (2011) Calculating determinants of block matrices. arXiv:1112.4379
268. Gómez S, Díaz-Guilera A, Gómez-Gardeñes J, Pérez Vicente CJ, Moreno Y, Arenas A (2013) Diffusion dynamics on multiplex networks. Phys Rev Lett 110:028701. https://doi.org/10.1103/physrevlett.110.028701
269. Solé-Ribalta A, De Domenico M, Kouvaris NE, Díaz-Guilera A, Gómez S, Arenas A (2013) Spectral properties of the Laplacian of multiplex networks. Phys Rev E 88(3):032807. https://doi.org/10.1103/physreve.88.032807
270. Tang L, Wu X, Lü J, Lu J, D'Souza RM (2019) Master stability functions for complete, intralayer, and interlayer synchronization in multiplex networks of coupled rössler oscillators. Phys Rev E 99(1):012304. https://doi.org/10.1103/physreve.99.012304
271. Omel'chenko OE, Wolfrum M, Maistrenko Y (2010) Chimera states as chaotic spatiotemporal patterns. Phys Rev E 81(6):065201(R). https://doi.org/10.1103/physreve.81.065201
272. Wolfrum M, Omel'chenko OE (2011) Chimera states are chaotic transients. Phys Rev E 84(1):015201. https://doi.org/10.1103/physreve.84.015201
273. Hizanidis J, Kanas V, Bezerianos A, Bountis T (2014) Chimera states in networks of nonlocally coupled Hindmarsh-Rose neuron models. Int J Bifurc Chaos 24(03):1450030. https://doi.org/10.1142/s0218127414500308
274. Omelchenko I, Omel'chenko OE, Hövel P, Schöll E (2013) When nonlocal coupling between oscillators becomes stronger: patched synchrony or multichimera states. Phys Rev Lett 110:224101. https://doi.org/10.1103/physrevlett.110.224101
275. Schmidt L, Schönleber K, Krischer K, García-Morales V (2014) Coexistence of synchrony and incoherence in oscillatory media under nonlinear global coupling. Chaos 24(1):013102. https://doi.org/10.1063/1.4858996

276. Suda Y, Okuda K (2015) Persistent chimera states in nonlocally coupled phase oscillators. Phys Rev E 92:060901. https://doi.org/10.1103/physreve.92.060901
277. Bera BK, Ghosh D (2016) Chimera states in purely local delay-coupled oscillators. Phys Rev E 93:052223. https://doi.org/10.1103/physreve.93.052223
278. Zakharova A, Kapeller M, Schöll E (2014) Chimera death: symmetry breaking in dynamical networks. Phys Rev Lett 112:154101. https://doi.org/10.1103/physrevlett.112.154101
279. Bick C (2017) Heteroclinic switching between chimeras, arXiv. https://doi.org/10.48550/arXiv.1703.03274
280. Bogomolov S, Slepnev A, Strelkova G, Schöll E, Anishchenko VS (2017) Mechanisms of appearance of amplitude and phase chimera states in a ring of nonlocally coupled chaotic systems. Commun Nonlinear Sci Numer Simul 43:25. https://doi.org/10.1016/j.cnsns.2016.06.024
281. Omelchenko I, Provata A, Hizanidis J, Schöll E, Hövel P (2015) Robustness of chimera states for coupled FitzHugh-Nagumo oscillators. Phys Rev E 91:022917. https://doi.org/10.1103/physreve.91.022917
282. Omel'chenko OE (2018) The mathematics behind chimera states. Nonlinearity 31(5):R121. https://doi.org/10.1088/1261-6544/aaaa07
283. Schöll E, Zakharova A (2018) Chasing chimeras: control of complex networks. Res Featur 123:34–37. https://doi.org/10.1038/s42254-018-0005-7
284. Omel'chenko OE, Knobloch E (2019) Chimerapedia: coherence-incoherence patterns in one, two and three dimensions. New J Phys 21(9):093034. https://doi.org/10.1088/1367-2630/ab3f6b
285. Sawicki J, Omelchenko I, Zakharova A, Schöll E (2019) Delay-induced chimeras in neural networks with fractal topology. Eur Phys J B 92:54. https://doi.org/10.1140/epjb/e2019-90252-0
286. Sawicki J, Ghosh S, Jalan S, Zakharova A (2019) Chimeras in multiplex networks: interplay of inter- and intra-layer delays. Front Appl Math Stat 5:19. https://doi.org/10.3389/fams.2019.00019
287. Schöll E, Zakharova A, Andrzejak RG (2019) Editorial on the research topic: Chimera states in complex networks. Front Appl Math Stat 5:62. https://doi.org/10.3389/fams.2019.00062
288. Larger L, Penkovsky B, Maistrenko Y (2013) Virtual chimera states for delayed-feedback systems. Phys Rev Lett 111:054103. https://doi.org/10.1103/physrevlett.111.054103
289. Gonzalez-Avella JC, Cosenza MG, Miguel MS (2014) Localized coherence in two interacting populations of social agents. Phys A 399:24–30. https://doi.org/10.1016/j.physa.2013.12.035
290. Gambuzza LV, Buscarino A, Chessari S, Fortuna L, Meucci R, Frasca M (2014) Experimental investigation of chimera states with quiescent and synchronous domains in coupled electronic oscillators. Phys Rev E 90:032905. https://doi.org/10.1103/physreve.90.032905
291. Bastidas VM, Omelchenko I, Zakharova A, Schöll E, Brandes T (2015) Quantum signatures of chimera states. Phys Rev E 92:062924. https://doi.org/10.1103/physreve.92.062924
292. Andrzejak RG, Rummel C, Mormann F, Schindler K (2016) All together now: analogies between chimera state collapses and epileptic seizures. Sci Rep 6:23000. https://doi.org/10.1038/srep23000
293. Banerjee T, Dutta PS, Zakharova A, Schöll E (2016) Chimera patterns induced by distance-dependent power-law coupling in ecological networks. Phys Rev E 94:032206. https://doi.org/10.1103/physreve.94.032206
294. Loos SAM, Claussen JC, Schöll E, Zakharova A (2016) Chimera patterns under the impact of noise. Phys Rev E 93:012209. https://doi.org/10.1103/physreve.93.012209
295. Kouvaris NE, Requejo RJ, Hizanidis J, Díaz-Guilera A (2016) Chimera states in a network-organized public goods game with destructive agents. Chaos Interdiscip J Nonlinear Sci 26(12):123108. https://doi.org/10.1063/1.4971974
296. Hizanidis J, Kouvaris NE, Zamora-López G, Díaz-Guilera A, Antonopoulos C (2016) Chimera-like states in modular neural networks. Sci Rep 6:19845. https://doi.org/10.1038/srep19845

297. Chouzouris T, Omelchenko I, Zakharova A, Hlinka J, Jiruska P, Schöll E (2018) Chimera states in brain networks: empirical neural vs. modular fractal connectivity. Chaos 28(4):045112. https://doi.org/10.1063/1.5009812
298. Bansal K, Garcia JO, Tompson SH, Verstynen T, Vettel JM, Muldoon SF (2019) Cognitive chimera states in human brain networks. Sci Adv 5(4):eaau8535. https://doi.org/10.1126/sciadv.aau8535
299. Kang L, Tian C, Huo S, Liu Z (2019) A two-layered brain network model and its chimera state. Sci Rep 9:14389. https://doi.org/10.1038/s41598-019-50969-5
300. Rybalova E, Bukh A, Strelkova G, Anishchenko V (2019) Spiral and target wave chimeras in a 2D lattice of map-based neuron models. Chaos Interdiscip J Nonlinear Sci 29(10):101104. https://doi.org/10.1063/1.5126178
301. Pournaki A, Merfort L, Ruiz J, Kouvaris NE, Hövel P, Hizanidis J (2019) Synchronization patterns in modular neuronal networks: a case study of C. elegans. Front Appl Math Stat 5(52):1. https://doi.org/10.3389/fams.2019.00052
302. Kundu S, Majhi S, Ghosh D (2019) From asynchronous to synchronous chimeras in ecological multiplex network. Eur Phys J Special Top 228(11):2429–2439. https://doi.org/10.1140/epjst/e2019-900013-1
303. Hagerstrom AM, Murphy TE, Roy R, Hövel P, Omelchenko I, Schöll E (2012) Experimental observation of chimeras in coupled-map lattices. Nat Phys 8:658–661. https://doi.org/10.1038/nphys2372
304. Tinsley MR, Nkomo S, Showalter K (2012) Chimera and phase cluster states in populations of coupled chemical oscillators. Nat Phys 8:662–665. https://doi.org/10.1038/nphys2371
305. Martens EA, Thutupalli S, Fourriere A, Hallatschek O (2013) Chimera states in mechanical oscillator networks. Proc Natl Acad Sci USA 110:10563. https://doi.org/10.1073/pnas.1302880110
306. Wickramasinghe M, Kiss IZ (2013) Spatially organized dynamical states in chemical oscillator networks: synchronization, dynamical differentiation, and chimera patterns. PLoS ONE 8(11):e80586. https://doi.org/10.1371/journal.pone.0080586
307. Hart JD, Bansal K, Murphy TE, Roy R (2016) Experimental observation of chimera and cluster states in a minimal globally coupled network. Chaos 26:094801. https://doi.org/10.1063/1.4953662
308. Totz J, Rode J, Tinsley MR, Showalter K, Engel H (2018) Spiral wave chimera states in large populations of coupled chemical oscillators. Nat Phys 14:282–285. https://doi.org/10.1038/s41567-017-0034-7
309. Majhi S, Perc M, Ghosh D (2016) Chimera states in uncoupled neurons induced by a multilayer structure. Sci Rep 6:39033. https://doi.org/10.1038/srep39033
310. Majhi S, Perc M, Ghosh D (2017) Chimera states in a multilayer network of coupled and uncoupled neurons. Chaos 27:073109. https://doi.org/10.1063/1.4991861
311. Bukh AV, Rybalova E, Semenova N, Strelkova G, Anishchenko V (2017) New type of chimera and mutual synchronization of spatiotemporal structures in two coupled ensembles of nonlocally interacting chaotic maps. Chaos 27:111102. https://doi.org/10.1063/1.5009375
312. Goremyko MV, Maksimenko VA, Makarov VV, Ghosh D, Bera B, Dana SK, Hramov AE (2017) Interaction of chimera states in a multilayered network of nonlocally coupled oscillators. Tech. Phys. Lett. 43(8):712–715. https://doi.org/10.1134/s1063785017080077
313. Andrzejak RG, Ruzzene G, Malvestio I, Schindler K, Schöll E, Zakharova A (2018) Mean field phase synchronization between chimera states. Chaos 28(9):091101. https://doi.org/10.1063/1.5049750
314. Winkler M, Sawicki J, Omelchenko I, Zakharova A, Anishchenko V, Schöll E (2019) Relay synchronization in multiplex networks of discrete maps. EPL 126:50004. https://doi.org/10.1209/0295-5075/126/50004
315. Dmitrichev AS, Shchapin D, Nekorkin VI (2019) Cloning of chimera states in a large short-term coupled multiplex network of relaxation oscillators. Front Appl Math Stat 5:9. https://doi.org/10.3389/fams.2019.00009

316. Drauschke F, Sawicki J, Berner R, Omelchenko I, Schöll E (2020) Effect of topology upon relay synchronization in triplex neuronal networks. Chaos 30:051104. https://doi.org/10.1063/5.0008341
317. De Domenico M (2017) Multilayer modeling and analysis of human brain networks. Gigascience 6(5):1–8. https://doi.org/10.1093/gigascience/gix062
318. Sieber J, Omel'chenko OE, Wolfrum M (2014) Controlling unstable chaos: Stabilizing chimera states by feedback. Phys Rev Lett 112:054102. https://doi.org/10.1103/physrevlett.112.054102
319. Bick C, Martens EA (2015) Controlling chimeras. New J Phys 17(3):033030. https://doi.org/10.1088/1367-2630/17/3/033030
320. Gambuzza LV, Frasca M (2016) Pinning control of chimera states. Phys Rev E 94:022306. https://doi.org/10.1103/physreve.94.022306
321. Isele TM, Hizanidis J, Provata A, Hövel P (2016) Controlling chimera states: the influence of excitable units. Phys Rev E 93(2):022217. https://doi.org/10.1103/physreve.93.022217
322. Gjurchinovski A, Schöll E, Zakharova A (2017) Control of amplitude chimeras by time delay in dynamical networks. Phys Rev E 95(4):042218. https://doi.org/10.1103/physreve.95.042218
323. Omelchenko I, Omel'chenko OE, Zakharova A, Wolfrum M, Schöll E (2016) Tweezers for chimeras in small networks. Phys Rev Lett 116:114101. https://doi.org/10.1103/physrevlett.116.114101
324. Omelchenko I, Omel'chenko OE, Zakharova A, Schöll E (2018) Optimal design of tweezer control for chimera states. Phys Rev E 97:012216. https://doi.org/10.1103/PhysRevE.97.012216
325. Ruzzene G, Omelchenko I, Schöll E, Zakharova A, Andrzejak RG (2019) Controlling chimera states via minimal coupling modification. Chaos 29:051103. https://doi.org/10.1063/1.5097570
326. Ruzzene G, Omelchenko I, Sawicki J, Zakharova A, Schöll E, Andrzejak RG (2020) Remote pacemaker-control of chimera states in multilayer networks of neurons. Phys Rev E 102:052216. https://doi.org/10.1103/physreve.102.052216
327. Kori H, Mikhailov AS (2004) Entrainment of randomly coupled oscillator networks by a pacemaker. Phys Rev Lett 93:254101. https://doi.org/10.1103/PhysRevLett.93.254101
328. Radicchi F, Meyer-Ortmanns H (2006) Entrainment of coupled oscillators on regular networks by pacemakers. Phys Rev E 73:036218. https://doi.org/10.1103/PhysRevE.73.036218
329. Gollo LL, Mirasso CR, Atienza M, Crespo-Garcia M, Cantero JL (2011) Theta band zero-lag long-range cortical synchronization via hippocampal dynamical relaying. PLoS ONE 6(3):e17756. https://doi.org/10.1371/journal.pone.0017756
330. Fischer I, Vicente R, Buldú JM, Peil M, Mirasso CR, Torrent MC, García-Ojalvo J (2006) Zero-lag long-range synchronization via dynamical relaying. Phys Rev Lett 97(12):123902. https://doi.org/10.1103/physrevlett.97.123902
331. Bergner A, Frasca M, Sciuto G, Buscarino A, Ngamga EJ, Fortuna L, Kurths J (2012) Remote synchronization in star networks. Phys Rev E 85:026208. https://doi.org/10.1103/physreve.85.026208
332. Moslonka-Lefebvre M, Gilligan CA, Monod H, Belloc C, Ezanno P, Filipe JAN, Vergu E (2016) Market analyses of livestock trade networks to inform the prevention of joint economic and epidemiological risks. J R Soc Interface 13(116):20151099. https://doi.org/10.1098/rsif.2015.1099
333. Grimm V, Revilla E, Berger U et al (2005) Pattern-oriented modeling of agent-based complex systems: lessons from ecology. Science 310(5750):987–991. https://doi.org/10.1126/science.1116681
334. Sprott JC (2004) Competition with evolution in ecology and finance. Phys Lett A 325(5):329–333. https://doi.org/10.1016/j.physleta.2004.06.014
335. Woolley-Meza O, Thiemann C, Grady D, Lee JJ, Seebens H, Blasius B, Brockmann D (2011) Complexity in human transportation networks: a comparative analysis of worldwide air transportation and global cargo-ship movements. Eur. Phys. J. B 84(4):589–600. https://doi.org/10.1140/epjb/e2011-20208-9

336. De Domenico M, Granell C, Porter MA, Arenas A (2016) The physics of spreading processes in multilayer networks. Nat Phys 12:901–906. https://doi.org/10.1038/nphys3865
337. Mucha PJ, Richardson T, Macon K, Porter MA, Onnela JP (2010) Community structure in time-dependent, multiscale, and multiplex networks. Science 328(5980):876–878. https://doi.org/10.1126/science.1184819
338. Nicosia V, Valencia M, Chavez M, Díaz-Guilera A, Latora V (2013) Remote synchronization reveals network symmetries and functional modules. Phys Rev Lett 110:174102. https://doi.org/10.1103/physrevlett.110.174102
339. Gambuzza LV, Cardillo A, Fiasconaro A, Fortuna L, Gómez-Gardeñes J, Frasca M (2013) Analysis of remote synchronization in complex networks. Chaos 23:043103. https://doi.org/10.1063/1.4824656
340. Gambuzza LV, Frasca M, Fortuna L, Boccaletti S (2016) Inhomogeneity induces relay synchronization in complex networks. Phys Rev E 93(4):042203. https://doi.org/10.1103/physreve.93.042203
341. Zhang L, Motter AE, Nishikawa T (2017) Incoherence-Mediated Remote Synchronization. Phys Rev Lett 118(17):174102. https://doi.org/10.1103/physrevlett.118.174102
342. Guillery RW, Sherman SM (2002) Thalamic relay functions and their role in corticocortical communication: generalizations from the visual system. Neuron 33(2):163–175. https://doi.org/10.1016/S0896-6273(01)00502-8
343. Bassett DS, Meyer-Lindenberg A, Achard S, Duke T, Bullmore ET (2006) Adaptive reconfiguration of fractal small-world human brain functional networks. Proc Natl Acad Sci USA 103(51):19518–19523. https://doi.org/10.1073/pnas.0604451103
344. Bassett DS, Bullmore ET (2006) Small-world brain networks. Neuroscientist 12(6):512–523. https://doi.org/10.1177/1073858406293182
345. Pecora LM, Sorrentino F, Hagerstrom AM, Murphy TE, Roy R (2014) Symmetries, cluster synchronization, and isolated desynchronization in complex networks. Nat Commun 5:4079. https://doi.org/10.1038/ncomms5079
346. Sawicki J, Omelchenko I, Zakharova A, Schöll E (2017) Chimera states in complex networks: interplay of fractal topology and delay. Eur Phys J Spec Top 226(9):1883–1892. https://doi.org/10.1140/epjst/e2017-70036-8
347. Sawicki J, Omelchenko I, Zakharova A, Schöll E (2018) Synchronization scenarios of chimeras in multiplex networks. Eur Phys J Spec Top 227:1161. https://doi.org/10.1140/epjst/e2018-800039-y
348. Schöll E, Zakharova A, Andrzejak RG (2020) Chimera states in complex networks. Res Top Front Appl Math Stat. https://doi.org/10.3389/978-2-88963-311-1
349. Rattenborg NC, Amlaner CJ, Lima SL (2000) Behavioral, neurophysiological and evolutionary perspectives on unihemispheric sleep. Neurosci Biobehav Rev 24:817–842. https://doi.org/10.1016/S0149-7634(00)00045-1
350. Rattenborg NC, Voirin B, Cruz SM, Tisdale R, Dell'Omo G, Lipp HP, Wikelski M, Vyssotski AL (2016) Evidence that birds sleep in mid-flight. Nat Commun 7:12468. https://doi.org/10.1038/ncomms12468
351. Mascetti GG (2016) Unihemispheric sleep and asymmetrical sleep: behavioral, neurophysiological, and functional perspectives. Nat Sci Sleep 8:221. https://doi.org/10.2147/NSS.S71970
352. Nikolaev AR, Gepshtein S, Gong P, van Leeuwen C (2010) Duration of coherence intervals in electrical brain activity in perceptual organization. Cerebral Cortex 20:365. https://doi.org/10.1093/cercor/bhp099
353. Ahn S, Rubchinsky LL (2013) Short desynchronization episodes prevail in synchronous dynamics of human brain rhythms. Chaos 23:013138. https://doi.org/10.1063/1.4770219
354. Ahn S, Rubchinsky LL, Lapish CC (2014) Dynamical reorganization of synchronous activity patterns in prefrontal cortex-hippocampus networks during behavioral sensitization. Cerebral Cortex 24:2553–2561. https://doi.org/10.1093/cercor/bht111
355. Jiruska P, de Curtis M, Jefferys JGR, Schevon CA, Schiff SJ, Schindler K (2013) Synchronization and desynchronization in epilepsy: controversies and hypotheses. J Physiol 591(4):787–797. https://doi.org/10.1113/jphysiol.2012.239590

356. Jirsa VK, Stacey WC, Quilichini PP, Ivanov AI, Bernard C (2014) On the nature of seizure dynamics. Brain 137:2210. https://doi.org/10.1093/brain/awu133
357. Sakaguchi H (2006) Instability of synchronized motion in nonlocally coupled neural oscillators. Phys Rev E 73(3):031907. https://doi.org/10.1103/physreve.73.031907
358. FitzHugh R (1961) Impulses and physiological states in theoretical models of nerve membrane. Biophys J 1:445–466. https://doi.org/10.1016/s0006-3495(61)86902-6
359. Watts DJ, Strogatz SH (1998) Collective dynamics of 'small-world' networks. Nature 393:440–442. https://doi.org/10.1038/30918
360. Arenas A, Díaz-Guilera A, Kurths J, Moreno Y, Zhou C (2008) Synchronization in complex networks. Phys Rep 469(3):93–153. https://doi.org/10.1016/j.physrep.2008.09.002
361. Barahona M, Pecora LM (2002) Synchronization in small-world systems. Phys Rev Lett 89(5):054101. https://doi.org/10.1103/physrevlett.89.054101
362. Hong H, Jun Kim B, Choi MY, Park H (2004) Factors that predict better synchronizability on complex networks. Phys Rev E 69:067105. https://doi.org/10.1103/PhysRevE.69.067105
363. Karakaya B, Minati L, Gambuzza LV, Frasca M (2019) Fading of remote synchronization in tree networks of stuart-landau oscillators. Phys Rev E 99(5):052301. https://doi.org/10.1103/physreve.99.052301
364. Sawicki J, Koulen JM, Schöll E (2021) Synchronization scenarios in three-layer networks with a hub. Chaos 31:073131. https://doi.org/10.1063/5.0053691
365. Anishchenko VS (1995) Dynamical Chaos-models and experiments. World Scientific, Singapore
366. Anishchenko VS, Astakhov V, Neiman AB, Vadivasova T, Schimansky-Geier L (2007) Nonlinear dynamics of chaotic and stochastic systems: tutorial and modern developments. Springer, Berlin. https://doi.org/10.1007/978-3-540-70889-3
367. Anishchenko VS, Vadivasova T, Strelkova G (2014) Deterministic nonlinear systems. Springer series in synergetics, Springer. https://doi.org/10.1007/978-3-642-41849-0
368. Semenova N, Zakharova A, Anishchenko VS, Schöll E (2016) Coherence-resonance chimeras in a network of excitable elements. Phys Rev Lett 117:014102. https://doi.org/10.1103/PhysRevLett.117.014102
369. Zakharova A, Semenova N, Anishchenko VS, Schöll E (2017) Time-delayed feedback control of coherence resonance chimeras. Chaos 27:114320. https://doi.org/10.1063/1.5000683
370. Shepelev IA, Vadivasova TE, Bukh AV, Strelkova GI, Anishchenko VS (2017) New type of chimera structures in a ring of bistable FitzHugh-Nagumo oscillators with nonlocal interaction. Phys Lett A 381:1398. https://doi.org/10.1016/j.physleta.2017.02.034
371. Shepelev IA, Strelkova G, Anishchenko V (2018) Chimera states and intermittency in an ensemble of nonlocally coupled Lorenz systems. Chaos 28(6):063119. https://doi.org/10.1063/1.5020009
372. Shepelev IA, Bukh AV, Muni SS, Anishchenko VS (2020) Role of solitary states in forming spatiotemporal patterns in a 2D lattice of van der Pol oscillators. Chaos Solitons Fractals 135:109725. https://doi.org/10.1016/j.chaos.2020.109725
373. Shepelev IA, Bukh AV, Strelkova GI, Anishchenko VS (2021) Anti-phase relay synchronization of wave structures in a heterogeneous multiplex network of 2D lattices. Chaos Solitons Fractals 143:110545. https://doi.org/10.1016/j.chaos.2020.110545
374. Shepelev IA, Bukh AV, Vadivasova TE, Anishchenko VS (2021) Synchronization effects for dissipative and inertial coupling between multiplex lattices. Commun Nonlinear Sci Numer Simul 93:105489. https://doi.org/10.1016/j.cnsns.2020.105489
375. Shepelev IA, Anishchenko VS (2021) Bistable labyrinth-like structures and chimera states in a 2D lattice of van der Pol oscillators. Commun Nonlinear Sci Numer Simul 93:105513. https://doi.org/10.1016/j.cnsns.2020.105513
376. Omelchenko I, Maistrenko Y, Hövel P, Schöll E (2011) Loss of coherence in dynamical networks: spatial chaos and chimera states. Phys Rev Lett 106:234102. https://doi.org/10.1103/physrevlett.106.234102
377. Semenova N, Zakharova A, Schöll E, Anishchenko VS (2015) Does hyperbolicity impede emergence of chimera states in networks of nonlocally coupled chaotic oscillators? Europhys Lett 112:40002. https://doi.org/10.1209/0295-5075/112/40002

378. Bogomolov S, Strelkova G, Schöll E, Anishchenko VS (2016) Amplitude and phase chimeras in an ensemble of chaotic oscillators. Tech Phys Lett 42(7):765–768. https://doi.org/10.1134/s1063785016070191
379. Vadivasova TE, Strelkova G, Bogomolov SA, Anishchenko VS (2016) Correlation analysis of the coherence-incoherence transition in a ring of nonlocally coupled logistic maps. Chaos 26:093108. https://doi.org/10.1063/1.4962647
380. Semenova N, Strelkova G, Anishchenko VS, Zakharova A (2017) Temporal intermittency and the lifetime of chimera states in ensembles of nonlocally coupled chaotic oscillators. Chaos 27:061102. https://doi.org/10.1063/1.4985143
381. Shepelev IA, Bukh AV, Vadivasova TE, Anishchenko VS, Zakharova A (May2017) Double-well chimeras in 2D lattice of chaotic bistable elements. Commun Nonlinear Sci Numer Simul 54:50–61. https://doi.org/10.1016/j.cnsns.2017.05.017
382. Strelkova G, Rybalova E, Anishchenko V, Zakharova A (2017) Effect of switchings and the lifetime of chimeras in an ensemble of nonlocally coupled chaotic maps. In: ICNAAM 2017. AIP conference proceedings
383. Shepelev IA, Bukh AV, Strelkova GI, Vadivasova TE, Anishchenko VS (2017) Chimera states in ensembles of bistable elements with regular and chaotic dynamics. Nonlinear Dyn 90:2317. https://doi.org/10.1007/s11071-017-3805-6
384. Semenova N, Vadivasova T, Anishchenko V (2018) Mechanism of solitary state appearance in an ensemble of nonlocally coupled Lozi maps. Eur Phys J Spec Top 227:1173. https://doi.org/10.1140/epjst/e2018-800035-y
385. Bukh AV, Strelkova GI, Anishchenko VS (2018) Synchronization of chimera states in coupled networks of nonlinear chaotic oscillators. Russ J Nonlinear Dyn 14(4):419. https://doi.org/10.20537/nd180401
386. Rybalova E, Strelkova GI, Anishchenko VS (2018) Mechanism of realizing a solitary state chimera in a ring of nonlocally coupled chaotic maps. Chaos Solitons Fractals 115:300–305. https://doi.org/10.1016/j.chaos.2018.09.003
387. Bukh AV, Schöll E, Anishchenko VS (2019) Synchronization of spiral wave patterns in two-layer 2D lattices of nonlocally coupled discrete oscillators. Chaos 29:053105. https://doi.org/10.1063/1.5092352
388. Rybalova E, Strelkova G, Schöll E, Anishchenko V (2020) Relay and complete synchronization in heterogeneous multiplex networks of chaotic maps. Chaos 30:061104. https://doi.org/10.1063/5.0008902
389. Sawicki J, Schöll E (2021) Influence of sound on empirical brain networks. Front Appl Math Stat 7:662221. https://doi.org/10.3389/fams.2021.662221
390. Berner R, Mehrmann V, Schöll E, Yanchuk S (2021) The multiplex decomposition: An analytic framework for multilayer dynamical networks. SIAM J Appl Dyn Syst 20(4):1752–1772. https://doi.org/10.1137/21m1406180
391. Zhang Y, Nishikawa T, Motter AE (2017) Asymmetry-induced synchronization in oscillator networks. Phys Rev E 95(6):062215. https://doi.org/10.1103/physreve.95.062215
392. Zhang Y, Motter AE (2020) Symmetry-independent stability analysis of synchronization patterns. SIAM Rev 62(4):817–836. https://doi.org/10.1137/19m127358x
393. Roelfsema P, Engel A, König P, Singer W (1997) Visuomotor integration is associated with zero time-lag synchronization among cortical areas. Nature 385:157–161. https://doi.org/10.1038/385157a0
394. Soteropoulos DS, Baker SN (2006) Cortico-cerebellar coherence during a precision grip task in the monkey. J Neurophysiol 95:1194. https://doi.org/10.1152/jn.00535.2005
395. Sherman SM (2016) Thalamus plays a central role in ongoing cortical functioning. Nat Neurosci 19(4):533. https://doi.org/10.1038/nn.4269
396. Rhodes PA, Llinas R (2005) A model of thalamocortical relay cells. J Physiol 565(3):765–781. https://doi.org/10.1113/jphysiol.2005.086287
397. Gollo LL, Mirasso CR, Villa AEP (2010) Dynamic control for synchronization of separated cortical areas through thalamic relay. NeuroImage 52(3):947–955. https://doi.org/10.1016/j.neuroimage.2009.11.058

398. Wang X, Vaingankar V, Sanchez CS, Sommer FT, Hirsch JA (2011) Thalamic interneurons and relay cells use complementary synaptic mechanisms for visual processing. Nat Neurosci 14(2):224. https://doi.org/10.1038/nn.2717
399. Halassa MM, Kastner S (2017) Thalamic functions in distributed cognitive control. Nat Neurosci 20(12):1669–1679. https://doi.org/10.1038/s41593-017-0020-1
400. de Curtis M, Paré D (2004) The rhinal cortices: a wall of inhibition between the neocortex and the hippocampus. Prog Neurobiol 74(2):101. https://doi.org/10.1016/j.pneurobio.2004.09.004
401. Prasad JA, Chudasama Y (2013) Viral tracing identifies parallel disynaptic pathways to the hippocampus. J Neurosci 33(19):8494–8503. https://doi.org/10.1523/JNEUROSCI.4809-12.2013
402. Nagumo J, Arimoto S, Yoshizawa S (1962) An active pulse transmission line simulating nerve axon. Proc IRE 50:2061–2070. https://doi.org/10.1109/JRPROC.1962.288235
403. Bassett DS, Zurn P, Gold JI (2018) On the nature and use of models in network neuroscience. Nat Rev Neurosci 19:566–578. https://doi.org/10.1038/s41583-018-0038-8
404. Kondo S, Miura T (2010) Reaction-diffusion model as a framework for understanding biological pattern formation. Science 329(5999):1616–1620. https://doi.org/10.1126/science.1179047
405. Bachmair CA, Schöll E (2014) Nonlocal control of pulse propagation in excitable media. Eur Phys J B 87(11):276. https://doi.org/10.1140/epjb/e2014-50339-2
406. Omelchenko I, Zakharova A, Hövel P, Siebert J, Schöll E (2015) Nonlinearity of local dynamics promotes multi-chimeras. Chaos 25:083104. https://doi.org/10.1063/1.4927829

Chapter 4
Modeling Physiological and Neuronal Systems

This chapter demonstrates the effective application of complex systems theory in modeling physiological and neuronal systems, underscoring the crucial roles of adaptation and synchronization as universal mechanisms. These phenomena are foundational to diverse systems, including physiological regulation, artificial intelligence learning processes, and socio-economic dynamics. In physiological systems, synchronization underpins neural network functionality and informs medical applications. Adaptivity, similarly, is pivotal in artificial intelligence, enabling neural networks to adjust to data and tasks, while in socio-economic systems, it governs the dynamics of social interactions and adaptive models for epidemiology and climate change. Together, these examples highlight the broad applicability and fruitfulness of complex systems theory across disciplines in Sect. 4.1.

In Sect. 4.2, a novel functional model for tumor disease and sepsis is introduced, emphasizing the dynamic interactions of cytokine activity between parenchymal and immune cells, regulated by adaptive coupling weights. This approach departs from biochemical-focused models, providing a unifying perspective on disease dynamics rooted in network-based interactions.

The study of neuronal systems in Sect. 4.3 integrates empirical brain network structures, derived from diffusion tensor imaging (DTI) and probabilistic tractography, with FitzHugh-Nagumo (FHN) oscillators to model neural activity. This framework is further applied in Sect. 4.4, where mesoscopic-scale brain models reveal partial synchronization patterns and hemispheric asymmetries in the dynamics, influenced by structural connectivity and controlled via inter-hemispheric coupling strength.

In Sect. 4.5, simulations of FitzHugh-Nagumo oscillators on empirical structural networks mimic synchronization phenomena observed in epileptic seizures. A direct comparison between simulation results and electroencephalography (EEG) data highlights the model's real-world applicability and predictive power. Additionally, the analysis underscores the role of structural features in shaping synchronization dynamics, providing avenues for further research into perception, sleep or various brain disorders through comparative studies of neuronal brain networks.

J. Sawicki, *Music Psychology—Balance of Relations*,
Current Research in Systematic Musicology 13,
https://doi.org/10.1007/978-3-031-97003-0_4

Overall, this Chapter illustrates the pivotal role of complex systems theory in bridging empirical data and theoretical models to advance our understanding of physiological and neuronal systems.

4.1 Adaptation and Synchronization as Modeling Tools

Adaptation and synchronization are fundamental phenomena observed across a diverse range of fields, where they play critical roles in shaping underlying processes. To enhance our understanding and predictive capabilities in these domains, modeling approaches leveraging these phenomena are widely employed. This section presents three application examples from three distinct areas:

Adaptation and synchronization are essential in physiological systems, particularly in understanding regulatory principles and mechanisms. Section 4.1.1 focuses on these phenomena within neural networks, exploring their foundational dynamics (see Sect. 4.1.1.1), their perspectives (see Sect. 4.1.1.2) and discussing specific medical applications (see Sect. 4.1.1.3).

Adaptivity lies at the core of machine learning, as "learning" fundamentally involves "adapting" to input data. Adaptive systems, such as neural networks, adjust based on the learning task, architecture, and algorithm. Section 4.1.2 examines various perspectives on adaptivity in artificial intelligence, with discussions on deep neural networks (see Sect. 4.1.2.1), recurrent neural networks (see Sect. 4.1.2.2), and reservoir computing (see Sect. 4.1.2.3).

Section 4.1.3 explores adaptation and synchronization in socio-economic systems, focusing on topics such as adaptive social interactions (see Sect. 4.1.3.2) and the integration of adaptive mechanisms in epidemiological (see Sect. 4.1.3.1) as well as climatic models (see Sect. 4.1.3.3).

Section 4.1.4 summarizes the main aspects of this section. Further details on these topics can be found in the publication [1, 2], which serves as the basis for this section.

4.1.1 Neural Adaptivity and Synchronization

This section highlights the critical role of adaptation and synchronization in physiological processes, particularly through neuronal plasticity. Evolution often converges on analogous solutions to related challenges, resulting in shared principles across distinct systems. The physiological properties of biological systems are best understood as complex networks of interactions, commonly referred to as regulatory networks. Under comparable conditions, these regulatory networks exhibit shared features known as adaptation motifs, each with distinct functional implications (Sect. 4.1.1.1).

Neuronal systems comprise millions of individual neurons, whose detailed dynamics are often mathematically intractable. However, methodologies have been developed to study the macroscopic collective dynamics emerging from such systems. One of the most effective approaches is the next-generation neural mass model, which enables low-dimensional representations of neuronal populations incorporating features such as frequency adaptation and short-term synaptic plasticity (Sect. 4.1.1.2).

Computational models have proven instrumental in elucidating the mechanisms underlying adaptation in neuronal systems. For instance, in medical applications, deep brain stimulation has become the gold standard for treating Parkinson's disease patients who exhibit abnormal neuronal synchrony and suffer from various motor and non-motor symptoms. Incorporating synaptic plasticity into computational models facilitates the design of targeted therapeutic stimulation protocols, improving treatment outcomes (Sect. 4.1.1.3).

4.1.1.1 Design Principles for Adaptation in Physiological Systems

In this section, we analyze motifs of adaptation within physiological regulatory networks. The physiological characteristics of biological systems arise from a complex interplay among their fundamental components. For example, the rate of protein synthesis from a gene depends on the presence of transcription factors–proteins whose production is, in turn, influenced by other transcription factors. Similarly, hormone secretion into the bloodstream is regulated by the concentrations of other circulating factors, which are themselves modulated by the presence of additional hormones. These intricate and interdependent interactions are collectively termed regulatory networks.

A pivotal concept in the study of regulatory networks is the observation that evolutionary processes often converge on similar solutions to analogous challenges. Under comparable conditions, regulatory networks across diverse biological systems exhibit recurring patterns or mathematical regularities known as regulatory motifs or *design principles* [3–5]. Identifying these design principles provides profound insights into the functional roles of specific regulatory interactions. This prompts a fundamental inquiry: what design principles enable biological systems to adapt? In this context, adaptation refers to the capacity of a system to self-regulate and maintain functionality despite fluctuations in internal parameters or external environmental conditions.

An illustrative example is the regulation of blood glucose homeostasis, where the glucose concentration (denoted as x) must be tightly controlled within a physiological range (5 mM). Deviations from this range can result in severe or life-threatening consequences. The human body has evolved a robust mechanism to lower elevated blood glucose levels through the action of pancreatic β-cells. These specialized cells detect glucose concentrations in the bloodstream and secrete insulin, a hormone that signals distant tissues–including adipose tissue, skeletal muscle, and liver cells–to reduce glucose levels.

This regulatory system sustains glucose concentrations at a steady state, despite being highly sensitive to multiple parameters, such as the number of β-cells, plasma volume, and tissue responsiveness to insulin. While these parameters can vary sig-

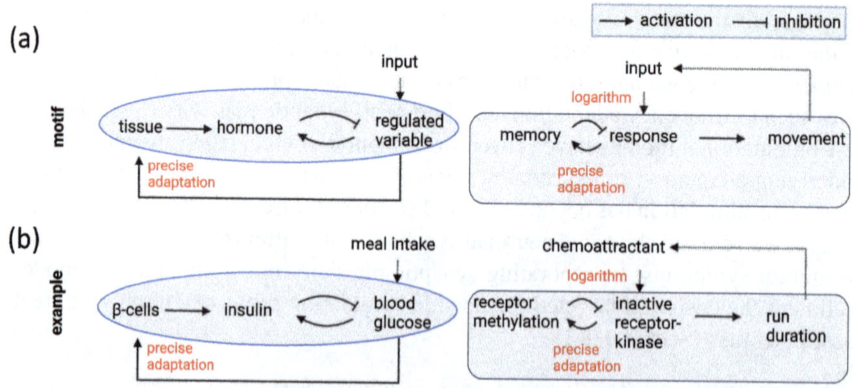

Fig. 4.1 Motifs for adaptation in physiological systems. **a** In hormone circuits, a hormone-regulated variable governs the growth rate of the tissue responsible for its secretion, enabling precise adaptation. This adaptation mechanism ensures that the dynamics of the regulated variable remain robust in the face of physiological variations. **b** Organisms employ a combination of logarithmic sensing, precise adaptation, coupled to movement regulation, to achieve robust sampling of an input field. This motif is observed in chemotaxis and potentially in the mammalian dopamine system. Figure adapted from [1]

nificantly among individuals, most people can nonetheless maintain blood glucose concentrations within the narrow physiological range [6].

A related problem arises in bacterial chemotaxis, where the bacterium Escherichia coli (*E. coli*) navigates its environment using a strategy akin to a random walk. This involves alternating between movement and reorientation at a characteristic rate, denoted as ϕ (the tumbling rate), which typically occurs every few seconds. The bacterium achieves directed navigation by modulating ϕ in response to sensed ligand molecules, known as attractants or repellents. For example, a sudden increase in the concentration of an attractant transiently reduces ϕ, resulting in a net drift towards regions with higher attractant concentrations.

Notably, at steady-state and for a fixed attractant concentration u, ϕ remains constant across a wide range of sensed attractant levels and is independent of u [7, 8]. This raises an intriguing question: how is ϕ maintained at a constant level despite variations in the input activity of the chemotactic signaling circuit?

This problem is closely related to the engineering challenge of disturbance rejection [9–11]. Disturbance rejection is exemplified in systems such as automotive cruise control, where the system maintains a constant speed despite changes in road slope, or thermostats, which stabilize room temperature under varying external conditions. The key to solving this challenge lies in the use of integral feedback control. In this mechanism, the feedback increases proportionally to the accumulated error (integrating the error signal), ensuring that the steady-state error converges to zero.

How is integral feedback implemented in biological circuits? In hormone circuits there appears to be a simple answer (Fig. 4.1a). Let x be the regulated variable and y to be its regulating hormone, with Z the mass of the tissue that secretes the hormone.

In the blood glucose system, x is blood glucose, y is blood insulin, and Z is β-cell mass. The following motif is observed across hormone systems: there is a slow negative feedback where the main regulator of the growth dynamics of Z is x, that is, x adjusts the death-, growth-, and replication-rates of the cells of Z. Thus:

$$\dot{Z} = f(x)Z, \tag{4.1}$$

where $f(x)$ is the x-dependent growth rate. The system will settle at the steady-state where $f(x) = 0$ (denoted x_0) regardless of variation in the other physiological parameters, including plasma volume, secretion rate, and the responses of remote cells.

The ubiquity of the motif suggests that it is uniquely advantageous. Why is it so prevalent? Beyond integral feedback, another intriguing phenomena occurs. Consider for example the following simple model for the glucose system:

$$\begin{aligned} \dot{u} &= u - sxy, \\ \dot{y} &= pZ - \gamma y, \end{aligned} \tag{4.2}$$

where s is the sensitivity to the response of the hormone, and p is the product of the per-cell secretion and (inverse) plasma volume. u is the time-dependent input, incorporating e.g. meal intake. Eq. (4.1) not only sets the steady-state of x to $x = x_0$, it makes the entire dynamics in response to any input u invariant of s, p [12]. These scale invariant dynamics are evident in clinical data from distinct hormonal systems [12–15]. Thus, in hormone systems, negative feedback from the regulated variable to its controlling tissue allows the system to adapt its dynamics to variability in key system parameters, which are uncertain and may be highly variable.

Scale invariance is also observed in bacterial chemotaxis, where the dynamics of the tumbling rate, $\phi(t)$, are modulated by the attractant input, $u(t)$, in a manner that depends on relative rather than absolute changes in $u(t)$. This phenomenon is known as fold-change detection [16]. Fold-change detection has been documented in the navigation systems of other simple organisms, such as worms and slime molds [17, 18].

What about more complex organisms? In vertebrates, including mice and humans, movement is regulated by dopamine transmission in the midbrain. Dopamine is released in response to surprise (or prediction errors) about rewards, such as food or drink. Specifically, better-than-expected outcomes lead to increased firing rates of dopaminergic neurons, while worse-than-expected outcomes transiently inhibit their firing [19]. Interestingly, the dopaminergic response is also scale-invariant [20]. Furthermore, during locomotion, dopamine levels change in a manner consistent with a response to the temporal derivative of a spatial input field [21].

On closer inspection, the dopamine system exhibits striking parallels to the chemotaxis system. In the case of dopamine, the input field corresponds to an animal's expectations about rewards [22]. This input field decreases spatially with distance from the locations where rewards are encountered, resembling the spatial decay of a chemical attractant from its source. Additionally, dopamine modulates movement vigor in a way that mirrors the influence of attractants on bacterial motion.

We therefore identified another regulatory motif: fold-change detection of an input field, which modulates movement statistics (Fig. 4.1b). What is the function of this motif? From the perspective of sensing, scale invariance allows us to remove uncertainty and retain sensitivity over a wide dynamic input range. An additional distinct advantage is apparent when we consider the coupling between sensing and movement. The fold-change detection circuit calculates the temporal logarithmic derivative of the input $u(t)$. In a spatial setting, we can consider a spatial input field $U(x)$; the movement dynamics of the organism over long time- and length- scales are captured by the stochastic dynamics:

$$\mathrm{d}x = \beta v^2 \phi \nabla \log U \ \mathrm{d}t + v^2 \sqrt{2\phi}\ \mathrm{d}W, \tag{4.3}$$

where v is the typical movement speed, and β depends on circuit parameters. The steady-state distribution of the organism location is $P(x) = U(x)^{\beta}$, which only depends on circuit parameters (rather than movement parameters); the motif thus provides a robust mechanism for sampling a power of the input field. This is again consistent with experimental observations on both chemotaxis and the dopamine system [22]. Thus, in these systems, a motif that appears to support adaptation of sensing in the background of uncertain input levels, in fact provides a mechanism for robust sampling of uncertain environments.

The examples discussed herein indicate that adaptation motifs enabling scale-invariant dynamics are widely observed in regulatory networks. Moreover, specific regulatory motifs, which recur under similar biological contexts, exhibit distinct functional roles that contribute to their physiological relevance. Identifying and characterizing these motifs, as well as comparing their behaviors across different systems, can significantly advance our understanding of the mechanisms through which adaptation is achieved in complex regulatory networks.

In conclusion, the identification of common regulatory motifs and the elucidation of their underlying design principles provide critical insights into the adaptive capabilities of physiological systems. These findings enhance our understanding of how such systems maintain functionality despite intrinsic variability and external environmental fluctuations.

4.1.1.2 A Next Generation Neural Mass Approach to Spike Frequency Adaptation and Short Term Plasticity

Neural mass models are mean-field approximations designed to replicate the dynamics of homogeneous neuronal populations. These models range from heuristic approaches, such as the well-known Wilson-Cowan model [23], to more sophisticated formulations derived from the eigenfunction expansion of the Fokker-Planck equation for membrane potential distributions [24, 25]. Recently, an exact derivation of a next-generation neural mass model was achieved for heterogeneous populations of spiking neurons [26–28]. This development leveraged analytic techniques from coupled phase oscillators [29] and was made possible for networks of Quadratic Integrate-and-Fire (QIF) neurons–representing Hodgkin's Class I excitable membranes [30]. These models describe the dynamics of neuronal networks using macroscopic variables such as the population firing rate and mean membrane potential, and have been applied across diverse neuroscientific contexts [31–40].

This section explores the emergent dynamics of next-generation neural mass models when neuronal populations adapt to environmental changes by modifying intrinsic or extrinsic properties. In particular, we examine collective dynamics, such as synchronization and bursting behavior, driven by spike-frequency adaptation and post-synaptic plasticity.

Spike-Frequency Adaptation Spike-frequency adaptation, a ubiquitous neurobiological phenomenon, occurs in neurons across vertebrate and invertebrate systems, central and peripheral alike. It is an intrinsic mechanism of adaptation resulting from slow negative feedback to cellular excitability. Experimental studies suggest that spike-frequency adaptation arises from balancing ionic currents triggered during action potential generation [41, 42]. Three principal ionic mechanisms contribute to adaptation:

- Voltage-gated potassium currents caused by high-threshold potassium channels [43].
- Calcium-dependent potassium currents resulting from calcium influx and intracellular calcium dynamics [44].
- Slow recovery from sodium channel inactivation [45].

These mechanisms collectively result in a characteristic firing behavior where neurons exhibit an initial increase in firing frequency followed by gradual reduction under prolonged stimulation.

The application of neural mass models to spike-frequency adaptation began with mean-field descriptions of networks of linear integrate-and-fire neurons [46]. Using Fokker-Planck formalism and adiabatic approximations for slow adaptation timescales, Gigante et al. identified distinct types of collective bursting dynamics. More recently, next-generation neural mass models incorporating spike-frequency adaptation mechanisms have been shown to produce complex collective behaviors, including nested oscillations with cross-frequency coupling [38, 47]. These studies highlight how spike-frequency adaptation shapes population-level dynamics, from promoting bursting in excitatory networks to suppressing tonic spiking in inhibitory ones.

Short-Term Plasticity Short-term plasticity (STP) reflects synaptic efficacy changes influenced by presynaptic activity history, providing an extrinsic adaptation mechanism [48–52]. STP manifests as two opposing effects:

- Short-term depression reduces synaptic efficacy through neurotransmitter depletion [53, 54], receptor desensitization [55, 56], or resource limitations [57, 58].
- Short-term facilitation enhances efficacy due to residual calcium influx that increases neurotransmitter release probability.

STP has been implicated in diverse cortical processes, including working memory, where facilitation and depression contribute to memory storage and retrieval, respectively [59–61]. Neural mass models equipped with STP have reproduced key working memory features, including sustained and transient oscillations across frequency bands [36]. These oscillations are thought to underpin memory operations in the brain, as evidenced by primate and human studies [62–65].

Integrated Neural Mass Models Next-generation neural mass models unify intrinsic (e.g., spike-frequency adaptation) and extrinsic (e.g., STP) adaptation mechanisms into a cohesive framework. These models offer precise descriptions of firing rates and mean membrane potentials, enabling insights into both spiking and subthreshold dynamics. Despite the varied timescales of these mechanisms–ranging from milliseconds for spike adaptation [41] to days for receptor density changes [56]–mean-field approaches remain robust.

Applications of these models extend to understanding rhythmic outputs in biological circuits, such as central pattern generators (CPGs) responsible for locomotion [66] and respiration [67]. By coupling adaptation-driven slow dynamics with fast synaptic timescales, these models elucidate the self-organization of rhythmic activity, even in the absence of rhythmic input. For example, spike-frequency adaptation has been shown to produce nested slow-fast oscillations and bursting behaviors across interacting neural populations [47].

In summary, the incorporation of spike-frequency adaptation and STP into exact neural mass models offers a powerful tool for exploring the mesoscopic dynamics of neural circuits. These advancements pave the way for more realistic neural architectures, potentially enabling the development of multilayer models that mimic cortical layer interactions [68].

4.1.1.3 Therapeutic Reshaping of Plastic Networks

Regular deep brain stimulation (DBS) is the gold standard treatment for patients with medically refractory Parkinson's disease [69–73]. For patients with advanced Parkinson's, studies have shown that combining regular DBS with medication is superior to medication alone [74]. Despite its therapeutic efficacy [75, 76], DBS is associated with side effects [77–80]. Notably, it may induce characteristic side effects collectively termed DBS-induced movement disorders [81, 82].

A key limitation of regular deep brain stimulation (DBS) is its inconsistent efficacy across different Parkinsonian symptoms. While effective in managing tremors and rigidity, standard DBS targeting the subthalamic nucleus (STN) or globus pallidus internus (GPi) offers limited or no improvement for gait and other axial symptoms such as balance and posture impairments. It may even exacerbate speech as well as affective and cognitive symptoms [83–86].

Addressing Abnormal Neuronal Synchrony Abnormal neuronal synchrony is a hallmark of Parkinson's disease [87]. Computational modeling has proposed innovative stimulation strategies to counteract this abnormal synchrony. Two key approaches include:

- Phase-dependent stimulus delivery, which applies desynchronizing stimulation based on neuronal phase [88].
- Compound stimuli, designed to induce desynchronization irrespective of initial dynamic conditions [89, 90].

Coordinated Reset (CR) stimulation was developed based on these principles. CR employs weak, phase-resetting stimuli that do not require sophisticated calibration [90]. This approach was selected for preclinical and clinical studies due to its theoretical potential to desynchronize pathological neuronal activity effectively. Initially, CR stimulation was envisioned for demand-controlled, closed-loop applications, where stimuli would be delivered dynamically in response to detected resynchronization or adjusted according to synchrony levels [90]. However, at that time, no implantable pulse generators were available for clinical testing [73, 91].

Subsequently, engineering-based advancements led to the development of closed-loop deep brain stimulation (DBS) systems capable of recording neuronal or muscular activity to suppress pathological patterns upon detection [92, 93]. Despite this progress, routine clinical applications of closed-loop DBS face several unresolved challenges [94].

A Novel Adaptive Stimulation Paradigm A qualitatively different stimulation approach, rooted in adaptive dynamical systems, was computationally developed to tackle these limitations [95]. Adaptivity, a fundamental feature of the nervous system, enables it to cope with complex physiological processes and environmental changes (see Sects. 4.1.1.1, 5.1.1, 4.1.1.2, 5.1.3, and 5.1.2). Adaptive mechanisms, however, can also become maladaptive in response to pathological changes, contributing to disease progression. In Parkinson's disease, dopamine depletion triggers a cascade of structural and functional changes [96].

To address these disease-related maladaptive changes, synaptic plasticity–particularly spike-timing-dependent plasticity (STDP)–was integrated into neuronal network models to design therapeutic stimulation [95]. STDP, a form of synaptic plasticity that depends on the precise timing of neuronal spikes [97–100], was incorporated to counteract pathological adaptations and facilitate therapeutic network remodeling (see also Sects. 5.1.1 and 4.1.1.2). This novel stimulation paradigm represents a radical departure from traditional DBS approaches, offering the potential for long-term restoration of healthy neuronal dynamics [95].

It was observed that Coordinated Reset stimulation can shift a network from an unfavorable, synchronized attractor to a more favorable, desynchronized attractor (Fig. 4.2) [95]. From then on, coordinated reset stimulation and further variants were computationally developed and optimized to robustly cause an "unlearn-

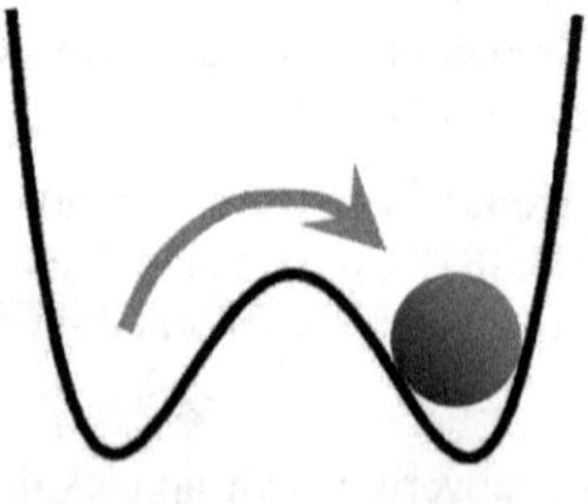

Fig. 4.2 Schematic illustrating how desynchronizing stimulation induces long-lasting therapeutic effects by leveraging plasticity. Spike-timing-dependent plasticity is a fundamental plasticity mechanism of the nervous system which adapts the synaptic strengths based on the relative timings of post- and presynaptic spikes [97–99]. Neural networks with spike timing dependent plasticity typically display bi- or multi-stability of stable states with stronger synchrony and synaptic connectivity and stable desynchronized states with weaker synaptic connectivity [95, 101, 103, 105, 109, 117], as illustrated by a simple double-well potential here. These states serve as models for pathological and physiological conditions. Coordinated reset stimulation may shift the network into the basin of attraction of a stable desynchronized state, in this way causing a long-lasting desynchronization [95]. Figure taken from [1]

ing" of pathological synchrony and synaptic connectivity, in this way causing long-lasting therapeutic effects [95, 101–105, 105–108]. A series of computational studies revealed *novel stimulus response characteristics* of neural networks with spike-timing-dependent plasticity:

- *Rebound of synchrony after cessation of stimulation*: Directly after cessation of coordinated reset stimulation, synchrony may reemerge and then spontaneously fade while further approaching the desynchronized attractor [101].
- *Cumulative effects*: Effects of coordinated reset stimulation may accumulate over time [109], and stimulation pauses may even improve the outcome [110].
- *Acute vs. long-term effects*: Acute stimulation effects (observed during stimulation) and long-term effects (emerging when the system relaxes into a stable state after cessation of stimulation) may differ substantially [103, 105]. One can even decouple neurons, i.e., reduce their synaptic weights, without desynchronization during stimulation [105]. In fact, acute effects do not necessarily serve as predictive markers for long-term outcome [103, 105].
- *Transition to non-invasive stimulation*: Long-term effects are favorable because they enable to reduce stimulation time and, hence, potentially reduce side effects. However, a profound advantage of this type of stimulation is that it does not require implants to permanently deliver stimulation. Rather, as predicted theoretically [111, 112], non-invasive stimulation can be delivered occasionally or regularly for a few hours. Non-invasive therapies are typically less risky and more appropriate for larger patient populations.
- *Functional restoration*: Not only stimulation-induced unlearning of abnormal synaptic connectivity and neuronal synchronization [95], but also reshaping net-

work connectivity by differentially up- or downregulating different synaptic connections [107] may contribute to a restoration of function.
- *Different plasticity mechanisms*: In Parkinson's disease pathophysiology, both spike-timing-dependent plasticity and structural plasticity [113, 114] are important [96] and may induce different stimulation responses [115, 116].

These computationally derived predictions and results enabled to design appropriate protocols for pre-clinical and clinical studies.

Invasive coordinated reset studies Coordinated reset deep brain stimulation was successfully tested in Parkinsonian monkeys [112, 118–120]. For instance, a few hours of coordinated reset deep brain stimulation led to therapeutic effects lasting for one month [112]. In addition, cumulative and long-lasting desynchronizing and therapeutic effects were observed in Parkinson's patients treated with coordinated reset deep brain stimulation [91].

Non-invasive coordinated reset studies Vibrotactile coordinated reset fingertip stimulation was developed to provide patients with a non-surgical and non-pharmacological treatment option [121]. To this end, instead of administering electrical bursts through depth electrodes, weak, non-painful vibratory bursts were non-invasively delivered in a coordinated reset mode to patients' fingertips [121]. A first in human study [122] as well as pilot studies [123] showed that vibrotactile Coordinated Reset stimulation is safe and tolerable and revealed a statistically and clinically significant reduction of Parkinson's disease symptoms off medication together with a significant reduction of high beta (21–30 Hz) power in the sensorimotor cortex. Remarkably, also axial symptoms, difficult to treat with regular deep brain stimulation, responded well to vibrotactile coordinated reset in these studies [122, 123]. For illustration, see patient videos in [123]. Of note, Parkinson's disease patients improved during a months-long vibrotactile coordinated reset treatment when evaluated after medication withdrawal, indicating a substantial improvement of the patients' conditions [123]. These findings indicate that vibrotactile coordinated reset treatment might even have an impact on metabolic and degenerative processes [123, 124], e.g., by slowing or even counteracting degeneration-related processes, e.g., vicious circles giving rise to oxidant stress and mitochondrial impairment, causing a bioenergetic crisis and the death of dopamine neurons in the substantia nigra [125–127].

In summary, rather than merely suppressing unwanted neuronal activity, stimulation techniques grounded in the principles of adaptive dynamical systems aim to induce lasting therapeutic effects. These techniques work by guiding affected neural systems toward more favorable attractor states, thereby addressing the root of pathological dynamics (Fig. 4.2).

4.1.2 Adaptation and Artificial Learning

In this section, we emphasize the critical role of adaptation in the realm of artificial learning, highlighting the symbiotic relationship between natural sciences and machine learning.

On one hand, theoretical physics provides powerful tools to illuminate the inner workings of machine learning algorithms, helping us move beyond the traditional "black box" perspective. For example, concepts from statistical physics are leveraged to address foundational questions, such as explaining the effectiveness of artificial learning despite the challenges posed by the curse of dimensionality (see Sect. 4.1.2.1). Additionally, physics-inspired models are used to generate controlled training datasets, enabling targeted investigations into machine learning capabilities, such as their capacity to extract and leverage memory from input sequences (see Sect. 4.1.2.2).

On the other hand, machine learning serves as a transformative tool for exploring and managing complex physical systems. This includes both investigating their behavior (see Sect. 4.1.2.2) and devising innovative control mechanisms (see Sect. 4.1.2.3). For instance, reservoir computing techniques provide novel approaches to controlling chaotic nonlinear systems (see Sect. 4.1.2.3). Another key challenge lies in deciphering the rules governing, and devising strategies to control, the collective behavior of self-organizing multi-agent systems.

This interplay between the natural sciences and machine learning underscores a bidirectional exchange of insights, advancing both fields and offering novel approaches to complex adaptive systems.

4.1.2.1 Adaptation and Neural Networks

Deep neural networks (DNNs) have driven numerous advancements in machine learning over the past decade. Their transformative impact spans various domains, beginning with groundbreaking progress in computer vision [128–132], extending to natural language processing [133–136], and excelling in complex strategic games like Go [137, 138] and Poker [139–141]. Their influence is also growing in the natural sciences [142], exemplified by their role in predicting the 3D structures of nearly all human proteins [143], a milestone in structural biology. Additional applications in physics are discussed in Sect. 4.1.2.2.

Despite their inspiration from biological neural circuits, as detailed in Sects. 5.1.1 and 4.1.1.2, machine learning neurons are far simpler. Interestingly, the adaptability of Deep neural networks (DNNs), distinct from biological mechanisms, underpins their success. For instance, the traditional task of classifying images as "cat" or "dog" exemplifies this adaptivity. Previously, models relied on manually crafted features like edge detection or patch correlations [144–146], requiring extensive domain expertise. In contrast, deep learning bypasses this by learning relevant features directly from raw data.

The key idea of deep learning is instead to learn the relevant features directly from data. So rather than computing a feature vector using a predefined set of transformations, we try to learn a function $f_\theta(x)$ that maps the raw images x directly to a "label" $y = \pm 1$ indicating whether the image shows a cat or a dog. A neural network is a particular functional form for $f_\theta(x)$, usually consisting of a series of alternating linear transformations and point-wise non-linear functions. There are different architectures which are appropriate for certain types of data or certain tasks [147]. The adjustable parameters θ, called weights, determine what the transformations compute exactly. They are found by maximizing the prediction accuracy of the network on a given set of images, which is called "training" the network [148]. In practice, simple first-order optimization methods such as stochastic gradient descent work best [149, 150]. Training a neural network is thus a general-purpose procedure to obtain features that are well-adapted to the input data and the task at hand.

Theoretical questions arise regarding the success of Deep neural networks (DNNs), especially given challenges like the curse of dimensionality, which predicts that the sample size required for accurate function estimation grows exponentially with input dimensionality [151]. Current research in statistical physics [142, 152–154] investigates how neural networks overcome this limitation.

A key insight is that real-world data, such as natural images, occupy a much smaller, lower-dimensional *manifold* within the high-dimensional input space, sketched as a two-dimensional curved surface in Fig. 4.3. While the manifold is not easily defined, it is tangible. This manifold, often estimated numerically to have a dimensionality 10–100 times smaller than the image space [155–159], concentrates meaningful data and mitigates the curse of dimensionality. The points that do represent real images tend to concentrate on a lower-dimensional *manifold* in input space.

Theoretical models of low-dimensional data, such as object manifolds [168], hidden manifolds [169, 170], or spiked covariate models [163, 171], offer controlled environments to study neural network adaptivity. Results suggest that neural networks adapt to these lower-dimensional structures more effectively than traditional methods like kernel methods [160–166].

This foundational research points toward broader inquiries: How do neural networks exploit low-dimensional structures in other modalities, such as language or amino acid sequences? Understanding these mechanisms unlocks new applications and further integrates DNNs into solving complex scientific and real-world challenges.

4.1.2.2 Machine Learning Applications in Complex Systems

Machine learning (ML) techniques have become increasingly prevalent in the investigation of physical systems [142]. While this section does not aim to provide an exhaustive overview of such applications, we highlight several notable instances within statistical physics. These include learning and sampling from equilibrium distributions [172], phase classification [173, 174], free energy difference estima-

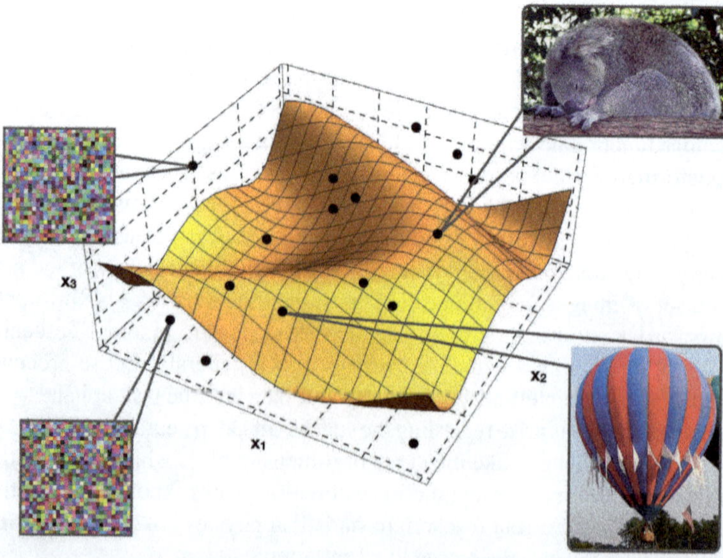

Fig. 4.3 Adaptivity to low-dimensional structures in real data is key to the practical success of neural networks: Image classification is a prototypical machine learning task with high-dimensional inputs, where each black point represents a possible input in a high-dimensional space. To a human observer, most of these inputs appear as random noise, like the "images" shown on the left. However, realistic images tend to concentrate on a lower-dimensional *manifold* within this high-dimensional input space, which is sketched here as a two-dimensional curved surface. The intrinsic dimension of this manifold in standard image datasets is much lower than the image dimension itself [155–159]. It is conjectured that neural networks efficiently learn in high-dimensional spaces by *adapting* to these manifolds, thus seemingly breaking the curse of dimensionality. Recent theoretical work has confirmed this conjecture for simple neural networks [160–166]. Figure taken from [1], images taken from the ImageNet [167] data set

tion [175], identifying the arrow of time [176], and entropy production estimation [177]. Readers interested in a more comprehensive discussion on the intersection of ML and physical sciences are referred to [142]. The interplay between complex systems and ML, however, is bidirectional: concepts from theoretical physics have deepened our understanding of ML algorithms [152] (see also Sect. 4.1.2.1). Here, we explore this duality, focusing on the role of adaptivity.

The Role of Adaptivity We first address the application of machine learning (ML) in solving physics problems, emphasizing adaptivity. Consider supervised learning tasks where the goal is to train a neural network to predict outputs given inputs. As detailed in Sect. 4.1.2.1, adaptivity is central to training networks. During optimization, network weights are adjusted to minimize prediction error, enabling the model to generalize accurately. Furthermore, network performance can be enhanced by incorporating historical input information, especially for sequential data. Recurrent neural networks (RNNs) are particularly suited for this purpose. These models maintain an internal state that evolves based on current and previous inputs, as formalized by the relation [147]:

$$\mathbf{h}_t = f(\mathbf{x}_t, \mathbf{h}_{t-1}; \boldsymbol{\theta}), \tag{4.4}$$

where f represents a neural network parameterized by θ. In the most basic form, the output of the network $\mathbf{y}_t$ can be calculated by applying another parameterized function to $\mathbf{h}_t$. While theoretically capable of capturing long-term dependencies, training RNNs is challenging due to issues like vanishing or exploding gradients [178]. Advanced architectures such as long short-term memory (LSTM) networks mitigate these problems by incorporating mechanisms for gradient preservation [179]. These architectures are foundational for tasks such as sequence-to-sequence (seq2seq) translation [134].

Recurrent neural networks (RNNs) have proven valuable in physics applications, such as inferring force fields and forecasting chaotic systems. Reference [180] examines the problem of inferring force fields from overdamped Brownian motion trajectories. Here, $\mathbf{x}_t$ corresponds to particle positions, and the network predicts parameters of the potential function. For instance, in a harmonic potential $U(x) = \frac{1}{2}kx^2$, the network infers the stiffness constant k. RNNs outperform conventional methods in data-limited scenarios and can handle non-conservative, time-dependent forces beyond the capabilities of traditional approaches.

Similarly, in [181] the focus is on chaotic time-series forecasting using the Kuramoto-Sivashinsky equation [182–184]. In this case, the input is a discretized scalar field, and the output predicts the field at the next time step. By employing reservoir computing–a variant of RNNs with fixed input-to-state mappings [185]–the authors achieve predictions far beyond the Lyapunov time. For a related discussion on controlling chaotic systems with reservoir computing, see Sect. 4.1.2.3. Both examples illustrate how RNNs leverage temporal dependencies via adaptive internal states to solve physics problems.

The reciprocal interaction between physics and machine learning (ML) is exemplified in [186], where a physical system aids in understanding seq2seq tasks. The study employs a stochastic switching-Ornstein-Uhlenbeck process, describing a Brownian particle in a harmonic potential with a stochastically alternating center. The non-Markovianity of the input is modulated via the waiting-time distribution between alternations. The task involves inferring the center's position from the particle trajectory, with ML models showing that increasing the model's memory enhances prediction accuracy. Interestingly, the non-Markovianity of inputs can have mixed effects on performance, and a correlation emerges between learning model performance and phase transitions in the stationary state of the stochastic process. This study underscores how memory depth in ML models relates to capturing long-term dependencies in sequential data.

In summary, the interplay between complex systems and machine learning (ML) remains a nascent but promising field of research. On one hand, ML tools are revolutionizing our ability to analyze and predict physical phenomena, with advancements in natural language processing poised to enhance human-algorithm communication [187]. On the other hand, principles from statistical physics and adaptive systems theory are providing foundational insights into the dynamics of learning and the influence of data structures on ML models [152]. This reciprocal relationship offers fertile ground for future exploration and innovation.

4.1.2.3 Adaptation and the Control of Dynamical Systems

In this section, we discuss the control of complex dynamical systems using closed-loop feedback, facilitated by a machine learning technique known as reservoir computing. The concept of adaptation emerges in two primary aspects: the dynamical system being controlled, referred to as the plant, and the controller itself. For effective control, the plant must generate measurable signals, typically through transducers, to infer its dynamical state. Additionally, there must be parameters available for adjustment to guide the plant's state toward the desired behavior, as depicted in Fig. 4.4.

The controller's role is to process the signals from the plant, infer its current state, compare it to the target behavior, and generate control perturbations applied to the adjustable parameters of the system. For complex systems, particularly those exhibiting chaotic dynamics, these control perturbations are inherently nonlinear functions of the plant's state and the desired behavior, categorizing the controller as nonlinear. Traditionally, implementing nonlinear controllers requires a precise mathematical model of the plant, demanding significant effort from control engineers and model developers.

A classic approach for controlling chaotic systems leverages the unstable sets inherent to chaos, such as unstable periodic orbits, which form the framework of chaotic dynamics in phase space [188, 189]. These systems naturally visit these sets, allowing linear control algorithms to stabilize the system within local neighborhoods. However, broader control tasks necessitate fully nonlinear controllers.

One solution for achieving fully nonlinear control involves employing machine learning to model the plant [190], a process known in control engineering as nonlinear system identification. Deep neural networks (DNNs), with their universal function

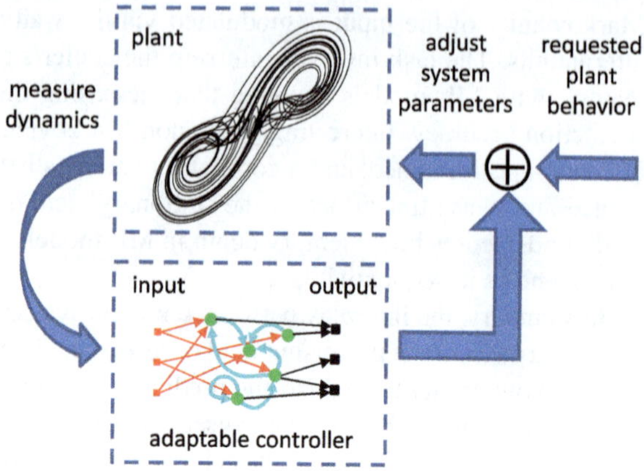

Fig. 4.4 A complex dynamical system controlled using closed-loop feedback. The controller is designed using an adaptive machine learning approach. Figure taken from [1]

approximation capabilities (see Sect. 4.1.2.1), can be trained to map measurements of the plant's state and target behavior to appropriate control perturbations. While this approach has demonstrated success, its reliance on extensive training data poses challenges for adaptivity, particularly when the plant dynamics evolve.

Reservoir computing offers a computationally efficient and data-efficient alternative for modeling dynamical systems [191]. This approach employs a recurrent neural network (RNN) structure, with neurons that exhibit finite temporal responses. The reservoir computer consists of three layers: an input layer (red squares), a reservoir of interconnected neurons (green dots), and an output layer (black squares), as shown in the lower dashed box of Fig. 4.4. The neuron dynamics are governed by differential equations driven by nonlinear functions of input signals and reservoir feedback, with simple exponential decay for memory.

This architecture provides short-term memory that can be tuned to match the plant dynamics. Unlike traditional neural networks, only the output layer's weights are trained, while the input and reservoir weights remain fixed after being initialized randomly. This significantly reduces both the size of the required training dataset and the computational overhead. Moreover, a single reservoir can support multiple tasks by coupling it with distinct output layers.

One application of reservoir computing in control involves training it to model the inverse dynamics of a plant under control [192]. By learning the perturbations needed to achieve a desired state at a future time, this method effectively guides systems with constrained, low-dimensional behavior, such as robotic arms. However, for high-dimensional, chaotic systems, a more sophisticated architecture, such as a parallel deep reservoir, may be necessary [193]. The training process for reservoir-computing-based inverse control typically requires on the order of 10 000 data points and modest computation time, making it feasible for real-time adaptation in response to changes in plant dynamics due to non-stationarity or damage.

A key challenge is further reducing the data requirements to enable deployment on resource-constrained devices, such as microprocessors in internet-of-things (IoT) systems. Recent work [194] reformulates the reservoir computer using delay lines for the plant signals and a nonlinear output layer, reducing training data requirements by an order of magnitude. However, it remains uncertain whether this approach compromises adaptivity. Current efforts aim to optimize this trade-off between rapid training and robust adaptivity, enabling broader applicability of reservoir computing in real-world control scenarios.

4.1.3 Adaptation and Synchronization in Socio-Economic Systems

This section explores the broad and complex field of socio-economic systems, where adaptation and synchronization naturally arise and play a pivotal role in modeling these systems. Adaptive networks are particularly crucial not only for the realistic examination of spreading dynamics but also for the development of strategies for disease containment, mitigation, and eradication. This concept is further elaborated in the final section of this fourth part, where an overview of adaptivity in epidemiology is presented (see Sect. 4.1.3.1). A related topic of significant interest is the interaction between social and epidemic systems, which focuses on the coevolutionary (adaptive) dynamics of interaction structures and their dynamical units (see Sect. 4.1.3.2).

Beyond the context of epidemiology, social systems themselves exhibit intrinsic adaptivity. In these systems, adaptivity can be understood as the process by which social structures evolve in response to external influences, enabling them to align with new circumstances. This adaptability is examined from various perspectives, addressing the challenges involved in achieving compatibility with changing conditions. Furthermore, anthropogenic influences on the Earth system also highlight the role of adaptivity in addressing environmental changes and disruptions (see Sect. 4.1.3.3). Lessons drawn from historical instances of adaptive mechanisms in this complex system provide valuable insights into the nature of perturbations and responses.

4.1.3.1 Adaptation and Synchronization in Epidemiology

Network epidemiology illustrates the critical role of adaptation and synchronization in understanding and controlling the spread of infectious diseases. Many pathogens propagate through direct contacts, which can be modeled using networks of social, transportation, and logistic interactions. These networks provide a mathematical foundation for formalizing individual-level interactions and identifying potential pathways of disease transmission. At the local scale, such as within populations or between groups, pathogen dynamics are often captured by compartmental models, including the susceptible-infected-recovered (SIR) model proposed by Kermack and McKendrick [195]. Adaptation becomes important when the network's structure responds dynamically to its state–such as the number of infections–aiming to mitigate outbreaks and synchronize efforts to contain the disease. This interaction forms a feedback loop as depicted in Fig. 4.5: (i) the network structure dictates the *dynamics* of disease spread, and (ii) the state of the system drives structural changes in the *network*.

Adaptivity in networks often functions as a control mechanism, optimizing the system's trajectory toward a desired state, such as minimizing infections [196]. Containment protocols, including social distancing or movement restrictions, exemplify adaptive responses driven by infection metrics. Notable applications of such adaptive

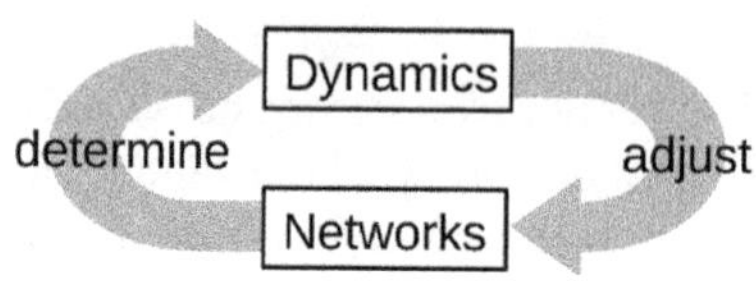

Fig. 4.5 Schematics of the interplay between dynamics and networks. Figure taken from [1]

network strategies include the H1N1 pandemic (2009), the Ebola outbreak (2014), and the ongoing SARS-CoV-2 pandemic [197–203]. Global airline transportation networks have been a recurring focus in these studies due to their significant role in disease transmission [204–206].

Adaptation also enables synchronization of interventions with disease dynamics. Numerical simulations highlight the limited delay caused by travel bans in slowing Ebola's spread in 2014 [199] and explore the feasibility of synchronized containment strategies, such as Zero-COVID or low-incidence goals [207, 208]. These models also shed light on suboptimal adherence to containment measures [203], underscoring the importance of adaptive and synchronized policies in ensuring effective responses.

The concept of adaptive networks in epidemiology predates recent public health emergencies. Early studies, such as those by Gross et al. [209, 210], investigated adaptive rewiring of susceptible individuals to avoid contact with infected nodes. These studies revealed complex dynamical patterns, including first-order transitions and hysteresis. Healthy nodes exhibited gradual evolution of their network connections, while infected nodes experienced abrupt isolation until recovery. However, the small-world structure of many social networks [211] complicates this adaptation, as rewiring can inadvertently create shortcuts that accelerate pathogen spread by linking distant infection clusters.

Synchronization also plays a critical role in surveillance and outbreak monitoring. Sentinel nodes, strategically chosen based on their network position, enable early detection and precise identification of outbreak origins. This approach has been validated in studies of animal disease networks [212, 213]. By synchronizing monitoring efforts with network dynamics, sentinel nodes provide actionable insights for targeted adjustments. Similarly, screening protocols in hospital-referral networks have proven effective in mitigating nosocomial infections [214, 215]. The COVID-19 pandemic's early stages in mainland China illustrate the power of synchronized, adaptive containment measures, where rapid depletion of the susceptible population led to subexponential infection growth [201]. After successful containment, synchronized relaxation of restrictions allowed the network to revert to its pre-epidemic state.

Adaptive networks are a subset of temporal or time-varying networks, where edges are active only during specific intervals [216, 217]. Synchronization of contact sequences with disease dynamics is critical, as only time-respecting paths contribute to transmission. Interactions occurring before an individual becomes infectious do not affect the spread. Temporal adaptation can be modeled mathematically using time-dependent adjacency matrices, facilitating frameworks like individual-based and pair-based models [218–221]. Furthermore, concepts such as network controllability [222] have been extended to temporal and multiplex networks [223, 224], enhancing our understanding of synchronized adaptive responses.

Summarizing, adaptive and synchronized network dynamics are fundamental for developing realistic models of disease transmission and designing effective containment strategies. The increasing availability of real-time data enhances these models' predictive capabilities, providing robust tools for synchronized interventions. Future challenges include integrating epidemiological models with frameworks that capture social behaviors and opinion dynamics. Such integration would facilitate a comprehensive understanding of disease propagation, offering a pathway to holistic strategies for disease containment, mitigation, and eradication.

4.1.3.2 Adaptation and Synchronization in Coevolutionary Networks

In the study of dynamical systems on networks, adaptive or coevolutionary networks exemplify the intricate interplay between adaptation and synchronization. In such networks, nodes represent dynamical entities, and edges depict the interactions among these entities. Beyond their static counterparts, adaptive networks incorporate dynamic topologies where the structure evolves in response to node dynamics, forming a feedback loop. This coupling between the dynamics on the network and the dynamics of the network creates a fully adaptive system, where the state of the nodes influences the topology and vice versa [210]. As such, we have a closed feedback loop in which the topology of the network influences the dynamics of the nodes and the state of the nodes influences the dynamics of the topology [225]. Combining the so-called dynamics on the network with the dynamics of the network in that way is what makes the system *fully adaptive* (see Fig. 4.5 in Sect. 4.1.3.1).

A classical example is the adaptive voter model, an extension of consensus and opinion-formation frameworks [226, 227]. In this model, a population with binary opinions interacts over a social network. Two key processes are at play: (i) individuals align their opinions with those of their neighbors, representing dynamics on the network, and (ii) individuals rewire connections away from those holding opposing opinions, forming new links with like-minded individuals, embodying dynamics of the network. This dual dynamic can lead to either:

- Dynamic equilibrium, where a stable coexistence of opposing opinions persists, or
- Static fragmentation, where the network disintegrates into components of homogeneous opinion [228–230].

The synchronization of rewiring and opinion adaptation determines the outcome, underscoring the critical role of coevolution in shaping the system's state.

Similarly, the adaptive susceptible-infected-susceptible (SIS) model [231] highlights the impact of adaptive dynamics on epidemic spreading. In this framework, susceptible individuals become infected through interactions with infected neighbors. On the other hand, adaptive rewiring allows susceptible individuals to sever ties with infected neighbors, forming connections with other susceptibles. Without adaptation, the system exhibits a supercritical transcritical bifurcation at a critical infection rate, transitioning smoothly to an endemic state. However, with adaptivity, this transition can become subcritical, introducing bistability. The interplay between epidemic spreading and rewiring synchronizes the dynamics, profoundly altering the bifurcation structure and, consequently, the system's behavior [231, 232].

These examples illustrate how adaptivity reshapes system behavior. In adaptive voter models, directed networks foster self-stabilizing structures, enabling fragmentation at lower critical thresholds [233, 234]. Extensions to continuous opinion spaces reveal the emergence of diverse communities rather than fragmentation [235, 236]. For adaptive epidemics, the focus has expanded to bifurcation analysis [237], thresholds [238], and early warning signs of outbreaks [239]. Studies of pandemic containment strategies have further explored the synchronization of quarantining and social distancing efforts [240].

Adaptivity also underpins synchronization phenomena in coupled phase oscillators, where adjustable coupling promotes explosive transitions into synchrony [241]. Adaptivity enables the control of cluster synchronization [196] and the emergence of frequency clusters through slow adaptation [242, 243]. These findings demonstrate how synchronization and adaptation mutually reinforce each other in diverse dynamical contexts.

Recent advancements have extended adaptive network models to higher-order structures, such as simplicial complexes and hypergraphs, which capture multi-node interactions. These frameworks have been applied to evolutionary games and adaptive voter models [244, 245]. The richer topology of higher-order networks introduces complex phenomenology, posing challenges for analysis while promising deeper insights into coevolutionary dynamics.

Concluding, adaptive and coevolutionary networks exemplify the intricate interplay between adaptation and synchronization, leading to emergent phenomena and new insights into dynamical systems. By integrating dynamics on and of the network, adaptive models reveal how structural changes synchronize with state dynamics to produce novel behaviors. Continued exploration of these systems, especially in the context of higher-order networks and real-world applications, will deepen our understanding of coevolutionary processes across disciplines.

4.1.3.3 Adaptability of the Earth System

The Earth system is a profoundly intricate and dynamic entity characterized by numerous interactions, encompassing both positive and negative feedback mechanisms. Due to its inherent complexity, it is occasionally referred to as a "horrendogram". Crucially, the Earth system is an open system, continuously exchanging energy and matter with its immediate and distant surroundings. These properties are fundamental to its capacity for adaptation, allowing the system to respond to both external shocks and internal perturbations.

Throughout its history, the Earth system has demonstrated remarkable adaptive capacity in the face of significant disruptions. Approximately 66 million years ago, a massive asteroid impact created the Chicxulub crater in the Yucatan Peninsula, Mexico. This event caused catastrophic changes, including megatsunamis, widespread wildfires, and a dramatic global temperature decrease. It is widely accepted that this impact triggered the Cretaceous–Paleogene extinction event, leading to the extinction of approximately 75% of Earth's plant and animal species, including all non-avian dinosaurs. Despite the magnitude of this event, the Earth system adapted, eventually stabilizing into a climate regime similar to its pre-impact state [246, 247].

Another example of intrinsic perturbation occurred with the Toba supervolcanic eruption approximately 74 000 years ago in Sumatra. This eruption led to significant climatic changes, including a global temperature drop of 3–5 °C. The Earth system adapted once more, transitioning through pronounced fluctuations to a stable climate regime, albeit with a temperature baseline distinctly lower than before the eruption [248, 249].

The Earth system has also demonstrated adaptive resilience to recurrent influences operating across diverse temporal scales. On longer timescales, Milanković cycles, driven by variations in Earth's orbital eccentricity, axial tilt, and precession, have significantly influenced glacial and interglacial cycles. These cycles are characterized by primary components of approximately 41 000 years and 95 000 years, among others. On shorter timescales, phenomena such as the El Niño–Southern Oscillation (ENSO), which recur every 3–7 years, affect global weather patterns, including monsoons and extreme climate events. The Earth system has consistently adapted to these periodic disturbances, maintaining stability across shifting climate regimes, including transitions between glacial and interglacial states.

In recent history, human activity has emerged as a dominant force impacting the Earth system. Anthropogenic emissions of greenhouse gases such as carbon dioxide (CO_2) and methane have driven global warming and induced profound changes in the Earth's climate. The system's ability to adapt to these human-induced perturbations remains uncertain, largely due to the unpredictability of future emissions and the challenges of mitigating their effects. International efforts, such as the United Nations Climate Change Conferences (COP), aim to curb emissions, but their long-term effectiveness is uncertain [251].

Efforts to predict the Earth system's response to anthropogenic influences often rely on models and observational data. However, significant challenges persist in accurately modeling these processes and understanding the full range of potential outcomes. A promising avenue for improving these predictions involves the study of tipping elements—large-scale subsystems of the Earth system that are sensitive to crossing critical thresholds. Examples include the Amazon rainforest and polar ice sheets, which, if destabilized, could trigger cascading effects across the system [250, 252].

The interactions among tipping elements, as well as their individual dynamics, remain only partially understood. Recent studies have begun to uncover connections between tipping elements, such as teleconnections between the Amazon and the Tibetan Plateau or the West Antarctic Ice Sheet as shown in Fig. 4.6. Quantitative analyses of near-surface air temperature fields have identified these teleconnections [253]. Conceptual models suggest that polar ice sheets could initiate tipping cascades, while elements like the Atlantic Meridional Overturning Circulation may act as mediators. Advances in machine learning offer promising tools for elucidating these complex interactions and improving predictions of tipping dynamics [254].

Concluding, the Earth system's history demonstrates its robust adaptive capacity in the face of both external shocks and intrinsic perturbations. However, the scale and nature of current anthropogenic influences represent an unprecedented challenge. While the system will undoubtedly adapt and eventually stabilize in response to these

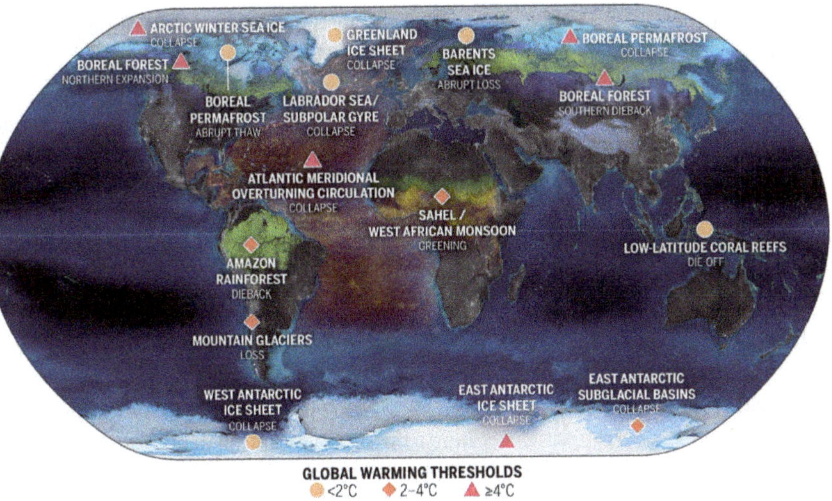

Fig. 4.6 The location of climate tipping elements in the cryosphere (blue), biosphere (green), and ocean/atmosphere (orange), and global warming levels at which their tipping points will likely be triggered. Pins are colored according to our central global warming threshold estimate being below 2 °C, i.e., within the Paris Agreement range (light orange, circles); between 2 and 4 °C, i.e., accessible with current policies (orange, diamonds); and 4 °C and above (red, triangles). Figure taken from [250]

changes, the resultant regime may be inhospitable for human life. Understanding and mitigating these impacts requires urgent scientific and societal efforts to enhance predictability and support sustainable interactions with this adaptive system.

4.1.4 Summary

In this section, we have demonstrated that adaptation and synchronization emerge as essential modeling tools across diverse complex systems, offering profound insights into their dynamics and potential applications.

Adaptive principles in physiological systems enable robustness against intrinsic variability and environmental fluctuations. Neural modeling incorporating spike-frequency adaptation and short-term plasticity (STP) provides mesoscopic insights, advancing multilayer neural architectures that replicate cortical interactions. Stimulation techniques based on adaptive dynamical systems reframe therapeutic approaches by steering neural systems toward favorable attractor states, addressing pathological behaviors at their core.

In machine learning (ML), adaptability underpins the success of deep neural networks (DNNs), which autonomously learn features from raw data, bypassing manual feature engineering. The reciprocal relationship between ML and complex systems is burgeoning: ML aids in predicting physical phenomena, while principles from statistical physics and adaptive systems theory elucidate learning dynamics and data structure effects on model performance.

Adaptation and synchronization in network models are crucial for understanding and managing disease transmission. By leveraging real-time data, these models enhance predictive accuracy and inform synchronized containment strategies. Future directions include integrating epidemiological models with frameworks capturing social and opinion dynamics to develop holistic containment and mitigation strategies.

Adaptive and coevolutionary networks highlight the interplay between structural evolution and synchronized state dynamics, revealing emergent behaviors. This perspective deepens our understanding of phenomena in higher-order networks and real-world applications, offering insights into coevolutionary processes across fields.

The Earth system exemplifies resilience through adaptation to both external shocks and intrinsic perturbations. However, unprecedented anthropogenic influences challenge this capacity, posing risks to human habitability. Adaptive modeling of tipping elements and synchronized interventions is essential for predicting outcomes and ensuring sustainable interactions with this adaptive system.

By identifying and unifying principles of adaptation and synchronization, researchers can better capture the intricate dynamics of complex systems, enabling realistic modeling, predictive insights, and effective interventions across disciplines.

4.2 Functional Modeling of Tumor and Sepsis

In this section, we adopt a dynamical systems framework to model pathological states arising from tumors or infections, with a specific emphasis on carcinogenesis and sepsis. Anchored by the innate immune system as a reference, we introduce a unified disease model. This model conceptualizes the interplay between parenchymal cells and immune cells through cytokine-mediated interactions and captures their co-evolutionary dynamics within a functional two-layer network. The objective is to illustrate how the intricate cooperation between parenchyma and stroma (the immune layer) in both physiological and pathological contexts can be effectively described by a paradigmatic phase oscillator model, in which adaptation and synchronization are of crucial importance.

The proposed framework interprets carcinogenesis, tumor progression, and sepsis as transitions driven by the destabilization of a synchronized healthy homeostatic state, leading to desynchronized pathological states or multifrequency clusters. In this duplex network representation, parenchymal cell metabolism and innate immune system responses are modeled as nodes within two interconnected layers. Cytokine-mediated interactions are characterized by adaptive coupling, with distinct time scales: a faster adaptation for immune cells and a slower adaptation for parenchymal cells. Bidirectional coupling between parenchymal and immune cells is modeled as fixed in this duplex network.

This approach provides critical insights into the emergence of organ-threatening pathological states driven by dysregulation in immune-parenchymal dynamics, particularly in sepsis. By identifying key interaction parameters, the model elucidates the stabilizing and destabilizing mechanisms underlying the parenchyma-stroma interplay. Consequently, the framework offers a functional explanation of carcinogenesis, sepsis-induced organ dysfunction, and recurrence risk within a realistic pathophysiological context.

Furthermore, we examine the robustness of these pathological states by analyzing their sensitivity to parameter variations. Numerical simulations reveal critical parameters that influence the dynamic interactions between the immune system and parenchymal cells, enhancing our understanding of how immune dysregulation can precipitate organ failure. Parenchyma, representing the functional tissue of an organ (e.g., epithelial cells in many organs), is contrasted with stroma, the structural connective tissue. This research highlights the mechanisms by which dysregulated immune responses can destabilize homeostasis, offering a deeper understanding of disease progression.

The structure of this section is as follows: Section 4.2.1 introduces the foundational concepts of a unified disease model and situates them within the emerging field of network physiology. Section 4.2.2 provides a concise overview of the physiological mechanisms underlying tumor development and sepsis. In Sect. 4.2.3, the functional network model is presented, detailing all variables, parameters, and their physiological interpretations. This section also describes the methods and metrics employed for subsequent numerical analyses, emphasizing their relevance to physiological contexts. Section 4.2.4 investigates various dynamical scenarios arising from model simulations of tumor progression in the presence of pathological cells and sepsis. In particular, computational simulations for identifying critical parameters in sepsis modeling are detailed in Sect. 4.2.5. Section 4.2.6 explores the relationship between clinical sepsis data and the functional modeling approach. Finally, the findings are synthesized in Sect. 4.2.7. Additional details and background information can be found in the publications [255, 256], which form the basis of this section.

4.2.1 Unified Disease Model

Tumors and sepsis are distinct diseases with different origins, progression time scales, and therapeutic approaches. Despite these differences, they exhibit notable similarities in age-related incidence, risk factors, clinical outcomes, and patterns of temporal recurrence. These shared characteristics warrant the development of a unified disease model capable of describing both tumor pathologies and sepsis. This model represents a paradigm shift by using the nonspecific innate immune system as a unifying reference point for both conditions [257].

The innate immune system plays a central role in maintaining organismal integrity by eliminating foreign pathogens, degrading damaged host cells, and coordinating wound healing processes. It is highly effective, preventing approximately 99% of infections [258]. This system, which emerged with multicellular life, remains a fundamental line of defense, characterized by its rapid response, decentralized regulation, and reliance on dynamic activator-inhibitor mechanisms [259, 260]. Structurally, the innate immune system comprises humoral and cellular components, the endothelium, and tissue stroma. Due to its complex organization and broad functional spectrum, the innate immune system is susceptible to dysregulation, which can manifest as tumor diseases or sepsis.

Although tumors and sepsis originate from distinct triggers–genetic mutations and infections, respectively–they share a relatively uniform disease trajectory following initial onset [261, 262]. Tumor progression is influenced by the genetic state of tumor cells, subsequent mutations, epithelial-mesenchymal transitions, and interactions with the innate immune system [263]. Recent studies have shed light on the intricate interplay between tumor and immune cells in cancer [264].

In sepsis, the innate immune system responds to infection, with the clinical course heavily dependent on the patient's baseline condition. Dysregulation of the innate immune system in both diseases can lead to shared clinical manifestations, including cachexia, coagulation disorders, and organ failure. In tumor diseases, these outcomes are further complicated by space-occupying lesions, invasive tissue destruction, and the nonphysiological production of cytokines, chemokines, and proliferation factors. It is suggested that not only the concentration and composition of cytokines but also the dynamics of their increase contribute to immune dysregulation and clinical outcomes [265, 266].

The Systemic Inflammatory Response Syndrome (SIRS) is a life-threatening condition characterized by organ dysfunction triggered by infection or endogenous antigens. As an induced disorder of the innate immune system, SIRS is highly complex and currently lacks a detailed model. Sepsis, the most severe subclass of SIRS, is defined as infection-induced organ failure, though a causative pathogen is identified in only 30-40% of cases. Sepsis-induced organ dysfunction often affects distant organs unrelated to the primary infection site [262]. Despite advancements in critical care, sepsis mortality remains high, with hospital mortality rates around 45% in intensive care units and up to 74% within four years [267].

Currently, there is no gold standard for sepsis diagnosis [268]. Clinical evaluation is based on infection-related organ dysfunction affecting systems such as the lungs, kidneys, liver, circulatory system, blood count, or central nervous system [267]. Even among patients with comparable risk factors, the trajectory of severe infections, such as pneumonia or peritonitis, remains unpredictable. While some recover, others rapidly progress to multi-organ failure. This variability reflects the highly individualized inflammatory response, for which no targeted therapy exists [269].

Dysregulated, pro-inflammatory cytokine release–commonly referred to as a cytokine storm–drives organ damage in sepsis, which may manifest sequentially or simultaneously across multiple organs. The severity ranges from mild dysfunction to severe organ failure. Clinically, organ function is assessed using the four-stage Sepsis-related Organ Failure Assessment (SOFA) score, which is updated daily. Not all organs are equally susceptible, and the extent of dysfunction varies among patients, reflecting differences in organ resilience to cytokine-induced damage. This section aims to model the conditions leading to organ failure, the resulting dysfunction, organ resilience, and the overall state of recovery.

The unified disease model uses the innate immune system as the central reference point and serves as the basis for modeling the nonlinear dynamics of complex networks. Rather than focusing on the biochemical, genetic, or tissue-specific mechanisms of carcinogenesis (reviewed extensively elsewhere [270]), this model provides a functional and dynamical framework to describe tumor progression, sepsis, and recurrence risk. By leveraging the principles of complex networks–ubiquitous in nature and technology across fields such as physics, chemistry, biology, neuroscience, engineering, and socio-economic systems–this approach offers a novel perspective on these diseases.

4.2.1.1 Physiological and Medical Framework

The parallels between cancer and inflammation, initially hypothesized by Rudolf Virchow in 1863, were later substantiated by the pioneering work of William Coley and Lloyd Old. Their findings culminated in the identification of Tumor Necrosis Factor (TNF, also known as "cachectin") as a critical cytokine in sepsis pathogenesis, an achievement credited to Bruce Beutler and colleagues in 1986 [271]. This discovery was subsequently honored with the Nobel Prize in 2011. Building on these insights, Hotchkiss and Moldawer in 2014 proposed disease-specific modulation of the adaptive immune system, specifically T-cell responses, as a therapeutic approach for cancer and sepsis [272].

The innate immune system's pro-inflammatory activation by cancer cells, pathogen-associated molecular patterns (PAMPs), and damage-associated molecular patterns (DAMPs) often remains clinically undetectable during the early stages of disease. Nonspecific risk factors such as age, smoking, and chronic inflammation contribute to systemic low-grade inflammation during this phase. Elevated cytokine levels, including TNF-α, IL-1, and IL-6, may be observed even in this early stage, arising from non-immunological sources. These perturbations disrupt the delicate equilibrium of immune homeostasis, potentially triggering a dysregulated host response that damages various organ systems.

In cancer, these chronically elevated cytokine levels manifest as symptoms such as fever, cachexia, fatigue, pain, anorexia, and anemia. Conversely, in sepsis and septic shock, a similarly uncontrolled immune response precipitates acute organ failure. The vascular endothelium emerges as a key target, with cytokine-induced disruptions in endothelial coagulation and microcirculatory function playing a pivotal role in the progression of organ dysfunction. Unfortunately, therapeutic strategies targeting this process have largely failed to achieve clinical success [273].

In advanced cancer stages, nonspecific inflammatory markers rise progressively, paralleling the acute inflammatory cascade observed in sepsis but over a more extended timescale. What occurs over hours in sepsis unfolds over weeks to months in cancer, earning the phenomenon the designation of "Slow-Motion Systemic Inflammatory Response Syndrome (SIRS)". Postmortem examinations of patients who succumb to metastatic cancer or sepsis often reveal no immediate morphological cause of death, leading to terms such as "septic multiorgan failure" or "tumor intoxication". These terms underscore the chaotic dysregulation of the nonspecific immune system.

The aberrant cytokine expression observed in these diseases has prompted the exploration of therapeutic interventions aimed at cytokine modulation, including blood purification techniques or the use of antibodies to neutralize cytokines or block their receptors [274]. However, clinical outcomes from such approaches have been largely disappointing, and they are not recommended for routine practice. In contrast, the inhibition of pro-inflammatory cytokines has shown efficacy in chronic inflammatory conditions and has become a cornerstone of treatment. Antibody therapies targeting cytokines in severe chronic inflammation can exert beneficial effects; however, these treatments also carry significant risks, including immunosuppression and adverse effects arising from the intricate pleiotropic roles of cytokines.

The development of personalized therapeutic strategies for cancer and sepsis is hindered by a lack of fundamental understanding of the innate immune system's dynamic communication network. Key gaps include knowledge of genetically determined cytokine polymorphisms, the temporal dynamics of cytokine interactions, the effects of cytokine pleiotropy in individual patients, the absence of clear thresholds for action, and the implications of cytokine short half-lives. Addressing these challenges is essential to advance therapeutic interventions and improve clinical outcomes for these complex diseases.

4.2.1.2 Framework of Network Physiology

Synchronization is a fundamental phenomenon in the dynamics of networks comprising coupled nonlinear oscillators [275–280]. Among the various synchronization patterns, cluster synchronization divides the network into groups of synchronous elements [281], while partial synchronization patterns such as chimera states exhibit coexisting domains of coherent (synchronized) and incoherent (desynchronized) states [282–287]. These synchronization phenomena have also been extensively studied in adaptive networks [243, 288–293], particularly in the context of adaptive two-layer networks of phase oscillators [294, 295].

In the framework of network physiology, synchronization plays a critical role, as multi-component physiological systems must interact within integrated networks to coordinate their functions [296–302]. Empirical studies on the structural organization and functional complexity of human physiological systems reveal the occurrence of phase synchronization and transitions between different synchronization regimes [303–306]. In complex diseases, the progression from a healthy state to a diseased state can involve abrupt critical transitions [307–311].

On the other hand, adaptive networks characterized by time-evolving connectivity are particularly relevant to modeling dynamical systems. Examples include biochemical systems with dynamically adjusting reaction rates [312], neuronal networks exhibiting synaptic plasticity [313–316], epidemic dynamics [209], and

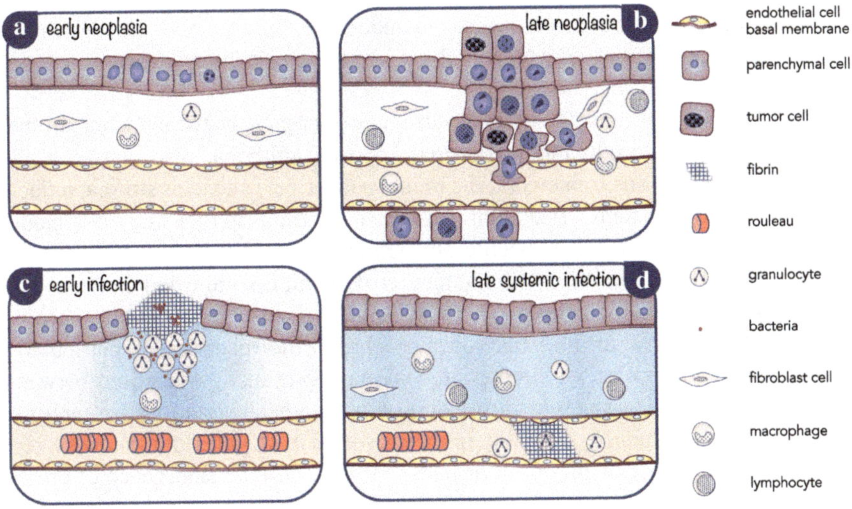

Fig. 4.7 Physiological setup: the main features of early and late neoplasia (**a**, **b**) and of early and systemic infection (**c**, **d**) in the micro-stromal bed. Panels **a**–**d** show characteristic tissue changes. Panel **a**: Carcinoma in situ, the stroma is not affected by the malignant cells. Panel **b**: Basal membrane rupture with a metastatic tumor, tumor cells and stroma interact directly with each other. Panel **c**: local inflammation after superficial injury, the innate immune system organizes the localization and repair of the parenchymal damage and invading pathogens are phagocytosed. Panel **d**: Parenchymal damage in systemic inflammation by disruption of microcirculation and cytokine action

biological or social systems [210]. Multilayer networks, which interconnect systems through distinct types of links, are a prominent area of focus in recent network science research [317–320]. One significant subclass is multiplex networks, where each layer contains an identical set of nodes, with interconnections limited to corresponding nodes across layers [321–332].

In this section, we employ a paradigmatic model of adaptively coupled phase oscillators [243, 289, 291, 333–336] to simulate the coupled dynamics of parenchymal and nonspecific immune cells. These entities are represented as nodes within a duplex network–the simplest form of a multiplex network [320]–which is known to exhibit complex synchronization phenomena [285, 325, 330, 337]. Cytokine interactions are modeled using adaptive coupling strengths: between parenchymal cell nodes (governed by a slower timescale) and immune cell nodes (faster timescale, though slower than the metabolic dynamics of phase oscillators).

The model emphasizes functional dynamics over biochemical specifics. Instead of modeling cytokine concentrations, cytokines are treated as mediators of information flow, capturing their activity in both intra-layer and inter-layer interactions. In the parenchymal layer, coupling includes a static, time-independent component and an adaptive, time-dependent component reflecting cytokine-mediated activity, while the immune layer coupling is entirely adaptive. The base topology for both layers is all-to-all coupling. Future extensions may include more sophisticated local dynamics, such as FitzHugh-Nagumo activator-inhibitor kinetics, and explore complex network topologies.

Parenchyma, the primary functional substance of organs, is distinct from stroma, which provides structural support. For many organs, epithelial cells constitute the parenchyma. Within this framework, the immune layer, referred to as stroma, reflects the innate immune system's structural and functional role. By focusing on dynamic interactions rather than organ-specific biochemical processes, the model abstracts cytokine activity as the driver of interaction dynamics within and between the parenchymal and immune layers.

The model allows for the investigation of synchronization phenomena, particularly frequency synchronization within the parenchyma, representing transitions between healthy and pathological states of the organ system. Parameter variations are analyzed to elucidate their influence on the interplay between a dysregulated immune system and the parenchyma. This approach provides insights into the emergence of organ-damaging interactions and the complex dynamics underlying synchronization in biological networks, contributing to a deeper understanding of disease progression in tumor pathologies and sepsis.

4.2.2 *Physiological Description*

This section provides an overview of the innate immune system, an evolutionarily conserved defense mechanism that relies on the coordinated activity of humoral and cellular components, mediated by cytokines, to neutralize pathogens and regulate inflammatory responses. The activity of the innate immune system exhibits interindividual variability, influenced by genetic predispositions and environmental factors, including inflammaging, chronic inflammation, and lifestyle-related elements. Dysregulation of this system plays a critical role in the development and progression of pathologies such as tumors, sepsis, and disease relapses.

4.2.2.1 Innate Immune System

The innate immune system operates in a site-independent manner, relying on the coordinated activity of humoral and cellular components, as well as cytokines, to maintain organismal integrity. Over the course of an individual's lifetime, the regulatory stability of this system is subject to changes due to the emergence of new cytokine sources. Several systemic factors contribute to these changes, including inflammaging, a chronic low-grade inflammatory state associated with aging [338, 339]; activation of the coagulation system [338, 340]; increased adiposity [341–343]; physical inactivity [344, 345]; age-related tissue fibrosis [346, 347]; chronic local inflammation [348]; and smoking-induced local hypoxia [349].

The innate immune system demonstrates varying levels of pro-inflammatory activation across individuals, with significant variability influenced by factors such as cytokine gene polymorphisms. The physiological production of cytokines, combined with additional pro-inflammatory cytokine activities, collectively modulates the activation level of the innate immune system. Key determinants of this activation include the pattern of cytokine release, their concentration, the types of cellular emitters involved, and the gradient of cytokine increase over time. These factors together shape the innate immune response and its ability to adapt to systemic and local challenges.

The innate immune system represents the most evolutionarily ancient component of immunity, with origins tracing back approximately 2.4 billion years. This co-evolution with multicellular organisms accounts for about 75% of the total evolutionary timeline [350, 351]. As the first line of defense, the innate immune system is highly effective, neutralizing over 99% of encountered pathogens, including bacteria, fungi, and viruses. In viral infections, it plays a critical role by targeting and eliminating infected cells, thereby halting viral replication.

Functionally, the innate immune system exhibits a consistent level of responsiveness throughout life, operating through spatially distributed humoral and cellular components. These components communicate via cytokines and other mediators, which travel through the bloodstream to detect infections, localize threats, initiate acute-phase responses, and regulate inflammation. Cells equipped with specific cytokine receptors respond to these signals, ensuring a coordinated immune response.

Sources of cytokines external to the immune system–such as adipose tissue, chronic inflammation, and comorbid diseases–can perturb the immune equilibrium, typically skewing it toward a pro-inflammatory state. Lifestyle factors, including smoking and physical inactivity, further modulate cytokine activity, influencing inflammatory dynamics. Cytokine gene polymorphisms are pivotal in shaping individual variations in the innate immune response, determining resilience to such perturbations [344, 352–358].

Inflammation is a highly complex process, given the diversity of pathogens–including bacteria, viruses, fungi, and endogenous materials–that can elicit an immune response. The innate immune system comprises multiple components and pathways, each contributing to inflammation's initiation and resolution. Redundant signaling pathways and the pleiotropy of cytokines, which exert multiple effects across different tissues, add to this complexity.

Typically, inflammation is localized, with the immune system containing and resolving it through phagocytosis and destruction of damaged cells. Locally released cytokines such as IL-1, IL-6, and TNF-α, produced by macrophages, fibroblasts, and endothelial cells, initiate multi-stage defense mechanisms. These cytokines stimulate cortisol synthesis via the anterior pituitary-adrenal axis. Cortisol, in turn, induces the production of cytokine receptors in hepatocytes, enabling the synthesis of acute phase proteins (APPs).

Acute phase proteins play diverse roles in modulating inflammation, being sequentially produced and regulated by feedback mechanisms. Locally, the balance between

APPs and pro-inflammatory cytokines ensures that the inflammatory response is neither excessively inhibited nor prolonged. In the bloodstream, APPs dominate, neutralizing circulating cytokines and preventing systemic inflammation.

When local inflammation overwhelms containment mechanisms, depletion of acute-phase resources in the bloodstream allows pro-inflammatory cytokines, mediators, and immune cells to disrupt distant organ function. Reactive oxygen species (ROS), along with pro-inflammatory cytokines released by activated polymorphonuclear leukocytes (PMNs) and macrophages, exacerbate tissue damage, particularly through endothelial interactions. These events can trigger Systemic Inflammatory Response Syndrome (SIRS) [353].

Tissue damage in SIRS and sepsis follows two principal pathways. First, microcirculatory disruption occurs due to intravascular coagulation initiated by inflammation, leading to reduced oxygen delivery to parenchymal tissues. Second, cytokines may inhibit mitochondrial respiration, forcing cells to rely on aerobic glycolysis for energy production. The relative dominance of these pathways varies across disease phases, affected organs, and individual patients. Furthermore, cytokines have cardiotoxic and neurotoxic effects, contributing to central nervous system dysfunction and systemic disease progression.

The understanding of these mechanisms highlights the complex interplay between localized and systemic immune responses, underscoring the critical role of the innate immune system in both health and disease.

4.2.2.2 Tumor Disease

In tumor diseases, mutant cells interact with macrophages of the innate immune system located within the stromal tissue. These mutant cells, characterized by genetic alterations and instability, represent a new cellular phenotype with modified regulatory behavior. In young, healthy individuals, mutant cells generally fail to thrive within a normal stromal microenvironment due to effective tissue surveillance mechanisms. Such cells often remain dormant or undergo apoptosis [359].

In contrast, chronic inflammation creates a conducive environment for mutant cells. Cytokines released by the activated stroma promote epithelial-mesenchymal transition (EMT) [360, 361], enabling these cells to proliferate, invade surrounding tissues, and reprogram normal macrophages into tumor-associated macrophages (TAMs) [362–364]. TAMs play a pivotal role in tumor progression by enhancing the supply of oxygen and nutrients, thereby supporting tumor growth and survival.

Cytokine signaling drives communication between tumor cells, TAMs, and normal stromal cells. The intensity of this signaling depends on the genetic profile of the tumor cells and the degree of stromal activation. As the tumor mass increases, the interaction between tumor cells and the innate immune system forms a self-reinforcing loop. This feedback mechanism establishes a cytokine-rich microenvironment around the tumor, which expands proportionally with tumor growth.

Cytokines released into the lymphatic system contribute to lymphogenic metastasis. These cytokines stimulate tumor cell proliferation within lymph nodes, initiating metastasis in a sequential manner through regional lymph nodes and forming organ-specific metastatic profiles [365]. Elevated circulating cytokine levels contribute to the systemic manifestations of tumor disease, such as cachexia [366], coagulopathies, and organ dysfunction. Advanced cases may present with life-threatening complications, including lung failure, space-occupying lesions, and tissue destruction.

Hematogenous metastasis becomes possible when disseminated tumor cells encounter sufficient growth factors, provided through further activation of the innate immune system. This process supports tumor cell survival and proliferation at distant sites, establishing secondary tumors and amplifying systemic disease progression.

Lewis Thomas' postulate [367] provides a framework for understanding tumor progression, suggesting that interactions initiated by tumor cells within the host organism drive key processes such as growth, invasion, metastasis, recurrence, and systemic symptomatology. This view highlights the role of the innate immune system not only in tumor containment but also in inadvertently facilitating tumor progression under certain conditions.

The intricate interplay between mutant tumor cells, TAMs, and the surrounding stroma underscores the complexity of tumor biology. These interactions shape the tumor microenvironment, influencing disease trajectory and presenting potential targets for therapeutic intervention.

4.2.2.3 Sepsis

The onset of sepsis mirrors conditions conducive to the survival of mutated cells in tissue: an activated innate immune system. Sepsis arises when the innate immune system fails to confine invading pathogens, enabling their systemic spread and interaction with the pre-activated immune response. This systemic activation initiates a complex inflammatory cascade. While antibiotics or surgical interventions may reduce the pathogen load, the intensity and outcome of the subsequent inflammatory response depend on host factors such as cytokine gene polymorphisms [356, 358, 368], which can either attenuate or escalate the pro-inflammatory reaction [369].

Pro-inflammatory cytokines play a pivotal role by targeting endothelial cells and the coagulation system, triggering the formation of microthrombi. These microthrombi not only impair blood flow but also release growth factors and additional cytokines, amplifying the inflammatory response [356]. This interaction compromises endothelial barrier integrity, leading to increased vascular permeability, tissue edema, and hypotension, which often necessitate aggressive fluid resuscitation [349]. Prolonged inflammation and impaired oxygen diffusion to tissues result in hypoxia, further promoting cytokine release and initiating metabolic disturbances in the parenchyma–a phenomenon referred to as a "second hit" [258]. These metabolic shifts contribute to the progression of sepsis and set the stage for multi-organ dysfunction.

The inflammatory cascade in sepsis predisposes critical organs–such as the kidneys, lungs, and liver–to failure. The release of endogenous damage-associated molecular patterns (DAMPs), including mitochondrial DNA [370], exacerbates the immune response. These molecules, released during trauma, burns, pancreatitis, or surgical injury, mimic the effects of pathogen-associated molecular patterns (PAMPs) and potentiate the septic response.

Sepsis exemplifies a maladaptive immune response, where the innate immune system's activation, intended for pathogen defense, inadvertently leads to systemic inflammation, vascular dysfunction, and organ damage. Understanding the interplay of cytokine dynamics, genetic predisposition, and the impact of endogenous danger signals provides critical insights into the pathogenesis of sepsis and potential therapeutic targets.

4.2.2.4 Relapse

Tumors and sepsis exhibit similar patterns that correlate with the progression of their primary manifestations. Both conditions initially require an activated innate immune system, which becomes dysregulated as the diseases advance. Following the removal of tumors or resolution of infections, the innate immune system's activation decreases but typically remains elevated compared to pre-disease levels. As a result, cytokine production persists, supporting the proliferation of residual tumor cells and the formation of tumor stroma [366, 371–373]. The kinetics of this process depend on factors such as tumor cell proliferation rate, cytokine availability, and the tumor cells' ability to develop their own stroma and establish blood vessel connections.

In the aftermath of a systemic infection, the innate immune system retains an activation status similar to pre-septic shock levels. Importantly, the infection may induce irreversible elements of trained immunity [258]. Survivors of septic shock face a challenging prognosis, with approximately 50% succumbing within the first two years post-sepsis due to persistent secondary infections, new tumor development, or cardiovascular issues [374]. Long-term outcomes for sepsis survivors are generally poor, marked by myopathy, neuropathy, cognitive impairments, exacerbation of pre-existing conditions, and elevated mortality. Regardless of prior health status, 74% of patients have died within two years of the illness, often due to re-infection with sepsis, cardiovascular complications, or tumor progression. The heightened vulnerability of post-sepsis survivors is largely attributed to dysregulated inflammation during the acute phase of the disease, compounded by tissue damage incurred during the initial inflammatory response and ongoing inflammation [267, 268, 374, 375].

4.2.3 Functional Model with Adaptation

In this approach, we simplify the complex dynamics of the innate immune system by focusing on the adaptive interactions between the parenchyma and stroma, modeling them as partially coupled oscillators. This conceptual framework allows for the representation of highly complex, self-organizing, and dynamic structures. The information exchange mediated by these couplings governs synchronization both within the parenchyma and stroma, as well as between them. Under physiological conditions, these components are in a well-synchronized state. In pathological states, however, desynchronization occurs, often due to tumors or failure of the acute-phase response (APR), leading to a disruption of the dynamic inflammatory equilibrium, which can progress to complete dysregulation.

In our unified model of cancer and sepsis, the underlying functional framework, represented by differential equations, remains the same; only the parameters differ between the two diseases. In the case of cancer, intraepithelial neoplasia represents a structural disruption within the parenchyma that progressively impacts the stroma, culminating in tumor formation. In contrast, in systemic infections, desynchronization begins in the stroma and, as the disease progresses, spreads to the parenchyma, resulting in the organ failures that are characteristic of sepsis. This framework represents a novel approach to qualitatively modeling two major disease categories and may potentially be extended to other conditions, such as atherosclerosis.

We propose a unified disease model centered around the innate immune system, mathematically represented through a two-layer dynamic network model of the parenchyma and stroma. The nonlinear communication between these components, modeled as adaptively coupled nonlinear oscillators, captures the interactions within cellular compartments involved in both cancer and sepsis. Our goal is to mathematically simplify and model the complex pathophysiological processes of the immune system, which are often difficult to utilize therapeutically in clinical settings. By applying model-based simulations of these diseases, we aim to improve individualized prognosis and therapy planning, ultimately leading to better clinical outcomes. Achieving these objectives requires interdisciplinary collaboration between clinicians, pathologists, mathematicians, and the pharmaceutical industry.

Given the distinct characteristics of cytokine release in cancer and sepsis, separate therapeutic strategies must be developed for each disease. The proposed method seeks to continuously monitor the individual progression of cellular reactions in metastatic cancer and sepsis, assess their course, and identify critical phases or tipping points in the disease's progression. This information would allow for the optimal timing and selection of personalized interventions. In metastatic cancer, the goal is to find therapeutic strategies that decouple tumor growth from the body's inflammatory response, improving quality of life by addressing symptoms such as cachexia, fatigue, and pain, or by reducing metastasis progression. This would involve inhibiting cytokine production from both the nonspecific immune system, which contributes to the pathogenic response to the tumor, and from the tumor cells themselves. In sepsis, a "reset" of the innate immune system–temporarily blocking pro-inflammatory driver cytokines–could provide an opportunity for the immunological feedback loop to re-synchronize and restore balance.

In the following, we introduce the functional model that we employ for the analysis of tumor and sepsis. We introduce all parameters and variables and provide details on the measures used to analyze the system.

4.2.3.1 Schematic Model for Tumor Disease and Sepsis

Organic tissue consisting of parenchymal cells and immune cells is shown schematically in Fig. 4.8. We depict the initial and final configurations for tumor disease and sepsis on a tissue element. The tissue element consists of the epithelial parenchyma, the basal membrane and the stroma. The parenchyma is the organ-specific functional layer. The basal membrane separates the parenchyma and stroma and is made of collagen of type IV. In the stroma, blood supply, lymphatic drainage, and immune response occur. The stroma consists of an extracellular matrix and embedded cells that do not form a solid association. The extracellular matrix is structurally composed of collagen, glycoproteins, proteoglycans, and water. Cells in the stroma are resident fibroblasts and fat cells, and mobile cells (macrophages, mast cells, granulocytes, and plasma cells).

The initial process of carcinogenesis is shown in Fig. 4.8a–c. In Fig. 4.8a, the parenchyma is normal, the stroma is inflammatorily activated. In Fig. 4.8b, a mutation of a tumor cell has occurred in the parenchyma, the inflammatory activation of the stroma continues. Finally in Fig. 4.8c, the mutant cells proliferate, break through the basal membrane and migrate into the blood and lymphatic vessels, and cytokines are emitted.

Figure 4.8d–f shows the systemic effects of the final process of tumor disease and sepsis: Figure 4.8d depicts a tissue element that is not directly affected by tumor, metastasis, or primary inflammation. The stroma exhibits inflammatory activation. In Fig. 4.8e, a systemic cytokine storm is occurring. Cytokine production by the

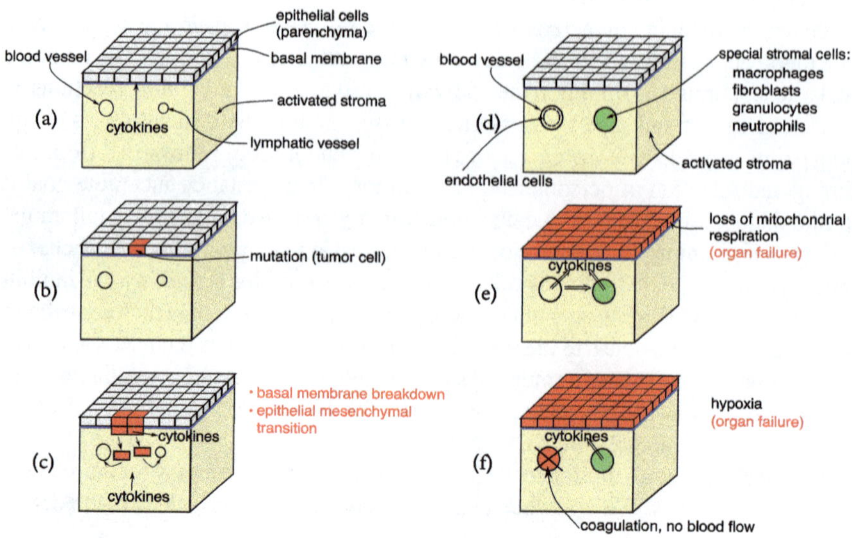

Fig. 4.8 Scheme of a tissue element showing the initial progression of tumor disease (**a**)–(**c**), the initial configuration for sepsis (**d**) and the final configurations found for both tumor disease and sepsis (**e**), (**f**). Figure taken from [255]

stromal cells is additionally stimulated. The parenchyma responds to the primary and secondary cytokine storm by uncoupling mitochondrial respiration and switching to aerobic glycolysis. The energy supply is no longer sufficient for organ-specific cellular functions and the organ fails. Figure 4.8f shows a variant of Fig. 4.8e. The cytokine storm activates the endothelium, blood coagulation is activated, oxygen transport breaks down, and hypoxia arises. The stromal cells respond with cytokine release. Both together, the hypoxia and the cytokine release, lead to the collapse of organ-specific cell functions in the parenchyma and the organ fails.

For sepsis modeling, a schematic illustration of organic tissue consisting of parenchymal cells and immune cells is shown in Fig. 4.9. Panel A depicts the initial configuration of a tissue element. The tissue element consists of the epithelial parenchyma, the basal membrane and the stroma. The parenchyma is the organ-specific functional layer. The basal membrane separates the parenchyma and stroma and is made of collagen of type IV which is a network-forming collagen underlying epithelial and endothelial cells. In the stroma, blood supply, lymphatic drainage and immune response occur. The stroma consists of an extracellular matrix and embedded cells that do not form a solid association. The extracellular matrix is structurally composed of collagen, glycoproteins, proteoglycans and water. Cells in the stroma are resident fibroblasts and fat cells, and mobile cells (macrophages, mast cells,

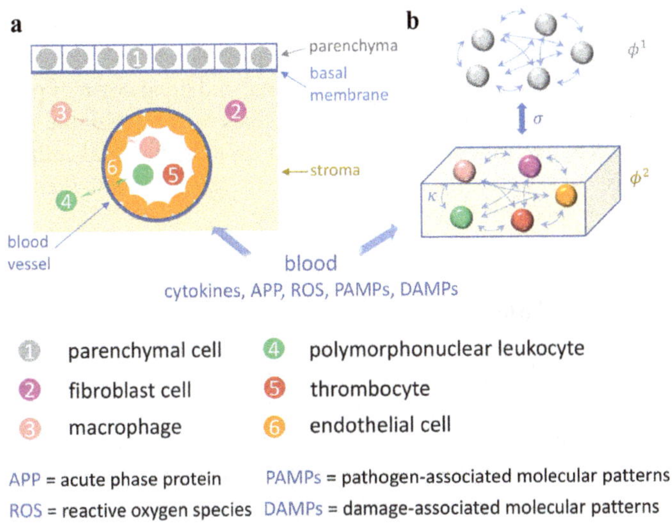

Fig. 4.9 Schematic illustration of the sepsis model. **a** A tissue element is depicted, in which the basic processes of sepsis take place: shown are the cells (colored) involved such as parenchymal, fibroblast, endothelial cells, and macrophages, polymorphonuclear leukocytes and thrombocytes in the parenchyma (grey), the stroma (yellow), and the capillary blood vessel. **b** depicts the functional interactions within and between the two corresponding network layers in our model, the parenchyma and the stroma (immune layer). Figure taken from [256]

granulocytes, and plasma cells). Panel B shows the functional interactions in the two-layer network model of the parenchymal layer and the immune layer.

Figure 4.9 shows the functional structure of the tissue element, in particular the reactants interacting during sepsis. With the blood supply via capillaries, pro- and inflammation-inhibiting molecules are delivered to the stroma of each organ. They originate from the primary focus of infection (pathogen-associated molecular patterns, damage-associated molecular patterns, cytokines), from the liver (acute phase proteins) and from the innate immune system (macrophages, polymorphonuclear leukocytes). The concentration of all reactants changes as the inflammatory response progresses. They initially interact with the endothelium of the capillaries. With the influx of pro- and inflammation-inhibiting reactants, the overall system (Fig. 4.9a) tries to maintain a local inflammation-inhibiting equilibrium. Blood flow and oxygen supply, especially to the parenchyma, must be ensured.

An ongoing blood flow and oxygen supply is achieved by the individual and locally adapted information processing of all cells of the innate immune system (macrophages, polymorphonuclear leukocytes), of the stroma (endothelial cells, fibroblasts), the specific activation of platelets, the pleiotropy of cytokines, i.e., their concentration- and pattern-dependent reaction patterns, and the acute phase proteins produced and released in the liver via cytokines in a time-delayed manner. All cells involved are potential cytokine sources.

The pathophysiological positive response pattern is the maintenance of the inflammation-inhibitory balance. The pathological situation is the initiation of disseminated intravascular coagulation, interruption of blood flow, oxygen diffusion pathways prolonged by fluid influx into the stroma, and breakdown of parenchymal oxygen supply. In parallel and in addition, cytokines interact with the parenchyma and reduce parenchymal function via impairment of mitochondrial cellular respiration. This process may develop an autocatalytic characteristic with the involvement of reactive oxygen species, ending in organ failure.

4.2.3.2 Functional Two-Layer Network Model

The unified disease model with the reference point given by the *innate immune system* is the basis for our model, which includes disease-specific initial conditions, mutant cells for tumor disease and infection-driven cytokine dysregulation. A volume element of tissue consisting of parenchyma, basal membrane and stroma is used as a model for tumor disease and sepsis, describing the functional interactions between parenchyma (organ tissue) and stroma (immune layer). We represent the network layer of parenchymal cells (superscript 1) by N phase oscillators ϕ_i^1, $i = 1, \ldots, N$, with partly fixed and partly adaptive coupling weights [376], and the network layer of immune cells (superscript 2) by N adaptively coupled phase oscillators ϕ_i^2. The communication through cytokines which mediate the interaction between the parenchymal cells is modeled by the coupling weights κ_{ij}^1, and those between the immune cells by coupling weights κ_{ij}^2.

The use of phase oscillators for the functional modeling of the interacting parenchymal cells and immune cells is motivated by the fact that phase oscillator networks are a paradigmatic model for collective coherent and incoherent dynamics. As discussed in detail in Sect. 4.2.2, healthy cells and tumor cells differ by their metabolic activity, i.e., tumor cells are less energy-efficient and thus have a faster cellular metabolism, which is reflected in our phase oscillator model by a higher frequency. Thus in the healthy homeostatic equilibrium state all parenchymal cells have the same lower frequency, while the pathological state splits into two clusters with different frequencies, healthy and unhealthy, i.e., a multifrequency cluster. The healthy state is assumed to be characterized by regular periodic, fully synchronized dynamics of the phase oscillators, i.e., all cells show the same collective frequency of cellular metabolism. The pathological state is described by multifrequency clusters with different frequencies, i.e., the pathological cells in the parenchyma attain a higher frequency, while the healthy cells are still frequency-synchronized with the "healthy" frequency. This loss of synchrony reflects the fact that in case of tumor disease, the malignant mutation basically leads to a loss of performance of the parenchymal cells. As a consequence, they are no longer fully coordinated, which leads to a loss of proliferation and apoptosis control [261, 263]. Such a pathological condition leads also to an alteration of the metabolic activity of the immune cells [361, 377–381]. In case of sepsis, i.e., a systemic inflammation, cells belonging to the innate immune system produce cytokines in an unregulated way affecting the parenchyma or microcirculation leading to organ failure [262, 382] which is also associated in our model with the loss of synchrony (Fig. 4.10).

We consider a general multiplex network with two layers each consisting of N identical adaptively coupled phase oscillators:

$$\dot{\phi}_i^1 = \omega^1 - \frac{1}{N}\sum_{j=1}^{N}(a_{ij}^1 + \kappa_{ij}^1)\sin(\phi_i^1 - \phi_j^1 + \alpha^{11}) - \sigma \sin(\phi_i^1 - \phi_i^2 + \alpha^{12}), \tag{4.5}$$
$$\dot{\kappa}_{ij}^1 = -\epsilon^1\left(\kappa_{ij}^1 + \sin(\phi_i^1 - \phi_j^1 - \beta)\right),$$

$$\dot{\phi}_i^2 = \omega^2 - \frac{1}{N}\sum_{j=1}^{N}\kappa_{ij}^2\sin(\phi_i^2 - \phi_j^2 + \alpha^{22}) - \sigma \sin(\phi_i^2 - \phi_i^1 + \alpha^{21}), \tag{4.6}$$
$$\dot{\kappa}_{ij}^2 = -\epsilon^2\left(\kappa_{ij}^2 + \sin(\phi_i^2 - \phi_j^2 - \beta)\right),$$

where $\phi_i^\mu \in [0, 2\pi)$ represents the phase of the ith oscillator ($i = 1, \ldots, N$) in the μth layer ($\mu = 1, 2$), ω^μ are the natural oscillator frequencies of the oscillators in the μth layer. The interaction between the oscillators within each layer is determined by the intralayer connectivity weights $a_{ij}^1 \in [0, 1]$ (fixed interaction within an organ) and $\kappa_{ij}^\mu \in [-1, 1]$ (adaptive interaction mediated by cytokines). We assume that the parenchymal layer has both fixed and adaptive couplings, while the immune layer has only adaptive coupling. Further the interactions within the layer depend on the phase lag parameters α^{11} and α^{22}.

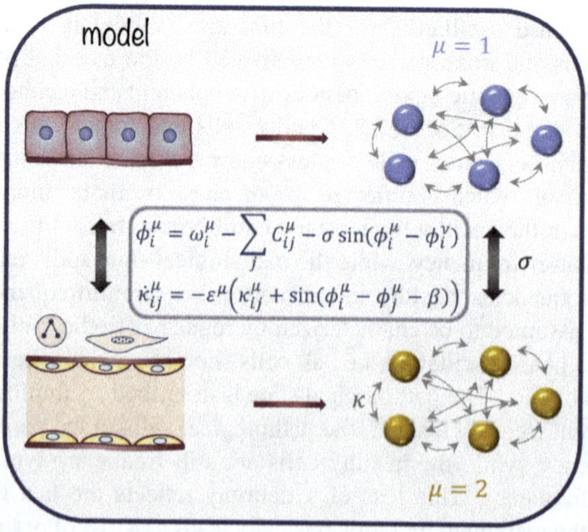

Fig. 4.10 Parenchyma and stroma are represented in the dynamic model as two communicating network layers. See physiological setup in Fig. 4.7. The figure shows the abstraction of the micro-stromal bed, which is divided into a parenchymal and a stromal information space. Both information spaces communicate within and with each other. They adapt to each other. Mathematically, this can be represented by a dynamic 2-layer network model [255, 256], as depicted in this figure. This method makes it possible to describe a wide variety of pathologies using a unified system of equations. The equations in the center represent Eqs. (4.5) and (4.6)

In this work our focus is on the interaction between the two layers and their synchronization. In particular, we analyze the onset of desynchronization in the parenchymal layer induced by an activated immune layer. The interaction of the layers is controlled by two main parameters, the interlayer coupling weight σ and the interlayer phase lag parameters α^{12} and α^{21}. Between the layers the interlayer coupling weights $\sigma \geq 0$ are fixed and symmetric for both directions of interaction. The phase lags can be considered to model interaction time delays [383, 384].

The adaptation rates $0 < \epsilon^\mu \ll 1$ separate the time scales of the slow dynamics of the coupling weights and the fast dynamics of the oscillatory system. The adaptation rate of the parenchymal layer ϵ^1 is assumed to be slow compared to the adaptation rate of the immune layer ϵ^2, i.e., $\epsilon^1 \ll \epsilon^2$ to account for the faster reaction of the immune cells, see also [255]. Thus we have two classes of adaptive coupling weights modeling two different cytokine mechanisms on two different timescales. Consequently by choosing two significantly different values for ϵ^1 and ϵ^2, a system with multiple times scale dynamics is obtained, i.e., "slow-fast-faster" dynamics ($\epsilon^1 \ll \epsilon^2 \ll 1$) [385, 386].

From a neuroscience perspective, the phase lag parameter β of the adaptation function $\sin(\phi_i^\mu - \phi_j^\mu - \beta)$ can also be called plasticity parameter [289] which accounts for different adaptation rules that may occur. Depending on the value of β the adaptation rule can be symmetric, i.e., with a cosine shape ($\beta = \pi/2$), or causal, i.e.,

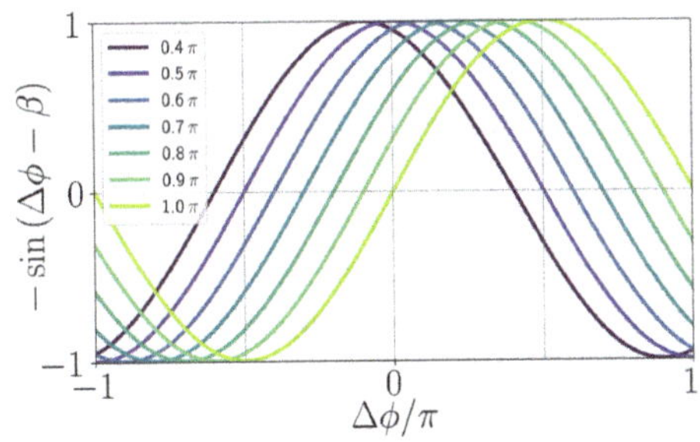

Fig. 4.11 Illustration of the adaptation function in dependence of the age parameter β. Figure taken from [256]

with a sine shape ($\beta = \pi$). Symmetric as well as causal relationship are well-known forms for spike timing-dependent plasticity in neuroscience [316, 387–390]. The shape of the adaptation function for different choices of the parameter β is provided in Fig. 4.11. By varying β from 0.4π to π, we can see that the maximum of the coupling term $-\sin(\Delta\phi - \beta)$, where $\Delta\phi \equiv \phi_i^\mu - \phi_j^\mu$, shifts from $\Delta\phi = -0.1\pi$ to $\Delta\phi = 0.5\pi$. Thus, for $\beta = 0.5\pi$ we have a Hebbian adaptation rule where the coupling term gives a maximum positive feedback for synchronization (*fire together, wire together*), while for $\beta \neq 0.5\pi$ the feedback is asymmetric, i.e., maximum positive feedback occurs for some phase lag $\phi_i^\mu - \phi_j^\mu = \beta - 0.5\pi$. Thus the adaptation lag $\beta = 0.5\pi$ seems to be most favorable for synchronization. For $\beta = \pi$ the coupling term is zero for synchronization, negative for $\phi_i^\mu < \phi_j^\mu$ and positive for $\phi_i^\mu > \phi_j^\mu$, i.e., the coupling weight κ_{ij}^μ, and hence the input from node j to node i, is increased if $\phi_i^\mu > \phi_j^\mu$, i.e., if the i-th oscillator is advancing the j-th, and vice versa. The parameter β plays an essential role in the model because it governs the adaptivity rule of the cytokines. It will be called *age parameter*, since it mimics a systemic sum parameter which accounts for different influences such as physiological changes due to age, inflammaging, systemic and local inflammatory baseline, adiposity, pre-existing illness, physical inactivity, nutritional influence, etc.

In the following we use a simplified model, where the natural frequencies of both layers are identical and set to zero in a co-rotating frame: $\omega^1 = \omega^2 = 0$. Further we assume phase lag parameters $\alpha^{11} = \alpha^{22}$, and $\alpha^{12} = \alpha^{21} = \alpha$ throughout the article. The matrix elements $a_{ij}^1 \in \{0, 1\}$ of the adjacency matrix A in the parenchymal layer are chosen as $a_{ij}^1 = 1$ if $i \neq j$ (global coupling).

4.2.3.3 Methodology and Measures

Heterogeneous dynamics, e.g., multifrequency clusters and tumor growth, may arise through the dynamic interaction of parenchymal cells, immune cells, and localized cytokine activity in a self-organized self-adaptive manner, even if the system parameters in the layers are chosen uniformly, i.e., homogeneous [242, 291, 391]. In addition to the plethora of dynamical states, adaptivity also induces a high degree of multistability [387]. Heterogeneity can enter the parameters of oscillator networks in various ways. Very prominent are heterogeneities in the natural frequencies of the oscillators or in the connectivity structure [392]. Heterogeneous frequencies ω_i are used to model pathological parenchymal cells which occur by spontaneous random mutations. We assume that all healthy cells possess the same natural frequency $\omega_i = \omega^h$, and a fraction r of randomly chosen pathological cells possess the natural frequency $\omega_i = \omega^p$. The characterization of tumor cells via their metabolic properties [261, 263] has revealed a difference in the metabolic activity between tumor and healthy cells [393]. Warburg demonstrated that cells after a malignant mutation obtain their energy via aerobic glycolysis, which provides only 4 mol ATP/mol glucose, in contrast to normal cells whose metabolism is based upon mitochondrial breathing yielding 36 mol ATP/mol glucose. Thus tumor cells have a metabolic efficiency of only about 10% of healthy cells, and therefore they need more glucose. Other general differences in metabolic activity between tumor cells and healthy cells include proliferation rate, apoptosis rate, initiation of angiogenesis, tissue invasion or metastatic capacity. Here, quantitative differences exist between different organ tumors. Ultimately, any change in metabolic performance, regardless of which metabolic branch is affected (structural, energetic, cell division, or special metabolism), can be conceived as a frequency change.

Using numerical simulations, we study whether the healthy state (frequency synchronized) is persistent also in the presence of a few pathological cells. Under certain conditions depending on various parameters (age, inflammaging, chronic inflammation, other basic diseases, obesity, smoking, lack of exercise), an unregulated cytokine expression and hence a possibly lethal tumor can occur. In these cases, the healthy (synchronized) state is not persistent anymore against the perturbation by the heterogeneity (tumor cells). For our study, the cytokine adaptivity parameter (which we call *age parameter*) β and the fraction r are considered as the main model parameters to account for the various system conditions (see Table 4.1).

In the case of sepsis, we assume that all cells possess the same natural frequency. To model the initial state for the potential occurrence of sepsis, we introduce a fixed initial perturbation of the cytokine activity in the immune layer representing a systemic immune response, see Fig. 4.16. We study the effect of this initial system perturbation on the emergence of the healthy state, i.e., synchronization, in dependence of the age parameter β. Under certain conditions depending on various parameters summarized by β (age, inflammaging, chronic inflammation, other basic diseases, obesity, smoking, lack of exercise, gene polymorphisms) the unregulated cytokine expression can progress into the parenchyma and desynchronize it. In these cases, the healthy (synchronized) state is not resilient anymore against the perturbation of the immune layer.

Further, we analyze how this dependency changes depending on other parameters that shape the interaction between the parenchyma (layer 1) and the immune system (layer 2), namely the interlayer coupling strength σ, the form of the initial immune layer activation expressed by the size of the perturbation $1 < C < N$, and the interlayer coupling phase lag $\alpha^{12} = \alpha^{21} = \alpha$. The latter parameter accounts for a delay in the layers' interaction where $\alpha = 0$ can be regarded as instantaneous.

Table 4.1 Physiological meaning of the dynamical variables, parameters and measures of the model (superscripts referring to layers $\mu = 1$ and $\mu = 2$ omitted). Table adapted from [255]

	Symbol	Name	Physiological meaning
Dynamical variable	ϕ_i	Phase	Metabolic activity
	κ_{ij}	Coupling weight	Cytokine activity
Parameter	α	Phase lag	Metabolic interaction delay
	β	Plasticity rule	Age, inflammaging, pre-existing diseases, etc
	ω	Natural frequency	Natural frequency of cellular metabolism
	ϵ	Time scale ratios	Time scales of cytokine activity
	C	Initial network perturbation	Local infection
	a_{ij}	Connectivity	fixed parenchymal cell-cell interaction
	σ	Interlayer coupling	Interaction between immune & parenchymal cells
Measure	$\langle\dot{\phi}_i\rangle$	Mean phase velocity	Collective frequency of cellular metabolism
	s	Standard deviation of frequency (see Eq. (4.9))	Pathogenicity (parenchymal layer), activation (immune layer)
	f	Frequency cluster ratio	Probability of a pathological state

In order to quantitatively characterize the dynamical collective state of the two-layer network, in particular its degree of frequency and phase synchronization, we introduce several measures. If the frequency of all adaptively coupled phase oscillators is the same, the phases may still be different. They may either be all the same (complete in-phase synchronization) or they may be phase-locked such that each phase oscillator oscillates with the same frequency but a fixed, time-independent phase difference. A special case is a splay state, where the phase differences of all oscillators average out, for instance if the phase of the j-th oscillator is $2\pi j/N$, $j = 1, \ldots, N$. In systems of the form Eqs. (4.5) and (4.6), it is possible to find in-phase synchronization and splay states, and they may be interpreted as different quality of synchronization [295]. In our set-up a splay state is interpreted as a more vulnerable collective state where small perturbations can quickly lead to partial or complete desynchronization.

First, we introduce the mean phase velocities of the oscillators j in both layers $\mu = 1, 2$

$$\langle\dot{\phi}_j^\mu\rangle = \frac{1}{T}\int_t^{t+T} \dot{\phi}_j^\mu(t')\mathrm{d}t' = \frac{\phi_j^\mu(t+T) - \phi_j^\mu(t)}{T} \tag{4.7}$$

with averaging time window T, and the spatially averaged mean phase velocity (frequency) for each layer $\bar{\omega}^\mu = \frac{1}{N}\sum_{j=1}^{N}\langle\dot{\phi}_j^\mu\rangle$. In case of frequency synchronized states within the layers, we further consider a classical measure for the phase coherence within each layer, namely, the Kuramoto-Daido order parameter [394, 395]. In particular, we look at the second moment of the order parameter R_2^μ as it is the most suitable characteristic for these kinds of patterns in adaptive networks as shown in [243, 295]. This measure of phase coherence is given by

$$R_2^\mu(t) = \frac{1}{N}\left|\sum_{j=1}^{N} e^{\mathrm{i}2\phi_j^\mu(t)}\right|. \tag{4.8}$$

It takes values $0 \leq R_2^\mu = 0 \leq 1$, where the lowest and the highest coherence correspond to 0 and 1 =, respectively. We recall that for $R_2^\mu = 0$ we call a state a splay state and for $R_2^\mu = 1$ an antipodal state [396]. A well-known example of a splay state is a state with fixed phase difference of $2\pi/N$ between neighboring oscillators on a ring network of N phase oscillators. Further we note that in-phase and anti-phase synchronized states are included in the class of antipodal states. We emphasize that splay states are still frequency synchronized, and hence are considered as healthy states, however, due to their weaker phase coherence properties, they may be considered as more vulnerable and less resilient than in-phase synchronized states.

Furthermore, for both layers $\mu = 1, 2$ we calculate the ensemble average s^μ (ensemble size N_E with ensemble elements E) of the standard deviation $\sigma_\chi(\bar{\omega}^\mu) = \sqrt{\frac{1}{N}\sum_{j=1}^{N}(\langle\dot{\phi}_j^\mu\rangle - \bar{\omega}^\mu)^2}$ of the mean phase velocities

$$s^\mu = \frac{1}{N_E}\sum_E \sigma_\chi(\bar{\omega}_E^\mu), \tag{4.9}$$

and the ensemble average of the corresponding normalized standard deviation $\frac{\sigma_\chi(\bar{\omega}_E^\mu)}{\bar{\omega}_E^\mu}$. If the latter quantities are non-zero, they indicate the formation of frequency clusters, where the respective layer splits into clusters with different frequencies, which is indicative of a pathological state. The ensemble average is necessary to account for the multistable nature of the system, i.e., for random initial conditions some of the simulations may give a pathological state, while some may still give a healthy state. This is similar to the real physiological situation where some patients will develop sepsis, while some will not.

We further introduce another complementary measure to quantify the occurrence of pathological states, which we call the frequency cluster ratio. The frequency cluster ratio f^μ is defined as the ratio between the number of frequency clusters N_f^μ in layer μ found for an ensemble of initial conditions and the size of the ensemble N_E, i.e., $f^\mu = N_f^\mu/N_E$. We consider an asymptotic state to be a frequency cluster (desynchronized, pathological state) if there exist one or more nodes $j \in \{1, \dots, N\}$ such that $\langle\dot{\phi}_j^\mu\rangle \neq \bar{\omega}^\mu$ (deviating frequencies).

Table 4.1 gives an overview of the dynamical variables, parameters and measures of the model. In the right column the physiological meaning of all quantities is given in a concise manner. In more detail, the quantities of the mathematical model which are listed in Table 4.1 have the following physiological correspondence:

(i) Variable ϕ_i mimics the metabolic activity of a parenchymal or immune cell i as a universal oscillatory phase variable. Here our aim is not to describe physiological processes on a detailed biochemical level, e.g., C-reactive protein (CRP) production after inflammatory activation, etc.
(ii) Variable κ_{ij} describes the cytokine activity mediating information flow between cells j and i. It stands for the susceptibility of communications between cells via cytokines. Cytokines have pleiotropic properties, i.e., they can exert different, even contrary, effects depending on the initial situation and in combination with other cytokines. Therefore we model them by an adaptive coupling strength. Reaction strength of cytokines is due to individually different gene polymorphisms.
(iii) Parameter α models the phase lag of the intralayer cell-to-cell coupling. Its origin is the time delay from the activation of a defined metabolic branch until the mediator release; within the inflammatory cascade, characteristic time delays exist for all reaction sub-steps starting from the triggering of the inflammatory reaction, for example, the reaction time of the individual liver cells starting from the input signal until the maximum blood concentration of interleukin IL-6 (2h) and CRP (12h) is reached.
(iv) Parameter β governs the adaptivity (or plasticity) rule of the cytokines. It mimicks a systemic sum parameter which may account for different influences such as physiological changes due to age in the extracellular matrix, inflammaging, systemic and local inflammatory baseline, adiposity, pre-existing illness, physical inactivity, nutritional influence, and others. In case of tumor disease, it can include the malignancy grade of tumor cells. For the sake of brevity, we call this parameter the age parameter.
(v) Parameter ω denotes the natural frequency of the basic metabolic activity of a single cell. For instance, it can stand for the minor CRP production in a normal healthy state.
(vi) Parameters ϵ^1 and ϵ^2 denote the inverse relaxation times (half-life) of the cytokines in the parenchymal and immune layer, respectively.
(vii) Parameter r denotes the fraction of mutant (pathological) parenchymal cells in the tissue element, and is implemented in the model as the fraction of cells with a deviating (pathological) natural frequency ω^p.
(viii) Parameter a_{ij} denotes the genetically fixed intercellular communication pathways between parenchymal cells, they are *wired* by fixed cell-to-cell connections. In contrast, communication of parenchymal cells via cytokines and of cells within the immune layer runs in an open communication channel, which is controlled self-adaptively, after [397].
(ix) Parameter σ denotes the interlayer coupling strength between the parenchymal and immune cells. It can be due to a mass transfer, e.g. cytokine expression of macrophages in the immune layer, and signal transfer into the parenchyma:

macrophages invade the extracellular matrix and penetrate the basal membrane to the parenchyma.

(x) The mean phase velocity $\langle \dot{\phi}_i \rangle$ is a measure which describes the collective frequency of the cellular metabolism of cell i due to all interactions with other cells in both layers. It denotes the system performance of a single metabolic branch of the parenchymal or immune cells of a tissue element according to their activation state (normal healthy state or pro-inflammatory activation with increased CRP production).

(xi) The ensemble-averaged standard deviation of the mean phase velocities s^1, s^2 is a measure which characterizes the inhomogeneity of metabolic activity within the parenchyma and the stroma (immune layer), respectively. It assumes non-zero values if the respective layer is not frequency-synchronized and splits into multifrequency clusters, e.g., a healthy cluster with one frequency and a pathological cluster with another frequency, and thus is a measure of pathogenicity in case of the parenchymal layer, or activation in case of the immune layer.

4.2.4 *Synchronization Scenarios and Clustering*

In this section, we present exemplary computer simulations of our model to demonstrate different dynamic scenarios which this model can produce already in its simplest form. For the parameter scans presented in the subsequent sections, we simulate system (4.5)–(4.6) for each set of parameters for the same ensemble of random initial conditions.

4.2.4.1 Tumor Disease

The model has not been refined or optimized with respect to the parameters, but our concern here is to display simulations which can describe principally different evolutions and outcomes of tumor disease. We assume $N = 200$, $\epsilon^1 = 0.03$, $\epsilon^2 = 0.3$, $a^1_{ij} = 1$ (global coupling in the parenchymal layer), $\sigma = 0.3$ (interlayer coupling), natural frequencies $\omega^h = \omega^2 = 0$ (co-rotating frame of healthy parenchymal cells), $\omega^p = 1$ (pathological cells describing spontaneous mutation, chosen randomly for R oscillators where $r = R/N$), and the coupling delay parameter α is chosen the same for both layers such that the frequencies of the healthy parenchymal and immune layers are similar. The parameters β is used to model the influence of age, inflammation status, environment etc. and also the degree of malignancy of the tumor.

Our system exhibits various dynamic patterns resembling healthy and pathological states. The results are depicted in Fig. 4.12, where the mean phase velocities $\langle \dot{\phi}^{\mu}_j \rangle$ are plotted in the right panel, and the absolute differences of the mean phase velocities with respect to a reference oscillator in the parenchyma $\langle \dot{\phi}^1_{100} \rangle$ and in the

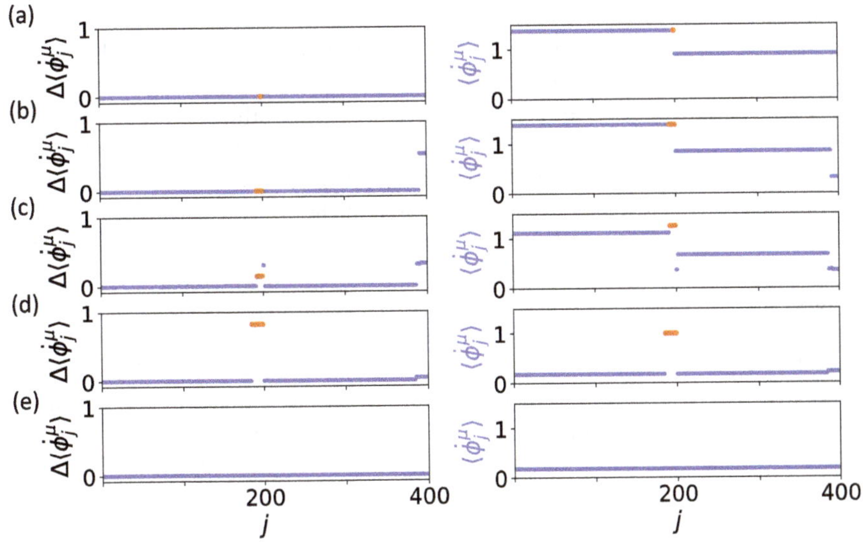

Fig. 4.12 Dynamical scenarios for healthy states without clusters and pathological states with multifrequency clusters for different values of β and r. The left panel shows the absolute differences of the mean phase velocities $\Delta\langle\dot{\phi}_j^\mu\rangle = |\langle\dot{\phi}_j^{1/2}\rangle - \langle\dot{\phi}_{100/300}^{1/2}\rangle|$, whereas the right panel shows mean phase velocities $\langle\dot{\phi}_j^\mu\rangle$. The parenchymal nodes are labeled $j = 1, ..., 200$, and the immune nodes $j = 201, ..., 400$ for $\beta = 0.45\pi$ and $r = 1\%$ (**a**), $\beta = 0.55\pi$ and $r = 4\%$ (**b**), $\beta = 0.60\pi$ and $r = 4\%$ (**c**), $\beta = 0.66\pi$ and $r = 7.5\%$ (**d**), and $\beta = 0.66\pi$ and $r = 0\%$ (**e**). Within each layer μ the nodes are sorted first by $\langle\dot{\phi}_j^1\rangle$, then by ϕ_j^1, respectively. Simulation parameters: $N = 200$, $\alpha = -0.28\pi$, $a_{ij}^1 = 1$, $\epsilon^1 = 0.03$, $\epsilon^2 = 0.3$, $\sigma = 0.3$, $\omega^h = \omega^2 = 0$, $\omega^p = 1$. The simulation time is 2 000 time units. For the mean phase velocities the last $T = 1\,000$ time units are taken for the temporal average, see Eq. (4.7). The mean phase velocities of nodes in layer $\mu = 1$ with $\omega^p = 1$ are marked red. **c**, **d** show pathological states. Figure taken from [255]

immune layer $\langle\dot{\phi}_{300}^2\rangle$, respectively, is plotted in the left panel. Figure 4.12a shows a healthy state with a low age parameter $\beta = 0.45\pi$ and only two tumor cells ($r = 1\%$) in the parenchymal layer. Although the two tumor cells in the parenchyma have different frequencies, the frequency-synchronized state with the healthy cells can be maintained due to the coupling. The coupled cooperative dynamics leads to a completely in-phase synchronized state in both layers. Note that the collective frequencies in the parenchymal layer and the immune layer are different (right panel) since the coupling terms are different. The same synchronized state of the parenchyma holds in Fig. 4.12b for increased age parameter $\beta = 0.55\pi$ and increased number of tumor cells ($r = 4\%$), but first indications of pathological behavior are visible in the changed activity of the immune layer, showing lower frequencies of the associated immune cells coupled to the tumor cells (right panel, $j = 393, \ldots, 400$). For higher age parameter $\beta = 0.60\pi$ and the same number of tumor cells ($r = 4\%$), Fig. 4.12c shows a scenario of an emergent pathological state, i.e., a two-frequency cluster with higher frequency of the pathological cells (red), in the parenchymal layer. A corresponding cluster of lower frequencies is observed in the immune layer (Fig. 4.12c as well. With increasing age parameter $\beta = 0.66\pi$ and tumor size $r = 7.5\%$, Fig. 4.12d shows an even more pathological state, where the frequency of the split-off pathological cluster (red) in the parenchyma is distinctly more different from the healthy cluster (blue), while the immune layer is no longer supporting a two-cluster state but has adjusted in frequency to the parenchyma (right panel). This indicates an essential change in the dynamic state which is frequency synchronized close to zero, which indicates the absence of strong coupling contributions to the frequency. This frequency synchronization between parenchymal and immune layer is maintained if the large age parameter $\beta = 0.66\pi$ is kept, but the number of tumor cells is chosen as zero (Fig. 4.12e), and a healthy completely frequency-synchronized state is obtained. In the following, we will characterize this new dynamic state in more detail.

Figure 4.13 shows details of these scenario for the same values of age parameter β and tumor size r. The left and right columns show snapshots of the cytokine matrices κ_{ij}^1 (parenchymal layer) and κ_{ij}^2 (immune layer), respectively. The second

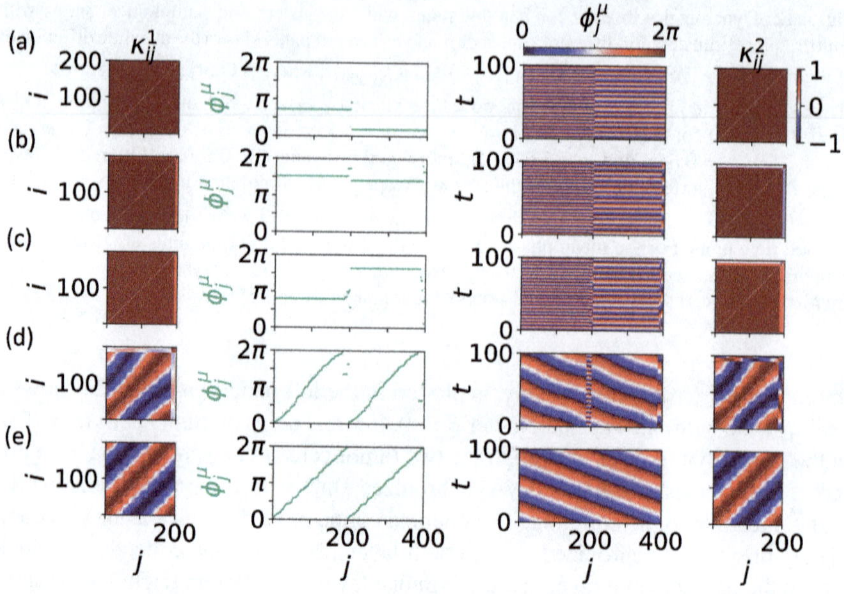

Fig. 4.13 Details of dynamical scenarios for healthy states without clusters (**a**), (**b**), (**e**) and pathological states with multifrequency clusters (**c**), (**d**) for different values of β and r chosen as in Fig. 4.12. The left and right columns show snapshots of cytokine matrices κ_{ij}^1 (parenchymal layer) and κ_{ij}^2 (immune layer), respectively (color coded). Second column: snapshots of phases ϕ_j^μ with the parenchymal nodes labeled $j = 1, ..., 200$, and the immune nodes labeled $j = 201, ..., 400$. Third column: space-time plot of phases $\phi_j^\mu(t)$ (color coded). Parameters: $\beta = 0.45\pi$ and $r = 1\%$ (**a**), $\beta = 0.55\pi$ and $r = 4\%$ (**b**), $\beta = 0.60\pi$ and $r = 4\%$ (**c**), $\beta = 0.66\pi$ and $r = 7.5\%$ (**d**), and $\beta = 0.66\pi$ and $r = 0\%$ (**e**). Within each layer μ the nodes are sorted first by $\langle\dot{\phi}_j^1\rangle$, then by ϕ_j^1, respectively. Other parameters as in Fig. 4.12. Figure taken from [255]

column shows snapshots of the instantaneous phases ϕ_j^μ, and the third column depicts space-time plots of the phases $\phi_j^\mu(t)$ visualizing the oscillations. In the healthy state in Fig. 4.13a, b the cytokine matrices are uniform and temporally constant within each cluster. The snapshots of the phases ϕ_j^μ (second column) and the space-time plots $\phi_j^\mu(t)$ (third column) show homogeneous in-phase oscillations in the parenchyma and the immune layer, respectively, though with different collective frequencies. In the pathological state in Fig. 4.13c the small pathological cluster breaks off at $j = 193, \ldots, 200$. For even larger age parameter $\beta = 0.66\pi$ and tumor size $r = 7.5\%$ in Fig. 4.13d the phase dynamics changes qualitatively. We obtain a pathological two-frequency cluster with strong frequency difference, where the healthy part is no longer in-phase synchronized, but becomes a splay state. With increasing age parameter β the frequency difference of the healthy cluster and the tumor cluster becomes larger, i.e., the tumor cells become autonomous. Thus larger age parameter can be associated with less favorable conditions of tumor disease (higher age, higher tumor malignity). A different scenario in dependence on the tumor size r with increased age parameter β leads from a mixed 2-frequency cluster to a healthy state, where the healthy cluster is a splay state (Fig. 4.13e). This state is different from the frequency clusters in Fig. 4.13a–c, where each frequency cluster is phase synchronized. In Fig. 4.13d, e the frequency of the healthy cluster in the parenchymal layer is fixed equal to that of the immune layer, but the phases of the parenchyma can no longer be kept in-phase, thus a strong overall perturbation of the healthy cluster is visible. The frequency of the pathological cluster is much more strongly separated from that of the healthy cluster. When the tumor size r is reduced for low age parameter β, below a certain r the splay state does no longer occur and the in-phase state is recovered Fig. 4.13(a). In contrast, for high age parameter β the splay state occurs even without tumor and multifrequency clustering in a healthy state (see Fig. 4.13e). The two different pathological scenarios in Fig. 4.13c and d are connected with different responses of the immune layer and might indicate different malignity of the tumor.

A map of regimes in the parameter plane of (β, r) is shown in Fig. 4.14a, b, where the ensemble average s^μ of the standard deviation of the mean phase velocities is plotted for the parenchymal and immune layer, respectively. Bright colors correspond to large deviations of the frequencies (mean phase velocities), and hence to pathological two-cluster states. Their enhancement with increasing age parameter and tumor size can be clearly seen. In Fig. 4.14c, d, a comparison between the ensemble average of the standard deviation and the normalized standard deviation is depicted. In comparison to the standard deviation, the normalized standard deviation shows higher values of the ensemble average s^1 in case of splay states (see Fig. 4.14c for $\beta > 0.6\pi$). On the contrary, the small bulge of s^1 at $\beta = 0.5\pi$ is less pronounced. Figure 4.15 depicts the comparison between the normalized standard deviation of the parenchymal and immune layer. It clearly shows the strong increase of the frequency deviation in the parenchyma associated with the pathogenicity of the tumor with increasing age parameter, while the much smaller bulge in the immune layer is associated with the immune layer activation before the tumor becomes clearly visible.

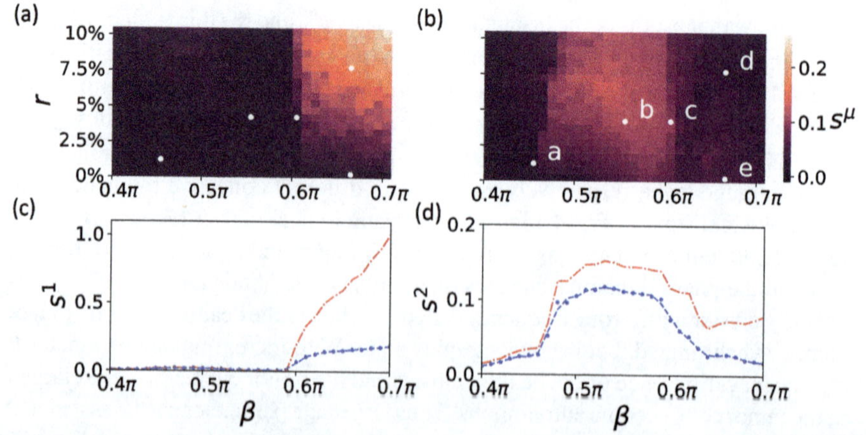

Fig. 4.14 Map of regimes: ensemble average s^μ of the standard deviation of the mean phase velocities in the parameter plane of age parameter β and tumor size r) (ensemble size is $N_E = 50$) for the parenchymal (**a**) and immune layer (**b**), respectively. Bright colors correspond to the formation of multifrequency clusters. The dynamics of the points marked (**a**)–(**e**) in the upper right panel are shown in Figs. 4.12 and 4.13. The lower panels depict the ensemble average s^μ (blue) in dependence of β for a fixed value of $r = 7.5\%$ for the parenchymal (**c**) and immune layer (**d**), respectively. The dashed line is a regression curve for an ensemble size of $N_E = 200$. For comparison the ensemble average s^μ of the standard deviation (blue curves) and the normalized standard deviation (red curves) of the spatially averaged mean phase velocities is shown in dependence of β for the parenchymal (**c**) and immune layer (**d**), respectively. Other simulation parameters as in Fig. 4.12. Figure taken from [255]

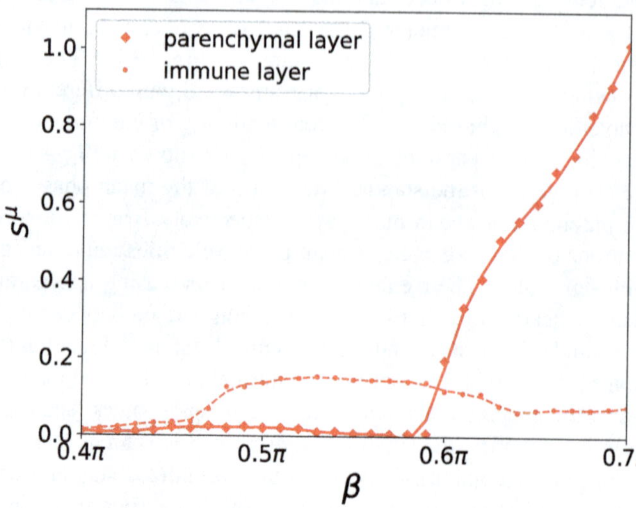

Fig. 4.15 Ensemble average s^μ of the normalized standard deviation of the mean phase velocities in dependence of β for a fixed value of $r = 7.5\%$ for the parenchymal (solid curve, diamonds) and immune layer (broken curve, dots), respectively for an ensemble size of $N_E = 200$. A cubic Savitzky–Golay filter has been applied to smooth the curves. Other simulation parameters as in Fig. 4.12. Figure taken from [255]

4.2.4.2 Sepsis

In our model, we focus on the stage of sepsis which is characterized by generalized inflammation. Depending on the individual subject and the local cellular situation of cytokine and fluid influx, either de-escalation of the inflammatory reaction and restoration of homeostasis, or organ failure occurs. Sepsis is usually preceded by a pre-septic perturbation of the parenchyma, e.g., by a wound which is infected by germs. This perturbation is terminated after a while by blocking off the wound by blood coagulation and eventually healing. Under "normal" conditions, the system returns to a healthy state; however, if sepsis occurs, an inflammatory immune response triggered by the infection spreads across the whole body (cytokine storm). As a consequence, the immune activity may invade large parts of the body through blood vessels and lead to severe organ failure and death. Whether sepsis terminates in a septic shock with severe consequences for the patient depends crucially on the ability of the immune and parenchymal system to regain homeostasis. As in the case of the tumor disease, the ability for returning to a healthy state, i.e., a frequency-synchronized state, is subsequently analyzed using the systemic sum parameter (age parameter) β.

For the simulations, we assume that a dysregulation of the cytokine activity in the immune system has already occurred due to a systemic immune response. The dysregulation of the cytokine activity is modeled by an initial perturbation of the cytokine activity matrix κ_{ij}^2 of the immune layer. For the latter, we consider a separation into two clusters with high activity between nodes from the same cluster ($\kappa_{ij}^2 = 1$) and no activity between nodes from different clusters ($\kappa_{ij}^2 = 0$). All other initial conditions are chosen randomly, as in the simulations for tumor disease in Sect. 4.2.4.1. An example initial condition is presented in Fig. 4.16. Note that we arbitrarily fix the cluster sizes of the initial cytokine activities in the immune layer for the rest of

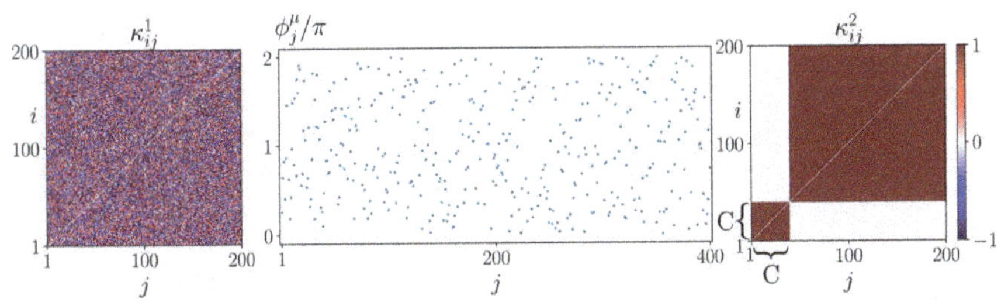

Fig. 4.16 Initial conditions of sepsis: Cytokine dysregulation expressed by a cluster structure of the cytokine activity matrix κ^2 imposes a systemic activation of the immune layer representing the beginning of sepsis. The figure shows an initial condition used for simulations of (4.5)–(4.6) with $N = 200$. The initial cytokine activities in the parenchymal layer κ_{ij}^1 and the initial phases in both layers are randomly drawn from a uniform distribution on the interval $[-1, 1]$ and the interval $[0, 2\pi]$, respectively. The cytokine activities in the immune layer κ_{ij}^2 are initially given by a two-cluster structure where the smallest cluster has size C. The cytokine activities κ_{ij}^2 are 1 within and 0 between the clusters. Figure taken from [255]

this paper. Furthermore, we increase the value of the interlayer coupling strength σ compared to our simulations of the tumor disease. This choice is natural in order to understand the mechanism in action during the progression of sepsis. As described in Sect. 4.2.2, pro-inflammatory cytokines act on endothelial cells and hence cause an increased blood vessel leakiness. As a result, more immune cells and cytokines enter the stroma which in consequence increases the immune-parenchymal interaction.

In the following, we present simulation results for different choices of the age parameter β showing that after an initial cytokine perturbation in the immune layer either the healthy frequency-synchronized state is restored, or the whole system goes to a desynchronized or multifrequency cluster state.

Figure 4.17 shows five representative dynamical scenarios that are induced by pathological initial cytokine activity in the immune layer. For an age parameter $\beta = 0.5\pi$ in Fig. 4.17a, we observe that the system relaxes to a healthy state, i.e., frequency synchronized state, after an initial immune layer perturbation. Hence, the pro-inflammatory response is stopped. A transient of 1 000 time units is discarded before the temporal average over $T = 1\,000$ is taken in the mean phase velocities $\langle \dot{\phi}_j^\mu \rangle$. In addition, Fig. 4.18a shows that in the final state the phases in both layers are in-phase synchronized representing a resilient healthy state. The same scenario can be observed as well for an increased value of the age parameter $\beta = 0.6\pi$ in Figs. 4.17b and 4.18b. While for $\beta = 0.5\pi$ it is unlikely to obtain a pathological

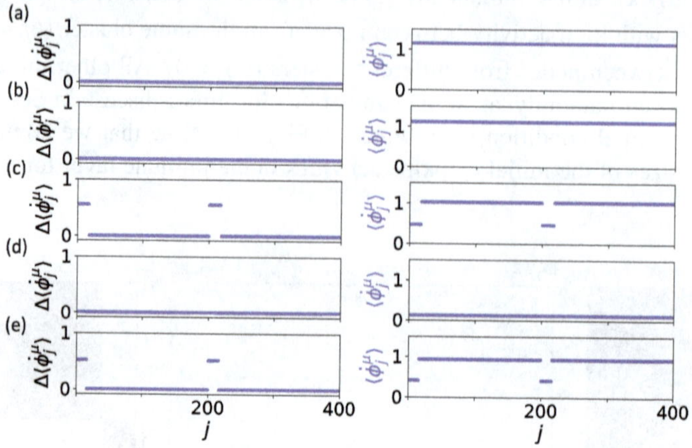

Fig. 4.17 Dynamical scenarios for healthy states without clusters and pathological states with multifrequency clusters for different values of β. The left panels show the absolute differences of the mean phase velocities $\Delta\langle\dot{\phi}_j^\mu\rangle = |\langle\dot{\phi}_j^{1/2}\rangle - \langle\dot{\phi}_{100/300}^{1/2}\rangle|$. The right panels show mean phase velocities $\langle\dot{\phi}_j^\mu\rangle$. The parenchymal nodes are labeled by $j = 1, \ldots, 200$, and the immune nodes by $j = 201, \ldots, 400$. The age parameter is $\beta = 0.5\pi$ (**a**), $\beta = 0.6\pi$ (**b**), $\beta = 0.6\pi$ (**c**), $\beta = 0.7\pi$ (**d**) and $\beta = 0.7\pi$ (**e**), where panels (**b**) and (**c**), and panels (**d**) and (**e**), respectively, only differ in the random initial conditions $\kappa_{ij}^1(0), \phi_j^\mu(0)$. Within each layer μ the nodes are sorted first by $\langle\dot{\phi}_j^1\rangle$, then by ϕ_j^1. Simulation parameters: $\sigma = 1, \omega^1 = \omega^2 = 0$. Other parameters as in Fig. 4.12 and the initial conditions as in Fig. 4.16. Figure taken from [255]

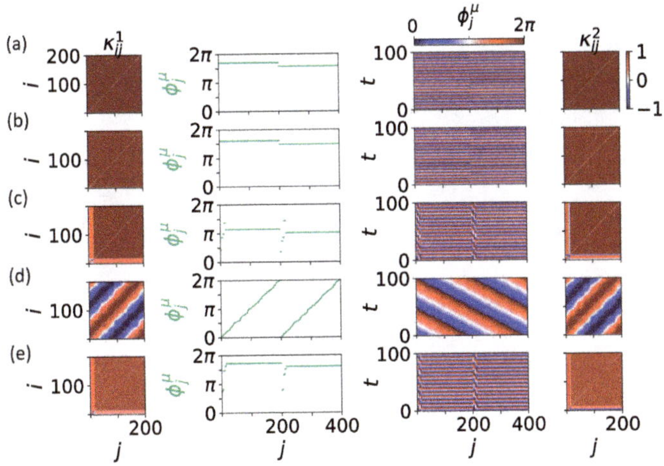

Fig. 4.18 Details of dynamical scenarios for healthy states without clusters (**a**), (**b**), (**d**) and pathological states with multifrequency clusters (**c**), (**e**) for different values of β chosen as in Fig. 4.17. The left and right columns show snapshots of cytokine activity matrices κ_{ij}^1 (parenchymal layer) and κ_{ij}^2 (immune layer), respectively (color coded). Second column: snapshots of phases ϕ_j^μ with the parenchymal nodes labeled $j = 1, ..., 200$, and the immune nodes labeled $j = 201, ..., 400$. Third column: space-time plot of phases $\phi_j^\mu(t)$ (color coded). The parameters β and all other parameters are chosen as in Fig. 4.17. Figure taken from [255]

state, i.e., a desynchronized state, from any random initial conditions of κ_{ij}^1 and the phases ϕ_j^μ, the probability of a pathological state for an ensemble of random initial conditions $\kappa_{ij}^1(0)$, $\phi_j^\mu(0)$ increases for $\beta = 0.6\pi$. In Figs. 4.17c and 4.18c, we depict a dynamical scenario where the initial immune layer perturbation induces a desynchronization of the parenchymal layer. We clearly observe the presence of a two-frequency cluster. In this situation, the initial activated immune response can not be compensated by the coupled system and pushes the parenchyma away from a homeostatic state that may have severe consequences for the organic tissue, compare with the discussion for tumor disease in Sect. 4.2.4.1. For even higher values of the age parameter $\beta = 0.7\pi$ the probability of desynchronization increases. Depending upon the random initial conditions of κ_{ij}^1 and the phases ϕ_j^μ, also for this value of β, we may obtain a frequency-synchronized (healthy) or a desynchronized (pathological) state. In Figs. 4.17d and 4.18d, we display a frequency-synchronized state for $\beta = 0.7\pi$. However, this particular state is not in-phase synchronized any more, as it is the case for the healthy states in Fig. 4.18a, b. Figure 4.18d shows this state which possesses a splay distribution of the phases. Splay distributions of the phases may be interpreted as more vulnerable and less resilient against further perturbations by pathological cells in the parenchymal layer or a "second hit" phenomena known for sepsis. The more likely dynamical scenario of an emergent pathological two-frequency cluster for the same value of $\beta = 0.7\pi$ is displayed in Figs. 4.17e and 4.18e.

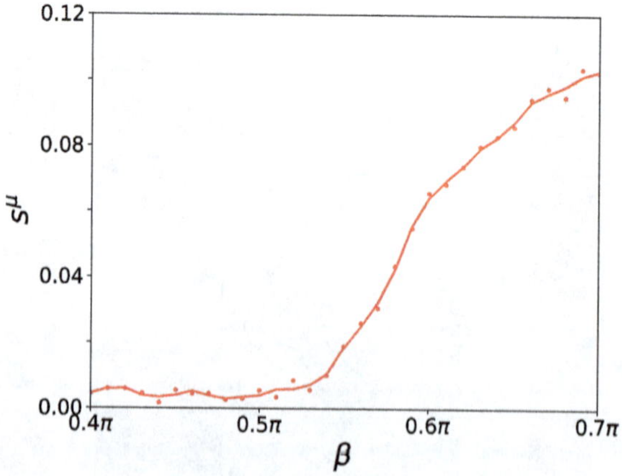

Fig. 4.19 Ensemble average s^μ of the normalized standard deviation of the mean phase velocities in dependence of β for a fixed value of $\sigma = 1$ for the parenchymal and immune layer (both curves coincide) for an ensemble size of $N_E = 200$. A cubic Savitzky-Golay filter has been applied to smooth the curve. Other simulation parameters as in Fig. 4.17. Figure taken from [255]

For all presented scenarios in Figs. 4.17 and 4.18, we note that the frequencies of the parenchymal and the immune layer are locked. The emergence of the dynamical phenomenon of locking [275] can be explained by the increased value of the interlayer coupling strength σ compared to the simulation of the tumor disease, see Sect. 4.2.4.1. Locking can be interpreted as the dynamical manifestation of the physiological observation that the immune system is taking control of the whole system in case of sepsis. For smaller interlayer coupling strength this locking cannot be observed, and hence desynchronization induced by a pathologically activated immune system is not possible.

The dynamical scenarios discussed in this section are very similar to the observations presented for tumor disease. Additionally, in agreement with the results found for tumor disease, the desynchronization of the parenchymal layer due to an initially activated immune layer is observed for higher age parameter β with increasing probability. We note that also here resilient and vulnerable healthy states may coexist with pathological desynchronized states. The increased desynchronization probability with increasing β is reflected in the increasing ensemble average of the normalized standard deviation of the mean phase velocities in Fig. 4.19.

4.2.5 *Critical Parameters for Modeling Sepsis*

This section is devoted to the numerical analysis of critical parameters controlling the interaction of the parenchyma with the immune system, i.e., σ and α, and the initial activation of the immune system, i.e., activation cluster size C, see Fig. 4.12.

In the following, we analyze the impact of these parameters in addition to the age parameter β that has been found to be crucial for the description of the patient's physiological condition.

4.2.5.1 Interlayer Interaction Strength as a Critical Parameter

In this subsection we investigate the influence of the interlayer coupling strength σ on the emergence of sepsis. The interlayer coupling strength appears naturally as an important parameter in order to understand the mechanism acting during the progression of sepsis. In fact, proinflammatory cytokines act on endothelial cells and hence cause an increased blood vessel leakiness [353]. As a result, more immune cells and cytokines enter the stroma, which consequently enhances the immune-parenchymal interaction.

In the following, we present simulation results in the (β, σ)-plane showing that after an initial cytokine perturbation in the immune layer either the healthy frequency-synchronized state is likely to be restored (dark shading), or the system is more likely to transition to a pathological desynchronized multifrequency cluster state (light shadings).

The top panels of Fig. 4.20 depict the ensemble average s^{μ} of the standard deviation of the spatially averaged mean phase velocities, which measures the average frequency desynchronization, corresponding to the amount of heterogeneous activity in the system. A high or low degree of desynchronization represents a pathological or healthy physiological condition, respectively. Splitting into frequency clusters corresponds to a pathological state of the parenchyma ($\mu = 1$) or activation of the immune layer ($\mu = 2$). Figure 4.20 shows three regimes of the coupling strength σ for which the system behaves qualitatively different. Within the first regime ($\sigma < 0.5$), the parenchyma, Fig. 4.20(top left), evolves independently of the immune system, Fig. 4.20(top right). This can be concluded from the different values of the average activity s^1 and s^2. In fact, for sufficiently large age parameter β the initial perturbation of the immune layer leads to persistent desynchronization (activation) of the immune layer. As shown in Sect. 4.2.4.1, the average activity s^2 increases with increasing age parameter. Up to the critical value $\sigma_c \approx 0.5$, the parenchyma synchronizes in most of the simulations independently of the age parameter, hence no organ-threatening desynchronization s^1 occurs. It is worth mentioning that below but near the critial value σ_c, the boundaries between low and high activity in the immune layer become more complex due to the increasing interaction of the immune system with the parenchyma.

In the second regime, i.e., in the interval of approximately $0.5 < \sigma < 0.8$, the systems starts to show interlayer phase locking, i.e., $\phi_i^1(t) - \phi_i^2(t) \approx \Delta_i \in [0, 2\pi)$ for all times t. We observe that beyond the threshold of σ_c, the parenchyma may also desynchronize depending on the age parameter β. The average desynchronization s^1 in the parenchyma and hence the potential for organ failure increases with increasing age parameter. For constant σ there always exists a threshold of the age parameter

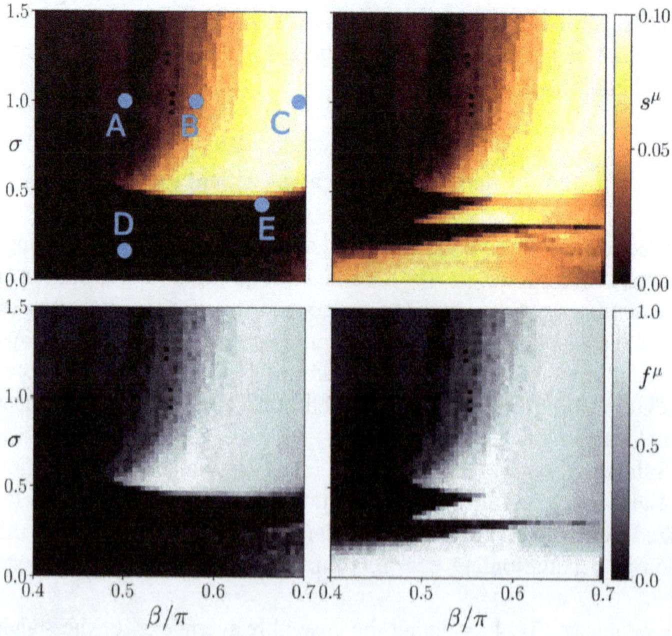

Fig. 4.20 Map of regimes: ensemble average s^μ of the standard deviation of the spatially averaged mean phase velocities (top panels) and the frequency cluster ratio (bottom panels) in the parameter plane of age parameter β and interlayer coupling strength σ for the parenchymal (left column) and immune layer (right column), respectively. Bright colors correspond to the formation of frequency clusters. Simulation parameters: $N = 200$, $\alpha^{11} = \alpha^{22} = -0.28\pi$, $\alpha^{12} = \alpha^{21} = 0$, $a_{ij}^1 = 1$, $\epsilon^1 = 0.03$, $\epsilon^2 = 0.3$, $\omega^1 = \omega^2 = 0$, $C = 40$. Ensemble size is $N_E = 50$. The simulation time is 2 000 time units, the averaging time window 1 000. Figure taken from [256]

above which the parenchyma is dynamically able to desynchronize. With increasing σ the threshold shifts to larger values of β.

In the third regime of the interlayer coupling strength ($\sigma > 0.8$), the threshold of β above which the parenchyma may desynchronize does not shift further to larger values, but remains approximately fixed. Hence, we observe a clear separation in terms of the age parameter between parameter regions with healthy and regions with pathological dynamics.

In order to support our conclusions drawn from the frequency desynchronization measure s^μ, we also plot the ratio f^μ of simulations yielding frequency clusters divided by the total number of simulations N_E for an ensemble of $N_E = 50$ random initial conditions in the bottom panels of Fig. 4.20. We observe that indeed a high value of s^μ correlates with a higher probability of finding a frequency cluster. Therefore, both measures can be used interchangeably.

In Fig. 4.21, we plot representative asymptotic states for different values of σ and β. They correspond to parameter values marked by letters A, B, C, D, E in Fig. 4.20. The left and right columns show snapshots of the cytokine activity matrices

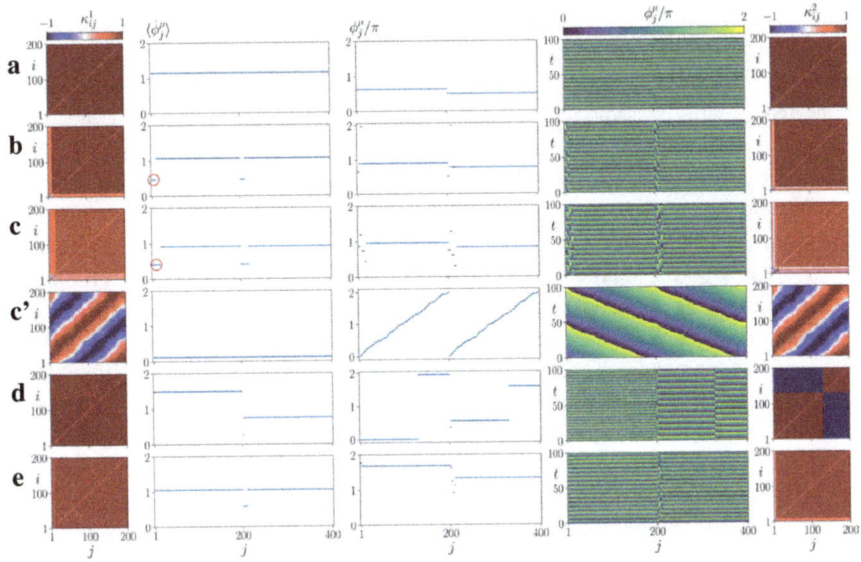

Fig. 4.21 Details of dynamics for healthy parenchymal states without clusters **a**, **c'**, **d**, **e** and a pathological parenchymal state with frequency clusters **b**, **c** for different values of β chosen as in Fig. 4.20. The states shown represent a healthy state in a ($\beta = 0.5\pi$, $\sigma = 1$), a pathological state in **b** ($\beta = 0.58\pi$, $\sigma = 1$) and **c** ($\beta = 0.7\pi$, $\sigma = 1$), where a red circle marks the small pathological cluster, a healthy but vulnerable state in **c'** ($\beta = 0.7\pi, \sigma = 1$) and two resilient states in **d** ($\beta = 0.5\pi, \sigma = 0.2$) and **e** ($\beta = 0.65\pi, \sigma = 0.45$). The left and right columns show snapshots of cytokine activity matrices κ^1_{ij} (parenchymal layer) and κ^2_{ij} (immune layer), respectively (color coded). Second column: mean phase velocities (average frequencies) $\langle\dot{\phi}^\mu_j\rangle$ of the oscillators. Third column: snapshots of phases ϕ^μ_j. The parenchymal nodes are labeled $j = 1, \ldots, 200$, and the immune nodes are labeled $j = 201, \ldots, 400$. Within each layer μ the nodes are sorted first by $\langle\dot{\phi}^1_j\rangle$, then by ϕ^1_j, respectively. Fourth column: space-time plot of phases $\phi^\mu_j(t)$ (color coded). All parameters are chosen as in Fig. 4.20. Figure taken from [256]

κ^1_{ij} (parenchymal layer) and κ^2_{ij} (immune layer), respectively. The second column shows the mean phase velocities (average frequencies) $\langle\dot{\phi}^\mu_j\rangle$ of the oscillators. The third column shows snapshots of the instantaneous phases ϕ^μ_j, and the fourth column depicts space-time plots of the phases $\phi^\mu_j(t)$ visualizing the oscillations.

We observe that depending on the choice of parameters different dynamical states emerge. In Fig. 4.21a an in-phase synchronized state is presented showing that the system is capable of evolving into a healthy state after an initial perturbation of the immune layer. All mean phase velocities in the parenchyma and in the immune layer (collective frequencies) are the same (second column), and the oscillators in each layer are in phase (third column). The space-time plot shows spatially homogeneous periodic oscillations. The adaptive coupling weights both in the parenchymal and the immune layer are homogeneous and all weights are equal to unity (left and right columns). Another completely healthy state is shown in Fig. 4.21c' where instead of an in-phase synchronized state a splay state is formed in both layers (third column), i.e., the order parameter $R_2^\mu = 0$ for both layers, but the frequencies are still the same (second column). The space-time plot (fourth column) shows traveling waves, rather than spatially homogeneous oscillations as in panel A. In [255], we have speculated that this type of synchronized states can be interpreted as a vulnerable state emerging in coexistence with pathological states. Indeed, Fig. 4.21c shows a pathological frequency cluster state for the same parameters but different initial conditions. Here both the parenchyma and the immune layer exhibit a two-frequency cluster state, where a smaller cluster with lower frequency splits off from the large cluster (marked by a small red circle in the second column). The small clusters can also be clearly seen in the snapshots of the phases (third column), in the perturbations of the space-time pattern (fourth column), and in the lighter red color in the cytokine matrices (left and right columns).

Figures 4.21d, e show states that can still be regarded as healthy from the perspective that the parenchymal nodes are in synchrony, while the immune layer remains activated after the initial perturbation and exhibits small clusters of deviating frequencies. This shows up in the mean phase velocity profiles (second column), in the snapshots of the phases (third column), in the space-time plot (fourth column) and in the cytokine matrix of the immune layer (right column). These states demonstrate a high degree of parenchymal resilience to the persistent activation of the immune layer. A pathological state is also presented in Fig. 4.21b. Here, the parenchymal layer shows desynchronization and a frequency cluster (small red circle), as well, which may be considered as the starting point of an organ failure.

The splay states with $R_2^\mu = 0$ for both layers (panel C') represent a special class of healthy states. In particular, due to their structure, the oscillators in this state effectively decouple and are potentially more vulnerable to external perturbations. Moreover, as also shown in [255], these states may coexist with frequency clusters. In order to quantify this observation, we plot the probability of finding a splay state in dependence on σ and β in Fig. 4.22. By comparing Fig. 4.20(top panels) with Fig. 4.22, we see that the regions for the existence of splay states have large overlap with the region of pathological cluster states of the paranchyma (yellow hatched area). It should, however, be noted that for intermediate values of the interlayer coupling strength and the age parameter there exists a large region in parameter space for which frequency clusters are very likely, whereas almost no splay states can be found.

Figure 4.23a presents a cut through the parameter plane of Fig. 4.20 (bottom left) at coupling strength $\sigma = 1$. It shows that the probability of a frequency cluster, i.e., a pathological sepsis state, sharply rises with age parameter β above approximately $\beta > 0.5\pi$. This curve compares favorably with empirical data of patients which gives the number of cases of sepsis per 100 000 inhabitants in Germany as a function of age, presented in Fig. 4.23b.

In this section, we have numerically analyzed the dependence of sepsis on the interlayer coupling strength and the age parameter after an initial perturbation of the immune system. We have identified three regimes with qualitatively different

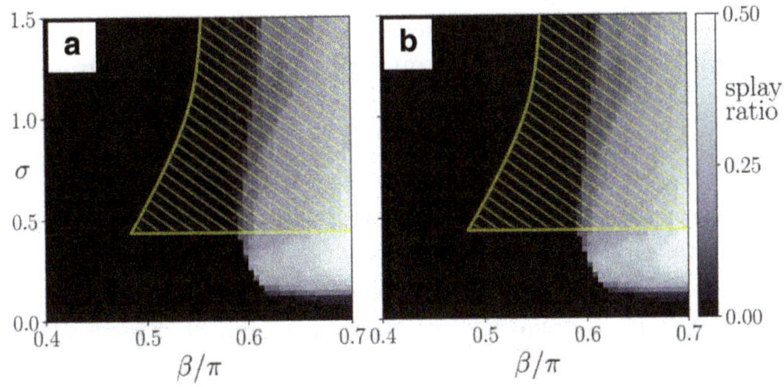

Fig. 4.22 Probability of finding a splay state (splay ratio), see Fig. 4.21c, from $N_E = 50$ random initial conditions, plotted in the parameter plane of interlayer coupling strength σ and the age parameter β. The yellow hatched area shows schematically the regime of pathological cluster states in the parenchyma. Data taken from simulation shown in Fig. 4.20. Figure taken from [256]

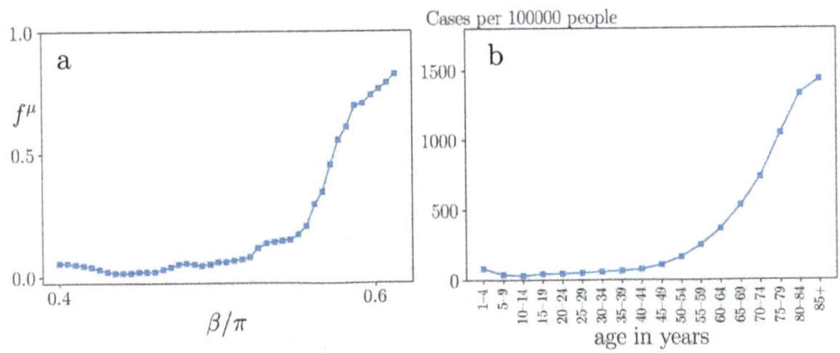

Fig. 4.23 Qualitative comparison of model prediction with empirical data. **a** Frequency cluster ratio for $\sigma = 1$ vs age parameter β, for the parameters in Fig. 4.20, where all data points were averaged over a sliding window of 4 neighboring data points. **b** Empirical data taken from [398] showing the hospitalization incidence of sepsis per 100 000 inhabitants in Germany by age group for the years from 2007 to 2013. Figure taken from [256]

dynamics. First, below a critical coupling strength, the healthy state is preserved for all values of the age parameter. Second, above the critical coupling strength the probability of sepsis sharply rises with increasing age parameter β above a threshold of β, and the threshold itself increases with increasing coupling strength. In the third regime this threshold saturates at a fixed value of β. This means that in a certain intermediate coupling range stronger coupling to the immune layer can preserve the healthy state even at larger age parameter, but eventually the age threshold cannot be shifted further, and the pathological state cannot be avoided. It also implies that an interlayer coupling weight slightly above a critical value could be potentially threatening for patients with a wide range of age parameters, in particular also "younger" patients, i.e., with smaller values of β. This threat, however, shifts to higher values of β as the coupling strength between the layers is increased. Remarkably, our simulations show that, depending upon the initial conditions, healthy states coexist with pathological states for the same parameter values, indicating that the outcome of sepsis after an initial perturbation of the immune system cannot be straightforwardly predicted.

4.2.5.2 Interlayer Phase Lag as a Critical Parameter

In this section, we analyze the dependence of sepsis on the interlayer phase lag parameter α. In particular, we investigate the robustness of our results from the previous subsection with respect to this parameter. Phase lags have been used to account for interaction delays [285, 384] and are known to be critical for the emergence of complex dynamics [213, 286, 399–403]. Motivated by the results presented in the previous subsection for the case $\alpha = 0$, we choose an interlayer coupling strength σ for which sepsis may occur. Therefore, we set $\sigma = 1$ throughout this subsection.

In Fig. 4.24, we show the ensemble average s^{μ} as a measure the average frequency desynchronization for both layers (top left and top right, respectively). In the bottom panels we plot the corresponding ratio f^{μ} of simulations yielding frequency clusters

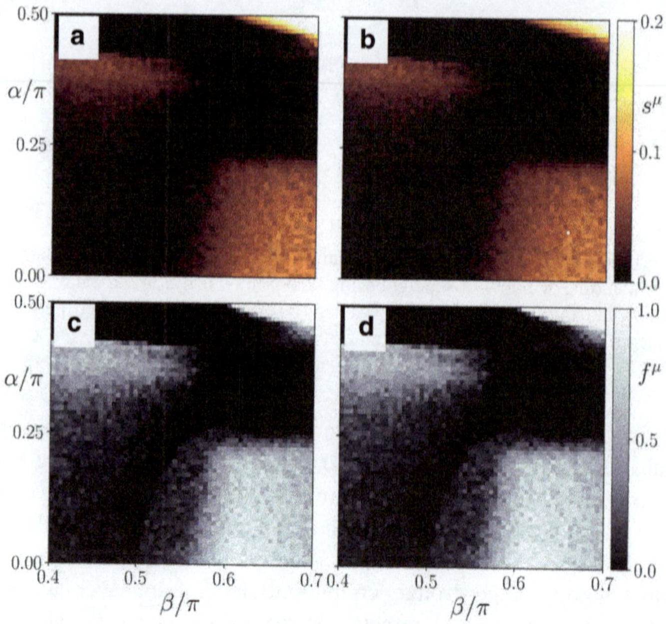

Fig. 4.24 Map of regimes: ensemble average s^{μ} of the standard deviation of the spatially averaged mean phase velocities (top panels) and the frequency cluster ratio (bottom panels) in the parameter plane of age parameter β and interlayer interaction phase lag α for the parenchymal (left column) and immune layer (right column), respectively. Bright colors correspond to the formation of frequency clusters. Ensemble size is $N_E = 50$. Simulation parameters: $\sigma = 1$, $\alpha \equiv \alpha^{12} = \alpha^{21}$; all other parameters as in Fig. 4.20. Figure taken from [256]

divided by the total number of simulations N_E for an ensemble of $N_E = 50$ random initial conditions. The behavior of the parenchyma ($\mu = 1$ in Fig. 4.24a, c) and the immune layer ($\mu = 2$ in Fig. 4.24b, d) is practically the same. From Fig. 4.24, we see that for small values of α, the threshold in the age parameter for the occurrence of sepsis is only slightly changed. It should be noted that with increasing but small interlayer layer phase lag the β threshold does not change much, but the transition from the healthy state to the pathological state becomes sharper, i.e., the frequency cluster ratio increases more sharply.

A dramatic change of the behavior occurs slightly below $\alpha = \pi/4$, which is also the value of the phase lag where in single-layer networks complex partial synchronization patterns of chimera-type are found [399, 400]. For larger values of $\alpha > \pi/4$, we observe that the dependence upon β flips, and the desynchronized (activated) state occurs with some probability for lower β, while for higher β the healthy synchronized state is observed. At approximately $\alpha \approx 0.42\pi$ another flip occurs, and with increasing β there is once more a pronounced transition from the synchronized state to a desynchronized frequency cluster state at a distinct threshold of β, which decreases with further increasing α. This alternating behavior is due to the periodic nature of the coupling function $\sin(\phi_i^1 - \phi_i^2 + \alpha)$. It indicates that the regime which corresponds to physiological conditions and to our interpretation of β as age parameter seems to be confined to $\alpha < \pi/4$, but within this interval the observed behavior is robust. Figure 4.27 depicts the map of regimes for a larger range of $\alpha \in [0, 2\pi]$. This clearly shows the structure of the tongues of two-cluster states (bright colors), which obeys a π-periodic pattern in α.

Figure 4.25a shows details of the dynamics for an exemplary parameter set $\alpha = 0.2\,\pi$, $\beta = 0.58\,\pi$, in a plot similar to Fig. 4.21. Comparing it with Fig. 4.21b where $\alpha = 0$, but the other parameters are the same, we see that our model is robust with respect to the parameter $\alpha = 0$.

4.2.5.3 Immune Activation as a Critical Parameter

This section is devoted to study the impact of the initial perturbation in the immune system corresponding to cytokine activation. For this, we vary the cluster size C of the initial condition of the adaptive coupling weight matrix presented in Fig. 4.12. Here, we choose the two other parameters as $\sigma = 1$ and $\alpha = 0$.

We see in Fig. 4.26 that independently of the initial cluster size a transition from the healthy synchronized state to the pathological desynchronized state in the parenchyma may be observed, and surprizingly the threshold β_c is insensitive to the size C of the initial perturbation in a wide range from only a few cells to half the immune system ($C/N = 0.5$). The behavior of the parenchyma ($\mu = 1$, left panel) and the immune layer ($\mu = 2$, right panel) is very similar.

Panels B, C of Fig. 4.25 show details of the dynamics for $C/N = 0.1$ ($C = 20$) and $C/N = 0.4$ ($C = 80$), respectively, in a plot similar to Fig. 4.21. Comparing it with Fig. 4.21b where $C = 40$, but the other parameters are the same, we see that the asymptotic state does not depend upon the size of the initial perturbation.

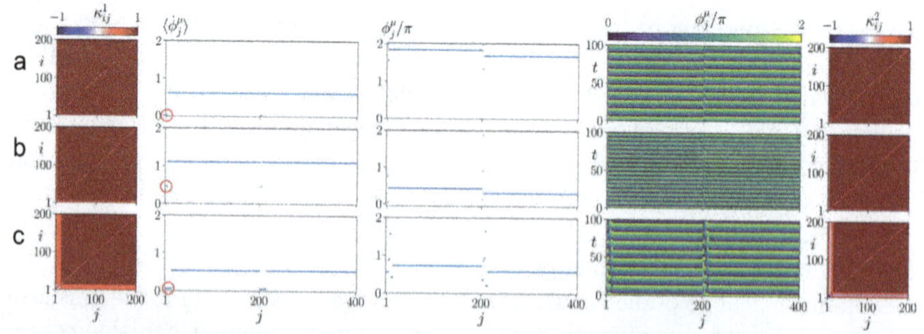

Fig. 4.25 Details of dynamics for different values of α and C. **a** $\alpha = 0.2\pi$, **b** $C/N = 0.1$, **c** $C/N = 0.4$. The left and right columns show snapshots of cytokine activity matrices κ_{ij}^1 (parenchymal layer) and κ_{ij}^2 (immune layer), respectively (color coded). Second column: mean phase velocities (average frequencies) $\langle\dot{\phi}_j^\mu\rangle$ of the oscillators. Third column: snapshots of phases ϕ_j^μ. The parenchymal nodes are labeled $j = 1, ..., 200$, and the immune nodes are labeled $j = 201, ..., 400$. Within each layer μ the nodes are sorted first by $\langle\dot{\phi}_j^1\rangle$, then by ϕ_j^1, respectively. Fourth column: space-time plot of phases $\phi_j^\mu(t)$ (color coded). A red circle marks the small pathological cluster in the parenchyma. Simulation parameters: $\sigma = 1$, $\beta = 0.58\pi$; all other parameters are as in Fig. 4.20. Figure taken from [256]

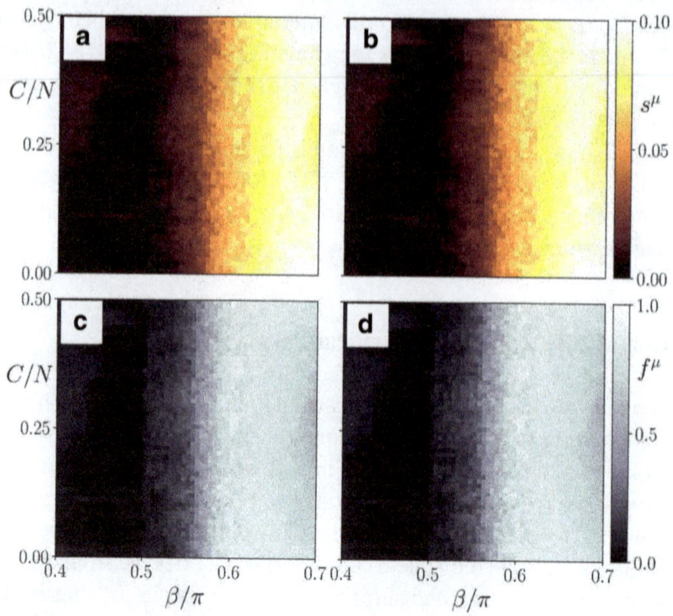

Fig. 4.26 Map of regimes: ensemble average s^μ of the standard deviation of the spatially averaged mean phase velocities (top panels) and the frequency cluster ratio (bottom panels) in the parameter plane of age parameter β and the initial immune layer perturbation expressed by the cluster size C/N for the parenchymal (left column) and immune layer (right column), respectively. Bright colors correspond to the formation of frequency clusters. Ensemble size is $N_E = 50$. Simulation parameters: $\sigma = 1$, $\alpha \equiv \alpha^{12} = \alpha^{21} = 0$; all other parameters as in Fig. 4.20. Figure taken from [256]

This finding seems to be in line with the medical observation that there is no direct relation between the cause and form of an inflammatory response and the frequency of occurrence of sepsis.

4.2.5.4 Analytic Approximations

The adaptive network model in Eqs. (4.5) and (4.6) can also be written in the form of two integral equations with an exponential kernel for the phases in the two layers ϕ_i^1 and ϕ_i^2 by using a Green's function technique to eliminate the differential equations for the adaptive coupling weights κ_{ij}^{μ}. The solution of the general inhomogeneous differential equation

$$\dot{\kappa}_{ij} = -\epsilon\left(\kappa_{ij} + \sin(\phi_i - \phi_j - \beta)\right) \tag{4.10}$$

is given by the integral

$$\kappa_{ij}(t) = -\epsilon \int_0^\infty dse^{-\epsilon s}\left(\sin(\phi_i(t-s) - \phi_j(t-s) - \beta)\right) \tag{4.11}$$

Hence the adaptive two-layer phase oscillator model in the co-rotating frame ($\omega^1 = \omega^2 = 0$) with $\alpha^{11} = \alpha^{22} = \alpha^0$ and $\alpha^{12} = \alpha^{21} = 0$ is reduced to:

$$\begin{aligned}\dot{\phi}_i^1 = -\frac{1}{N}\sum_{j=1}^{N}\Big(a_{ij}^1 - \epsilon^1 \int_0^\infty & dse^{-\epsilon^1 s}\sin(\phi_i^1(t-s) - \phi_j^1(t-s) - \beta)\Big) \\ & \times \sin(\phi_i^1 - \phi_j^1 + \alpha^0) - \sigma \sin(\phi_i^1 - \phi_i^2),\end{aligned} \tag{4.12}$$

$$\begin{aligned}\dot{\phi}_i^2 = \frac{1}{N}\sum_{j=1}^{N}\Big(\epsilon^2 \int_0^\infty & dse^{-\epsilon^2 s}\sin(\phi_i^1(t-s) - \phi_j^1(t-s) - \beta)\Big) \\ & \times \sin(\phi_i^2 - \phi_j^2 + \alpha^0) - \sigma \sin(\phi_i^2 - \phi_i^1),\end{aligned} \tag{4.13}$$

The adaptation function $\sin(\phi_i^{\mu} - \phi_j^{\mu} - \beta)$ shown in Fig. 4.11 now enters as a distributed time delayed feedback which contains the whole history. For the completely synchronized (healthy) state $\phi_i^{\mu} = \phi_j^{\mu} = \phi^{\mu}$ this term can be integrated out, using $\epsilon \int_0^\infty dse^{-\epsilon s} = 1$ and $\frac{1}{N}\sum_{j=1}^{N} = 1$, and setting $a_{ij}^1 = 1$ (for $N - 1 \approx N$):

$$\dot{\phi}^1 = -(1 + \sin\beta)\sin\alpha^0 - \sigma \sin(\phi^1 - \phi^2), \tag{4.14}$$

$$\dot{\phi}^2 = -\sin\beta \sin\alpha^0 - \sigma \sin(\phi^2 - \phi^1), \tag{4.15}$$

The condition for frequency synchronization $\langle\dot{\phi}^1\rangle = \langle\dot{\phi}^2\rangle$ yields a condition for the phase lag between the two layers 1 and 2

$$\sin(\phi^1 - \phi^2) = -\frac{\sin\alpha^0}{2\sigma} \tag{4.16}$$

which agrees with the numerical simulations in Fig. 4.21a ($\phi^1 - \phi^2 = 0.126\pi$). It follows from Eq. (4.16) that $\sigma > \frac{|\sin\alpha^0|}{2}$ is a condition for the existence of the fully in-phase synchronized state in both layers, e.g., $\sigma > 0.385$ for $\alpha^0 = -0.28\pi$.

For cluster states in either the immune layer, or in both layers, the situation is more complicated. If a large synchronized cluster with $\langle\dot{\phi}_i^1\rangle = \omega_L$ coexists with a smaller cluster of different frequency $\langle\dot{\phi}_j^1\rangle = \omega_L - \Delta\omega$ and desynchronized phases θ_j, Eqs. (4.14) and (4.15) must be supplemented for the large cluster ($i \in L$) and the small cluster ($j \in S$) by correction terms. These are complicated temporally oscillating functions, and the condition in Eq. (4.16) is modified for frequency synchronization of the large cluster at frequency ω_L and for the small cluster at $\omega_L - \Delta\omega$. By temporal averaging over trigonometric functions, one may obtain rough approximations. Assuming slow adaptation ϵ^1 and ϵ^2, and inserting the phases $\phi_i^1 = \omega_L t$ for $i \in L$ and $\phi_j^1 = (\omega_L - \Delta\omega)t + \theta_j$ for $j \in S$, for non-synchronous solutions i, j the fast oscillating terms in the integrals average out to zero. Thus more detailed expressions for the regime of existence of frequency cluster states as a function of σ and β may be derived (Fig. 4.27).

4.2.6 *Functional Model and Sepsis Prediction*

This section explores the relationship between clinical sepsis data and the dynamic systems approach to sepsis modeling introduced in Sect. 4.2.3. The primary objective is to identify critical parameters within the *Sepsis Occurrence in Acutely Ill Patients* (SOAP) study, a European-wide observational study from 2002, that influence the temporal progression of sepsis and patient outcomes at hospital discharge. The SOAP data set is introduced in Sect. 4.2.6.1.

Simultaneously, insights from dynamic systems theory are utilized to model the pathological progression of sepsis and its organ-damaging effects. In this framework, sepsis is characterized as a dysregulation of the homeostatic state (frequency synchronized), transitioning to a pathological state (desynchronized or multifrequency clustering) in the parenchyma. This approach identifies the critical parameters driving these transitions (see Sect. 4.2.5).

Section 4.2.6.2 provides a conceptual framework for quantitatively and reliably estimating the prognosis of individual sepsis patients and phenotypically categorizing patients to inform data-driven therapeutic strategies. A key focus is the evaluation

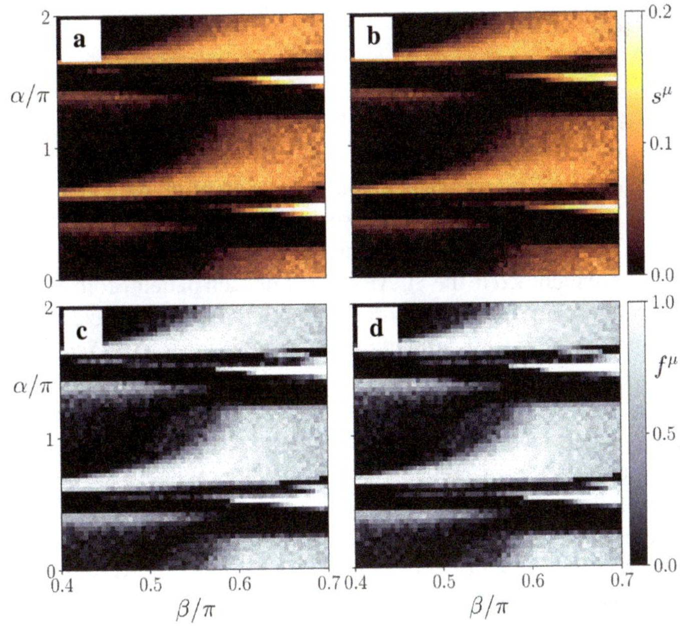

Fig. 4.27 Map of regimes: ensemble average s^{μ} of the standard deviation of the spatially averaged mean phase velocities (top panels) and the frequency cluster ratio (bottom panels) in the parameter plane of age parameter β and interlayer interaction phase lag α for the parenchymal (left column) and immune layer (right column), respectively. Bright colors correspond to the formation of frequency clusters. Ensemble size is $N_E = 50$. Simulation parameters: $\sigma = 1$, $\alpha \equiv \alpha^{12} = \alpha^{21}$; all other parameters as in Fig. 4.20. Figure taken from [256]

of SOFA scores to identify criteria that influence the temporal progression and outcomes of sepsis. These criteria form the basis for constructing mathematical models that capture the functional dynamics of sepsis, enabling personalized and effective treatment plans.

The insights gained from this analysis have significant implications for modeling approaches such as those proposed in [255, 256, 404] or Sect. 4.2.3. By integrating critical clinical parameters identified in the SOAP study with dynamic systems models, we can enhance the predictive power and explanatory depth of sepsis models. This integration allows for the simulation of lethality outcomes and progression trajectories, providing a robust framework for understanding the complex interplay between systemic dysregulation and organ-specific damage in sepsis.

Furthermore, this combined approach offers the potential to refine prognostic tools and therapeutic interventions by aligning clinical observations with theoretical predictions. These advancements could significantly improve patient stratification, treatment planning, and outcome prediction in clinical settings.

4.2.6.1 Clinical Sepsis Study

The *Sepsis Occurrence in Acutely Ill Patients* (SOAP) study was a multicenter, cohort-based, observational study. In this study, laboratory, hemodynamic, and diagnostic data were prospectively collected until the occurrence of one of the following: death, hospital discharge, or a maximum observation period of 60 days. The study population consisted of patients admitted between May 1 and May 15, 2002, to one of 198 Intensive Care Units (ICUs) located in 24 European countries. A total of 3147 adult patients were included in the SOAP study. For each patient, core data, including age and clinical outcomes, were recorded, along with daily data to capture the progression of mortality risk. Organ function was assessed using the *Sequential Organ Failure Assessment (SOFA)* score.

The *Sequential Organ Failure Assessment* (SOFA) score, previously known as the *Sepsis-related Organ Failure Assessment* score, is designed to monitor a patient's status during their Intensive Care Unit (ICU) stay by evaluating the degree of organ dysfunction or failure. The score is calculated every 24 hours and provides valuable insight into the patient's mortality risk over the course of intensive care treatment. The SOFA score evaluates six organ systems: respiratory (*SOFA-res*), cardiovascular (*SOFA-cir*), hepatic (*SOFA-hep*), coagulation (*SOFA-coa*), renal (*SOFA-ren*), and neurological (*SOFA-cns*). For each system, points ranging from 0 (normal function) to 4 (severely impaired function) are assigned based on specific parameters as described in Table 4.2. The abbreviations for the organ systems are:

- *res*: respiratory system,
- *cir*: circulation,
- *hep*: hepatic function,
- *coa*: blood coagulation,
- *ren*: renal function, and
- *cns*: central nervous system.

By summing the individual *SOFA-organ* scores, a *SOFA-total* score is obtained, which ranges from 0 to 24. This total score serves as a measure of a patient's overall organ function. Both the individual scores and their progression over time provide critical information for clinical decision-making.

The SOAP study has been the subject of extensive research and analysis [405–411]. Key clinical findings with regard to mortality derived from this study are summarized below [411]:

(i) Increased mortality is observed in patients with unchanged or rising SOFA values during the first two days of Intensive Care Units (ICU) admission.
(ii) Single organ failure associated with sepsis is linked to a higher mortality risk compared to single organ failure without sepsis.
(iii) In cases of multiple organ failure (MOF), mortality rates are independent of the presence or absence of sepsis.

Table 4.2 Sequential Organ Failure Assessment (SOFA) Score: The SOFA score evaluates six organ systems: respiratory (*SOFA-res*), cardiovascular (*SOFA-cir*), hepatic (*SOFA-hep*), coagulation (*SOFA-coa*), renal (*SOFA-ren*), and neurological (*SOFA-cns*). For each system, points ranging from 0 (normal function) to 4 (severely impaired function) are assigned based on specific parameters. Table adapted from [262]

SOFA	**0**	**1**	**2**	**3**	**4**
res PaO_2 [mmHg]	≥400	<400	<300	<200 +support	<100 +support
cir MAP [mmHg] dopamine adrenaline noradrenaline [μg/kg/min]	≥ 70	< 70	≤ 5 or dobutamine	>5 ≤0.1 ≤0.1	>15 >0.1 >0.1
hep bilirubin [μmol/l]	<20	20–32	33–101	102–204	>204
coa platelets [$\times 10^3/\mu l$]	≥ 150	<150	<100	<50	<20
ren creatinine [mg/dl] urine output [ml/day]	< 1.2	1.2–1.9	2.0–3.4	3.5–4.9 <500	>5.0 <200
cns Glasgow Coma Scale	15	13–14	10–12	6–9	< 6

(iv) The time to reach the maximum SOFA score (*SOFA-max*) is typically 2–3 days longer in decedents than in survivors (1–2 days).
(v) Early progression of organ dysfunction is a defining feature of the clinical trajectory in critically ill patients.
(vi) Dysfunction of the central nervous system (CNS) in systemic inflammatory response syndrome (SIRS) without infection is more frequently associated with mortality than CNS dysfunction in patients with sepsis.

Mortality is defined as the number of deaths for a certain area and for a certain period in relation to the mean total population. In contrast, lethality L can be defined as the number of deaths in relation to the number of diseased persons.

4.2.6.2 Link Between Clinical Sepsis Data and Model

To establish a connection between our functional model and clinical reality, statistical methods are employed, specifically the coefficient of determination (R^2). The R^2 value for a data set $y_1, \ldots, y_n$ and its modeled values $f_1, \ldots, f_n$ is defined as

$$R^2 = 1 - \frac{SS_{\text{res}}}{SS_{\text{tot}}}, \tag{4.17}$$

with

$$SS_{\text{res}} = \sum_i (y_i - f_i)^2, \tag{4.18}$$

$$SS_{\text{tot}} = \sum_i (y_i - \overline{y})^2, \tag{4.19}$$

where SS_{res} is the residual sum of squares, SS_{tot} the total sum of squares, and $\overline{y} = \sum_i y_i$ the mean of the observed data.

R^2 quantifies the proportion of variance in the dependent variable that can be predicted from the independent sets of variables. In statistical modeling, R^2 serves as a critical metric for assessing the predictive capability of a model or testing hypotheses based on associated datasets. It provides a measure of how well the observed outcomes are replicated by the model, based on the proportion of the total outcome variation explained by the model.

Despite the limited availability of data in the studies [405–411], an effort was made to align the model curve for the ensemble average s^μ from Fig. 4.19 with the lethality L obtained from the SOAP study to facilitate predictive analysis. In Fig. 4.28, lethality values are shown as blue, green, and black dots, representing lethality rates for each of three age groups (green: 15-49 years, blue: 50-69 years, black: 70-89 years) as a function of the *SOFA-total* score measured at patient admission (SOFA_{init}). Corresponding colored curves were generated by adjusting (shifting and scaling) the red model curve. The model curve (red) describes the measured value $s^\mu(\beta)$ as a function of the parameter β, establishing the following functional relationship:

$$s^\mu(\beta) = F(\beta), \tag{4.20}$$

where the function F can be obtained numerically. The adjustment of the model curve, involving both shifting and scaling, is performed based on the parameters SOFA_{init} and age:

$$L = F'\big(\beta_{\text{eff}}(\text{SOFA}_{init}, \text{age})\big) \tag{4.21}$$

$$= \gamma_1 F(\beta + \gamma_2) + \gamma_3, \tag{4.22}$$

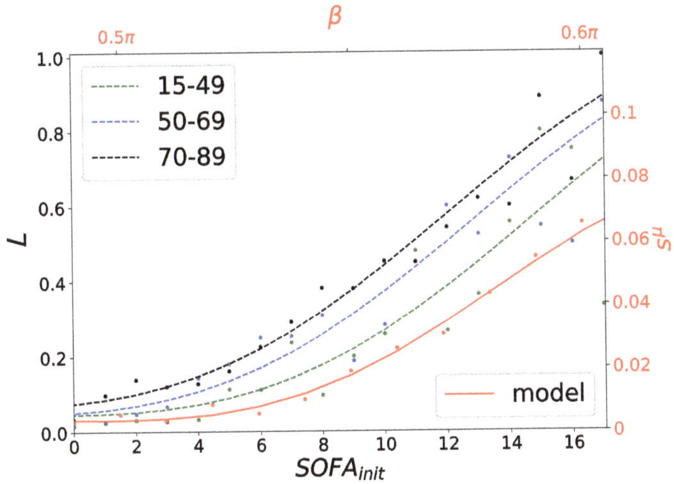

Fig. 4.28 Link between clinical sepsis data and functional model: Shown are empirical lethality L curves (dashed lines) as a function of $\text{SOFA}_{\text{init}}$ for each of three age groups (green: 15-49 years, blue: 50-69 years, black: 70-89 years). The data points represent the empirical data from the SOAP study [405–411], while the compensation curves were generated based on the model curve (red solid line) from Eq. (4.21). The model curve is taken from Fig. 4.15 and shows s^{μ} as a function of β and results in an agreement of $R^2 = 95\%$ (black), 86% (blue), and 81% (green) for the empirical lethality curves

where $\gamma_1, \gamma_2, \gamma_3$ are parameters derived through an optimization process utilizing regression techniques and artificial intelligence methods for the quantitative translation of SOFA_{init} and age to β_{eff}, a quasi "effective" version of β.

The R^2 values indicate an excellent agreement between clinical and model data, with $R^2 = 95\%$ (black), 86% (blue), and 81% (green), underscoring the strong potential of this functional modeling approach. As expected, lethality L generally increases with age. However, with our model, this qualitative observation can be quantitatively specified, enabling predictions of lethality based on parameters such as the SOFA score. Similarly to the β parameter, other critical parameters (refer to Sect. 4.2.5), such as the interlayer coupling strength σ, the interlayer coupling phase lag α, or the size C of the initial perturbation in the activated immune layer, can also be incorporated into the predictive modeling framework using the same methodology. This approach allows for the exploration of additional factors influencing sepsis progression and outcomes, thereby enhancing the precision and applicability of the model. For more robust predictions and enhanced reliability, comprehensive datasets are essential, as the promising results obtained here may lead to high expectations.

Concluding, a strong correlation has been observed between the critical parameters derived from the SOAP study and those identified in our dynamic sepsis model. These findings suggest that parameters extracted from clinical observations can be integrated into the dynamic model of sepsis introduced in this section. The identified parameters can subsequently undergo further testing and validation by comparing the lethality rates observed in the clinical data with those predicted by the model simulations.

4.2.7 Summary

In the field of network physiology, we present a functional model of coupled dynamical systems that elucidates both healthy and pathological conditions associated with carcinogenesis and sepsis. This model focuses on the interaction between parenchymal cells and immune cells, mediated by cytokines, encapsulating the co-evolutionary dynamics of these components. While parallels have been drawn between cancer and infectious diseases in medical sciences, the underlying mechanisms that trigger similar pathological states in tumors and sepsis remain poorly understood [272].

This section has introduced a novel, unified functional model for tumor disease and sepsis within the context of complex network dynamics, using the innate immune system as a common reference point for both conditions. Unlike organ-specific models, our approach emphasizes dynamic interactions over biochemical specifics. In particular, adaptation and synchronization are of crucial significance. Specifically, cytokine activity is modeled as an information flow between the parenchymal and immune layers, with adaptive coupling weights governing the communication between parenchymal cells and immune cells. This approach distinguishes itself from models that treat cytokine concentrations as isolated dynamical variables, focusing instead on the activation dynamics of cytokines [412].

We introduce a basic form of a two-layer network model using adaptively coupled phase oscillators, which, despite its simplifications, captures essential similarities between tumor disease and sepsis. Two critical parameters are identified: tumor size r, representing the fraction of mutant parenchymal cells with altered metabolic activity frequencies, and an adaptivity parameter β (the "age parameter"), which integrates factors such as age, inflammaging, adiposity, pre-existing conditions, physical inactivity, nutritional status, and malignancy in the context of tumor disease.

In the healthy state, the model reflects a fully frequency-synchronized condition of metabolic activity across both the parenchyma and immune layers. In pathological scenarios, desynchronization of the parenchymal layer occurs, resulting in the formation of two distinct frequency clusters: one corresponding to healthy tissue and the other to dysfunctional tissue, as seen in organ failure. The tendency for desynchronization becomes more pronounced with higher values of the age parameter β, which increases the probability of desynchronization in both tumor disease and sepsis.

To characterize the dynamical patterns resulting from this model, we introduce temporally averaged mean phase velocities as collective frequencies of metabolism, and the ensemble-averaged standard deviation s^{μ} as a measure of pathogenicity in the parenchymal layer ($\mu = 1$) and metabolic activation in the immune layer ($\mu = 2$). A nonzero value of $s^1 \neq 0$ immediately indicates a pathological state. The initial perturbation for both tumor disease and sepsis is considered: for tumors, we assume heterogeneity in the natural frequency distribution of parenchymal cells, accounting for the faster metabolism of mutated cells, while for sepsis, we consider a homogeneous network of oscillators with an initially activated cytokine matrix representing a systemic immune response. We also assume high coupling strength between the parenchymal and immune layers due to the pro-inflammatory immune response in sepsis.

For tumor disease, our simulations show that at small values of the age parameter β, which correspond to younger, healthier individuals, nearly all simulations maintain a healthy state even with a relatively high fraction r of mutated cells. This indicates that the system is robust to random mutations in parenchymal cells. As β increases, reflecting aging or other physiological stresses, the immune layer becomes more activated to maintain synchronization within the parenchymal layer. However, only a low fraction of mutated cells r can still maintain a healthy state under these conditions. Once r exceeds a certain threshold, the immune layer fails to preserve synchronization, leading to a desynchronized pathological state. At very high values of β, the immune layer becomes completely dysregulated, resulting in a severe pathological state of tumor disease. In cases of in-phase synchronization, tumor removal may restore the healthy state, whereas in scenarios with splay synchronization, the non-tumor cluster becomes significantly perturbed, making it difficult to return to a healthy state even after tumor removal.

Similarly, for sepsis, we demonstrate various dynamical scenarios for different age parameters. The age parameter β proves crucial in determining whether the system can return to a healthy state. By selecting specific parameter values, close to those used in tumor disease examples, we visualize the emergence of both synchronized and desynchronized states. Additionally, we identify parameter regimes where healthy states coexist with pathological desynchronized states, with the final outcome depending on the initial conditions. Further systematic studies should explore the dependence on initial perturbations of the immune layer cytokine matrix, including the initial cluster size and interlayer coupling strength.

Using a simple model of phase oscillators in a two-layer system, we explore how dysregulated interactions between the immune system and parenchyma can lead to organ-threatening conditions. Our analysis shows that the dynamic cooperation between the parenchymal and immune layers can result in either a healthy, synchronized state or a pathological state marked by desynchronization or multifrequency clusters within the parenchyma. This framework offers insights into how sepsis arises from the disruption of homeostasis, highlighting the complex dynamics that either stabilize or destabilize interactions between the parenchyma and the immune system.

The model represents parenchymal cells (metabolism) and immune cells (innate immune response) as phase oscillators in a duplex system, capturing indirect communication mediated by cytokines. Adaptive coupling weights are used to model the interaction between immune cells, which adapt on a faster timescale, and parenchymal cells, which adapt on a slower timescale. The system also includes fixed bidirectional coupling between parenchymal and immune cells.

In a pathophysiological context, the different scenarios derived from our model can be interpreted as follows: inflammation without organ failure (the parenchyma remains in-phase synchronized, Fig. 4.21a), organ failure (the parenchyma forms a two-frequency cluster state, Fig. 4.21b, c), systemic spreading (large-scale desynchronization, large frequency clusters), healing or resilience of the parenchyma despite persistent immune activation (synchronization of the parenchyma with a two-frequency immune layer, Fig. 4.21d, e), and relapse from a vulnerable healthy state (splay-synchronized state, Fig. 4.21c'). Critical interaction parameters include the adaptation phase lag β, which determines the adaptation law and acts as a physiological sum parameter (referred to as the "age parameter"), the interlayer coupling strength σ, the interlayer coupling phase lag α, and the size C of the initial perturbation in the activated immune layer cytokine coupling matrix, representing the immune system's initial activation due to inflammation.

An adaptation phase lag β around $\pi/2$ corresponds to a cosine-like adaptation function, maximizing for a healthy (synchronized) state, while larger phase lags β indicate delayed adaptability (see Fig. 4.11). Thus, $\beta \approx \pi/2$ reflects fast adaptability, typical of young age and good physical health, favoring the healthy state, while larger values of β hinder maintaining a healthy state. The size C of the initial perturbation is another critical factor. We use special initial conditions–random initial phase distributions in the parenchyma and immune layer, as well as random perturbations in the immune layer coupling matrix–designed to explore the entire dynamic state space. This approach reveals multistability between healthy and pathological states, where the probability of observing a pathological state depends on initial conditions. By choosing a fully phase-synchronized healthy state as the initial condition, the probability of observing healthy states increases.

Through extensive simulations, we analyze the dynamics of the sepsis model with respect to these critical parameters and find that β and σ are essential in describing the emergence of pathological states. Next to the critical role of the age parameter β in tumor disease, we show that variations in β and σ lead to different dynamical regimes with clear pathophysiological implications. We have mapped out parameter regimes where initial inflammation can (i) be regulated, leading to a completely healthy state (healing), (ii) persist without regulation by the immune system, keeping the parenchyma healthy (chronic inflammation), or (iii) lead to immune-parenchymal dysregulation, resulting in a pathological state and eventual organ failure. Furthermore, we compare the probability of pathological states in our simulations with empirical data on sepsis hospitalization rates in Germany, revealing a striking similarity, suggesting the robustness of our functional modeling approach.

Moreover, we have outlined the integration of clinical sepsis data from the *Sepsis Occurrence in Acutely Ill Patients* (SOAP) study with a dynamic systems-based sepsis model. It emphasizes identifying critical parameters, such as those derived from SOFA scores, to model sepsis progression and patient outcomes. By leveraging statistical methods like the coefficient of determination (R^2), the study demonstrates high predictive accuracy in linking model simulations to observed lethality data across age groups. Parameters such as interlayer coupling strength and initial immune activation are also highlighted for their potential to refine predictive models. This approach underscores the necessity of comprehensive datasets for robust predictions and aligns clinical observations with theoretical modeling to improve prognosis estimation and therapeutic strategies.

In conclusion, we propose a unified disease model based on the innate immune system, successfully describing tumorigenesis, organ dysfunction in sepsis, and disease recurrence within a coherent framework. By varying three key parameters–age, the fraction of mutated cells, and initial cytokine activation–we capture the dynamics of both conditions. Our findings provide a foundation for future dynamic disease modeling efforts, paving the way for improved predictive capabilities and clinical outcomes. This approach aligns with recent advancements in network physiology [413] introduced in Sect. 1.4.2. Network physiology represents an interdisciplinary field that integrates physiological modeling across scales, offering a promising framework for understanding complex diseases like sepsis. Future studies should focus on extending this model to explore disease progression and systemic interactions between organ systems, ultimately improving our understanding and prediction of clinical outcomes.

4.3 Modeling Brain Dynamics

This section presents a modeling framework for brain dynamics that integrates empirical brain network structures with neural oscillator models. The empirical network is introduced in Sect. 4.3.1 using Diffusion Tensor Imaging (DTI) and probabilistic diffusion tractography, which together provide connection strength data between brain regions, ensuring the model accurately reflects realistic brain connectivity.

The framework also incorporates experimental data on brain electrical activity, particularly through electroencephalography (EEG) in Sect. 4.3.2, which plays a crucial role in understanding various brain states such as epilepsy, sleep, and perception. EEG data are vital for both clinical diagnostics and research, offering insights into brain function under both normal and pathological conditions.

To simulate brain dynamics neuronal models are investigated in Sect. 4.3.3. A widely recognized population-level model is the FitzHugh-Nagumo (FHN) model, originally designed for individual neuron dynamics. It can be adapted as a coarse-grained representation of excitable media, making it suitable for large-scale simulations of neuronal networks (see Sect. 4.3.4).

Finally, the analysis of network properties and synchronization is crucial for understanding complex brain dynamics. By introducing analytical methods and metrics in Sect. 4.3.5, emergent behaviors from collective dynamics can be explored, providing deeper insights into the mechanisms underlying both normal and disordered brain activity. Section 4.3.6 summarizes the key steps of our modeling framework for brain dynamics.

4.3.1 Empirical Structural Brain Networks

In this section, we introduce an empirical structural brain network, as illustrated in Fig. 4.29a. The connection strengths between brain areas (nodes) are derived using a combination of Diffusion Tensor Imaging (DTI) and probabilistic diffusion tractography. DTI determines the bulk water motion within each voxel of a brain scan, while tractography estimates the connections between different brain regions [414]. The coupling strengths between the oscillators are based on experimental data and parameterized by two coefficients, ζ and σ. Inter-hemispheric coupling ζ controls the connection strength between the left and right hemispheres, while intra-hemispheric coupling strength σ controls the connection strength within a hemisphere.

The synchronization behavior of the system is determined by these parameters. For instance, the system achieves chimera states with relatively strong intra-hemispheric coupling and weak inter-hemispheric coupling (see also Sect. 3.3.1). A full analysis of the effects of these coupling strengths on the synchronization regime has been conducted by Ramlow et al. [415]. The connectivity matrix and a spatial representation of the connectivity are shown in Figs. 4.29a and 4.29b, respectively.

Structural connectivities are derived from data collected from 20 individuals, with results averaged across subjects. Notably, although it may not be apparent in Fig. 4.29a, there is a slight asymmetry between hemispheres in the structural connectivity of the network.

The brain network is obtained from diffusion-weighted magnetic resonance imaging (dMRI) data measured in healthy human subjects as part of a larger study on connectivity changes in schizophrenia. For details of the measurement procedure, including acquisition parameters, refer to [414, 417–419], where the authors provide structural brain connectivity matrices from diffusion Magnetic Resonance

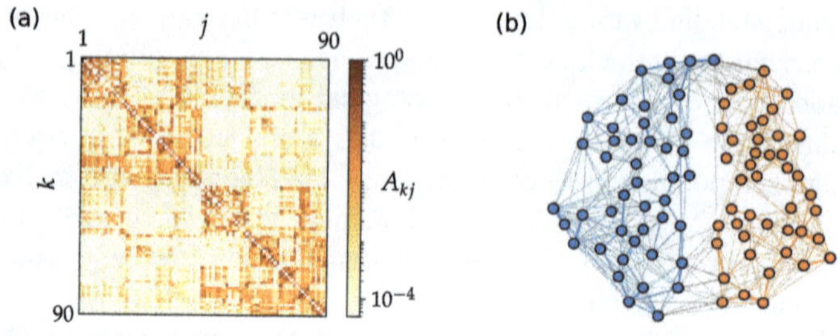

Fig. 4.29 Model for the hemispheric brain structure: **a** Weighted adjacency matrix $\mathbf{A} = \{A_{kj}\}$ of the averaged empirical structural brain network derived from twenty healthy human subjects by averaging over the coupling between two brain regions k and j. The brain regions k, j are taken from the Automated Anatomic Labeling (AAL) atlas [416], but re-labeled such that $k = 1, ..., 45$ and $k = 46, ..., 90$ correspond to the left and right hemisphere, respectively. **b** Schematic representation of the graph of the brain structure with highlighted left (dark blue) and right (light orange) hemisphere. The position of every node is determined by a so-called force atlas, i.e. connected nodes are attracted to one another. Figure modified from [415]

Imaging (dMRI) in 88 healthy subjects, along with raw imaging and demographic data, enabling broad use in computational neuroscience. For prior use of these structural networks in analyzing chimera states, see [420]. The data were analyzed using probabilistic tractography [419], implemented in the FMRIB Software Library (www.fmrib.ox.ac.uk/fsl), where FMRIB stands for Functional Magnetic Resonance Imaging of the Brain.

The cortex and subcortex are divided into 90 predefined regions based on the Automated Anatomical Labeling (AAL) atlas [416]. Each node of the network corresponds to a brain region. Indirect information about white matter fibers connecting different brain regions is provided by dMRI, which measures the preferred diffusion direction in each voxel of the brain. Probabilistic tractography [419] generates for each voxel a set of $n_s = 5\,000$ streamlines simulating possible white matter fiber tracts. A coefficient P_{kj} gives the connectivity probability from region k to region j, calculated as the proportion of streamlines connecting voxels in region k to those in region j, conditional on originating in region k.

Thus, a weighted adjacency matrix of size 90×90, with node indices $k \in N = \{1, 2, \ldots, 90\}$, is constructed. Each entry in this adjacency matrix, representing connectivity between two regions, is averaged over 20 subjects (mean age 33 years, standard deviation 5.7 years, 10 females, 2 left-handed), yielding the average empirical structural brain network $\mathbf{A} = \{A_{kj}\}$. For further details, refer to [420].

The pipeline for constructing the structural network follows prior studies of connectivity differences between healthy subjects and schizophrenia patients [421]. However, obtaining connectivity information using diffusion tractography faces several challenges [422, 423]. Although estimates of structural connection strength and direction from brain activity measurements are theoretically possible, their relationship varies with local dynamics and coupling parameters [423].

In contrast to the original AAL indexing, where sequential indices correspond to homologous brain regions, Fig. 4.29 reorders indices such that $k \in N_L = \{1, 2, \ldots, 45\}$ corresponds to the left hemisphere and $k \in N_R = \{46, \ldots, 90\}$ to the right hemisphere. This highlights the hemispheric structure, specifically the stronger intra-hemispheric coupling compared to inter-hemispheric coupling (see Fig. 4.29a). Additionally, there is a slight asymmetry in structural connectivity between the hemispheres, observed in all 20 networks [415]. For example, differences can be seen by comparing matrix elements at $k = j \approx 20$ with counterparts at $k = j \approx 65$. Such asymmetries, widely reported in neuroscience literature, are linked to functional lateralization, such as the prevalence of language functions in the left hemisphere. These asymmetries also vary with gender [424, 425] and age [426].

A weak but systematic asymmetry between hemispheres, i.e., a difference in the strength of corresponding connections in the left and right hemispheres (Wilcoxon rank-sum test, $p_W < 0.01$, see Sect. 4.3.5.8), was observed in over 20% of connections, most prominently in the temporal cortex. However, no significant variation in asymmetry was found with respect to gender, age, or handedness. This is likely due to the small sample size, which provides insights into the general structure and asymmetry of the brain but limits conclusions about individual variability.

4.3.1.1 Cortical and Subcortical Regions

The human brain is a highly complex organ, with the cortex and subcortex playing distinct but interconnected roles essential for cognition, behavior, and physiological regulation. The cortex, or cerebral cortex, is the outermost layer of the brain and is responsible for higher-order functions such as perception, reasoning, decision-making, language, and voluntary motor control. It is divided into lobes–frontal, parietal, temporal, and occipital–each specializing in specific processes like sensory interpretation, memory, and planning.

Beneath the cortex lies the subcortex, comprising structures such as the basal ganglia, thalamus, and limbic system, which support fundamental functions. The subcortex is involved in motor control, emotional processing, memory formation, and autonomic functions like heart rate and respiration. Together, the cortex and subcortex form a dynamic network, integrating sensory input with internal states to produce coordinated responses and adaptive behavior. This intricate interplay underpins both conscious experiences and subconscious processes, highlighting the brain's remarkable capacity for complex and flexible functioning.

The structural connectivity matrices are utilized as realistic inputs for modeling, rather than as precise representations of the existence and strength of each connection in the human brain. The process of constructing such connectivity information using diffusion tractography is known to encounter various methodological challenges [427]. Although it is theoretically possible to estimate the strength and direction of structural connections from measurements of brain activity, the relationship between these estimates and the underlying structural connectivity can vary significantly depending on (experimentally unmeasured) parameters of local dynamics and coupling functions [423].

The weighted adjacency matrix $\mathbf{A} = \{A_{kj}\}$, of size 90×90, with node indices $k \in N = \{1, 2, \ldots, 90\}$, was derived from averaged diffusion-weighted magnetic resonance imaging data (see Fig. 4.29). Each node of the network corresponds to a specific brain region within the cortex or subcortex. The anatomical labels of these regions are provided in Table 4.3, which outlines the correspondence between network nodes and their associated brain regions.

4.3.2 *Activity in the Brain*

The brain's electrical activity is a fundamental aspect of its function, enabling communication within and between neural circuits. This activity arises from the electrochemical processes of neurons, the primary cells responsible for information transmission in the brain. Neurons communicate through electrical signals known as action potentials, which are generated by the movement of ions across the neuronal membrane through specialized ion channels.

In addition to action potentials, neurons also produce postsynaptic potentials during synaptic transmission, where the release of neurotransmitters at synapses triggers

Table 4.3 Cortical and subcortical regions, according to the Automated Anatomical Labeling atlas (AAL). Note that the numbering of the brain regions is different from the original numbering [416]

Label L/R	Region	Lobe
1/46	Precentral	Central region
2/47	Frontal Sup	Frontal lobe
3/48	Frontal Sup Orb	Frontal lobe
4/49	Frontal Mid	Frontal lobe
5/50	Frontal Mid Orb	Frontal lobe
6/51	Frontal Inf Oper	Frontal lobe
7/52	Frontal Inf Tri	Frontal lobe
8/53	Frontal Inf Orb	Frontal lobe
9/54	Rolandic Oper	Central Region
10/55	Supp motor area	Frontal lobe
11/56	Olfactory	Frontal lobe
12/57	Frontal Sup medial	Frontal lobe
13/58	Frontal Med Orb	Frontal lobe
14/59	Rectus	Frontal lobe
15/60	Insula	Insula
16/61	Cingulum Ant	Limbic lobe
17/62	Cingulum Mid	Limbic lobe
18/63	Cingulum Post	Limbic lobe
19/64	Hippocampus	Limbic lobe
20/65	ParaHippocampal	Limbic lobe
21/66	Amygdala	Sub cort. gray nuc.
22/67	Calcarine	Occipital lobe
23/68	Cuneus	Occipital lobe
24/69	Lingual	Occipital lobe
25/70	Occipital Sup	Occipital lobe
26/71	Occipital Mid	Occipital lobe
27/72	Occipital Inf	Occipital lobe
28/73	Fusiform	Occipital lobe
29/74	Postcentral	Central region
30/75	Parietal Sup	Parietal lobe
31/76	Parietal Inf	Parietal lobe
32/77	Supramarginal	Parietal lobe
33/78	Angular	Parietal lobe
34/79	Precuneus	Parietal lobe
35/80	Paracentral Lobule	Frontal lobe
36/81	Caudate	Sub cort. gray nuc.
37/82	Putamen	Sub cort. gray nuc.
38/83	Pallidum	Sub cort. gray nuc.
39/84	Thalamus	Sub cort. gray nuc.
40/85	Heschl	Temporal lobe
41/86	Temporal Sup	Temporal lobe
42/87	Temporal Pole Sup	Limbic lobe
43/88	Temporal Mid	Temporal lobe
44/89	Temporal Pole Mid	Limbic lobe
45/90	Temporal Inf	Temporal lobe

localized changes in membrane potential. These collective electrical activities generate electric fields that can propagate through brain tissue. When populations of neurons are synchronized in their activity, the resultant fields can be detected externally, forming the basis for non-invasive recording techniques such as electroencephalography (EEG).

The brain's electrical dynamics underlie a wide range of cognitive and physiological processes, from sensory perception and motor control to memory and emotion. Understanding these activities provides critical insights into both normal brain function and pathological states, such as epilepsy, neurodegenerative disorders, and brain injuries.

Electroencephalography (EEG) is a widely used medical diagnostic and research method for recording the brain's electrical activity by measuring voltage fluctuations on the scalp. These voltage fluctuations arise from the synchronized activity of large populations of neurons, primarily pyramidal neurons in the cerebral cortex. Synaptic activity during neuronal communication generates postsynaptic potentials, which produce localized electric fields.

When these electric fields are sufficiently synchronized across a region, they create currents that propagate through the brain and surrounding tissues to the scalp, where they can be detected. EEG electrodes placed on the scalp capture these signals, which are influenced by the anatomical structure of the brain, skull, and scalp, as well as the conductivity properties of the intervening tissues. The recorded signals, therefore, represent the summed electrical activity of neurons beneath each electrode.

EEG signals are shaped by complex interactions between excitatory and inhibitory neural processes and are further modulated by volume conduction, which attenuates and spatially smears the original neural signals. Despite these challenges, EEG remains a valuable tool due to its high temporal resolution, allowing for the real-time monitoring of dynamic brain processes.

EEG measurements have broad applications, ranging from clinical diagnosis to research. They are instrumental in distinguishing epileptic seizures, evaluating head injuries, and studying brain dynamics during different states such as sleep (see Sect. 4.4.1) and perception (see Sect. 5.4). This versatility underscores the importance of EEG in advancing our understanding of brain function in both health and disease [428, 429].

4.3.2.1 Electroencephalography and Labeling Atlas

To enable a meaningful comparison between EEG-recorded and simulated epileptic seizure-related synchronization phenomena, we aimed to estimate which regions of the brain, as defined by the Automated Anatomical Labeling (AAL) atlas, correspond to the different electrode pairs used in the EEG recordings. This was achieved by selecting the AAL regions that are situated directly beneath the electrode pairs, in the vicinity of the skull, as shown in Table 4.4. The assignment of electrode locations

Table 4.4 Mapping of the EEG electrodes to the AAL. Electrodes were referenced in bipolar montage. This means that the EEG time series for AAL area 7 in Fig. 4.56d, for example, shows the measured voltage difference of electrodes F7 and T3. The AAL area covered by their locations were approximately identified and assigned. The electrode locations are shown in Fig. 4.56b. Table modified from [403]

Left/right electrodes	AAL number	Brain area
Fp1-F3/Fp2-F4	2/47	Front Sup
Fp1-F7/Fp2-F8	4/49	Front Mid
F3-C3/F4-C4	10/55	Supp Mot area
C3-Cz/Cz-C4	1/46	Precentral
F7-T3/F8-T4	7/52	Front Inf Tri
FT9-T3/T4-FT10	41/86	Temp Sup
T3-T5/T4-T6	43/88	Temp Mid
T3-C3/C4-T4	29/74	Postcentral
C3-P3/C4-P4	30/75	Par Sup
T5-O1/T6-O2	26/71	Occ Mid
P3-O1/P4-O2	25/70	Occ Sup

to specific AAL brain areas was informed by the method outlined in [430], which provides a standard mapping between EEG electrode positions and corresponding cortical regions.

It is important to emphasize, however, that a perfect or one-to-one correspondence between EEG electrodes and AAL brain areas is not feasible. The reasons for this limitation stem from the inherent complexity of brain anatomy and the non-invasive nature of EEG measurements. EEG signals are not recorded from a single, localized brain region but instead represent the electrical activity of a broader cortical area, with each electrode capturing the sum of the electrical activity from multiple underlying sources. Moreover, the spatial resolution of EEG is limited by the number and placement of the electrodes, which complicates the precise identification of the brain regions responsible for the observed activity. Therefore, while the electrode-to-brain area assignment offers valuable insight, it is important to acknowledge that this mapping is an approximation rather than an exact representation of brain activity.

In Sect. 4.5, EEG data were recorded at a sampling rate of 256 Hz (16-bit A/D conversion) within a frequency range of 0.3–70 Hz from 19 electrodes arranged in a bipolar montage. The locations and nomenclature of these electrodes follow the standardized guidelines established by the American Electroencephalographic Society [431]. To analyze the EEG signals, we apply a Morlet wavelet transform, which allows us to filter the signals and focus on frequencies in the range $f_i \in \{1\,\text{Hz}, \ldots, 32\,\text{Hz}\}$.

This filtering process enables the extraction of phase information corresponding to the different frequency bands. The phase data derived from the Morlet wavelet transform are subsequently used to apply various analytical measures, as outlined in Sect. 4.3.5. These measures provide a deeper understanding of the underlying dynamical mechanisms in the brain's activity, allowing for the investigation of synchronization phenomena and other critical aspects of brain function during epileptic seizures. By analyzing the extracted phases, we can gain insights into the temporal evolution of neural oscillations and their potential role in seizure dynamics.

4.3.3 Modeling Neuronal Activity

Neural oscillations refer to rhythmic or repetitive patterns of neuronal activity in the central nervous system, representing a fundamental aspect of brain function. These oscillations manifest across a broad spectrum of temporal and spatial scales, ranging from the activity of individual neurons to the coordinated dynamics of extensive neural networks. Commonly categorized by frequency bands (see Sect. 5.4.5), each band is associated with distinct cognitive and behavioral functions.

The emergence of neural oscillations is primarily attributed to the interaction between excitatory and inhibitory neuronal populations and the intrinsic biophysical properties of neurons, such as membrane potential and ion channel dynamics. Oscillations are crucial for organizing information processing within and across brain regions, supporting functions such as sensory integration, memory, attention, and perception. Dysregulated oscillatory activity is implicated in various neurological and psychiatric conditions, including epilepsy, Parkinson's disease, schizophrenia, and autism spectrum disorder.

Investigating the mechanisms of neural oscillations is central to neuroscience, as these dynamics offer critical insights into sensory processing, cognitive function, and behavior generation. Computational models, including Hodgkin-Huxley and FitzHugh-Nagumo frameworks, are widely employed to study these dynamics, providing a robust theoretical basis for interpreting experimental data.

Following the identification of cortical and subcortical regions in Sect. 4.3.1, this section focuses on the neuronal dynamics within individual regions. A variety of models exist to simulate neuronal activity, differing in their governing equations and operational scale. While single-neuron models effectively replicate biological characteristics [432], their high resolution imposes significant computational demands, limiting their feasibility for large-scale simulations [433]. Additionally, it is unclear whether large-scale systems of such models yield actionable insights into overarching brain computations [434, 435].

Conversely, population models are better suited for resource-constrained simulations and have successfully replicated cognitive tasks [436, 437]. Since neuronal states often oscillate in quasi-periodic rhythms, oscillator models offer an efficient representation of these dynamics [438]. In Chaps. 4 and 5, the FitzHugh-Nagumo oscillator, a well-established population-level oscillator model, is employed to study the system's dynamics.

4.3.3.1 Hodgkin-Huxley Model

The Hodgkin-Huxley (HH) axon model was originally developed to describe the membrane voltage dynamics underlying the generation of an action potential in the axon of a giant squid. This model represents the membrane voltage V as a dynamic equilibrium governed by four key currents: a leak current, sodium (I_{Na}) and potassium (I_K) ion currents, and an external input current (I_{ext}).

$$C\dot{V} = I_{\text{ext}} - I_{\text{K}} - I_{\text{Na}} - I_{\text{leak}} \tag{4.23}$$

All currents, except for the external input, are modeled as conductance currents. These currents are functions of the membrane voltage V_m and the differences between V_m and their respective equilibrium potentials: V_{Na} for sodium, V_K for potassium, and V_{leak} for the leak current. Importantly, V_{leak} is also defined as the resting potential of the cell ($V_{\text{leak}} = V_{\text{rest}}$), which varies among neurons but is commonly around −65 mV [439].

The individual currents are expressed as follows:

$$I_{\text{K}} = \hat{g}_K n^4(V - V_K) \tag{4.24}$$
$$I_{\text{Na}} = \hat{g}_{Na} m^3 h(V - V_{Na}) \tag{4.25}$$
$$I_{\text{leak}} = \hat{g}_{\text{leak}}(V - V_{\text{leak}}) \tag{4.26}$$

Here, $\hat{g}_X$ represents the maximum conductance for the respective ion channels. The dynamic conductance fraction at any moment is determined by gating variables n, m, and h, which describe the state of ion gates and vary based on the membrane voltage V.

To avoid confusion in terminology, the gating variables are presented with simplified notation. A complete mathematical description of the conductance dynamics is available in [439]. These gating variables dictate the rates of ion flow and are governed by specific dynamic equations that depend on the voltage. By adjusting the parameters α_x and β_x for the gating dynamics, the HH model reproduces the behavior of biological action potentials with high fidelity. The dynamics of gating variables $x \in \{n, m, h\}$ are described as:

$$\begin{aligned}\dot{x} &= \alpha_x(V)(1-x) - \beta_x(V)x \\ &= \frac{x_\infty(V) - x}{\tau_x(V)}\end{aligned} \tag{4.27}$$

$$\tau_x(V) = \frac{1}{\alpha_x(V) + \beta_x(V)} \tag{4.28}$$

$$x_\infty(V) = \frac{\alpha_x(V)}{\alpha_x(V) + \beta_x(V)} \tag{4.29}$$

The action potential initiation begins with an external input current (I_{ext}) that depolarizes the neuron, reducing the negativity of the membrane voltage. This depolarization activates the sodium channel opening variable m, which responds rapidly due to the high speed of sodium channel dynamics. The influx of positively charged sodium ions (Na^+) increases, further depolarizing the cell in a positive feedback loop.

As the membrane depolarizes, the sodium inactivation variable h and potassium channel opening variable n also begin to respond. The inactivation of sodium channels (via h) reduces Na^+ inflow, while the activation of potassium channels (n) allows an outward flow of potassium ions (K^-). Together, these processes hyperpolarize the membrane, bringing it closer to its resting potential (V_{rest}). As the membrane voltage stabilizes, the gating variables x_∞ settle to their steady-state values.

This sequence resets the neuron, making it ready for subsequent excitatory stimuli. For comprehensive details on the formulation and implications of the model, readers are encouraged to consult the foundational work by Hodgkin and Huxley [432].

4.3.3.2 FitzHugh-Nagumo Oscillator

FitzHugh-Nagumo (FHN) oscillators are a simplified representation of the Hodgkin-Huxley (HH) neuron model. By consolidating the ionic dynamics into a single *recovery* variable, the FHN reduces the 4-dimensional HH system in Eqs. (4.23)–(4.26) to a more tractable 2-dimensional model. However, the FHN model is not directly derived from the HH model; rather, it originates from the van der Pol relaxation oscillator (see Sect. 3.1.1.1). The van der Pol oscillator is governed by the following equation:

$$\ddot{x} + c(x^2 - 1)\dot{x} + x = 0 \tag{4.30}$$

In this formulation, the nonlinear damping term $x^2 - 1$ enables self-sustained oscillations, with the coefficient c determining the nonlinearity's intensity. To convert this into the FHN form, one applies the Lienard transformation, yielding two first-order differential equations:

$$\dot{x} = c\left(y + x - \frac{x^3}{3}\right) \tag{4.31}$$

$$\dot{y} = -\frac{x}{c} \tag{4.32}$$

The equations can be further extended into the Bonhoeffer-van der Pol (BVP) model by including additional terms, as described in [440]:

$$\dot{x} = c\left(y + x - \frac{x^3}{3} + z\right) \tag{4.33}$$

$$\dot{y} = \frac{-(x - a + by)}{c} \tag{4.34}$$

In this framework, x (the activator) represents the neuron's membrane voltage, while y (the inhibitor) serves as the recovery variable, which lacks a direct biological counterpart in real neurons. Similar to the HH model, the FHN model can exhibit either excitable behavior or self-sustained oscillations, depending on the choice of parameters. The dynamics of the two state variables operate on different time scales, determined by c, fulfilling the two essential properties of relaxation oscillators: nonlinearity and time-scale separation, which together enable the characteristic buildup and release of energy.

The FHN model retains its structure as a 2-dimensional oscillator with a single fixed point, similar to the van der Pol system. For the purposes of this study, we adopt the slightly modified form of the FHN equations presented in [415]. In this version, the input term z is neglected, reflecting the brain's autonomous dynamics. This simplification reduces the y-nullcline to a vertical line whose position depends solely on a. Furthermore, the nonlinear parameter c is replaced by a time-scale separation parameter ϵ. The modified equations are expressed as follows:

$$\epsilon\dot{u} = u - \frac{u^3}{3} - v \tag{4.35}$$

$$\dot{v} = u + a \tag{4.36}$$

Here, the parameter a determines the existence of a stable limit cycle, with a limit cycle present for $a < 1$. The parameter ϵ controls the time-scale difference between the dynamics of u and v, a critical feature for relaxation oscillator behavior. In this paper, we set $a = 0.5$ to ensure the system operates in the oscillatory regime. A comprehensive analysis of the model, including stability and bifurcation behavior, is provided in [415]. The phase plane, illustrating the model dynamics, is depicted in Fig. 4.30. Although originally designed to model single-neuron action potentials, the FHN model can also serve as a quasi mean-field approximation, representing the collective activity of neuronal populations.

4.3.3.3 Dynamical Features

The FitzHugh-Nagumo (FHN) system is fully described by its two state variables: the membrane voltage u and the recovery variable v. Consequently, its state space can be visualized as a 2-dimensional phase plane, as previously discussed and mathematically defined by Eqs. (4.35)–(4.36). In this context, the nullclines as introduced in Sect. 2.4.1 correspond to the values of u where $\dot{v} = 0$ and values of v where $\dot{u} = 0$. Near these nullclines, the dynamics in the respective directions are relatively slow.

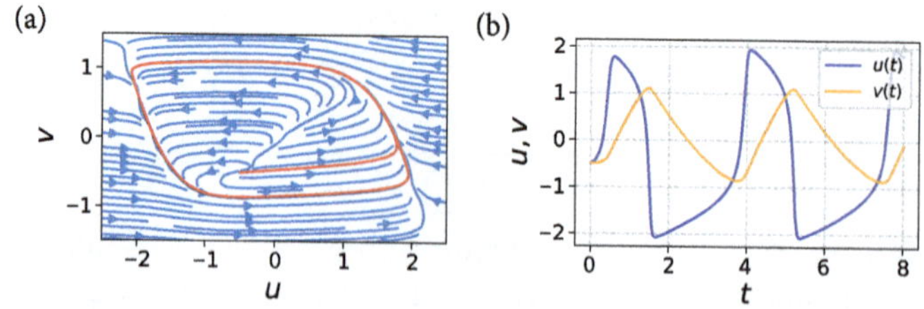

Fig. 4.30 Illustrations of the dynamics of the FitzHugh-Nagumo oscillator in Eqs. 4.35–4.36: **a** vector flow field illustrating the limit cycle in the FHN phase plane, **b** time-series oscillations of the u (blue) and v (orange) states of the FHN. Parameters given by $\epsilon = 0.08$, $a = 0.5$

For instance, close to the u-nullcline, the dynamics of the u state are slow, while the v state dynamics are unaffected. These nullcline conditions can be rearranged from Eqs. (4.35)–(4.36) as follows:

$$u\text{-nullcline: } v = u - \frac{u^3}{3} \quad (4.37)$$

$$v\text{-nullcline: } u = -a \quad (4.38)$$

Thus, in the FHN oscillator, the v-nullcline is represented as a vertical line, while the u-nullcline is described by a third-order polynomial (cubic). In the context of the FHN network, the shapes of the nullclines are dictated by the local dynamics, whereas nonlocal inputs result in shifts in the nullclines. Figure 4.31 illustrates the nullclines for a typical FHN oscillator. It can be observed that the nullclines intersect at a single point, indicating a unique fixed point in the system $\mathbf{f}^*$. This fixed point can be derived from the nullcline equations, as it represents the only state that satisfies both nullcline conditions simultaneously:

$$\mathbf{f}^* = (u^*, v^*) = (-a, \frac{a^3}{3} - a) \quad (4.39)$$

An important consideration in this description is that it assumes an isolated FHN oscillator. In a networked configuration, time-varying nonlocal inputs I_{nonlocal} influence both the shape of the vector field and the positions (though not the shapes) of the nullclines. As a result, the vector field for each oscillator becomes time-dependent. The u-nullcline undergoes vertical shifts, and the v-nullcline experiences horizontal shifts in response to these inputs. These shifted nullclines, referred to as the *effective nullclines*, are defined as follows:

$$v = u - \frac{u^3}{3} + I_{\text{nonlocal}} \quad (4.40)$$

$$u = -a + I_{\text{nonlocal}} \quad (4.41)$$

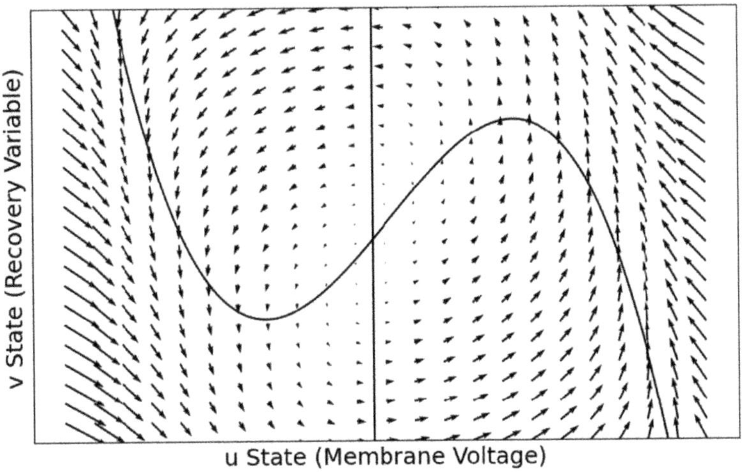

Fig. 4.31 The phase plane of a FitzHugh-Nagumo (FHN) oscillator illustrates its dynamical behavior. In the typical phase plane, the vector field is depicted using arrows positioned at regular intervals, while the nullclines are represented as continuous black lines. The u-nullcline exhibits a cubic shape, whereas the v-nullcline is a vertical line. These nullclines can be identified by observing the points in the phase plane where the vector field has zero length in either the u- or v-direction. A trajectory within the phase plane is constructed by following the vector directions at each point in the space. The relative contributions of the vector field components in the u- and v-directions are governed by the FHN time separation parameter ϵ, where $\epsilon < 1$. In this analysis, ϵ is chosen to be small, leading to dynamics in the u-direction dominating the behavior. Consequently, the vector field components are predominantly horizontal across most regions of the phase plane. This representation highlights the distinct time-scale separation intrinsic to the FHN model, with the fast u-direction dynamics interacting with the slower v-direction dynamics to produce the characteristic relaxation oscillations

4.3.4 Brain Network Dynamics

In this section, we introduce a modeling approach for brain dynamics that integrates the empirical brain structure introduced in Sect. 4.3.1 combined with neuronal models outlined in Sect. 4.3.3. Specifically, we employ the FitzHugh-Nagumo (FHN) model (see Sect. 4.3.3.2), a paradigmatic framework for simulating neuronal spiking dynamics [440–442]. While originally formulated as a simplified model of single-neuron activity, the FHN model is frequently utilized as a coarse-grained representation of excitable media [443, 444].

In this modeling approach, the 90 regions of the human brain, as defined by the Automated Anatomical Labeling (AAL) atlas [416], are represented as a network of $N = 90$ nodes. Each region is modeled by an FHN oscillator, described by an activator variable (u_k, representing the membrane potential) and an inhibitor variable (v_k, representing recovery). The brain regions $k = 1, 2, \dots, 90$ are divided into two sets, corresponding to the left ($k \in N_L = \{1, \dots, 45\}$) and right ($k \in N_R = \{46, \dots, 90\}$) hemispheres as can be seen in Fig. 4.29.

The dynamics of the network are governed by the equations:

$$\begin{aligned}\epsilon \dot{u}_k &= u_k - \frac{u_k^3}{3} - v_k \\ &+ \sigma \sum_{j \in N_H} A_{kj} \left[B_{uu}(u_j - u_k) + B_{uv}(v_j - v_k) \right] \\ &+ \varsigma \sum_{j \notin N_H} A_{kj} \left[B_{uu}(u_j - u_k) + B_{uv}(v_j - v_k) \right],\end{aligned} \tag{4.42a}$$

$$\begin{aligned}\dot{v}_k &= v_k + a \\ &+ \sigma \sum_{j \in N_H} A_{kj} \left[B_{vu}(u_j - u_k) + B_{vv}(v_j - v_k) \right] \\ &+ \varsigma \sum_{j \notin N_H} A_{kj} \left[B_{vu}(u_j - u_k) + B_{vv}(v_j - v_k) \right],\end{aligned} \tag{4.42b}$$

where $k \in N_H$ denotes nodes in either the left (N_L) or right (N_R) hemisphere, and $\epsilon = 0.05$ captures the separation of timescales between the fast activator variable (u_k) and the slower inhibitor variable (v_k) [440]. Depending on the parameter a, the FHN model exhibits either excitable behavior ($|a| > 1$) or self-sustained oscillations ($|a| < 1$). For this study, we focus on the oscillatory regime by setting $a = 0.5$, sufficiently far from the Hopf bifurcation point. The isolated oscillator dynamics are illustrated in Fig. 4.30. For all simulations we use initial conditions randomly distributed on the circle $u_k^2 + v_k^2 = 4$.

Intra- and inter-hemispheric connections are modeled with coupling strengths σ and ς, respectively. For simplicity, unless otherwise stated, these coupling strengths are set equal ($\varsigma = \sigma$). The interaction between nodes is defined by a rotational coupling matrix:

$$\mathbf{B} = \begin{pmatrix} B_{uu} & B_{uv} \\ B_{vu} & B_{vv} \end{pmatrix} = \begin{pmatrix} \cos\phi & \sin\phi \\ -\sin\phi & \cos\phi \end{pmatrix}, \tag{4.43}$$

where $\phi = \frac{\pi}{2} - 0.1$ represents the coupling phase. This configuration introduces strong cross-coupling between activator (u) and inhibitor (v) states [400], a mechanism frequently observed in biological systems [445]. In our macroscopic representation, activator-inhibitor coupling replaces the microscopic complexity of neuronal interactions [446]. This results in strong interdependence between u and v, while interactions within the same state (u to u or v to v) are weaker.

The coupling phase ϕ introduces a phase lag or time delay, enabling complex synchronization patterns, including intermittent high and low synchrony. This dynamic is characteristic of the brain's operation at the critical edge of different dynamical regimes [447]. The coupling phase is analogous to the phase-lag parameter in the Kuramoto model, which has been instrumental in describing synchronization phenomena in oscillator networks [399]. In our simulations, purely activator-activator coupling fails to produce partial synchronization, highlighting the importance of cross-coupling for generating nontrivial dynamics, such as chimera states, in ring topologies [400].

4.3.5 Network and Synchronization Measures

In this section, we will introduce a set of analytical methods and metrics employed to characterize the dynamics of complex systems, with a particular emphasis on the study of network properties and synchronization phenomena within networks. These approaches are integral for understanding the collective behavior of interconnected systems, where the interactions between individual components lead to emergent dynamics that cannot be fully understood by examining isolated elements. The analysis of network properties provides insight into the structure and function of complex systems, enabling the identification of key topological features.

4.3.5.1 Mean Phase Velocity

The mean phase velocity ω_k serves as a key frequency synchronization metric, providing a measure of the average frequency over the entire simulation time, excluding the initial transient period. By comparing the mean phase velocities of different oscillators, one can assess whether the oscillators are frequency synchronized on average. This measure is particularly useful for identifying stationary chimera states, where oscillators may exhibit desynchronized dynamics while still maintaining a common frequency. Numerically, ω_k is computed by determining the number of full oscillations M_k completed within a specified time window ΔT, and is given by the equation:

$$\omega_k = 2\pi \frac{M_k}{\Delta T} \tag{4.44}$$

For consistency, throughout this work we set $\Delta T = 10\,000$. In an uncoupled system ($\sigma = 0$), the mean phase velocity (or natural frequency) of each node is approximately $\omega_k = \omega_{\text{FHN}} \approx 2.6$.

In addition to the individual oscillator metrics, we introduce **hemispheric measures** to characterize the degree of synchronization within the sub-networks of the left and right hemispheres. These provide complementary insights into the global synchronization dynamics. One such measure is the spatially averaged mean phase velocity $\overline{\omega}$, defined as:

$$\overline{\omega} = \frac{1}{90} \sum_{k=1}^{N} \omega_k, \tag{4.45}$$

Here, $\overline{\omega}$ represents the mean phase velocity averaged across both hemispheres. The average of all mean phase velocities within a specific hemisphere, denoted as $\langle\omega\rangle_H$, provides a measure of synchronization for each hemisphere individually.

Further, we consider the **temporal mean** $\langle\Omega(t)\rangle$ of the collective frequency Ω of the mean field, as defined in the literature [448]. The collective frequency $\Omega(t)$ is the time derivative of the mean phase $\psi(t)$, and is calculated as:

$$\Omega(t) \equiv \dot{\psi}(t), \quad R(t)e^{i\psi(t)} = \frac{1}{90}\sum_{k=1}^{N} \exp[i\theta_k(t)] \tag{4.46}$$

The temporal mean $\langle\Omega(t)\rangle$ can then be compared with the spatially averaged mean phase velocity to assess the overall synchronization of the network.

Finally, we define the instantaneous mean phase velocity ω_k^1 for each individual oscillation, which captures the frequency at each moment in time. This provides a time-resolved measure of the phase velocity, allowing for the detection of more complex phenomena such as moving chimera states [400, 449, 450]. The instantaneous mean phase velocity is computed for each kth oscillator over each oscillation, with the formula:

$$\omega_k^1 = 2\pi \frac{1}{\Delta T_k} \tag{4.47}$$

Here, ΔT_k refers to the time duration of the k-th oscillation. This metric resolves the phase velocity at each moment in time and is particularly useful for systems exhibiting multiple timescales in their dynamics. The superscript 1 distinguishes this instantaneous measure from the mean phase velocity ω_k, which is averaged over the entire simulation period.

4.3.5.2 Kuramoto Order Parameter

To quantify the degree of phase synchronization within a neural network, the global *Kuramoto order parameter* is employed. Denoted as $r(t)$ or $R(t)$, this parameter is defined as:

$$R(t) = \frac{1}{N}\left|\sum_{k=1}^{N} e^{i\phi_k(u_k, v_k)}\right|. \tag{4.48}$$

where $\phi_k(t)$ represents the *dynamical phase* of the k-th oscillator. This phase is not taken directly from the standard geometric phase, which is defined as

$$\tilde{\phi}_k(t) = \arctan\left(\frac{v_k}{u_k}\right),$$

due to inherent issues in systems exhibiting *slow-fast dynamics*, such as the *FitzHugh-Nagumo (FHN) model*. Specifically, the nonlinear nature of the geometric phase introduces inhomogeneities in phase velocity, resulting in artificial periodic fluctuations in $R(t)$, which do not reflect actual changes in synchronization.

To mitigate these artifacts, a transformation from the geometric phase $\tilde{\phi}_k(t)$ to a *dynamical phase* $\phi_k(t)$ (also denoted as $\theta_k(t)$ is employed. This transformation ensures a *constant phase velocity* $\dot{\phi}_k$, effectively linearizing the phase evolution across the FHN oscillation cycle. For uncoupled oscillators, the mapping $t(\tilde{\phi}_k)$ is computed numerically, assigning a time value to each geometric phase such that:

$$\phi_k = \frac{2\pi t(\tilde{\phi}_k)}{T},$$

where T denotes the period of oscillation. This transformation homogenizes the time scale, ensuring that for uncoupled oscillators, the phase evolves uniformly and the phase relationship remains consistent.

As illustrated in Fig. 4.32, this transformation maps the nonlinear trajectory of the geometric phase onto the unit circle, stretching or compressing time intervals to eliminate distortions caused by the slow-fast dynamics. The resulting *dynamic phase* allows for a more accurate computation of $R(t)$, with fluctuations in the order parameter now reflecting genuine changes in synchronization rather than artifacts of nonlinear phase evolution.

The Kuramoto order parameter $R(t)$ takes values in the range ([0, 1]), where:

- $R(t) = 1$ indicates *complete phase synchronization*,
- $R(t) \approx 0$ corresponds to *complete desynchronization*,
- intermediate values represent *partial synchronization*.

Because $R(t)$ is computed at each time step, it serves as a *time-resolved measure* of network synchrony, unlike spatially distributed metrics such as the average oscillator frequency $\overline{\omega}_k$. The use of the dynamic phase ensures that $R(t)$ reliably captures the evolving synchronization state of the network, free from distortions introduced by nonlinear phase trajectories.

Additionally, the temporal mean of the Kuramoto order parameter is calculated as:

$$\langle R(t) \rangle = \frac{1}{\Delta T} \int_0^{\Delta T} R(t)\, \mathrm{d}t, \tag{4.49}$$

which provides an estimate of the overall dynamical behavior of the system over time.

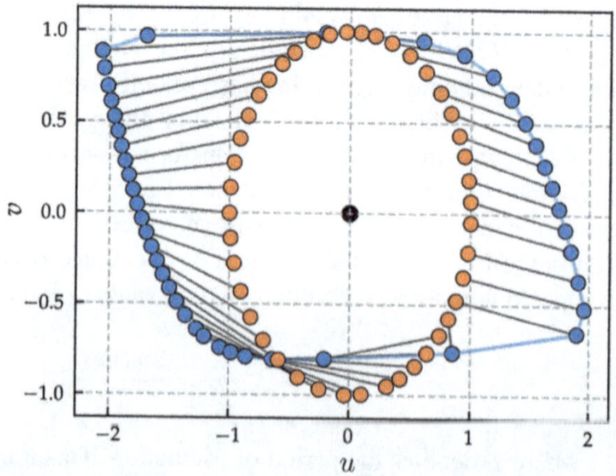

Fig. 4.32 Mapping the dynamic phase: The conversion from the geometric phase to the dynamic phase requires the elimination of inhomogeneities in the time scale. This process involves mapping the phase plane oscillations onto a unit circle, which effectively stretches or compresses the time scales, thereby homogenizing them. This transformation results in the linearization of the dynamic phase. The figure is adapted from [415]

4.3.5.3 Global Phase Coherence

Additionally, we introduce a measure known as the global phase coherence, R_Δ [403], which generalizes the mean phase coherence between two oscillators (i, j), previously studied in [291, 444, 451–453], to an arithmetic mean over all pairs (i, j):

$$R_\Delta = \frac{1}{N(N-1)/2} \sum_{\substack{i,j=1,\\ i>j}}^{N} \left| \frac{1}{T} \sum_{t=1}^{T} e^{i\Delta\Phi_{ij}(t)} \right|, \tag{4.50}$$

where $\Delta\Phi_{ij}(t) = \phi_i(t) - \phi_j(t)$ represents the dynamic phase difference between two oscillators (i, j) at time t. The complex exponential $e^{i\Delta\Phi_{ij}(t)}$ is averaged over a time window of duration T. If the phase relationship between two oscillators remains constant throughout the time window T (frequency-locking), the modulus of the time-averaged $e^{i\Delta\Phi_{ij}(t)}$ approaches 1. Conversely, if the phases are decoupled, the modulus decreases.

The choice of T influences the resolution and stability of R_Δ. For small T, the time resolution of the mean phase coherence is high, but in the limit $T \to 0$, $R_\Delta \to 1$, as any two oscillators appear phase-locked over infinitesimally short intervals, which provides no meaningful information about temporal evolution. Conversely, for large T, short-term fluctuations are smoothed, enabling observation of significant changes in coherence over time. However, in the limit $T \to \infty$, time resolution is entirely lost. To balance these effects, we select a time window corresponding to three oscillation periods for each frequency band, ensuring high temporal resolution while allowing for notable fluctuations in R_Δ.

The normalization of R_Δ ensures a value of 1 under conditions of perfect coherence. This measure was employed to analyze global phase coherence during epileptic seizures across various frequency bands. Optimal seizure identification was achieved at $f = 3$ Hz, a frequency characteristic of absence seizures [454, 455]. For this frequency, a time window of three periods corresponds to $T = 1$ s. It is important to note that $R_\Delta = 1$ for phase-locked states with a fixed phase difference, distinguishing it from the Kuramoto order parameter, which equals 1 only for complete in-phase synchronization of all oscillators.

4.3.5.4 Master Stability Equation

In Sect. 1.3.2, synchronization was introduced as a fixed relation between dynamical objects. From a mathematical perspective, synchronization can be defined in various ways. This monograph primarily focuses on oscillatory systems, where a phase ϕ can be defined. Consequently, phase-lag synchronization serves as a suitable framework: two oscillators, i and j, are phase-lag synchronized if

$$x_i(t) = x_j(t + \tfrac{\phi}{2\pi}T), \tag{4.51}$$

where T represents the period of oscillations in such systems. The term *in-phase synchronization* applies when $\phi = 0$, and *anti-phase synchronization* corresponds to $\phi = \pi$. In the study of chimera states, the mean phase velocity of an oscillator is often employed instead of tracking its precise time series. In these cases, the more general concept of *frequency synchronization* is utilized, where synchronization occurs when the mean phase velocities of two oscillators coincide.

The network with diffusive coupling introduced in Eq. (2.10) is considered under the assumption of in-phase synchronization, $\mathbf{x}_i = \mathbf{x}_j = \mathbf{x}_s$. The stability of this synchronized state is then analyzed. Section 2.4.1 presents a stability analysis for dynamical systems, which can be applied here. Linearizing the network dynamics around the synchronous state yields

$$\dot{\mathbf{x}}_s = \mathbf{F}(\mathbf{x}_s) + \sigma \sum_{j=1}^{N} A_{ij}\mathbf{H}[\mathbf{x}_s - \mathbf{x}_s]. \tag{4.52}$$

The largest Lyapunov exponent, $\Lambda_{\max}$, can be computed to assess stability. However, this computation can become computationally expensive for large networks. A more efficient approach involves the master stability equation, developed in [456] and later extended to delay-coupled networks [457]. Equation (2.10) can be reformulated using matrix notation by defining $\mathbf{X} = \{\mathbf{x}_i\}$:

$$\dot{\mathbf{X}} = \mathbf{F}(\mathbf{X}) + \sigma \mathbf{A} \otimes \mathbf{H}(\mathbf{X}), \tag{4.53}$$

where $\mathbf{F}(\mathbf{X}) = \{\mathbf{f}(\mathbf{x}_i)\}$, $\mathbf{H}(\mathbf{X}) = \{\mathbf{H}(\mathbf{x}_i)\}$, and $\otimes$ represents the tensor product. The variational equation for a small perturbation around the synchronized state, $\mathbf{X}_s + \delta\mathbf{X} = \mathbf{x}_s + \delta\mathbf{x}_i$, is expressed as

$$\delta\dot{\mathbf{X}} = [D\mathbf{F}(\mathbf{X}_s) + \sigma \mathbf{A} \otimes \mathbf{H}]\delta\mathbf{X}. \tag{4.54}$$

The diagonalization of the adjacency matrix, $\mathrm{diag}(\mathbf{A}) = \mathbf{S}^{-1}\mathbf{A}\mathbf{S}$, where $\mathbf{S}^{-1}$ is the matrix that diagonalizes $\mathbf{A}$ ($\mathbf{S}^{-1}\mathbf{S} = \mathbb{1}$), enables simplification. Using the transformation $\delta\tilde{\mathbf{X}} = (\mathbf{S}^{-1} \otimes \mathbb{1}_m)\delta\mathbf{X}$, where m denotes the local phase space dimension, the master stability equation is obtained:

$$\delta\dot{\tilde{\mathbf{x}}}_i = [D\mathbf{f}(\mathbf{x}_s) + \sigma\lambda_i\mathbf{H}]\delta\tilde{\mathbf{x}}_i, \tag{4.55}$$

where λ_i are the eigenvalues of the coupling matrix $\mathbf{A}$. Instead of using Eq. (4.52), the master stability Eq. (4.55) can be employed to compute $\Lambda_{\max}$ and evaluate the stability of the synchronized state. The master stability function separates the effects of local dynamics from those of the network topology. This derivation assumes identical local dynamics $\mathbf{f}$ and coupling schemes $\mathbf{H}$, although such assumptions are not strictly necessary, as demonstrated in studies of delay-coupled networks [281, 458]. Additionally, the row sum of the coupling matrix $\mathbf{A}$ must remain constant ($\sum_j A_{kj} = \text{const.}$), a condition fulfilled by circulant matrices, where each row is a rotated version of the preceding row. In this monograph, the master stability approach is utilized in several sections, including Sects. 3.2 and 4.4.

4.3.5.5 Root Mean Square of Inputs

The root mean square (RMS) of the total nonlocal input to each oscillator is a way to quantify the magnitude of the input, neglecting whether it is positive or negative. Here, it is calculated over each individual oscillation I^1_{RMS}. Note that in this analysis, all nonlocal inputs for which the RMS is calculated contribute to the state u. Similar patterns are observed for inputs to v, but are excluded for brevity. The equation for I^1_{RMS} of the kth node is calculated as follows:

$$I^1_{\mathrm{RMS},k} = \sqrt{\frac{\sum_{t=0}^{T_{\mathrm{OSC}}}(I_k^t)^2}{T_{\mathrm{OSC}}}}, \tag{4.56}$$

where T_{OSC} is the number of time steps in an oscillation and I_k^t is the input to the kth nodes at time t.

For further analysis, the difference between the RMS input for each oscillation I^1_{RMS} and for the full simulation time I_{RMS} can be calculated. This difference shows how the RMS inputs fluctuate over the course of several oscillations. The following Eq. (4.57) shows how the RMS deviation is calculated for one oscillation of the kth node:

$$\Delta I^1_{\mathrm{RMS},k} = I^1_{\mathrm{RMS},k} - I_{\mathrm{RMS},k} \quad , \tag{4.57}$$

where the superscript 1 indicates the RMS input is calculated for an individual oscillation, and its absence indicates the RMS input for the full simulation duration.

4.3.5.6 Fast Fourier Transform

The Fast Fourier Transform (FFT) is an efficient algorithm for computing the Discrete Fourier Transform (DFT) and its inverse. The DFT is a fundamental tool in signal processing, data analysis, and other fields where the transformation of data from the time domain to the frequency domain is required.

The DFT of a discrete sequence $x[n]$ of length N is defined as:

$$X[k] = \sum_{n=0}^{N-1} x[n]e^{-j\frac{2\pi}{N}kn}, \quad k = 0, 1, \ldots, N-1, \tag{4.58}$$

where $x[n]$ represents the input sequence in the time domain, $X[k]$ is the resulting sequence in the frequency domain, j is the imaginary unit ($j^2 = -1$), and N is the number of points in the sequence.

The inverse DFT (IDFT) transforms data back from the frequency domain to the time domain and is given by:

$$x[n] = \frac{1}{N}\sum_{k=0}^{N-1} X[k]e^{j\frac{2\pi}{N}kn}, \quad n = 0, 1, \ldots, N-1. \tag{4.59}$$

Although the Discrete Fourier Transform (DFT) is a powerful tool, its direct computation has a computational complexity of $O(N^2)$ due to the nested summation over N terms for each output frequency bin k. This high computational cost makes it impractical for large datasets.

The Fast Fourier Transform (FFT) significantly reduces this computational burden by exploiting the periodicity and symmetry properties of the complex exponential term $e^{-j\frac{2\pi}{N}kn}$. Specifically, the FFT algorithm divides the DFT computation into smaller sub-problems, recursively reducing the complexity to $O(N \log N)$. This improvement is particularly impactful in real-time signal processing and large-scale data analysis.

One of the most widely used Fast Fourier Transform (FFT) algorithms is the Cooley-Tukey algorithm, which decomposes the Discrete Fourier Transform (DFT) of a sequence into a combination of smaller DFTs of even-indexed and odd-indexed elements. For example, the DFT can be expressed as:

$$X[k] = \sum_{n=0}^{N/2-1} x[2n]e^{-j\frac{2\pi}{N/2}k} + \sum_{n=0}^{N/2-1} x[2n+1]e^{-j\frac{2\pi}{N}k}. \tag{4.60}$$

This recursive decomposition is applied iteratively until the computation reaches trivial cases of $N = 2$ or $N = 1$, dramatically improving the efficiency.

The Fast Fourier Transform (FFT) is particularly advantageous for studying complex systems and nonlinear oscillators, where signals often exhibit intricate patterns and a mixture of frequency components. The benefits include:

- **Efficient Frequency Analysis**: Complex systems often generate signals with multiple interacting frequencies. The FFT allows rapid identification and characterization of these frequency components, which is essential for understanding synchronization, bifurcations, and resonance phenomena in nonlinear oscillators.
- **Resolution of Coupled Dynamics**: In systems with coupled oscillators, the FFT provides insights into the distribution of energy across modes and helps identify dominant frequencies and harmonics, revealing underlying mechanisms of interaction.
- **Real-Time Processing**: Many complex systems, such as physiological signals or neural dynamics, require real-time analysis. The FFT's computational efficiency makes it feasible to analyze large datasets in real time, enabling applications like real-time monitoring of brain activity or adaptive control of oscillatory systems.
- **Quantitative Measures of Nonlinearity**: The FFT is integral to methods like spectral analysis, power spectral density estimation, and phase synchronization, which are used to quantify nonlinearity and chaotic behavior in oscillatory systems.
- **Robustness in Noise-Prone Environments**: In complex systems where signals are often noisy, the FFT's ability to filter and isolate frequencies enhances the clarity and interpretability of data.

In summary, the Fast Fourier Transform (FFT) transforms the computational landscape of frequency analysis, providing a fast, scalable, and practical method for implementing the Discrete Fourier Transform (DFT). Its application extends beyond general signal processing to the intricate analysis of complex systems and nonlinear oscillators, making it an indispensable tool in modern scientific research.

4.3.5.7 Spectral Clustering

Network analysis can be simplified by clustering nodes with strong interconnections while separating groups with weaker connections. This approach, which leverages the topological structure of a network, can be visualized using adjacency matrices with block structures. Spectral clustering is a widely used technique for identifying such structures. It maps an adjacency matrix to an embedding space, condensing the representation to dimensions with the highest informational content, and then applies clustering algorithms, such as K-means, to group data points [459].

K-means clustering assigns data points to one of K cluster centers by minimizing Euclidean distances, following these steps [460, 461]:

(i) Initialize K cluster centers (randomly or as specific data points).
(ii) Assign each data point to the nearest cluster center.
(iii) Recalculate cluster centers to minimize intra-cluster errors.
(iv) Repeat steps 2 and 3 until convergence criteria are met.

Spectral clustering combines K-means with adjacency matrix analysis. Connection strengths, such as those derived from diffusion tensor imaging (DTI), determine the topological relationships between network nodes. The process involves:

(i) Symmetrizing the adjacency matrix.
(ii) Solving the eigenvalue problem for the symmetrized matrix.
(iii) Determining the number of clusters (K).
(iv) Computing the spectral embedding of the data.
(v) Performing K-means clustering on the embedded data.

Numerous variations exist in spectral clustering, including methods for selecting K, symmetrizing matrices, and determining embedding spaces [459, 462]. When ground truth labels are unavailable, clustering performance is evaluated using metrics such as the Silhouette Coefficient, Calinski-Harabasz Index, and Davies-Bouldin Index, implemented in the scikit-learn library [463]. These metrics assess cluster compactness and separability:

- **Silhouette Coefficient**: Balances intra- and inter-cluster distances ($[-1, 1]$, where $+1$ is optimal).
- **Calinski-Harabasz Index**: Compares between- and within-cluster dispersion (higher values are better).
- **Davies-Bouldin Index**: Measures cluster similarity (lower values are better).

Since K-means initialization is randomized, multiple iterations are performed to identify the best clustering using scikit-learn's inertia attribute, which minimizes distance error [460]. Clustering instances are ranked by normalizing the three metrics and averaging their scores.

In this Chapter, we employ Atev's method for spectral clustering, which demonstrated superior performance across several metrics [462]. The approach involves specific symmetrization and embedding steps:

- **Symmetrization**: Uses the adjacency matrix A and the diagonal degree matrix D, where degrees represent the sum of a node's weighted connections.
- **Spectral Embedding**: Embeds the network into a lower-dimensional space using eigenvalues and eigenvectors. The "eigengap" heuristic is applied to determine K, based on the gap between consecutive eigenvalues [459].

In more detail, Atev's method consists of the following symmetrization and embedding steps [462]:

$$A_{\text{sym}} = D_{\text{sym}}^{-\frac{1}{2}} \left(\frac{1}{2}(A + A^T) \right) D_{\text{sym}}^{-\frac{1}{2}} \tag{4.61}$$

$$D_{\text{sym}}^{-\frac{1}{2}} = \frac{1}{2}(D_{\text{in}} + D_{\text{out}}) \tag{4.62}$$

$$Y_{\text{embed}} = D_k^{-\frac{1}{2}} V_k^T, \tag{4.63}$$

where A is the $N \times N$ adjacency matrix. D is the diagonal in- or out-degree matrix of A, consisting of the row sum (in-degree) or column sum (out-degree) of A. A node's degree is the sum of all of its weighted connections to other nodes. Y is the spectral embedding of the system, with Dk representing the $k \times k$ diagonal matrix of the first k nonzero eigenvalues and V_k ($N \times k$) the columns of the right eigenvector matrix associated with those eigenvalues.

The resulting clusters reflect the network's topology, with their evaluation confirming Atev's method as an effective choice for the tested configurations ($K = 2, 3, 4$). This provides a robust framework for analyzing complex networks through spectral clustering.

4.3.5.8 Wilcoxon Rank Sum Test

The Wilcoxon rank-sum test, also known as the Mann-Whitney U test, is a nonparametric statistical method used to compare two independent samples. Unlike parametric tests, such as the t-test, it does not assume a normal distribution of the data, making it suitable for analyzing datasets with unknown or non-normal distributions. The test evaluates whether the distributions of the two groups differ significantly in terms of their central tendency.

The test begins by combining and ranking all observations from both groups. Let R_1 and R_2 denote the sums of ranks for groups 1 and 2, respectively. The test statistic U for each group can be computed as:

$$U_1 = n_1 n_2 + \frac{n_1(n_1 + 1)}{2} - R_1, \tag{4.64}$$

where n_1 and n_2 are the sample sizes of groups 1 and 2. Similarly, U_2 is derived from R_2.

The smaller of U_1 and U_2 is used as the test statistic, and its significance is assessed against critical values from the Mann-Whitney distribution. For larger sample sizes ($n_1, n_2 > 20$), the distribution of U approximates a normal distribution, allowing for a z-score transformation:

$$z = \frac{U - \mu_U}{\sigma_U}, \tag{4.65}$$

where μ_U and σ_U are the mean and standard deviation of U, calculated as:

$$\mu_U = \frac{n_1 n_2}{2}, \quad \sigma_U = \sqrt{\frac{n_1 n_2 (n_1 + n_2 + 1)}{12}}. \tag{4.66}$$

The null hypothesis (H_0) posits that the two groups come from the same population, with no difference in distribution. The alternative hypothesis (H_a) suggests a difference in the distributions. Rejection of H_0 is based on the significance level (α), often set to 0.05.

The Wilcoxon rank-sum test is particularly valuable in scenarios involving ordinal data, skewed distributions, or small sample sizes. Its robustness and simplicity make it a preferred choice in various scientific disciplines, ranging from biology to social sciences.

4.3.6 *Summary*

In this section, we have presented a modeling framework for brain dynamics, which integrates empirical brain network structures with neural oscillator models. The empirical brain network is constructed using Diffusion Tensor Imaging (DTI) and probabilistic diffusion tractography, which together provide information on the connection strengths between brain regions. This approach is grounded in experimental data, ensuring that the model reflects realistic brain connectivity.

Equally important is the integration of experimental data related to the electrical activity of the brain, particularly through the use of electroencephalography (EEG). Electroencephalography (EEG) plays a critical role in understanding brain dynamics, offering insights into various brain states, including epilepsy, sleep, and perception. These measurements are indispensable for clinical diagnostics and research, providing valuable information about brain function under both normal and pathological conditions.

To model brain dynamics within individual regions, we use the FitzHugh-Nagumo (FHN) model, a widely recognized population-level model that simulates neuronal dynamics. The FHN model, initially designed to capture the behavior of individual neurons, is adapted here as a coarse-grained representation of excitable media, making it suitable for large-scale brain simulations. Unlike single-neuron models, which are computationally demanding and often impractical for large systems, the FHN model offers an efficient way to simulate the quasi-periodic oscillatory activity typical of neuronal networks.

The analysis of network properties and synchronization within these models is essential for understanding complex brain dynamics. By focusing on the interactions between interconnected oscillators, we explore emergent behaviors that arise from the collective dynamics of the system. Analytical methods for characterizing network topology and synchronization phenomena allow for a deeper understanding of how brain regions function together, shedding light on the mechanisms underlying both normal and disordered brain activity.

4.4 Partial Synchronization and Unihemispheric Sleep

Synchronization is a critical aspect of neural system dynamics. In many cases, only a part of a neuronal system exhibits synchronization, while other regions remain desynchronized. This phenomenon of partial synchronization can have both beneficial and detrimental effects on an organism's health. This section examines the specific phenomenon of unihemispheric sleep, characterized by the synchronization of activity in only one hemisphere of the brain (see Sect. 4.4.1). Among many animal species, this is termed unihemispheric sleep, whereas in humans, an analogous phenomenon is observed, known as the first night effect [415]. The discussion will particularly emphasize the contributions of distinct brain regions to the emergence of unihemispheric sleep. Furthermore, a potential relationship is proposed between the dynamics of three *eccentric* oscillators and the coupling inputs they receive from other oscillatory systems (see Sect. 4.4.2). Furthermore, the mechanism underlying additional oscillations of the eccentric nodes is identified, revealing a relationship between the network topology and its dynamic behavior (see Sect. 4.4.3). Section 4.4.4 briefly summarizes the key findings. This section is partly based on [415, 464, 465].

4.4.1 Unihemispheric Sleep

Humans and many animal species require a daily period of rest, commonly referred to as *sleep*. Despite its prevalence across mammalian evolution, the fundamental purpose and mechanisms of sleep remain only partially understood [466]. Sleep in healthy individuals can be categorized into distinct stages, typically identified through electroencephalograms (EEG), as introduced in Sect. 4.3. These stages alternate cyclically throughout a single sleep period.

Table 4.5 outlines the distinct stages of sleep. Broadly, sleep can be divided into rapid-eye-movement (REM) sleep and non-rapid-eye-movement (NREM) sleep. NREM sleep is further subdivided into three stages: N1, N2, and N3. The N3 stage, often referred to as deep sleep or slow-wave sleep, is particularly significant. Each stage is characterized by unique EEG patterns, especially the bandwidth of EEG signals, which provides a distinct marker for classifying these stages.

The right column of Table 4.5 illustrates the EEG time series corresponding to various sleep stages, including wakefulness. Notably, slow waves in the δ-regime, which exhibit the highest amplitude, are of particular interest. These waves are character-

Table 4.5 Classification of sleep stages as described by [467–469]: Sleep is broadly categorized into rapid-eye-movement (REM) and non-rapid-eye-movement (NREM) stages. NREM sleep is further divided into N1, N2, and N3, with N3 commonly referred to as deep sleep or slow-wave sleep. Each sleep stage is characterized by dominant electroencephalogram (EEG) wave patterns. The right column of the table presents exemplary EEG time series for the characteristic waves observed in each sleep stage [470]. Table taken from [285]

			(Sleep) stages					EEG
Waves		Frequency (in Hz)	Awake	R	N1	N2	N3	
Beta	β	15–25	•		•			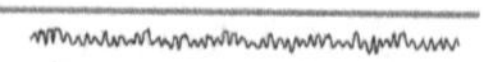
Alpha	α	8–12	•	•				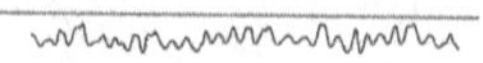
Theta	θ	4–7		•	•	•		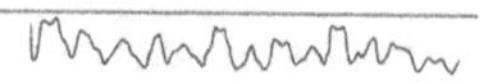
Delta	δ	0.5–4					•	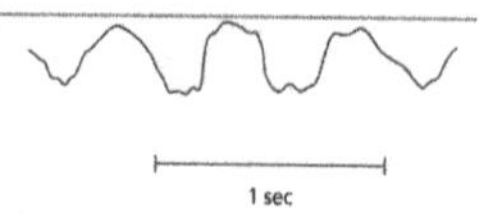

istic of the slow-wave sleep stage, reflecting a significant degree of neural activity synchronization and the coordinated firing of large neuronal populations [471–473]. For a comprehensive description of sleep stages, readers are referred to [469, 474].

Unihemispheric slow-wave sleep is commonly observed in certain birds and aquatic mammals, such as dolphins [475–479]. This state is marked by a dynamic asymmetry between the two hemispheres: one hemisphere exhibits δ-wave activity characteristic of sleep, while the other operates in the α or β regime indicative of wakefulness. This reflects a spatially distinct coexistence of high and low synchronization, allowing one part of the brain to sleep while the other remains awake.

In humans, unihemispheric sleep is not observed, but a related phenomenon, the first-night effect, demonstrates the brain's ability to exhibit interhemispheric asymmetries. The first-night effect is characterized by heightened vigilance in one hemisphere during sleep in unfamiliar surroundings, possibly as a protective mechanism [480]. Both unihemispheric sleep and the first-night effect can be explained in terms of partial synchronization.

Dynamically, unihemispheric sleep is hypothesized to result from the spontaneous symmetry breaking of chimera states, wherein regions of synchronization and desynchronization coexist spatially within oscillator networks [283, 481–483]. Neurophysiological mechanisms underlying this phenomenon remain unclear, but structural interhemispheric separation appears to be a prerequisite. Computational models employing two-community networks of FitzHugh-Nagumo (FHN) oscillators–where intra-hemispheric coupling is stronger than inter-hemispheric coupling–have successfully reproduced dynamics reflective of unihemispheric sleep [285, 415]. These models highlight the sensitivity of partial synchronization patterns, such as chimera states, to system parameters and initial conditions.

4.4.2 Analysis of Network Dynamics

The following section presents simulations of the dynamics described by Eq. (4.42). Throughout these simulations, the parameters are set to $a = 0.5$, $\varepsilon = 0.05$, $\varphi = \frac{\pi}{2} - 0.1$. The coupling strength σ is chosen such that it is as high as possible while still avoiding full synchronization for long simulations ($\approx 10\,000$ time units). For the empirical connectivities, we choose $\sigma = \varsigma = 0.6$. In order to compare our simulations with real data (electroencephalography (EEG) recordings of absence seizures; see Sect. 4.5), we transform the dimensionless time units of the FitzHugh-Nagumo (FHN) oscillator model to real time units by comparing the FHN oscillation period of a single FHN oscillator $T = 2.56$ to the dominant frequency of an absence seizure at about $f = 3\,\text{Hz}$ [454, 455, 484, 485]. Therefore, the simulation time is converted to real time by 1 second $= 2.56/3 = 0.85$ simulation time units (see Sect. 4.5).

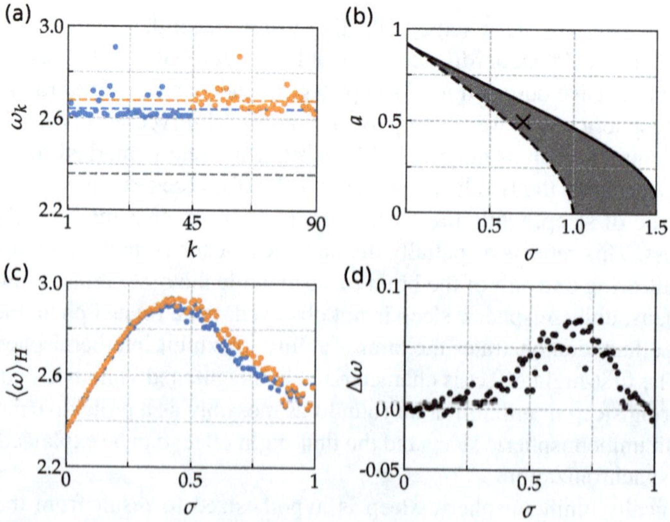

Fig. 4.33 Asymmetry scenarios in the brain network: **a** Mean phase velocity ω_k for each node k (dots) and spatially averaged hemispheric mean phase velocity $\langle\omega\rangle_H$ (dashed lines) for coupling strength $\sigma = \varsigma = 0.7$. The color code highlights the left (dark blue) and right (light orange) hemisphere. The gray dashed line at $\omega \approx 2.4$ denotes the mean phase velocity for the uncoupled system. **b** Numerically calculated critical coupling strength in the (σ, a) plane for the transition between incoherence and frequency synchronization using the average brain network (solid line, $\sigma = \varsigma$) and isolated hemispheres (dashed line, $\varsigma = 0$). The cross denotes the parameters ($\sigma = 0.7$, $a = 0.5$) used in panel (**a**). **c** Hemispheric mean phase velocities $\langle\omega\rangle_H$ as a function of the coupling strength $\sigma = \varsigma$, color code as in panel (**a**). **d** Difference between left and right hemispheric mean phase velocity $\Delta\omega$ as a function of the coupling strength $\sigma = \varsigma$. The difference assumes a maximum at $\sigma \approx 0.7$. Other parameters: $\epsilon = 0.05$, $a = 0.5$, $\phi = \frac{\pi}{2} - 0.1$. Figure taken from [415]

4.4.2.1 Dynamical Asymmetry and Partial Synchronization

As described in [415], dynamical asymmetries in the system emerge due to the inherent structural asymmetry in the empirical connectivity of the brain. This phenomenon can be quantified through the mean phase velocity ω of oscillators within each hemisphere, as depicted in Fig. 4.33a. The figure illustrates that the system remains desynchronized across all oscillators for the range of coupling strengths considered. However, it also reveals a significant difference in the average mean phase velocities $\langle\omega\rangle_H$ between the hemispheres, which constitutes the system's dynamical asymmetry (see Fig. 4.33a). Notably, the mean phase velocity $\langle\omega\rangle$ of the uncoupled system ($\varsigma = 0$) is indicated as a dashed gray line at $\omega \approx 2.4$, demonstrating a considerably lower value compared to the coupled system. This observation underscores that inter-hemispheric coupling increases the overall mean frequency of the system.

A specific region of intra- and inter-hemispheric coupling strengths exists where the system exhibits dynamical asymmetry. This region, determined numerically, is influenced by the excitation parameter (a) and the coupling strengths σ and ς. It is

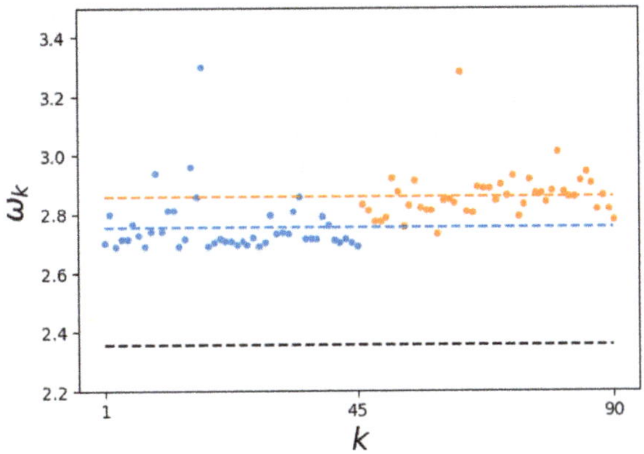

Fig. 4.34 Asymmetric mean phase velocity between left and right hemispheres showing the mean phase velocity ω_k for each oscillator k in the left (blue) and right (orange) hemispheres. A reproduction of Fig. 4.33a with different initial conditions. Other parameters as in Fig. 4.33. Figure adapted from [464]

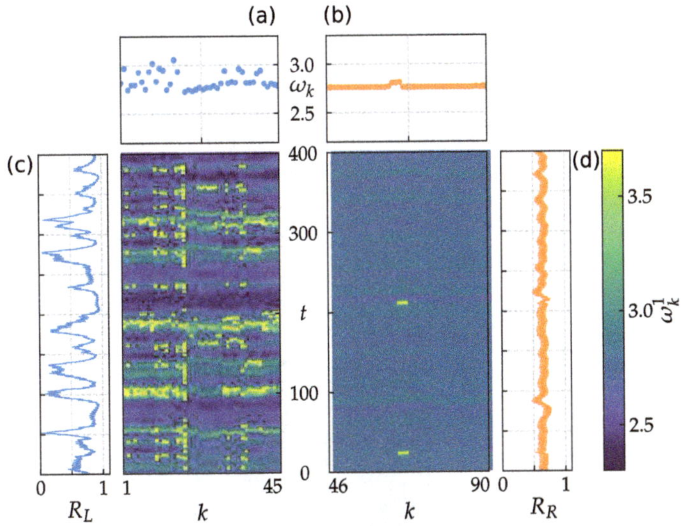

Fig. 4.35 Partial synchronization pattern for $\sigma = 0.70$, $\varsigma = 0.15$ with low and high degree of synchronization in the left (**a, c**) and right (**b, d**) hemisphere, respectively. **a, b** Mean phase velocity profiles ω_k. **c, d** inner panels: space-time plots of node-wise phase velocity ω_k^1 averaged over a single oscillation, outer panels: hemispheric Kuramoto order parameter $R_{L,R}$ as a function of time t. Other parameters as in Fig. 4.33. Figure adapted from [415]

highlighted as the gray area in Fig. 4.33b. The parameter set corresponding to the system illustrated in Fig. 4.33a is marked with a black X. Figure 4.33b and c further elucidate the behavior of dynamical asymmetry, showing its dependence on the intra-hemispheric coupling strength σ. Specifically, the parameter ranges in these figures are represented by a horizontal line in Fig. 4.33b, extending from $a = 0.5, \sigma = 0$ to $a = 0.5, \sigma = 1$.

A limitation of the mean phase velocity ω is that it represents a time-averaged value. While effective for illustrating oscillatory behavior with spatial resolution over long time periods, ω is less suited for capturing temporal variations in synchronization. To analyze how synchronization evolves over time, a time-resolved metric is necessary. The instantaneous mean phase velocity ω^1 provides this capability, offering insights into changes in synchronization both temporally and spatially (across oscillators). This metric is essential for identifying chimera states and partial synchronization (Fig. 4.34).

Figure 4.35 illustrates the partial synchronization behavior in the two-hemisphere oscillator network. The left hemisphere demonstrates desynchronization across both time and space, whereas the right hemisphere primarily exhibits synchronized behavior over time and among oscillators, with the exception of three oscillators displaying distinct dynamics. This behavior arises from a combination of relatively high intra-hemispheric coupling ($\omega = 0.7$) and weak inter-hemispheric coupling ($\varsigma = 0.15$). We will take a closer look at this eccentric behavior in the next section.

4.4.2.2 Eccentric Behavior

In the bottom-right plot of Fig. 4.36, three nodes exhibit what will hereafter be referred to as *eccentric* behavior. This plot presents the instantaneous phase velocity

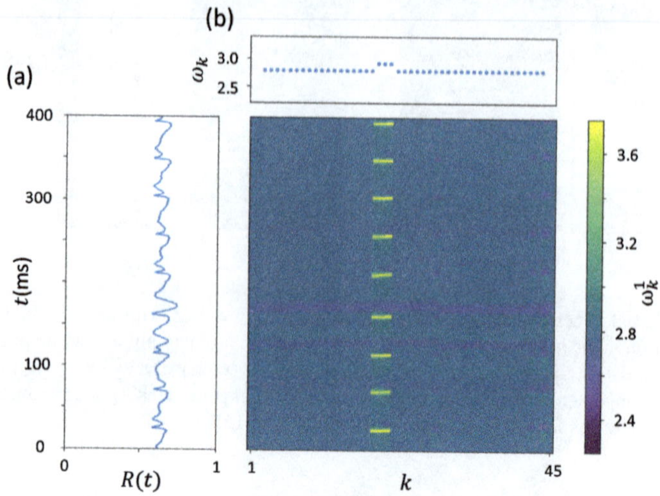

Fig. 4.36 Eccentric Behavior: The synchronization dynamics of the isolated right hemisphere ($\varsigma = 0$) reveal the eccentric behavior exhibited by nodes 19, 20, and 21. This behavior is evident in the Kuramoto order parameter (R), shown in panel (**a**), and the mean phase velocity (ω), displayed in the top of panel (**b**). In the bottom of panel (**b**), the space-time plots of node-wise phase velocity ω_k^1 averaged over a single oscillation is shown. The system as a whole demonstrates a high degree of synchronization, as indicated by a consistently elevated R. However, the three eccentric nodes deviate from this pattern, maintaining distinct dynamics that set them apart from the synchronized behavior of the other nodes. The number of nodes here is set from $46 \ldots 90$ to $k = 1 \ldots 45$. Other parameters as in Fig. 4.33. Figure adapted from [464]

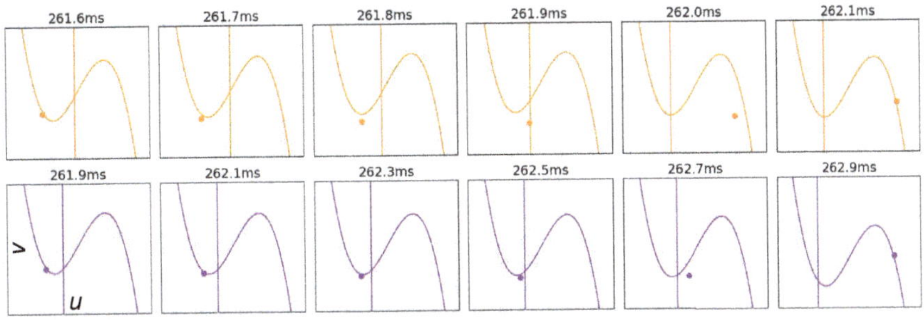

Fig. 4.37 Series of snapshots illustrating the effect of nullcline position and motion on the oscillator dynamics. Top: the nullcline moves the fixed point away from the eccentric node during the extra oscillation, allowing it to accelerate horizontally at the bottom of the oscillation. Bottom: reference node nullclines display less motion and their position locates the fixed point closer to the limit cycle of the node, decelerating the dynamics and resulting in relatively slow motion at the bottom of the oscillation. Figure adapted from [464]

ω_k^1 of each node over the final 500 ms of the simulation. The eccentric nodes, located in the right hemisphere, are characterized by quasi-periodic episodes of relatively high instantaneous phase velocity compared to other nodes. We will only focus on the right hemisphere of the brain and set the number of nodes in this section from $46\ldots90$ to $k = 1\ldots45$ as shown in Fig. 4.36.

In the state space representation, this behavior indicates that these nodes complete an additional oscillation relative to the rest of the system, termed the *reference* nodes. The eccentric nodes, identified as nodes 19, 20, and 21, correspond to the amygdala, hippocampus, and parahippocampal regions, respectively, as defined by the Automated Anatomical Labeling (AAL) atlas [416].

In the case of an isolated right hemisphere ($\varsigma = 0$), the eccentric behavior not only persists but becomes more regularly timed. This observation suggests that the desynchronized left hemisphere exerts a disruptive influence on the eccentric behavior. To investigate the mechanism underlying this phenomenon, we focus on the dynamics of the isolated hemisphere.

An analysis of the time evolution of oscillator trajectories, alongside their nullclines in the phase plane, reveals distinct differences in motion between the nullclines of eccentric nodes and those of reference nodes. Specifically, leading up to and during the additional oscillation, the nullclines of the eccentric nodes exhibit pronounced movement. This behavior is depicted in Fig. 4.37.

The position of a nullcline influences the location of the system's fixed point, where the dynamics slow down. Consequently, the marked nullcline motion of the eccentric nodes serves as a defining characteristic. As shown by the FitzHugh-Nagumo effective nullcline equations (Eqs. (4.41)), the nullcline position for each node is uniquely determined by its nonlocal inputs, further emphasizing the individualized dynamics of the eccentric nodes.

In Fig. 4.38, the phase plane trajectories of the three eccentric nodes are illustrated. The plot spans a time range that includes the period leading up to and a few oscillations

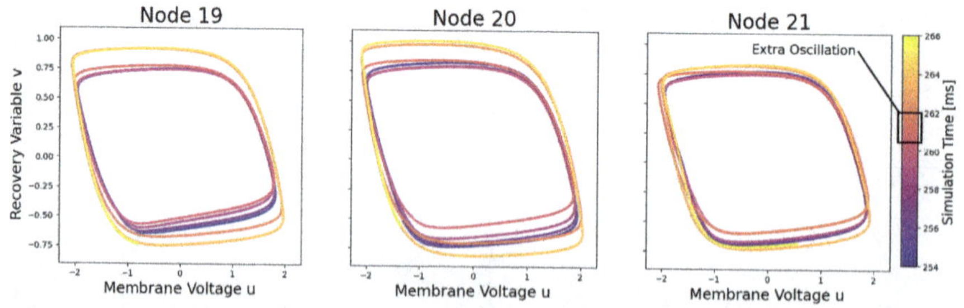

Fig. 4.38 Phase plane trajectories of the three eccentric nodes: This plot illustrates the phase plane trajectories of the three eccentric nodes over a time range encompassing the lead-up to, and several oscillations following, the extra oscillation. The timing of the extra oscillation is marked on the color bar with a black box. During this process, the eccentric nodes transition between a large and a small trajectory path, highlighting the distinctive dynamical behavior associated with their eccentricity. Other parameters as in Fig. 4.36. Figure adapted from [464]

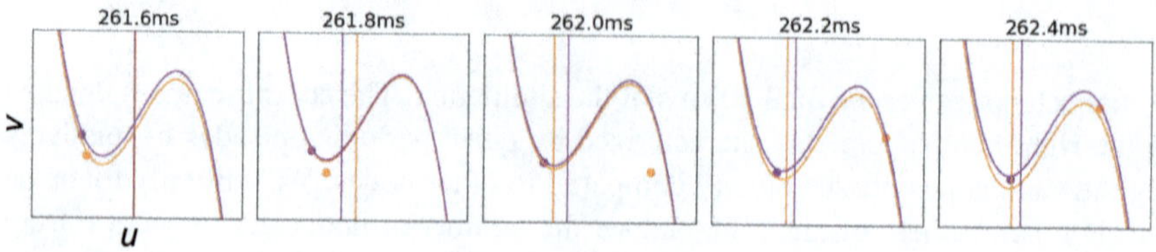

Fig. 4.39 Comparison of effective nullclines and fixed point positions during an eccentric node extra oscillation: This analysis compares the positions of the effective nullclines and the fixed point relative to the oscillator states during an extra oscillation of an eccentric node. The behavior of one of the three eccentric nodes is shown in orange, while a representative reference node is depicted in purple. For the reference node, the oscillator state remains close to the nullclines and the fixed point as it approaches the lower branch of the u-nullcline. This proximity results in relatively slow dynamics during this phase. In contrast, for the eccentric node, the nullclines and fixed point shift away from the oscillator state, enabling it to accelerate. This dynamic interaction between the nullclines, the fixed point, and the oscillator state constitutes the mechanism driving the extra oscillation characteristic of the eccentric nodes. Figure adapted from [464]

following the occurrence of an extra oscillation by the eccentric nodes. The timing of the extra oscillation is marked on the color bar by a black box. During this process, the eccentric nodes exhibit a transition between large and small trajectory paths, highlighting their distinctive dynamical behavior.

In Fig. 4.39, a comparison is made between the positions of the effective nullclines and the fixed point relative to the oscillator states during an eccentric node's extra oscillation. An eccentric node is depicted in orange, while a representative reference node is shown in purple. For the reference node, the oscillator state remains proximate to the nullclines and fixed point as it approaches the lower branch of the u-nullcline. This proximity results in relatively slow dynamics at that point in time. Conversely, for the eccentric oscillator, the nullclines and fixed point shift away from the oscillator state, enabling it to accelerate. This dynamic interaction explains the occurrence of the extra oscillation in the eccentric nodes.

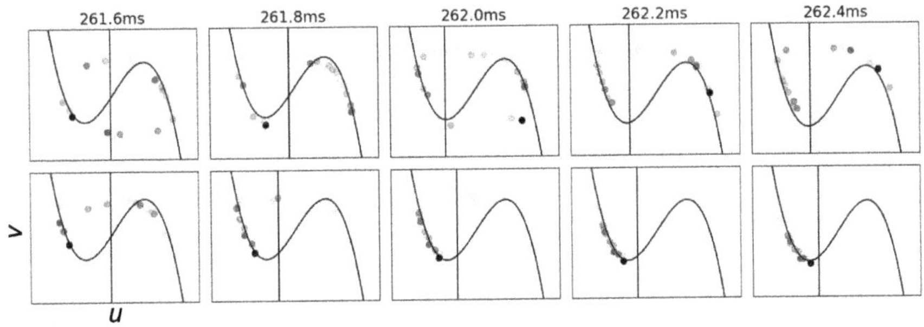

Fig. 4.40 Illustration of the relation between connection strength and relative position in phase space for eccentric and reference nodes: This figure depicts the connection strength and phase space positions of typical eccentric (top row) and reference (bottom row) nodes. The nodes of interest (black dots) correspond to those in Fig. 4.39 and are shown here alongside their effective nullclines. Nonlocal connection strengths, as defined by the adjacency matrix, are represented by opacity (transparent = no connection, opaque = strongest connection). In keeping with the color scheme of previous figures, purple denotes reference nodes, while orange represents eccentric nodes. Top: The eccentric node has strong connections to reference nodes located antipodally to it along the v-axis. This alignment results in a significant nonlocal input that accelerates the eccentric node in the positive u-direction. Bottom: In contrast, the representative reference node is clustered with the nodes it is most strongly connected to. This configuration yields relatively weak nonlocal inputs, leading to more uniform and less eccentric dynamics compared to the top scenario. Figure adapted from [464]

Figure 4.40 shows the relationship between connection strength and the relative position in phase space for typical eccentric (top row) and reference (bottom row) nodes. The nodes of interest (black dots) are the same as those in Fig. 4.39, and they are depicted here with their effective nullclines. Nonlocal connection strengths, as determined by the adjacency matrix, are represented by opacity, with transparent indicating no connection and opaque indicating the strongest connection. As in Fig. 4.39, orange and purple represent eccentric and reference nodes, respectively.

In Fig. 4.40a, the eccentric node demonstrates strong connections to reference nodes located antipodally along the v-axis, which results in an acceleration in the positive u-direction. In Fig. 4.40b, the representative reference node is clustered with nodes to which it has the strongest connections, leading to relatively weak nonlocal inputs and more stable, uniform dynamics.

4.4.3 Pacemaker Role of Eccentric Network Nodes

The eccentric synchronization behavior is characterized by the tendency of three network nodes to pseudo-periodically perform an additional oscillation relative to their neighboring nodes. In this section, several insights into the system's dynamics are developed, and a relationship between its topology and dynamics is identified.

In Sect. 4.4.3.1, the frequency spectrum of the sub-network consisting of all non-eccentric nodes is analyzed. While this approach provides valuable information, it requires the system to be operational for predictions, which limits its practical utility. More critically, we focus on examining the root mean square (RMS) input changes over extended timescales in Sect. 4.4.3.2, specifically investigating the variance in the inputs to the eccentric nodes compared to the other nodes in the system.

Additionally, brain regions are grouped in two distinct ways: by spectral clustering and by their phase relationships. Spectral clustering is a topological measure derived from the empirical connectivity matrix, whereas phase clustering represents the system's dynamic behavior. Comparing these two clustering methods in Sect. 4.4.3.3 offers potential insights into the mechanisms driving the dynamics of the eccentric nodes, shedding light on the interplay between network topology and dynamics.

4.4.3.1 Natural Frequencies of Populations

The Fourier transform of each node's oscillatory time series provides a detailed representation of the dominant frequencies present within the signal, referred to as the signal's power spectrum or frequency spectrum (see Sect. 4.3.5.6). Given that both FitzHugh-Nagumo (FHN) state variables–u (membrane voltage) and v (recovery variable)–oscillate at approximately the same frequency, their power spectra exhibit qualitative similarities. Consequently, either variable can be used for the Fourier transformation. In this analysis, the u-state is utilized taking advantage of the Fast Fourier Transform (FFT) introduced in Sect. 4.3.5.6.

To compute the power spectrum for a population of oscillators, the Fourier transform is first applied to the time series of individual nodes, followed by averaging the resulting spectra across all nodes within the population. In cases where oscillators exhibit heterogeneous behavior, variations in their individual frequency spectra tend to smooth or flatten the population's average power spectrum. Conversely, if dominant frequencies are shared across oscillators in the population, these frequencies manifest as prominent peaks in the average power spectrum.

Given that the majority of oscillators in the system exhibit frequency synchronization, it is anticipated that certain dominant frequencies will emerge, effectively governing the overall dynamics of the system. In Fig. 4.41a, the power spectrum of an isolated FitzHugh-Nagumo (FHN) oscillator is compared with those of randomly selected nodes from the connected reference and eccentric populations. By contrast, Fig. 4.41b focuses exclusively on a subset of nodes within the separated reference population, showing that all nodes in this group exhibit qualitatively similar frequency spectra.

Given this consistency, we investigate the average power spectra of separated populations in greater detail. Figure 4.42a displays two characteristic examples of average power spectra observed in separated reference populations, compared to the distribution of an isolated FHN oscillator. While higher harmonics are present, they are excluded from this analysis.

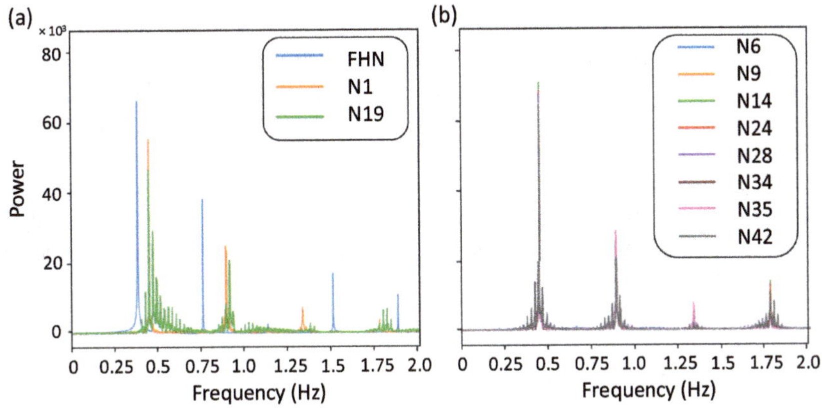

Fig. 4.41 Prototypical examples of the power spectra for individual oscillators calculated by the Fast Fourier Transform (FFT) introduced in Sect. 4.3.5.6: In panel (**a**), the spectrum of an isolated FitzHugh-Nagumo (FHN) oscillator is compared with those of randomly selected nodes from both the connected reference and eccentric populations. In panel (**b**), a subset of nodes from the separated reference population is shown. All nodes within the reference population exhibit qualitatively similar frequency spectra. Figure adapted from [465]

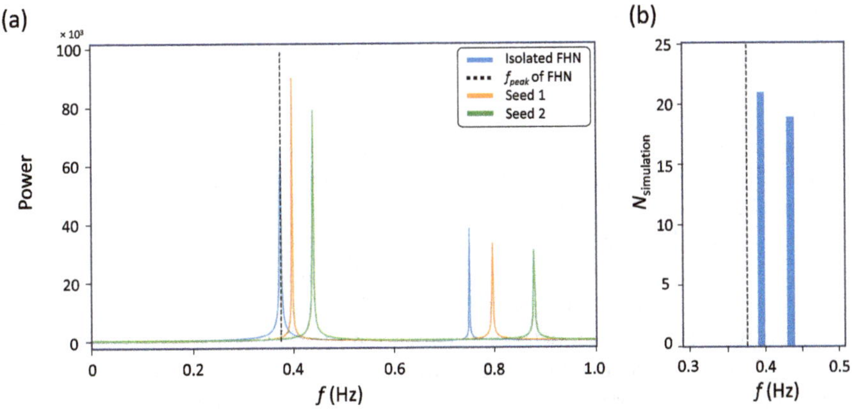

Fig. 4.42 The average power spectra of separated reference populations are presented for 40 different initializations. In panel (**a**), two representative examples of average power spectra from separated reference populations are shown, compared to the distribution of an isolated FHN oscillator. While higher harmonics are present, they are excluded from this analysis. In panel (**b**), a histogram illustrates the distribution of peak frequencies f_{peak} across the reference population for all forty sets ($N_{\text{simulation}}$) of initial conditions. The dashed black line indicates the peak frequency f_{peak} of an isolated FHN neuron. Notably, there is a clear difference between the frequency peak f_{peak} of the isolated oscillator and those of the two characteristic reference populations. Figure adapted from [464]

Figure 4.42b provides a histogram illustrating the distribution of peak frequencies within the reference population across forty different sets of initial conditions. The dashed black line indicates the frequency peak of an isolated FHN neuron, highlighting the discrepancy between this isolated frequency peak and those of the two characteristic reference populations. This difference underscores the influence of network coupling on the frequency dynamics of the reference populations.

It was determined that the eccentric behavior can be anticipated by analyzing the frequency spectrum of the sub-network composed exclusively of non-eccentric nodes. While this finding is intriguing and provides valuable insight into the system's dynamics, it is important to note that this approach requires the system to already be operational to make such predictions. Consequently, its practical utility is limited, as it does not offer a means for a priori identification of eccentric behavior.

4.4.3.2 Root Mean Square of Inputs

The root mean square (RMS) of the total nonlocal input to each oscillator provides a quantitative measure of the input magnitude, independent of its polarity (positive or negative). In this section, the RMS is computed for each individual oscillation and is represented as I^1_{RMS} as introduced in Eq. (4.56) of Sect. 4.3.5.5. An example of this calculation is depicted for a subset of nodes in Fig. 4.43.

Figure 4.43a shows the instantaneous mean phase velocity of the oscillators, with the rows corresponding to the eccentric nodes identifiable by yellow coloration, marking the periods during which they undergo an additional oscillation. Figure 4.43c presents the root mean square (RMS) of the input to the eccentric nodes (orange) and three reference nodes (purple). It is evident that the RMS input for the eccentric nodes increases and peaks prior to the extra oscillation, while the reference nodes exhibit a flat RMS input progression. Although only three reference nodes are shown, this trend holds for all other nodes in the system.

Figure 4.43b illustrates the root mean square (RMS) input deviation as introduced in Eq. (4.57), where zero represents the average RMS input over the entire simulation. Both the RMS and mean phase velocity metrics are calculated on a per-oscillation basis, enabling a direct comparison between the two. The data clearly shows that the RMS input peak occurs before the extra oscillation and drops below the average afterward. The analysis reveals that the RMS inputs consistently peak prior to the onset of an extra oscillation and subsequently fall below the average following its occurrence. Furthermore, eccentric nodes exhibit significantly greater variation in their RMS deviation compared to other nodes. The observed connection between fluctuations in RMS input and the changes in frequency of the eccentric nodes provides a compelling explanation for the mechanism underlying the eccentric behavior.

Figure 4.44a illustrates a positive monotonic relationship between I^1_{RMS} and ω_1 for the selected nodes. To validate this observation and generalize it across all nodes, the Pearson correlation coefficient r_p was computed for each node, utilizing all oscillation data as measurements. The resulting correlation coefficients are presented in Fig. 4.44b, with non-significant correlations ($r_p < 0.05$) highlighted in red. These results indicate a statistically meaningful relationship between the root mean square (RMS) input to a node and its frequency, represented by the mean phase velocity.

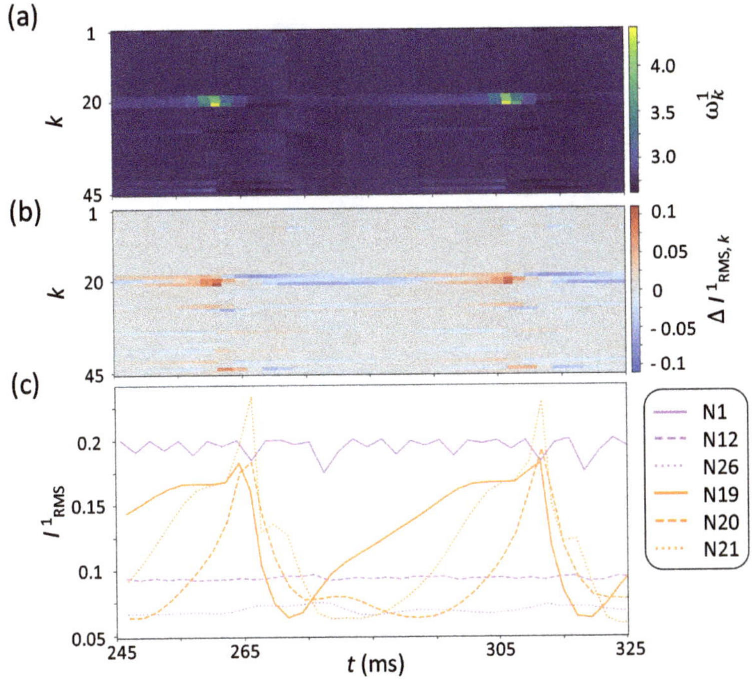

Fig. 4.43 Root mean square of inputs: Top panel (**a**): The instantaneous mean phase velocity of the oscillators is depicted, with rows corresponding to individual oscillators. The eccentric nodes are distinguished by their yellow coloration, indicating intervals when they undergo an additional oscillation. Middle panel (**b**): The RMS input deviation is plotted (see Eq. (4.57)), where a value of zero represents the average RMS input over the entire simulation. This metric is calculated on a per-oscillation basis, enabling a direct comparison with the mean phase velocity. The RMS input deviation demonstrates that the input peak precedes the extra oscillation, followed by a drop below the average level post-oscillation. This analysis underscores the temporal relationship between RMS input dynamics and the extra oscillatory behavior of the eccentric nodes. Bottom panel (**c**): The root mean square input I^1_{RMS} is shown for the eccentric nodes N19, N20, N21 (orange) and three representative reference nodes N1, N12, N26 (purple). The RMS input for eccentric nodes builds progressively and peaks just before the extra oscillation occurs, contrasting sharply with the flat progression observed for the reference nodes. This pattern holds true across all non-eccentric nodes in the system, although only three are explicitly shown. Figure adapted from [465]

Summarizing, we have identified the mechanism underlying the extra oscillations of the eccentric nodes. These oscillations are driven by nodes exhibiting significant root mean square (RMS) input changes over extended timescales. This behavior is characterized by a pronounced peak in the RMS input leading up to the extra oscillation, followed by a subsequent decline. As a result, eccentric nodes are distinct from other nodes in the system, demonstrating notably higher variance in their inputs.

The identified mechanism can be described from two perspectives: through the temporal dynamics of the input time series or via the motion of the nodes' nullclines

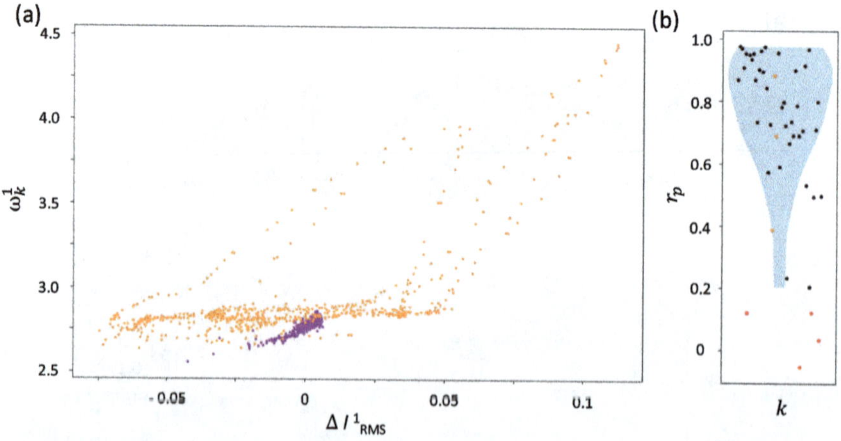

Fig. 4.44 Relation between input and phase velocity: In panel (**a**), a scatter plot of the instantaneous mean phase velocity ω_k^1 against the single oscillation root mean square input strength I^1_{RMS} is shown. The same nodes are plotted as in Fig. 4.43, with the same color correspondence (orange = eccentric, purple = reference). Each point represents a single oscillation. Panel (**b**): Pearson correlation coefficients r_p for the relationship between ω_k^1 and I^1_{RMS}. Correlations are calculated individually for each node, taking the individual oscillation values as measurements. The eccentric node correlations are shown in orange. Red points represent nodes whose correlations are not significant and are not included in the distribution envelope (light blue area). Figure adapted from [464]

within the phase plane. Both approaches provide complementary insights into the system's behavior and are illustrated in this section.

4.4.3.3 Spectral Clustering and Dynamic Clustering

Spectral clustering can be interpreted from several perspectives, as outlined by Luxburg [459] and introduced in Sect. 4.3.5.7. One approach views it as identifying clusters within a similarity matrix such that points within clusters have high similarity (high weights), and points between clusters exhibit low similarity (low weights). This corresponds to a graph cut of the similarity matrix, optimizing the partitioning of the graph.

Another interpretation involves random walks on the graph. Spectral clustering identifies groupings such that a random walker remains within clusters for extended periods, rarely transitioning between clusters. This interpretation is closely tied to the commute time (or resistance distance) between network nodes. Unlike the shortest path distance, the commute time incorporates multiple paths between nodes and their respective lengths, effectively measuring the information bandwidth between nodes. Given the established significance of nonlocal inputs in predicting eccentric node

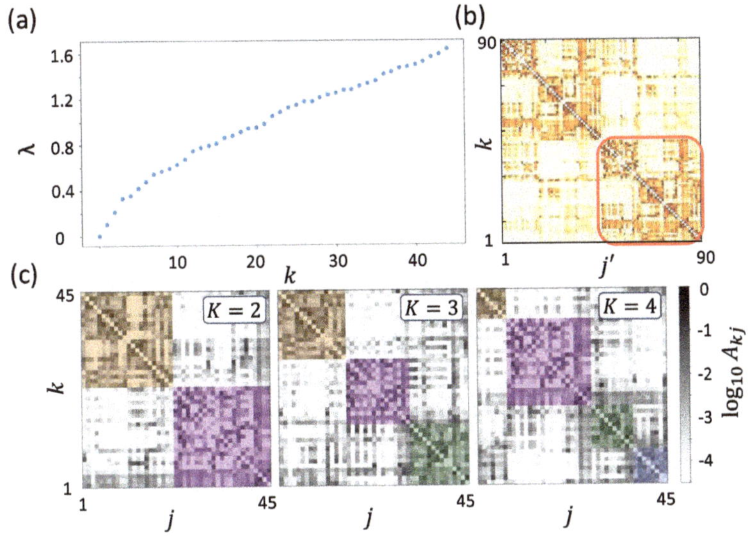

Fig. 4.45 Spectral clustering: Panel (**a**) depicts the eigenvalue λ distribution of the right hemisphere's adjacency matrix ($k = 1 \ldots 45$). In panel (**b**), the adjacency matrix for both hemispheres ($k = 1 \ldots 90$) are shown (see also Fig. 4.29a), where the red border marks the right hemisphere. Panel (**c**) shows the comparison of the sorted adjacency matrices of the right hemisphere. The clusters are denoted by color, with the cluster containing the eccentric nodes always shown in orange. Left: $K = 2$. Middle: $K = 3$. Right: $K = 4$. Figure adapted from [465]

behavior, the commute time perspective provides a particularly relevant framework for analyzing spectral clustering results.

Figure 4.45 displays the eigenvalue distribution of the network's adjacency matrix. The eigengap heuristic, which identifies the first significant gap between consecutive eigenvalues (excluding zero eigenvalues), suggests that the optimal number of clusters is $K = 2$, as the first prominent gap occurs between the second and third nonzero eigenvalues. Nevertheless, results for $K = 3$ and $K = 4$ are also analyzed for comparison.

The sorted adjacency matrices corresponding to the top clusterings for $K = 2, 3, 4$, are illustrated in Fig. 4.45. The clusters are denoted by color, with the cluster containing eccentric nodes consistently shown in orange. Clustering performance is evaluated visually to determine how well the block structure of the adjacency matrix is preserved, as metrics for comparing clusterings across different K values are not directly applicable.

For $K = 2$, the block structure of the sorted adjacency matrix is most distinct, with minimal strong connection weights between clusters (i.e., outside the colored blocks). At $K = 3$, the block structure is less defined, with increased connections between the green cluster and the others. For $K = 4$, inter-cluster connections become more pronounced, particularly between the green and blue clusters and the orange and blue clusters.

While spectral clustering is based solely on the adjacency matrix, network nodes can also be grouped dynamically by their phase relationships. This process, referred to as dynamical clustering, categorizes oscillators based on their relative phases. The relative phase ϕ_r of each oscillator is calculated by comparing its phase to a baseline node at each time point. The choice of the baseline node does not influence the results, as it merely shifts the relative phase distribution.

The relative phase distributions of individual nodes reveal that reference nodes generally maintain a stable relative phase, whereas eccentric nodes exhibit a broader distribution due to their variable frequencies. Node 1 is chosen as the baseline for calculating relative phases, resulting in a constant relative phase of zero for this node.

In Fig. 4.46, the relative phase distributions of eccentric (orange), in-phase reference (purple), and antiphase reference (green) nodes are compared between dynamical clustering and spectral clustering results ($K = 3$). Eccentric nodes display a wide phase distribution, whereas reference nodes form multimodal distributions. The division between in-phase and antiphase reference groups is marked by a vertical red line

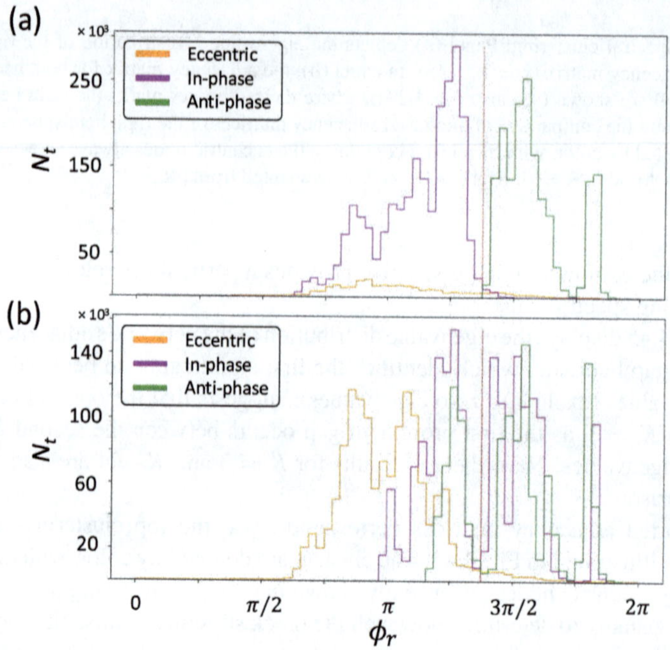

Fig. 4.46 Comparison between dynamic and spectral clustering: The relative dynamic phase clustering (**a**) and the spectral clustering (**b**) versus the number of time points N_t in the simulations. The relative phase ϕ_r distributions of eccentric (orange), in phase reference (purple) and antiphase reference (green) nodes, compared between dynamic (**a**) and spectral (**b**) clustering ($K = 3$) methods. Eccentric node relative phases are distributed over a wide range. Reference node phases form a multimodal distribution. The division between in phase and antiphase reference node groups is denoted by a vertical red line at $11\pi/8$. Node 1 (the relative reference node) is excluded from this plot for clarity. Figure adapted from [465]

at $11\pi/8$. Node 1 is excluded from this plot for clarity. This analysis highlights the distinct dynamic and topological features of eccentric nodes compared to reference populations.

Summarizing, in this section brain regions have been classified using two distinct approaches: spectral clustering, which is based on the empirical connectivity matrix, and clustering by phase relationships. Spectral clustering serves as a topological measure, identifying clusters that optimize the block structure of the connectivity matrix. In contrast, phase clustering captures the dynamic interactions within the system by grouping regions based on their phase relationships.

A comparative analysis of these two clustering methods reveals a relationship between the system's underlying topology and its dynamic behavior. However, the precise nature of this relationship remains ambiguous, warranting further investigation to elucidate the mechanisms linking structural connectivity and dynamical patterns.

4.4.4 Summary

The field of complex networks seeks to understand and describe a variety of natural and technological systems by analyzing their statistics, topology, and dynamics. This approach aims to elucidate how the overall behavior of a network arises from the interactions between its components. Among these systems, the brain stands out as a quintessential example of a complex network, given its significance to medicine, psychology, and broader human understanding.

This section has investigated the dynamics of a mesoscopic-scale brain model. The model incorporates an empirical structural connectivity matrix derived from diffusion tensor imaging and employs a FitzHugh-Nagumo oscillatory model to represent the mean-field neural activity of individual brain regions. The regions are defined based on the Automated Anatomical Labeling atlas. Previous work with this model identified three brain regions exhibiting eccentric synchronization behavior, characterized by a pseudo-periodic extra oscillation in these network nodes (vertices) relative to their neighbors.

While this section does not conclusively identify the topological features that define the network's eccentric brain regions, it has advanced the understanding of system dynamics and uncovers a relationship between network topology and dynamics. Key findings include the following:

Firstly, eccentric behavior can be anticipated from the frequency spectrum of the sub-network comprising all non-eccentric nodes. However, as this prediction requires the system to already be running, its practical utility is limited.

Secondly, the mechanism driving the extra oscillations in eccentric nodes is established. These oscillations occur in nodes whose root mean square input exhibits significant variations over long time scales, characterized by a peak preceding the oscillation and a subsequent drop. Eccentric nodes thus demonstrate a greater variance in their inputs compared to other nodes. This mechanism is described both through input time series and the motion of nullclines in the phase plane, as illustrated herein.

Finally, a method for reducing the network while preserving eccentric behavior has been developed. This reduction involves removing nodes whose inputs to their neighbors fall below a specified threshold. Notably, the eccentric nodes in the reduced network differ from those in the full network. This simplified network may facilitate future studies in identifying definitive links between topology and dynamics.

Additionally, the brain regions have been grouped using two approaches: spectral clustering, which leverages the empirical connectivity matrix, and clustering based on phase relationships. Spectral clustering is a topological method that identifies clusters by maximizing the block structure of the connectivity matrix. Phase clustering, on the other hand, reflects the system's dynamics. Comparing these two clustering methods reveals a relationship between the system's topology and dynamics, though the precise nature of this relationship remains to be determined.

4.5 Synchronization and Epileptic Seizures

Synchronization within the human brain is fundamental for normal physiological functioning [486]; however, it is also intricately linked to seizures, the hallmark symptom of epilepsy [487–489]. Epilepsy, a complex neurological disorder, is increasingly conceptualized as a network disease [490]. Gaining a deeper understanding of how the topology of epileptic networks influences seizure initiation and termination is of paramount importance. In this context, we employ complex networks of coupled oscillators to simulate synchronization phenomena observed in the human brain. Specifically, we use FitzHugh-Nagumo oscillators, a well-established model for neural dynamics [442]. By integrating empirical structural brain connectivity data derived from human subjects as the coupling matrix, we observe spontaneous episodes of heightened synchronization resembling epileptic seizures.

To investigate the network properties contributing to these pathological synchronization events, we simulate dynamics across various paradigmatic network topologies. These include networks with links rewired in a small-world fashion, networks with fractal connectivity patterns, and networks where uniform coupling weights are replaced by empirical weights derived from diffusion-weighted magnetic resonance imaging. Furthermore, we evaluate the impact of global network properties–such as the average clustering coefficient and the mean shortest path length–on the dynamics of seizure-like synchronization phenomena. Our model simulations are compared with electroencephalographic (EEG) recordings of human epileptic seizures to strengthen the validity of our findings. Understanding the interplay between network structure and the resulting complex synchronization dynamics is crucial for advancing the study of seizure mechanisms.

This section delves into patterns of partial synchronization in networks of FitzHugh-Nagumo oscillators (see Sect. 4.3). Specifically, we explore the topological conditions necessary for the emergence of synchronization patterns in the system's dynamics. Long-term simulations of oscillatory dynamics are conducted across several paradigmatic network structures, including random networks, regular nonlocally coupled ring networks, fractally connected ring networks, and small-world networks with varying rewiring probabilities. Among these configurations, small-world networks with intermediate rewiring probabilities most closely replicate the dynamics observed with empirical structural connectivity. For other network topologies, either no spontaneous epileptic-seizure-related synchronization phenomena emerge, or the system remains consistently highly synchronized throughout the simulation. This suggests that a topology balancing regularity and randomness is optimal for the spontaneous initiation and termination of seizure-like synchronization episodes. Importantly, systems with empirically derived structural connectivity exhibit synchronization dynamics that closely mirror those observed during human epileptic seizures.

The objective of this section is to elucidate the influence of neural network coupling structure on the emergence of seizure-like synchronization phenomena. For readability, the term “seizure” will be used interchangeably with epileptic-seizure-related synchronization phenomena in the following discussion. To achieve this goal, we analyze a range of network topologies relevant to neuroscience to understand how specific structures facilitate spontaneous, prolonged synchronization in predominantly desynchronized systems.

This section is organized as follows: In Sect. 4.5.1, we provide an overview of the physiological and clinical aspects of epilepsy. Section 4.5.2 describes the model parameters introduced in Sect. 4.3 and presents the simulation results across various network topologies, emphasizing the influence of coupling structure on the spontaneous emergence of synchronization phenomena. In Sect. 4.5.3, we compare these simulated seizure-like events with electroencephalographic (EEG) recordings of generalized epileptic seizures in humans. Section 4.5.4 concludes with a summary of the key findings. For more detailed information, we recommend referring to the original publication [403] on which this section is based.

4.5.1 Epilepsy

Epilepsy is a neurological disorder affecting nearly 70 million individuals globally [491]. It is characterized by seizures, defined as a “transient occurrence of signs and/or symptoms due to abnormal excessive or synchronous neuronal activity in the brain” [492]. Seizures are broadly categorized into generalized and focal types. Generalized seizures involve nearly the entire brain [455], while focal seizures are confined to a specific region. Generalized seizures are further classified based on their observable symptoms, such as muscle contractions, limb shaking, spasms, or sudden loss of muscle activity.

A notable type of generalized seizure, distinguished by the absence of gross muscular activity, is the absence seizure. These seizures are particularly suitable for EEG studies as they minimize movement-related artifacts. During an absence seizure, an individual may lose consciousness, abruptly halt any ongoing activity, remain still, blink repeatedly, and then resume previous behavior immediately after the seizure ends. Notably, the person is typically unaware of the event and may have no memory of it due to the transient loss of consciousness.

The development of the concept of the epileptic network [493–495] has been significantly influenced by network-theoretical approaches. An epileptic network consists of anatomically and functionally interconnected cortical and subcortical regions of the brain. Seizures may originate, propagate, and terminate through components of these networks, which otherwise sustain normal brain activity during seizure-free intervals [493].

To advance the understanding of epileptic networks and their temporal dynamics, the study of seizure dynamics can benefit from research on synchronization in complex networks. This area is of significant interest due to its broader relevance to synchronization phenomena in both natural and technological systems [275, 279, 496–498]. It is well established that the propensity for synchronization in a system depends on the dynamics of individual oscillators, the nature of their coupling, and the network's structural connectivity. This section focuses on the structural connectivity of neural networks and its role in seizure emergence. Specifically, we investigate synchronization events that are (i) generalized (affecting the entire network), (ii) of prolonged duration relative to system dynamics, and (iii) self-initiating and self-terminating. While external and internal stimuli play critical roles in triggering seizures [499–501], our primary goal is to identify network structures that inherently promote such events.

Prior studies have highlighted characteristic changes in network properties associated with generalized seizures [502, 503] and emphasized the critical role of coupling topology in shaping seizure dynamics [504–510]. For networks of neurons modeled using the FitzHugh-Nagumo paradigm, seizure-like dynamics have been explored in two primary topologies: empirical structural brain connectivity derived from diffusion-weighted magnetic resonance imaging, and mathematically constructed modular fractal networks [420].

Moreover, the role of partial synchronization phenomena [285, 287, 511–513] has been investigated in the context of seizure initiation [514] and termination [515]. These studies collectively underscore the importance of network topology in understanding the mechanisms underlying epileptic seizures.

4.5.2 *Epileptic-Seizure-Related Synchronization Phenomena*

In order to gain deeper insight into the interplay of dynamics and network topology, especially regarding the occurrence of seizures, we consider different networks. In this section, we present simulations of the dynamics described by Eq. (4.42) in Sect. 4.3.4 for various network topologies A_{kj}.

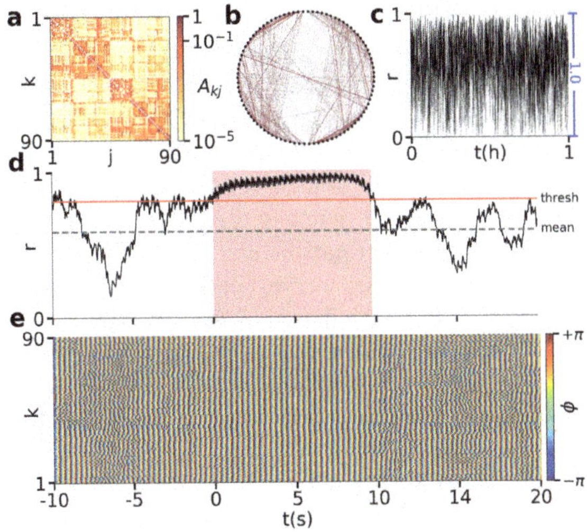

Fig. 4.47 Epileptic-seizure-like synchronization phenomena in a FitzHugh-Nagumo network with empirical connectivity. **a** Weighted adjacency matrix obtained from empirical human diffusion tensor imaging (DTI), averaged over 20 subjects [417]. Indices $k, j = 1, \ldots, 45$, and $k, j = 46, \ldots, 90$ label the left and right hemisphere, respectively. The dense intra-hemispheric connectivities (first and second diagonal blocks) and the sparse inter-hemispheric connectivities can be clearly seen. **b** Schematic plot of the network structure. The left (right) semicircle corresponds to the left (right) hemisphere, the nodes are numbered clockwise sequentially $1, \ldots, 90$ starting from the bottom of the circle. The link thickness is proportional to the weights A_{kj}. **c** Global Kuramoto order parameter r vs time (time interval one hour). It fluctuates strongly in the range from 0 to almost 1 (blue vertical bar). **d** Global Kuramoto order parameter r vs time relative to the onset of a seizure (time interval 30 s). The horizontal dashed grey line denotes the time average over 3 h. The horizontal full red line marks the threshold of $r = 0.8$. If $r > 0.8$ for more than 8 s, we define this as a seizure (pink shaded region). **e** Space-time plot of the dynamical phases corresponding to panel (**d**). The left (right) hemisphere is shown in the lower (upper) half. Simulation parameters: $a = 0.5$, $\varepsilon = 0.05$, $\varphi = \frac{\pi}{2} - 0.1$, $N = 90$, $\sigma = \varsigma = 0.6$. Figure taken from [403]

4.5.2.1 Empirical Brain Network

First, we consider an empirical structural brain network. In Fig. 4.47b, the network topology is schematically represented on a ring, where the left and right hemispheres correspond to the left and right half-circle, respectively. Most links are intra-hemispheric, and only very few inter-hemispheric connections can be seen. Note that the width of the links in Fig. 4.47b is proportional to their weight.

Following previous studies on synchronization phenomena [400, 415, 420], we fix the parameters $a = 0.5$, $\varepsilon = 0.05$, and $\varphi = \frac{\pi}{2} - 0.1$ in our simulations for all topologies. The coupling strength σ is chosen such that it is as high as possible while still avoiding full synchronization for long simulations ($\approx 10\,000$ time units). For the empirical connectivities, we choose $\sigma = \varsigma = 0.6$. In order to compare our simulations with real data (EEG recordings of absence seizures; see Sect. 4.5.3), we transform the dimensionless time units of the FitzHugh-Nagumo (FHN) oscillator model to real time units by comparing the FHN oscillation period of a single FHN oscillator $T = 2.56$ to the dominant frequency of an absence seizure at about $f = 3$ Hz [454, 455, 484, 485]. Therefore, the simulation time is converted to real time by 1 second $= 2.56 \cdot 3 = 7.68$ simulation time units.

The results of the simulation are shown in Fig. 4.47c, d, e. In panels (c) and (d), we show the global Kuramoto order parameter $r(t)$, which measures the degree of synchronization. Panels (c) and (d) also reveal periods of very high and of very low synchronization of the system as a function of time, varying in a range from 0 to almost 1 (panel (c)). The temporal average of the order parameter $\langle r \rangle$ (horizontal dashed grey line in (d)) and its standard deviation δ are given by $\langle r \rangle \pm \delta = 0.59 \pm 0.21$ for the full simulation of 164 min. We define a threshold of high synchrony as $r_{\text{th}} = \langle r \rangle + \delta = 0.8$ (horizontal red line in (d)). This threshold value is kept at 0.8 for all simulations in this section, even if the mean and the standard deviation differ for other topologies. In the simulation presented in Fig. 4.47, the order parameter is found to be in high synchrony with $r > 0.8$ during 17% of the simulation time. Only if the synchronization remains above the threshold for at least 8 s, we define this time interval as a seizure.

In Fig. 4.47d, the order parameter is shown versus time for one exemplary seizure. Approximately 6 s prior to the start of the seizure, the order parameter drops to a low value of $r \approx 0.2$. Such an apparent desynchronization can often be observed prior to the onset of focal epileptic seizures [452, 514, 516, 517], and it is not yet clear whether it can be observed prior to generalized seizures in humans, although it has been observed in animal studies [518]. The order parameter then increases above $r > 0.8$ (onset of seizure) and remains in high synchrony for almost 10 s. The seizure interval is shown as a pink shaded region; it marks the time of high synchronization without interruption. In the full simulation of 164 min., 11 seizures were detected, giving an average of 4 seizures per hour. Their average duration was 10.8 s, with a standard deviation of 1.3 s. In Fig. 4.47e, the dynamic phases of the oscillators are shown as space-time plot for the same time interval as in (d). The lower half of the panel corresponds to the left hemisphere ($k \leq 45$), the upper half to the right hemisphere ($k \geq 46$). Since both hemispheres synchronize strongly, this resembles a generalized seizure.

4.5.2.2 Random Surrogate Network

In order to gain deeper insight into the interplay of dynamics and network topology, especially regarding the occurrence of seizures, we consider different artificially constructed networks. First, we study a surrogate network with all links of the empirical connectivity matrix randomly rewired (Fig. 4.48). Note that the set of weights of all links is the same as in Fig. 4.47. However, the graph in panel (b) looks much denser due to the larger number of inter-hemispheric long-range connections. The simulation of the global Kuramoto order parameter $r(t)$, for one hour, shows that, on average,

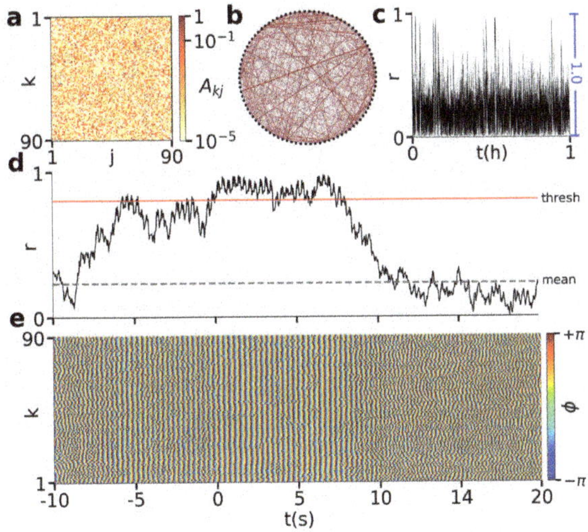

Fig. 4.48 Same as Fig. 4.47 with all links randomly rewired. Simulation parameters as in Fig. 4.47. Figure taken from [403]

the system is less synchronized, see Fig. 4.48c. Note that there are very short intervals of strong synchronization despite the very low average degree of synchronization. However, high synchrony $r > 0.8$ is observed only in 1% of the simulation time, and one such event is shown in Fig. 4.48d. The global synchronization at $t \approx 7$ s is similar to the global synchronization in Fig. 4.47d at $t \approx 7$ s. However, the dynamic phases of the space-time plot in Fig. 4.48e appear less coherent since the connectivity of neighboring nodes is, on average, much smaller in the random network, preventing local synchronization. Since $r > 0.8$ never holds for more than 8 s, according to our definition, not a single seizure is found in the simulation. The average degree of synchronization $\langle r \rangle = 0.23$ decreases significantly as compared to the one seen for the empirical connectivity (Fig. 4.47). The links in this random network are the same as in the empirical network. Therefore, also the average node strength is equal, see Table 4.6. However, the weighted clustering coefficient [519] and the average weighted shortest path length [520] decrease by 41 and 42%, respectively. It is difficult to assess which network measure is an appropriate characteristic quantity related to the decrease in the average degree of synchronization. We address this question later in Sect. 4.5.2.5.

Interestingly, by increasing σ to 0.7 in the random surrogate network, the system attains an average Kuramoto order parameter of $\langle r \rangle = 0.60$, similar to the empirical connectivity. However, even if high synchrony $r > 0.8$ is observed 47% of the time (see Fig. 4.49), only 4.7 seizures per hour were detected, compared to 4 seizures per hour and 17% high synchrony with the empirical connectivity, see Table 4.7.

The random surrogate network of Fig. 4.48 shows smaller synchronization values compared to the empirical network, even though it consists of the same set of weights. Therefore, no epileptic-seizure-related synchronization phenomena (seizures) were found. If the coupling strength is increased to $\sigma = 0.7$, the average order parameter

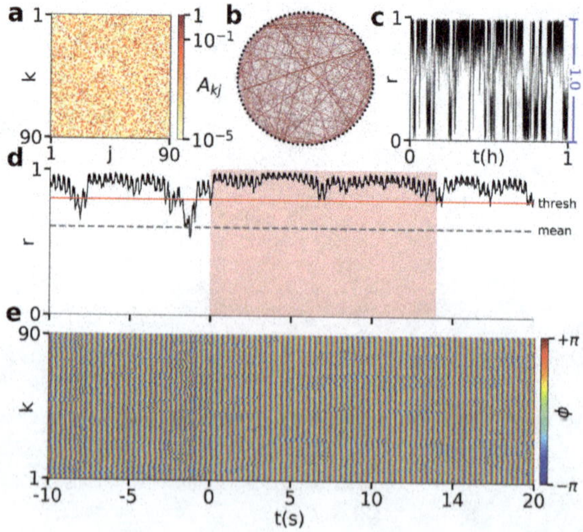

Fig. 4.49 Same as Fig. 4.48 (random surrogate network) with a coupling strength of $\sigma = 0.7$. Note that $\langle r \rangle = 0.60$ both for the random surrogate network and the empirical network. However, high synchrony $r > 0.8$ makes up 47% of the time in the random network and only 17% of the time in the empirical network. Interestingly, the number of seizures seen on the random network (4.7 per hour) is not much larger than the number of seizures seen on the empirical network (4 per hour, visual edge width $2 \cdot w_{ij}$). Figure taken from [403]

$\langle r \rangle$ becomes 0.6, which corresponds to the average order parameter of the empirical network, see Fig. 4.49. Notably, despite high synchrony $r > 0.8$ during 47% of the time, only 4.7 seizures per hour are found (compared to 17% high synchrony and 4 seizures for the empirical network). In addition, the time series in Fig. 4.49c shows time intervals of both extreme synchronization (shown in (d) and (e)) and extreme desynchronization. Medium synchronization, at about $r \approx 0.5$, which we consider as corresponding to a healthy state, is rarely apparent in the time series. This supports our claim that the clustering coefficient in brain networks must not be too small.

4.5.2.3 Fractal Connectivity

In mathematics, a *fractal* is a self-similar structure with a non-integer Hausdorff dimension. In nature, structures similar to fractals appear frequently [521]; examples are the vegetable Romanesco, snowflakes, proteins [521], and blood vessels [522]. The white matter tracts in the human brain were reported to have a quasi-fractal structure [523, 524]. This inspired simulations of networks of FHN oscillators with a one-dimensional [328, 449, 525, 526] or two-dimensional [420, 527] fractal coupling structure, and also for other dynamical models [450, 528–532]. To create a (one-dimensional) ring network with fractal connectivity, we follow the procedure described in [527]. Choose a base pattern b_{init} that consists of a string of ones and

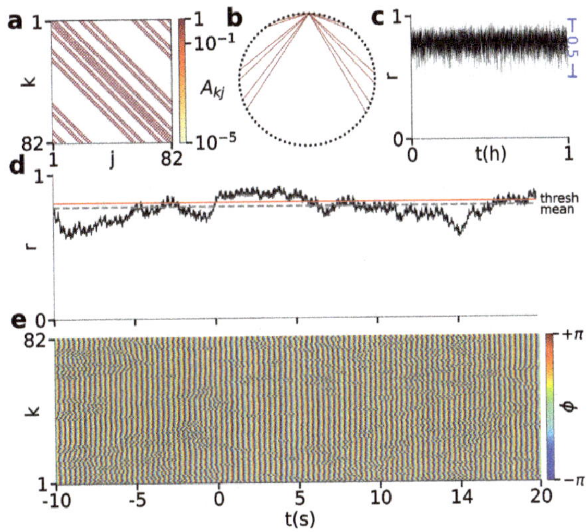

Fig. 4.50 Same as Fig. 4.47 for a ring network with fractal connectivity. Simulation parameters as in Fig. 4.47, except for $\sigma = 0.01$ and $N = 82$. In **b** for clarity, only links emanating from one representative node are shown. All other nodes have the same coupling topology. Figure taken from [403]

zeros. In this section, $b_{\text{init}} = (101)$ is used. Then iterate this base n times: For each 1, substitute the initial base pattern b_{init}, for each 0, substitute a string of zeros of size b with $b = 3$ corresponding to the length of the initial base pattern. The nth hierarchy level is reached after $n - 1$ iterations. A mathematical fractal is obtained in the limit of an infinite number of iterations $n \to \infty$. Since a network is finite, only a finite number of iterations can be performed, and the resulting string is a *quasi-fractal*. Put 0 in front of the string to exclude self-coupling [529]. The resulting string of binary digits becomes the first row of the fractal ring adjacency matrix, where 1 represents a link, 0 represents no link. For every following row, the string is shifted by one element to the right. Via this procedure, a circulant matrix is obtained. After n hierarchical steps, the obtained network consists of $N = b^n + 1$ nodes. In this section, $n = 4$, so the network has 82 nodes. The adjacency matrix is shown in Fig. 4.50a and the connections for one exemplary node are sketched in Fig. 4.50b.

The mean node strength $S = \sum_i s_i/N$, where $s_i = \sum_j a_{ij}$ is the i-th node strength or node degree, is very large ($S = 16$) for this network. Consequently, it synchronizes completely at a relatively small coupling strength σ. Therefore, in order to avoid permanent complete synchronization, we reduce σ to 0.01. Figure 4.50c shows the order parameter for the (unweighted) fractal ring. The average order parameter is $\langle r \rangle = 0.77$, high synchrony $r > 0.8$ is observed during 32% of the time, but the range of r-values in (c) is comparatively small (0.5). Overall, the synchronization varies little, and no clearly defined seizures were found (d). The space-time plot (e) shows an overall moderately synchronized pattern without distinct bursts of synchrony. Thus it seems that fractal connectivities are not appropriate to model realistic epileptic seizures. Either different fractal connectivities or a larger network size or a particular weight distribution is necessary. The latter point is addressed next.

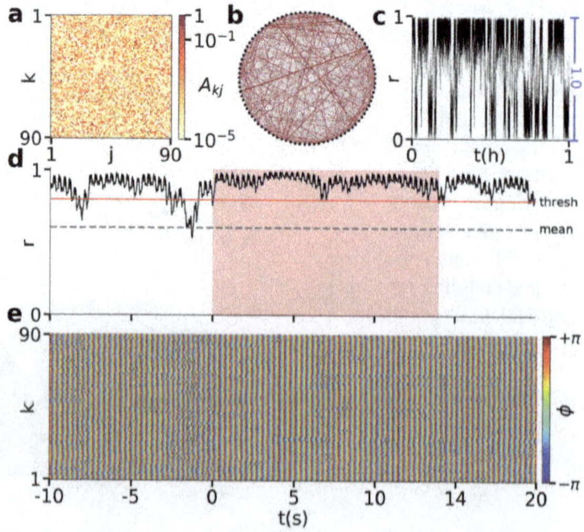

Fig. 4.51 Same as Fig. 4.47 for a ring network with fractal connectivity and weights selected randomly from the empirical connectivity matrix in Fig. 4.47. Simulation parameters as in Fig. 4.47, except for $\sigma = 6.1$ and $N = 82$. Figure taken from [403]

4.5.2.4 Weighted Fractal Connectivity

To achieve a more realistic weight distribution than just 1 and 0, we replace all 1312 non-zero links of the fractal ring in Fig. 4.50 by randomly chosen weights of the 7793 links of the empirical connectivity matrix in Fig. 4.51a, b. Due to the much smaller weights, we have to increase σ from 0.01 to $\sigma = 6.1$ in order to obtain partial synchrony. Now the dynamics resembles the empirical one, as shown in Fig. 4.51c. The order parameter varies strongly in time. However, despite a few short events of high synchrony $r > 0.8$ during 2% of the total time, not a single seizure was detected, Fig. 4.50d, e. For higher values of σ, still, no seizure can be found because the system starts to stabilize at a fixed value of the order parameter of about $r \approx 0.7$. Overall, the fractal connectivity with empirical weights shows rich, brain-like synchronization behavior. However, despite short phases with high synchrony, no well-defined seizures were found.

4.5.2.5 Small-World Networks

Next, we consider small-world-like networks, which can be constructed according to the Watts-Strogatz algorithm [211] by starting from a nonlocally coupled ring and randomly rewiring links with a probability p. With increasing p, these networks are characterized by decreasing average clustering coefficient (which quantifies the strongly coupled neighborhoods) and decreasing mean shortest path length. In some intermediate regime of p and for large enough average node degree S, the small-world property of large clustering coefficient and short path length is found. For the largest probability $p = 1$, we obtain an Erdős-Rényi random network [533]. The simulations are repeated 10 times for each probability p with different random initial conditions. The coupling strength σ is chosen for each p such that the simulations give a sufficiently high degree of synchronization while avoiding complete synchrony during the whole simulation.

In the nonlocally coupled ring at $p = 0$, each of the 90 nodes is connected to its three nearest neighbors on each side; thus, the node degree (strength) is $S = 6$. The links are nondirected and nonweighted. This nonlocally coupled ring is shown in Fig. 4.52a, b. It is necessary to tune the coupling parameter $\sigma = 0.0506$ very carefully. For slightly larger values of σ, the system would fully synchronize during the whole simulation time, while for lower σ, it would not synchronize at all. This underlines the system's sensitivity to small parameter changes [534, 535]. For a long shortest path length and maximum clustering in the nonlocally coupled ring with $p = 0$, the system fully synchronizes ($\langle r \rangle = 0.99$) in 8 simulations. In 2 simulations, it is completely desynchronized ($\langle r \rangle = 0.01$), leading to an arithmetic mean of $\overline{\langle r \rangle} = 0.79$ for all simulations. One typical simulation, with $\langle r \rangle = 0.99$, is shown in Fig. 4.52c–e.

If the rewiring probability is increased slightly to $p = 0.006$, we obtain a small-world network. It still has 95% of the clustering coefficient of the nonlocally coupled ring, but its mean shortest path length is reduced by 36%. The topology is shown in Fig. 4.53a and b. With an increase of p and the resulting reduction of the average shortest path length, the synchronization decreases, see Fig. 4.53c. While all other parameters are kept constant, and the clustering coefficient remains high, $\overline{\langle r \rangle} = 0.80$ is reduced to $\overline{\langle r \rangle} = 0.47$ averaged over all 10 simulations. For $p = 0.006$, $\langle r \rangle$ is for some simulations as low as $\langle r \rangle = 0.03$ and for some simulations as high as $\langle r \rangle = 0.76$, showing a higher sensitivity to initial conditions. In both scenarios, $p = 0$, and

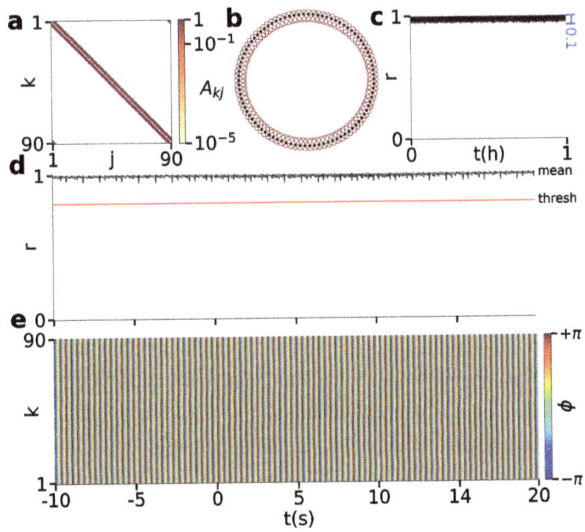

Fig. 4.52 Same as Fig. 4.47 for a nonlocally coupled ring network, corresponding to the Watts-Strogatz model with rewiring probability $p = 0$ and $N = 90$, $S = 6$ (every node is coupled to 3 neighbors on each side). No seizures were found, the system fully synchronizes. Average clustering coefficient $C = 0.60$, mean shortest path length $L = 7.92$. Simulation parameters as in Fig. 4.47, except for $\sigma = 0.0506$. Figure taken from [403]

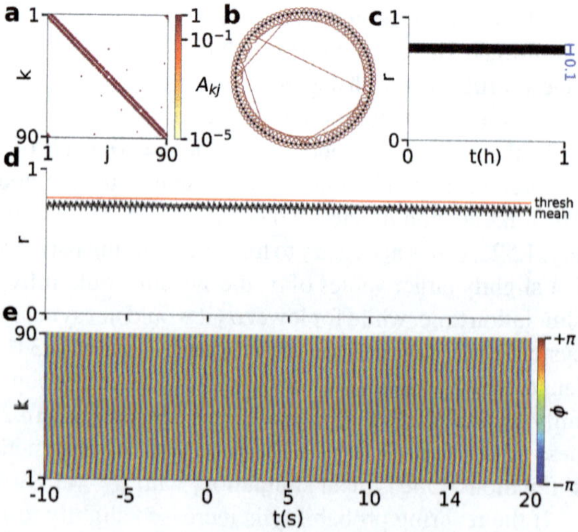

Fig. 4.53 Same as Fig. 4.47 for a Watts-Strogatz network with rewiring probability $p = 0.006$. No seizures were found. Average clustering coefficient $C = 0.57$, mean shortest path length $L = 5.07$. Simulation parameters as in Fig. 4.47, except for $\sigma = 0.0506$, mean node degree $S = 6$. Figure taken from [403]

$p = 0.006$, and for all simulations, the instantaneous degree of synchronization is approximately constant in time and varies in a very small range of 0.1. In fact, by inspecting Figs. 4.52d and 4.53d closely, one can recognize small periodic amplitude fluctuations of $r(t)$. This seems to be related to the slow-fast nature of the FHN system. Since the range of these fluctuations is very small, when full synchronization does not occur, no seizures are detected. The average degree of synchronization, however, depends sensitively upon the random initial conditions. All values are listed in Table 4.8.

By comparing Figs. 4.52 and 4.53, the impact of the shortest path length upon the network dynamics becomes apparent. In general, after decreasing the network's shortest path length, the synchronization reduces significantly. In theory, this should reduce the risk of epileptic seizures; it contradicts earlier findings of an, on average, shorter path lengths in the functional networks of subjects with epilepsy [503]. However, our results should be taken with caution: The range of the order parameter is smaller than 0.1 in both cases and does not allow for realistic brain modeling. This indicates that brain networks must not have too large clustering coefficients, which questions the hypothesis that the human brain is a small-world network [536–541].

Next, we increase the rewiring probability to $p = 0.232$, see Fig. 4.54. The clustering and shortest path are reduced to 41% and 38% of their original values. At $p = 0.232$, the mean shortest path length is very close to its minimum, and the clustering is still significant. The order parameter is $\langle r \rangle \approx 0.52$ for all simulations, proving its independence of initial conditions. We find high synchrony during 14% of the time, and 0.6 seizures per hour with an average duration of 14.7 s with a standard deviation of 3.5 s. One seizure is shown in Fig. 4.54d and e. The dynamics is similar to the one for the empirical connectivity.

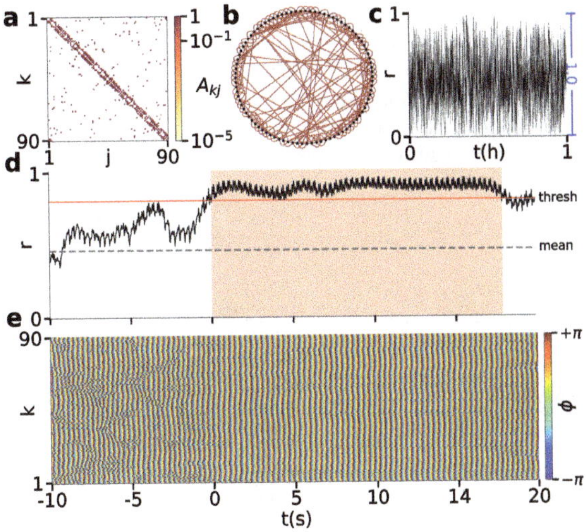

Fig. 4.54 Same as Fig. 4.47 for a Watts-Strogatz network with rewiring probability $p = 0.232$. Average clustering coefficient $C = 0.25$, mean shortest path length $L = 2.97$. Simulation parameters as in Fig. 4.47, except for $\sigma = 0.0506$, mean node degree $S = 6$. Figure taken from [403]

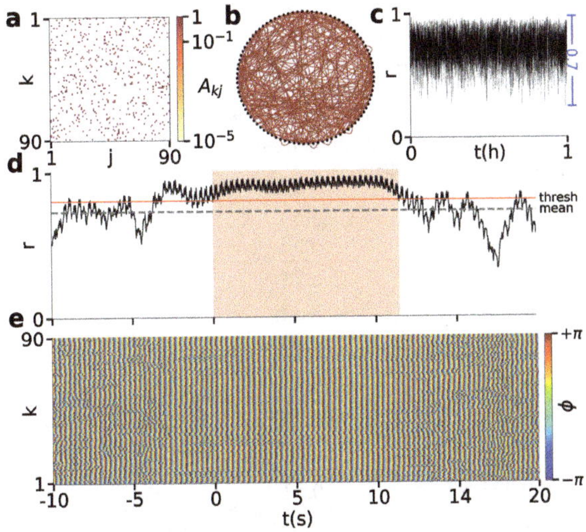

Fig. 4.55 Same as Fig. 4.47 for a random network (Watts-Strogatz network with rewiring probability $p = 1$). Average clustering coefficient $C = 0.053$, mean shortest path length $L = 2.67$. Simulation parameters as in Fig. 4.47, except for $\sigma = 0.0506$, mean node degree $S = 6$. Figure taken from [403]

For $p = 1$, corresponding to a random network (Fig. 4.55), the mean shortest path length reduces further by only 11%, whereas the clustering reduces further by 79%. Now we find a very high Kuramoto order parameter of $\langle r \rangle = 0.73$ and a high degree of synchrony during 25% of the time. Note the very high average of r. Strikingly though, despite the strong synchronization in general, fewer seizures (only 0.5 per hour) are observed as compared to our findings reported in Fig. 4.54. One such seizure is shown in Fig. 4.55d, e. The values of the network measures for all topologies are summarized in Table 4.6; and the simulation results are summarized in Table 4.7.

Table 4.6 Network measures for various topologies. Empirical: empirical Diffusion Tensor Imaging (DTI) network, Random: Random surrogate network with DTI weights, Fractal Unw: Fractal connectivity (unweighted), Fractal DTI: Fractal connectivity with randomly selected DTI weights, SW: Small-world Watts-Strogatz model with rewiring probabilities p; Nodes: number of nodes; Links: number of (non-zero) links; Weights: weighted/unweighted; S: average node strength; C: average (weighted) clustering coefficient; L: mean (weighted) shortest path length. Table taken from [403]

Network	Nodes	Links	Weights	S	C	L
Empirical	90	7793	weight.	1.3	1.7×10^{-3}	59
Random Surr.	90	7793	weight.	1.3	1.0×10^{-3}	34
Fractal Unw.	82	1312	unw.	16	0	2.1
Fractal DTI	82	1312	weight.	0.11	0	270
SW $p = 0$	90	270	unw.	6	0.60	7.9
SW $p = 0.006$	90	270	unw.	6	0.57	5.1
SW $p = 0.232$	90	270	unw.	6	0.25	3.0
SW $p = 1$	90	270	unw.	6	0.05	2.7

Table 4.7 Comparison of seizures for different networks. σ: coupling strength; $\overline{\langle r \rangle}$: average order parameter, $\Delta r = r_{\max} - r_{\min}$: range of r; $r > 0.8$: percentage of time with high synchrony; Num: number of seizures per 1 simulation hour; Duration: average duration of seizures. For each of the last four topologies, 10 simulations with an average of 2.9 h were performed. For $p = 0$, $r > 0.8$ was true 0% of the time for 2 simulations and 100% for 8 simulations. Minimum duration of seizures 8 s. Table taken from [403]

Topology	σ	$\overline{\langle r \rangle}$	Δr	$r > 0.8$ (%)	Num	Duration
Empirical	0.6	0.59	0.99	17	4.0	10.8 ± 1.3 s
Random	0.6	0.23	0.99	1	0	–
Random	0.7	0.60	0.99	47	4.7	10.2 ± 2.6 s
Fractal Unw	0.01	0.77	0.47	32	0	–
Fractal DTI	6.1	0.37	0.95	2	0	–
$p = 0$	0.05	0.79	0.07	80	0	–
$p = 0.006$	0.05	0.47	0.10	0	0	–
$p = 0.232$	0.05	0.52	0.98	14	0.6	14.7 ± 3.5 s
$p = 1$	0.05	0.73	0.81	25	0.5	9.0 ± 0.1 s

By comparing Figs. 4.54 and 4.55, it becomes apparent that seizures are more likely for a higher average clustering coefficient of the network than for a smaller clustering coefficient. This is in line with [503]. Remarkably, this is the case even though the average degree of synchronization decreases significantly for a larger average clustering coefficient. This shows that the occurrence of seizures is not proportional to the average global Kuramoto order parameter of the network, as one might expect. For a reduction of the clustering coefficient, the synchronization might increase while the probability of seizures decreases.

Table 4.8 For the small-world networks, we repeat the simulation 10 times with different random initial conditions for each rewiring probability p. The table shows the global Kuramoto order parameter $\langle r \rangle$, averaged over the simulation time. Each simulation has a duration of approximately 3 h in real time. Table taken from [403]

Simulation	1	2	3	4	5	6	7	8	9	10
$p = 0$	0.99	0.99	0.99	0.99	0.99	0.99	0.02	0.99	0.01	0.99
$p = 0.006$	0.13	0.46	0.03	0.76	0.75	0.21	0.34	0.46	0.75	0.75
$p = 0.232$	0.46	0.47	0.46	0.46	0.47	0.47	0.47	0.47	0.47	0.46
$p = 1$	0.73	0.72	0.73	0.73	0.73	0.73	0.73	0.73	0.73	0.73

When comparing Fig. 4.54c with Fig. 4.55c, the impact of clustering on seizure probability shows up (since the mean shortest path length changes little). For rich synchronization dynamics, including both very low and very high synchronization, the clustering coefficient needs to be sufficiently high: For $p \leq 0.006$, the difference between maximum and minimum of $r(t)$ is $\Delta r = 0.1$, whereas for $p \geq 0.232$ it is $\Delta r \geq 0.8$.

In conclusion, our simulations indicate that the human brain seems to effectively function in a specific window of medium clustering. If the clustering is too large, the neural synchronization is approximately constant in time ($\Delta r \approx 0.1$). The brain, however, shows both low and high synchronization values on the EEG during different tasks and mental states such as sleep. Moreover, epileptic brains, which function normally most of the time, appear to synchronize during generalized seizures fully. This shows that the brain is capable of sustaining both very coherent and very incoherent oscillatory states, which is not possible for too large clustering coefficients.

On the other hand, if the clustering coefficient is too small, the synchronization fluctuates rapidly in time and does not resemble the dynamics of simulations with an empirical brain network. Furthermore, the range of the degree of synchronization ($\Delta r \approx 0.7$) is decreased compared to medium clustering ($\Delta r \approx 1$). One might speculate, based on these simulations, that the difference between healthy and epileptic brains might show up in the network's slightly altered clustering coefficient [542, 543].

For Figs. 4.52, 4.53, 4.54 and 4.55, we have repeated all simulations 10 times to average out dynamic differences that might be caused by different random initial conditions. Table 4.8 summarizes the mean global Kuramoto order parameter $\langle r \rangle$ for each simulation. Note that for $p = 0.232$ and $p = 1$, the dynamics is independent of the initial conditions. For $p = 0$ and $p = 0.006$, on the other hand, the initial conditions determine the entire time series.

4.5.3 *Comparison with Absence Seizures*

We have obtained the electroencephalography (EEG) recordings from the Department of Epileptology of the University of Bonn from a 12 years old subject who suffered from absence seizures. The study was approved by the ethics committee of the University of Bonn, and a parent gave written informed consent that the clinical data might be used and published for research purposes.

Figure 4.56 presents a comparison of our simulations for an empirical connectivity with an EEG recording of an absence seizure. Electrodes and brain areas are depicted in Fig. 4.56a–c, measured EEG data in Fig. 4.56d simulated and EEG-related global phase coherence and global Kuramoto order parameter are shown in Fig. 4.56e and f, respectively, and the simulated and EEG-related space-time plot of the phases in Fig. 4.56g and h, respectively.

Figure 4.56a shows the 90 AAL brain areas underlying the simulations in (e) and (g) as red and blue dots. In (b), the approximated electrode locations are shown as black dots. The EEG time series are obtained by subtracting the voltages from two electrode pairs. For each of the 22 bipolar recording channels, we assign the closest AAL areas, which are colored in (b). These colored brain areas are marked blue in (a), and the ones which are not associated with electrodes in red. Fig. 4.56c provides the bottom view of (b) to clarify that no subcortical brain areas are assigned to the comparison. The EEG recording is presented in (d). For three exemplary electrode pairs, denoted by grey arrows in (c), grey lines connect them to the data in (d). The seizure start and end are marked by vertical red lines and are determined by visual inspection of the EEG by an expert.

In Figure 4.56e and f, both the global Kuramoto order parameter $r(t)$ (black line) and the global phase coherence $R_\Delta(t)$ (blue curve) are shown for simulation and EEG recording, respectively. In the simulation, the coupled dynamics of all 90 brain areas is computed. However, only the 22 areas that were accessible with the EEG are used to calculate the simulated measures R_Δ and r for better comparison.

The phases underlying the computation of R_Δ and r are shown in the space-time plots of Fig. 4.56g and h. We extracted phases from the EEG data shown in Fig. 4.56d after filtering for $f = 3$ Hz. The frequency synchronization during the seizure can easily be distinguished from the pre- and post-seizure activity. The left and the right y-axis show the label of each of the 22 assigned AAL brain areas in Table 4.4.

Let us examine the EEG-recorded seizure in more detail. The value of the global phase coherence in panel (f) before and after the seizure is $R_\Delta \approx 0.5$. At the (electrical) onset of the seizure, R_Δ increases quickly and reaches almost $R_\Delta \approx 1$ within one second. Applying our threshold condition $R_\Delta > 0.8$ to the global phase coherence, we mark the onset and termination of the seizure, shown as a pink shaded region. Note that the start of the seizure (vertical red line) slightly precedes the pink shaded region. This is expected since we obtain R_Δ by averaging over the time window $T = 1$ s, resulting in a lower time resolution of R_Δ. For the same reason, the end of the seizure, according to our threshold definition, slightly precedes the marked seizure termination. However, in both cases, the measured absence seizure lasts for approximately 10 s. Overall, 4 of the 5 seizures from the data set were identified correctly by our seizure detection criterion.

Note that during the recorded seizure temporal fluctuations of the Kuramoto order parameter $r(t)$ are stabilized, in accordance with our simulations. However, contrasting our simulations, $r(t)$ remains at a low value of $r \approx 0.5$. Therefore, if we apply our previous seizure detection criterion $r > 0.8$ to the Kuramoto order parameter,

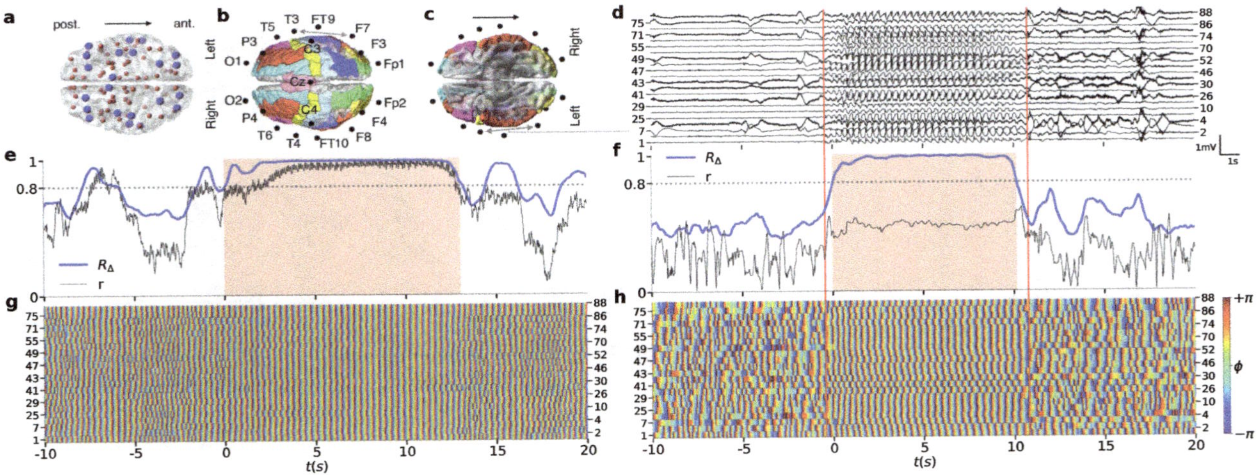

Fig. 4.56 Comparison of simulated and measured epileptic-seizure-related synchronization phenomena. **a** The 90 AAL brain areas marked by red and blue dots. The simulations of the network shown in panels (**e**), (**g**) are based on all areas (red and blue). However, the global phase coherence and global Kuramoto order parameter are calculated, for better comparison with the EEG-recorded seizure, for the areas shown in blue only. **b** Top view on brain: Electrode locations on the brain (black dots). All AAL brain areas corresponding to the blue dots in panel (a) are assigned to the EEG locations and highlighted in color. **c** Bottom view on brain: the grey line connects the exemplary electrode pair F7-T3 to their EEG time series shown in panel (**d**). **d** EEG recording of an exemplary absence seizure. Red vertical lines mark the onset and termination. **e** Simulated global Kuramoto order parameter r vs time (time interval 30 s) in black during one simulated seizure. Global phase coherence R_Δ vs time in blue. The horizontal grey dashed line marks the threshold of $r = 0.8$. In contrast to the previous figures, here the threshold $r > 0.8$ is applied to R_Δ instead of r to enable a comparison between simulated and measured data. The resulting duration of the simulated seizure is shown as a pink shaded region. **f** Same as in panel (**e**) for the EEG recording in (**d**). **g** Simulated space-time plot of the dynamic phases corresponding to panel (**e**). The dynamical phases of the left (right) hemisphere are shown in the lower (upper) half. The dynamics results from all 90 brain areas, but only the phases of the 22 areas sampled with the EEG are shown. Simulation parameters: $a = 0.5$, $\varepsilon = 0.05$, $\varphi = \frac{\pi}{2} - 0.1$, $N = 90$, $\sigma = 0.6$. **h** Same as panel (**g**) for EEG recording in panels (**d**) and (**f**). **a–c** were created using BrainNet Viewer [544]. Figure taken from [403]

we would not detect a seizure from the EEG recording. This highlights the difference between both synchronization measures. Due to time delays in the real system (brain), we cannot expect complete phase synchronization but only synchronization with respect to relative phases, which is measured by R_Δ. For this reason, R_Δ is more appropriate than r for quantifying brain synchrony.

When examining the simulated data in Fig. 4.56e, we notice that R_Δ and r correlate strongly, since delay is not included in the simulated Eq. (4.42). Due to the small averaging window of $T = 1$ s, the signal fluctuates strongly before and after the simulated seizure. We show the same plot with $T = 3$ s in Fig. 4.57 and with $T = 5$ s in Fig. 4.58. For these longer averaging windows, the fluctuations are successively averaged out, and substantial similarity between EEG recording and simulation can be seen. However, the time resolution decreases, prohibiting the precise determination of the start and end of the simulated seizure. For a better comparison of our simulation with the EEG recording, we have also applied our seizure detection criterion to R_Δ in the simulation. The simulated seizure lasts 13 s and is shown as a pink shaded region.

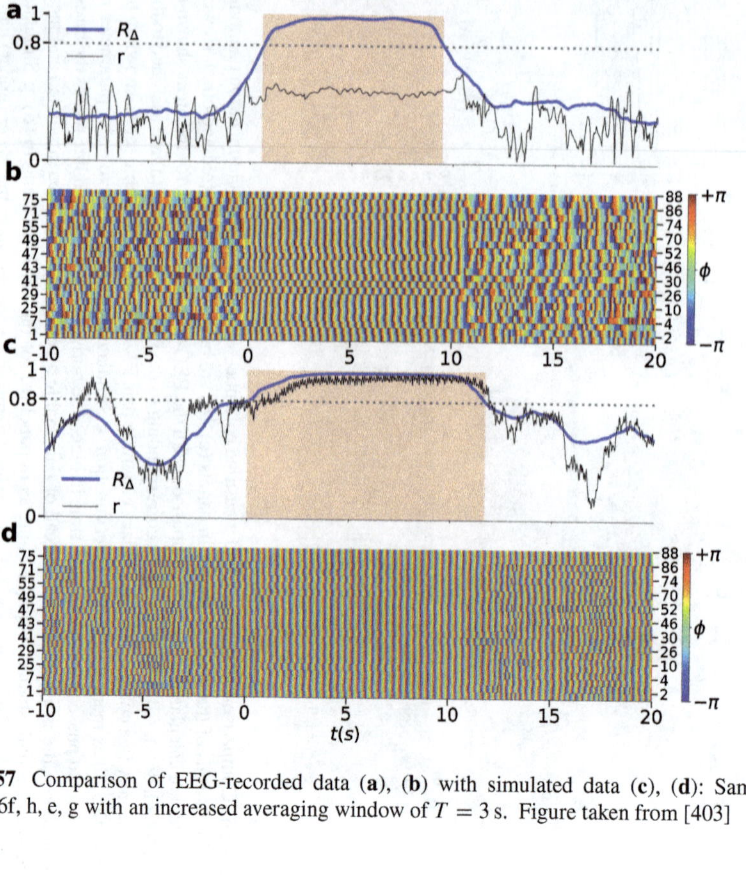

Fig. 4.57 Comparison of EEG-recorded data (**a**), (**b**) with simulated data (**c**), (**d**): Same as Fig. 4.56f, h, e, g with an increased averaging window of $T = 3$ s. Figure taken from [403]

In general, R_Δ and r are similar for the simulation, but due to the weaker definition of synchronization, the global phase coherence is, on average, larger than the order parameter. Overall, we find both for the simulated and empirically observed seizures (i) a global phase coherence of $r \approx 0.6$ (simulation) and $r \approx 0.5$ (empirical) before the seizure, (ii) $r \approx 1$ (simulation and empirical) during the seizure, and (iii) seizure durations of 10.8 ± 1.3 s (simulation) and 10.2 ± 2.4 s (empirical). Due to the quantitative similarity of the simulated events of excessive synchronization with empirical seizures [545], we term them epileptic-seizure-related synchronization phenomena.

For the calculation of the global phase coherence R_Δ, the average phase-locking is calculated over a time window T. For visualizing R_Δ, it is of great importance whether the averaging window T is chosen rather small or rather large. In Figs. 4.57 and 4.58, we re-plot data from Fig. 4.56 for different values of the averaging window T. Note that for longer averaging windows, the simulation data reproduces the data from the EEG recording better. However, due to the lower time resolution, the precise onset and end of the seizure cannot be resolved. Figures 4.59, 4.60, 4.61 and 4.62 show additional EEG recorded data for other seizures of the same subject with epilepsy. Figures 4.63 and 4.64 show very long simulated time series for the empirical connectivity and the random surrogate connectivity, respectively, and the corresponding histograms of the distribution of inter-seizure intervals (Figs. 4.65

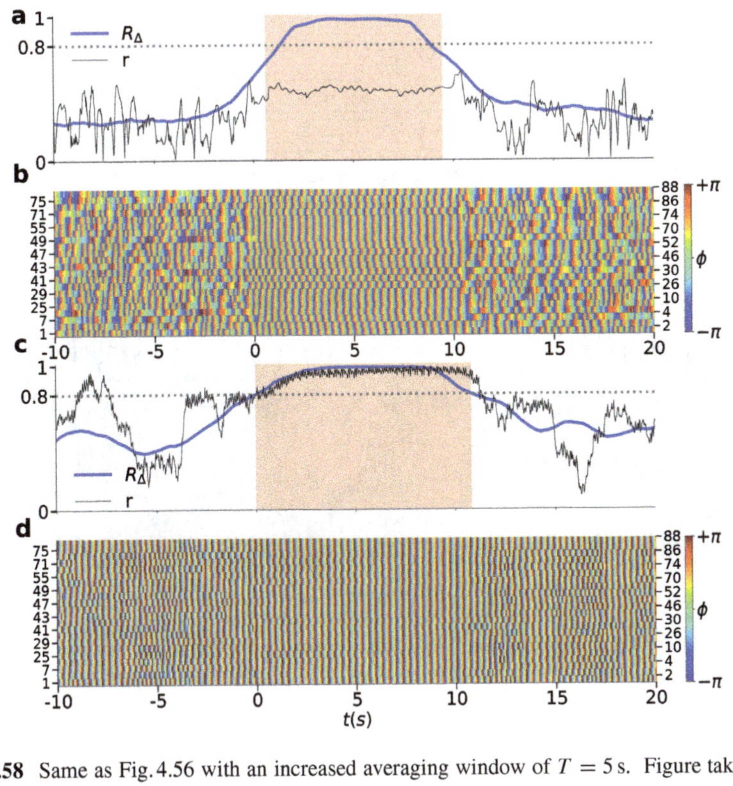

Fig. 4.58 Same as Fig. 4.56 with an increased averaging window of $T = 5$ s. Figure taken from [403]

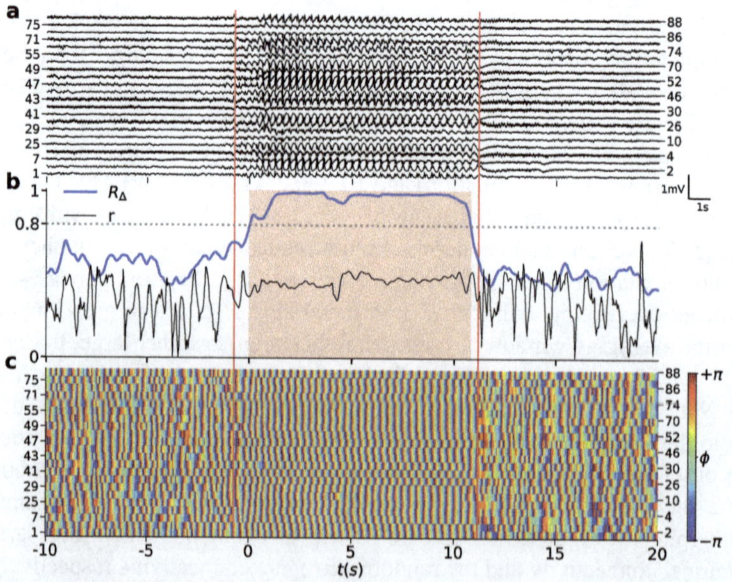

Fig. 4.59 EEG recorded data: Same as Fig. 4.56d, f, h, but for another seizure of the same subject with epilepsy. Figure taken from [403]

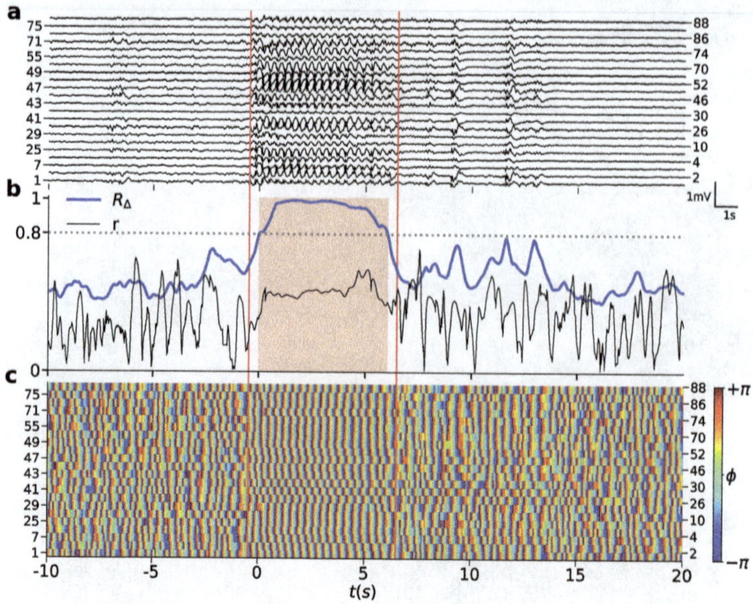

Fig. 4.60 EEG recorded data: Same as Fig. 4.56d, f, h, but for another seizure of the same subject. Figure taken from [403]

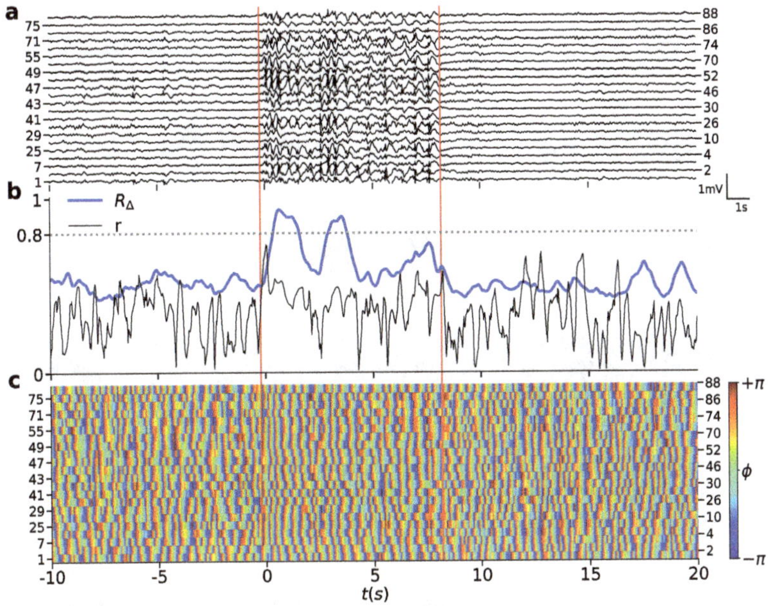

Fig. 4.61 EEG recorded data: Same as Fig. 4.56d, f, h, but for another seizure of the same subject. For this seizure, our threshold definition, applied to the global phase coherence, did not identify the seizure correctly. Figure taken from [403]

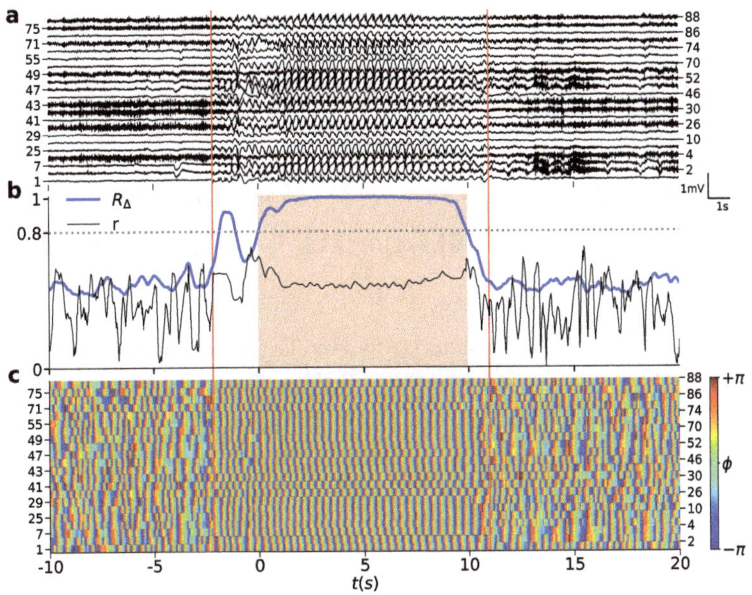

Fig. 4.62 EEG recorded data: Same as Fig. 4.56d, f, h, but for another seizure of the same subject. Figure taken from [403]

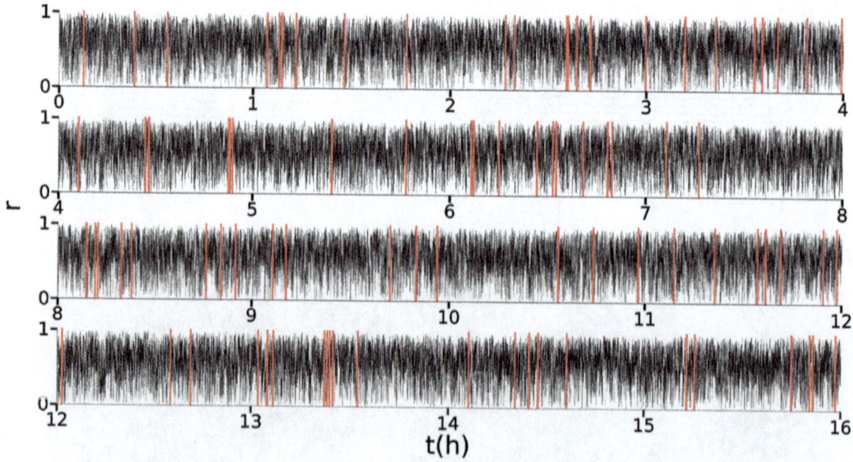

Fig. 4.63 Global Kuramoto order parameter $r(t)$ for a very long simulation of 16 h with the empirical connectivity with a coupling strength of $\sigma = 0.6$. Vertical red lines indicate seizures. Figure taken from [403]

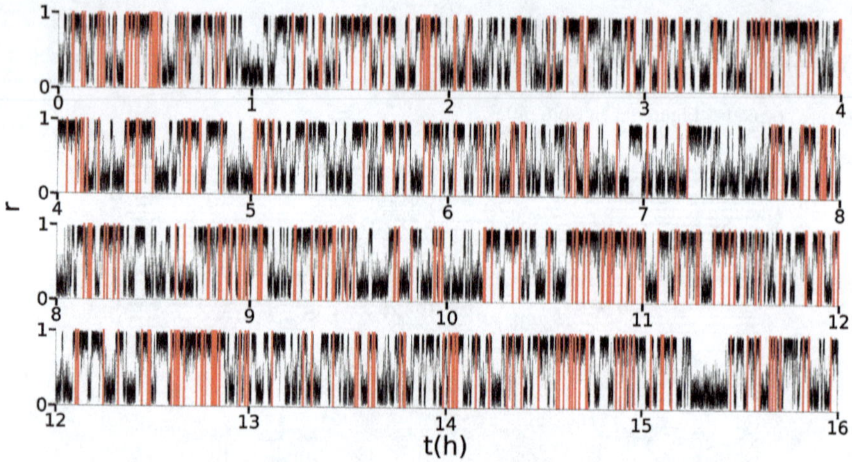

Fig. 4.64 Global Kuramoto order parameter $r(t)$ for a very long simulation of 16 h with the random surrogate connectivity with a coupling strength of $\sigma = 0.7$. Vertical red lines indicate seizures. Figure taken from [403]

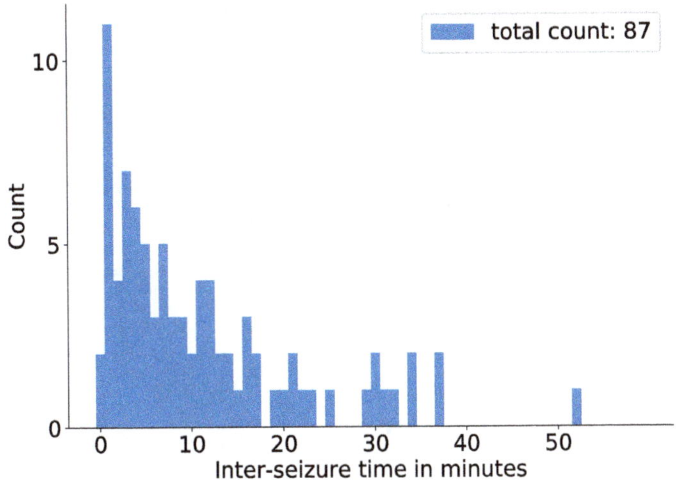

Fig. 4.65 Histogram of the inter-seizure intervals of Fig. 4.63. Bin size = 1 min. Figure taken from [403]

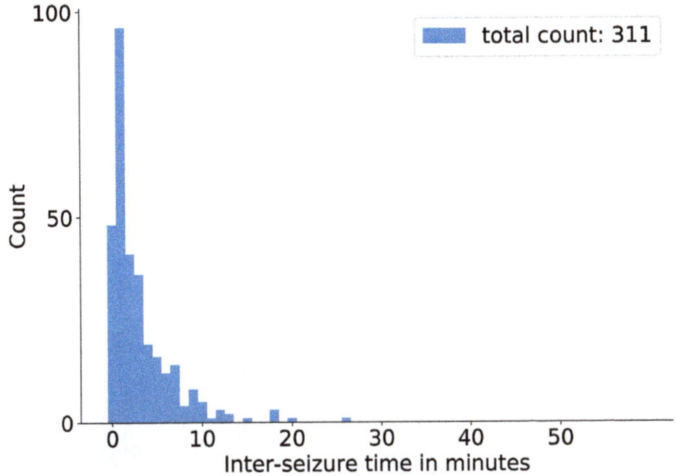

Fig. 4.66 Histogram of the inter-seizure intervals of Fig. 4.64. Bin size = 1 min. Figure taken from [403]

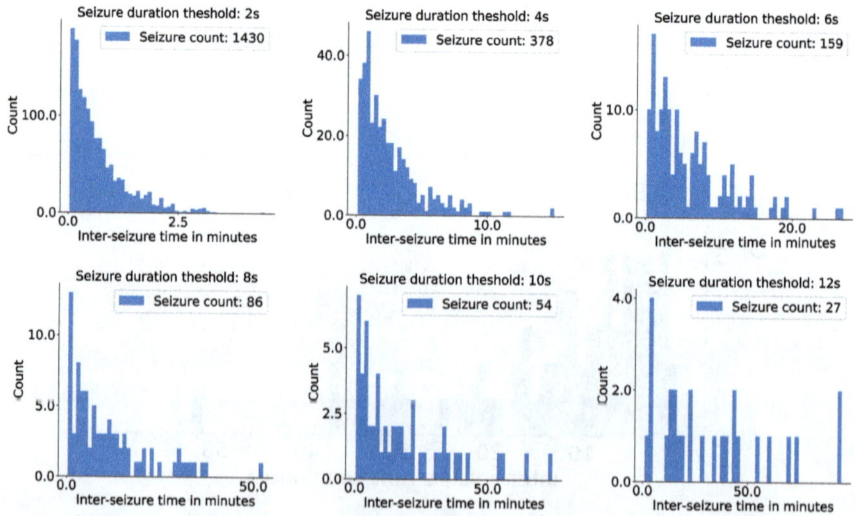

Fig. 4.67 Same as Fig. 4.65 but for various seizure duration thresholds varying between 2 s and 12 s. The distributions of the inter-seizure intervals of the empirical connectivity are independent of the seizure duration threshold. Parameters: $a = 0.5$, $\varepsilon = 0.05$, $\varphi = \frac{\pi}{2} - 0.1$, $N = 90$, $\sigma = 0.6$. Figure taken from [403]

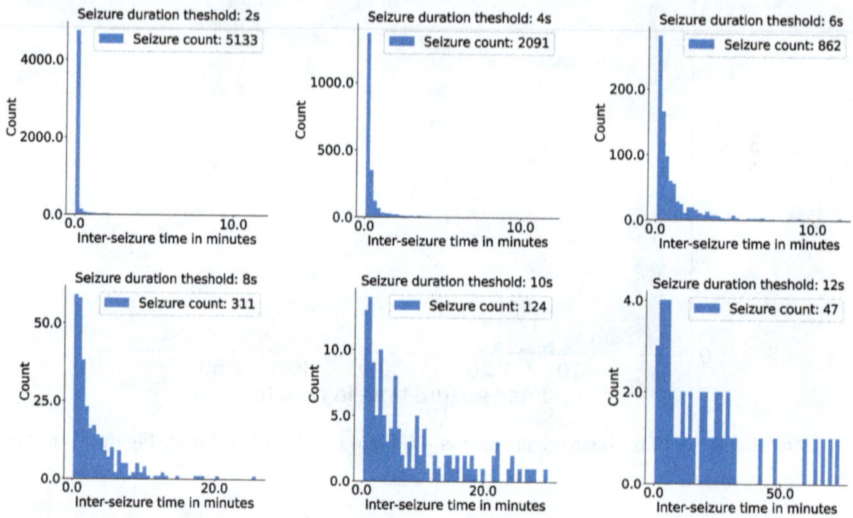

Fig. 4.68 Same as Fig. 4.66 but for various seizure duration thresholds varying between 2 s and 12 s. The distributions of the inter-seizure intervals of the empirical connectivity are independent of the seizure duration threshold. Parameters: $a = 0.5$, $\varepsilon = 0.05$, $\varphi = \frac{\pi}{2} - 0.1$, $N = 90$, $\sigma = 0.7$. Figure taken from [403]

and 4.66). These distributions are independent of the seizure duration threshold (see Figs. 4.67 and 4.68).

4.5.4 Summary

This section has demonstrated that FitzHugh-Nagumo oscillators, coupled via empirical structural connectivities derived from human subjects, exhibit synchronization phenomena closely resembling those observed during epileptic seizures. A comparison between long-term simulations using empirical connectivity data and EEG recordings of epileptic seizures reveals a remarkable similarity between the simulated and real-world data. By exploring FitzHugh-Nagumo oscillators on various network topologies, we have gained valuable insights into the relationship between network structure and synchronization dynamics. Specifically, our findings emphasize the critical role of weight distribution and clustering coefficient in shaping synchronization behavior. To advance the understanding of epilepsy, future investigations could compare these and other network metrics in both structural and functional brain networks of individuals with epilepsy and healthy controls [490].

To probe the interaction between dynamics and network topology, we conducted simulations across a range of artificial network structures: random networks, regular nonlocally coupled ring networks, ring networks with fractal connectivities, and small-world networks with varying rewiring probabilities. Specifically, we analyzed the effects of progressively disrupting the organized structure of the empirical connectivity matrix through random rewiring while maintaining the weight distribution and average node strength. Additionally, we examined quasi-fractal connectivity on a ring network, transforming it into a weighted network with an empirical weight distribution for comparison with empirical connectivity.

Further analysis assessed the impact of average clustering coefficient and average shortest path length on the frequency of epileptic-seizure-related synchronization episodes. Using the Watts-Strogatz small-world algorithm with varying rewiring probabilities, we found that a small-world network with intermediate rewiring probability showed the closest agreement with simulations based on empirical structural connectivity. For other network topologies, the simulations either lacked spontaneously occurring seizure-like synchronization events or exhibited persistent high synchronization throughout. These findings suggest that a balance between regularity and randomness in network topology promotes the self-initiation and self-termination of seizure-like episodes of high synchronization.

Critically, our results highlight that the clustering coefficient plays a significant role: it should not be excessively high (as in regular ring networks) or too low (as in purely random networks). Thus, optimal synchronization dynamics are associated with intermediate rewiring probabilities between 0 and 1. The interplay between regularity and randomness, as well as potential contributions from community structures, may be a fruitful avenue for future research.

Macroscopic alterations in brain network structure, such as lesions caused by stroke, are known contributors to epilepsy [546]. Moreover, epilepsy surgery intentionally modifies the brain's network structure to treat certain epilepsy types [547–549]. Future research could investigate whether there are systematic differences in network structure between the brains of individuals with epilepsy and those of healthy subjects, shedding light on structural factors that predispose to or mitigate the occurrence of epileptic seizures.

References

1. Sawicki J, Berner R, Loos SAM, Anvari M, Bader R, Barfuss W, Botta N, Brede N, Franović I, Gauthier DJ, Goldt S, Hajizadeh A, Hövel P, Karin O, Lorenz-Spreen P, Miehl C, Mölter J, Olmi S, Schöll E, Seif A, Tass PA, Volpe G, Yanchuk S, Kurths J (2023) Perspectives on adaptive dynamical systems. Chaos Interdiscip J Nonlinear Sci 33(7):071501. https://doi.org/10.1063/5.0147231
2. Johnson-Groh M (2023) A look at adaptive systems from biology to machine learning. Scilight 2023(30):301105. https://doi.org/10.1063/10.0020288
3. Tyson JJ, Chen KC, Novak B (2003) Sniffers, buzzers, toggles and blinkers: dynamics of regulatory and signaling pathways in the cell. Curr Opin Cell Biol 15(2):221–231. https://doi.org/10.1016/s0955-0674(03)00017-6
4. Alon U (2019) An introduction to systems biology an introduction to systems biology, 2 edn. Chapman & Hall/CRC Computational Biology Series, Taylor & Francis, Philadelphia, PA. https://doi.org/10.1201/9780429258771
5. Ferrell J (2021) Systems biology of cell signaling. CRC Press, Boca Raton, FL. https://doi.org/10.1201/9780429322366
6. Kahn SE, Prigeon RL, McCulloch DK, Boyko EJ, Bergman RN, Schwartz MW, Neifing JL, Ward WK, Beard JC, Palmer JP (1993) Quantification of the relationship between insulin sensitivity and beta-cell function in human subjects. Evidence for a hyperbolic function. Diabetes 42(11):1663–1672. https://doi.org/10.2337/diab.42.11.1663
7. Macnab RM, Koshland DEJ (1972) The gradient-sensing mechanism in bacterial chemotaxis. Proc Natl Acad Sci U S A 69(9):2509–2512. https://doi.org/10.1073/pnas.69.9.2509
8. Berg HC, Tedesco PM (1975) Transient response to chemotactic stimuli in escherichia coli. Proc Natl Acad Sci U S A 72(8):3235–3239. https://doi.org/10.1073/pnas.72.8.3235
9. Yi TM, Huang Y, Simon MI, Doyle J (2000) Robust perfect adaptation in bacterial chemotaxis through integral feedback control. Proc Natl Acad Sci U S A 97(9):4649–4653. https://doi.org/10.1073/pnas.97.9.4649
10. El-Samad H, Goff JP, Khammash M (2002) Calcium homeostasis and parturient hypocalcemia: an integral feedback perspective. J Theor Biol 214(1):17–29. https://doi.org/10.1006/jtbi.2001.2422
11. Briat C, Gupta A, Khammash M (2016) Antithetic integral feedback ensures robust perfect adaptation in noisy biomolecular networks. Cell Syst 2(1):15–26. https://doi.org/10.1016/j.cels.2016.01.004
12. Karin O, Swisa A, Glaser B, Dor Y, Alon U (2016) Dynamical compensation in physiological circuits. Mol Syst Biol 12(11):886. https://doi.org/10.15252/msb.20167216
13. Karin O, Raz M, Tendler A, Bar A, Korem Kohanim Y, Milo T, Alon U (2020) A new model for the hpa axis explains dysregulation of stress hormones on the timescale of weeks. Mol Syst Biol 16(7):e9510 (2020). https://doi.org/10.15252/msb.20209510
14. Korem Kohanim Y, Milo T, Raz M, Karin O, Bar A, Mayo A, Mendelson Cohen N, Toledano Y, Alon U (2022) Dynamics of thyroid diseases and thyroid-axis gland masses. Mol Syst Biol 18(8):e10919. https://doi.org/10.15252/msb.202210919
15. Tendler A, Bar A, Mendelsohn-Cohen N, Karin O, Kohanim YK, Maimon L, Milo T, Raz M, Mayo A, Tanay A, Alon U (2021) Hormone seasonality in medical records suggests circannual endocrine circuits. Proc Natl Acad Sci 118(7). https://doi.org/10.1073/pnas.2003926118
16. Lazova MD, Ahmed T, Bellomo D, Stocker R, Shimizu TS (2011) Response rescaling in bacterial chemotaxis. Proc Natl Acad Sci 108(33):13870–13875. https://doi.org/10.1073/pnas.1108608108
17. Larsch J, Flavell SW, Liu Q, Gordus A, Albrecht DR, Bargmann CI (2015) A circuit for gradient climbing in c. elegans chemotaxis. Cell Rep 12(11):1748–1760. https://doi.org/10.1016/j.celrep.2015.08.032
18. Kamino K, Kondo Y, Nakajima A, Honda-Kitahara M, Kaneko K, Sawai S (2017) Fold-change detection and scale invariance of cell–cell signaling in social amoeba. Proc Natl Acad Sci 114(21). https://doi.org/10.1073/pnas.1702181114

19. Schultz W, Dayan P, Montague PR (1997) A neural substrate of prediction and reward. Science 275(5306):1593–1599. https://doi.org/10.1126/science.275.5306.1593
20. Tobler PN, Fiorillo CD, Schultz W (2005) Adaptive coding of reward value by dopamine neurons. Science 307(5715):1642–1645. https://doi.org/10.1126/science.1105370
21. Kim HR, Malik AN, Mikhael JG, Bech P, Tsutsui-Kimura I, Sun F, Zhang Y, Li Y, Watabe-Uchida M, Gershman SJ, Uchida N (2020) A unified framework for dopamine signals across timescales. Cell 183(6):1600–1616. https://doi.org/10.1016/j.cell.2020.11.013
22. Karin O, Alon U (2022) The dopamine circuit as a reward-taxis navigation system. PLOS Comput Biol 18(7):1–24. https://doi.org/10.1371/journal.pcbi.1010340
23. Wilson HR, Cowan JD (1973) A mathematical theory of the functional dynamics of cortical and thalamic nervous tissue. Kybernetik 13(2):55–80. https://doi.org/10.1007/BF00288786
24. Mattia M, Del Giudice P (2002) Population dynamics of interacting spiking neurons. Phys Rev E 66(5):051917. https://doi.org/10.1103/PhysRevE.66.051917
25. Schaffer ES, Ostojic S, Abbott LF (2013) A complex-valued firing-rate model that approximates the dynamics of spiking networks. PLoS Comput Biol 9(10):e1003301. https://doi.org/10.1371/journal.pcbi.1003301
26. Luke TB, Barreto E, So P (2013) Complete classification of the macroscopic behavior of a heterogeneous network of theta neurons. Neural Comput 25(12):3207–3234. https://doi.org/10.1162/NECO_a_00525
27. Laing CR (2014) Derivation of a neural field model from a network of theta neurons. Phys Rev E 90(1):010901. https://doi.org/10.1103/PhysRevE.90.010901
28. Montbrió E, Pazó D, Roxin A (2015) Macroscopic description for networks of spiking neurons. Phys Rev X 5(2):021028. https://doi.org/10.1103/PhysRevX.5.021028
29. Ott E, Antonsen TM (2008) Low dimensional behavior of large systems of globally coupled oscillators. Chaos Interdiscip J Nonlinear Sci 18(3):037113. https://doi.org/10.1063/1.2930766
30. Ermentrout GB, Kopell N (1986) Parabolic bursting in an excitable system coupled with a slow oscillation. SIAM J Appl Math 46(2):233–253. https://doi.org/10.1137/0146017
31. Byrne A, Brookes MJ, Coombes S (2017) A mean field model for movement induced changes in the beta rhythm. J Comput Neurosci 43(2):143–158. https://doi.org/10.1007/s10827-017-0666-6
32. Schmidt H, Avitabile D, Montbrió E, Roxin A (2018) Network mechanisms underlying the role of oscillations in cognitive tasks. PLoS Comput Biol 14(9):e1006430. https://doi.org/10.1371/journal.pcbi.1006430
33. Byrne Á, O'Dea RD, Forrester M, Ross J, Coombes S (2020) Next-generation neural mass and field modeling. J Neurophysiol 123(2):726–742. https://doi.org/10.1152/jn.00314.2019
34. Ceni A, Olmi S, Torcini A, Angulo-Garcia D (2020) Cross frequency coupling in next generation inhibitory neural mass models. Chaos Interdiscip J Nonlinear Sci 30(5):053121. https://doi.org/10.1063/5.0017527
35. Bi H, Segneri M, Di Volo M, Torcini A (2020) Coexistence of fast and slow gamma oscillations in one population of inhibitory spiking neurons. Phys Rev Res 2(1):013042. https://doi.org/10.1103/PhysRevResearch.2.013042
36. Taher H, Torcini A, Olmi S (2020) Exact neural mass model for synaptic-based working memory. PLoS Comput Biol 16(12):e1008533. https://doi.org/10.1371/journal.pcbi.1008533
37. Segneri M, Bi H, Olmi S, Torcini A (2020) Theta-nested gamma oscillations in next generation neural mass models. Front Comput Neurosci 14:47. https://doi.org/10.3389/fncom.2020.00047
38. Gast R, Schmidt H, Knösche TR (2020) A mean-field description of bursting dynamics in spiking neural networks with short-term adaptation. Neural Comput 32(9):1615–1634. https://doi.org/10.1162/neco_a_01369
39. Gerster M, Taher H, Škoch A, Hlinka J, Guye M, Bartolomei F, Jirsa V, Zakharova A, Olmi S (2021) Patient-specific network connectivity combined with a next generation neural mass model to test clinical hypothesis of seizure propagation. Front Syst Neurosci 15:79. https://doi.org/10.3389/fnsys.2021.691147

40. Gast R, Knösche TR, Schmidt H (2021) Mean-field approximations of networks of spiking neurons with short-term synaptic plasticity. Phys Rev E 104(4):044310. https://doi.org/10.1103/PhysRevE.104.044310
41. Fuhrmann G, Markram H, Tsodyks M (2002) Spike frequency adaptation and neocortical rhythms. J Neurophys 88(2):761–770. https://doi.org/10.1152/jn.2002.88.2.761
42. Benda J, Herz AV (2003) A universal model for spike-frequency adaptation. Neural Comput 15(11):2523–2564. https://doi.org/10.1162/089976603322385063
43. Brown D, Adams P (1980) Muscarinic suppression of a novel voltage-sensitive K^+ current in a vertebrate neurone. Nature 283(5748):673–676. https://doi.org/10.1038/283673a0
44. Madison D, Nicoll R (1984) Control of the repetitive discharge of rat CA1 pyramidal neurones in vitro. J Physiol 354(1):319–331. https://doi.org/10.1113/jphysiol.1984.sp015706
45. Fleidervish IA, Friedman A, Gutnick MJ (1996) Slow inactivation of Na^+ current and slow cumulative spike adaptation in mouse and guinea-pig neocortical neurones in slices. J Physiol 493(1):83–97. https://doi.org/10.1113/jphysiol.1996.sp021721
46. Gigante G, Mattia M, Del Giudice P (2007) Diverse population bursting modes of adapting spiking neurons. Phys Rev Lett 98(14):148101. https://doi.org/10.1103/PhysRevLett.98.148101
47. Ferrara A, Angulo-Garcia D, Torcini A, Olmi S (2023) Population spiking and bursting in next-generation neural masses with spike-frequency adaptation. Phys Rev E 107(2):024311. https://doi.org/10.1103/PhysRevE.107.024311
48. Stevens CF, Wang Y (1995) Facilitation and depression at single central synapses. Neuron 14(4):795–802. https://doi.org/10.1016/0896-6273(95)90205-2
49. Markram H, Tsodyks M (1996) Redistribution of synaptic efficacy between neocortical pyramidal neurons. Nature 382(6594):807–810. https://doi.org/10.1038/382807a0
50. Abbott LF, Varela J, Sen K, Nelson S (1997) Synaptic depression and cortical gain control. Science 275(5297):221–224. https://doi.org/10.1126/science.275.5297.221
51. Zucker RS, Regehr WG (2002) Short-term synaptic plasticity. Annu Rev Physiol 64:355–405. https://doi.org/10.1146/annurev.physiol.64.092501.114547
52. Abbott L, Regehr WG (2004) Synaptic computation. Nature 431(7010):796–803. https://doi.org/10.1038/nature03010
53. Jones MV, Westbrook GL (1996) The impact of receptor desensitization on fast synaptic transmission. Trends Neurosci 19(3):96–101. https://doi.org/10.1016/0166-2236(96)80018-7
54. Wong AY, Graham BP, Billups B, Forsythe ID (2003) Distinguishing between presynaptic and postsynaptic mechanisms of short-term depression during action potential trains. J Neurosci 23(12):4868–4877. https://doi.org/10.1523/JNEUROSCI.23-12-04868.2003
55. Turrigiano GG (2008) The self-tuning neuron: synaptic scaling of excitatory synapses. Cell 135(3):422–435. https://doi.org/10.1016/j.cell.2008.10.008
56. Pozo K, Goda Y (2010) Unraveling mechanisms of homeostatic synaptic plasticity. Neuron 66(3):337–351. https://doi.org/10.1016/j.neuron.2010.04.028
57. Virkar YS, Shew WL, Restrepo JG, Ott E (2016) Feedback control stabilization of critical dynamics via resource transport on multilayer networks: How glia enable learning dynamics in the brain. Phys Rev E 94:042310. https://doi.org/10.1103/PhysRevE.94.042310
58. Huang YT, Chang YL, Chen CC, Lai PY, Chan C (2017) Positive feedback and synchronized bursts in neuronal cultures. PloS one 12(11):e0187276. https://doi.org/10.1371/journal.pone.0187276
59. Mongillo G, Barak O, Tsodyks M (2008) Synaptic theory of working memory. Science 319(5869):1543–1546. https://doi.org/10.1126/science.1150769
60. Tsodyks M, Uziel A, Markram H et al (2000) Synchrony generation in recurrent networks with frequency-dependent synapses. J Neurosci 20(1):RC50–RC50. https://doi.org/10.1523/JNEUROSCI.20-01-j0003.2000
61. Luccioli S, Ben-Jacob E, Barzilai A, Bonifazi P, Torcini A (2014) Clique of functional hubs orchestrates population bursts in developmentally regulated neural networks. PLoS Comput Biol 10(9):e1003823. https://doi.org/10.1371/journal.pcbi.1003823

62. Tallon-Baudry C, Bertrand O, Peronnet F, Pernier J (1998) Induced γ-band activity during the delay of a visual short-term memory task in humans. J Neurosci 18(11):4244–4254. https://doi.org/10.1523/JNEUROSCI.18-11-04244.1998
63. Howard MW, Rizzuto DS, Caplan JB, Madsen JR, Lisman J, Aschenbrenner-Scheibe R, Schulze-Bonhage A, Kahana MJ (2003) Gamma oscillations correlate with working memory load in humans. Cerebral cortex 13(12):1369–1374. https://doi.org/10.1093/cercor/bhg084
64. Van Vugt MK, Schulze-Bonhage A, Litt B, Brandt A, Kahana MJ (2010) Hippocampal gamma oscillations increase with memory load. J Neurosci 30(7):2694–2699. https://doi.org/10.1523/JNEUROSCI.0567-09.2010
65. Roux F, Wibral M, Mohr HM, Singer W, Uhlhaas PJ (2012) Gamma-band activity in human prefrontal cortex codes for the number of relevant items maintained in working memory. J Neurosci 32(36):12411–12420. https://doi.org/10.1523/JNEUROSCI.0421-12.2012
66. Lindén H, Berg RW (2021) Why firing rate distributions are important for understanding spinal central pattern generators. Front Hum Neurosci 15:719388. https://doi.org/10.3389/fnhum.2021.719388
67. Rubin JE, Bacak BJ, Molkov YI, Shevtsova NA, Smith JC, Rybak IA (2011) Interacting oscillations in neural control of breathing: modeling and qualitative analysis. J Comput Neurosci 30(3):607–632. https://doi.org/10.1007/s10827-010-0271-2
68. Miller EK, Lundqvist M, Bastos AM (2018) Working memory 2.0. Neuron 100(2):463–475. https://doi.org/10.1016/j.neuron.2018.09.023
69. Benabid A, Pollak P, Gross C, Hoffmann D, Benazzouz A, Gao D, Laurent A, Gentil M, Perret J (1994) Acute and long-term effects of subthalamic nucleus stimulation in Parkinson's disease. Stereotact Funct Neurosurg 62(1–4):76–84. https://doi.org/10.1159/000098566
70. Pinter M, Alesch F, Murg M, Seiwald M, Helscher R, Binder H (1999) Deep brain stimulation of the subthalamic nucleus for control of extrapyramidal features in advanced idiopathic Parkinson's disease: one year follow-up. J Neural Transm 106(7):693–709. https://doi.org/10.1007/s007020050201
71. Rodriguez-Oroz MC, Obeso JA, Lang AE, Houeto JL, Pollak P, Rehncrona S, Kulisevsky J, Albanese A, Volkmann J, Hariz MI, Quinn NP, Speelman JD, Guridi J, Zamarbide I, Gironell A, Molet J, Pascual-Sedano B, Pidoux B, Bonnet AM, Agid Y, Xie J, Benabid AL, Lozano AM, Saint-Cyr J, Romito L, Contarino MF, Scerrati M, Fraix V, Van Blercom N (2005) Bilateral deep brain stimulation in parkinson's disease: a multicentre study with 4 years follow-up. Brain 128(Pt 10):2240–2249. https://doi.org/10.1093/brain/awh571
72. Benabid AL, Chabardes S, Mitrofanis J, Pollak P (2009) Deep brain stimulation of the subthalamic nucleus for the treatment of Parkinson's disease. Lancet Neurol 8(1):67–81. https://doi.org/10.1016/S1474-4422(08)70291-6
73. Lozano AM, Lipsman N, Bergman H, Brown P, Chabardes S, Chang JW, Matthews K, McIntyre CC, Schlaepfer TE, Schulder M, Temel Y, Volkmann J, Krauss JK (2019) Deep brain stimulation: current challenges and future directions. Nat Rev Neurol 15(3):148–160. https://doi.org/10.1038/s41582-018-0128-2
74. Deuschl G, Schade-Brittinger C, Krack P, Volkmann J, Schäfer H, Bötzel K, Daniels C, Deutschländer A, Dillmann U, Eisner W, Gruber D, Hamel W, Herzog J, Hilker R, Klebe S, Kloss M, Koy J, Krause M, Kupsch A, Lorenz D, Lorenzl S, Mehdorn HM, Moringlane JR, Oertel W, Pinsker MO, Reichmann H, Reuss A, Schneider GH, Schnitzler A, Steude U, Sturm V, Timmermann L, Tronnier V, Trottenberg T, Wojtecki L, Wolf E, Poewe W, Voges J (2006) A randomized trial of deep-brain stimulation for Parkinson's disease. N Engl J Med 355(9):896–908. https://doi.org/10.1056/NEJMoa060281
75. Weaver FM, Follett K, Stern M, Hur K, Harris C, Marks WJ, Rothlind J, Sagher O, Reda D, Moy CS et al (2009) Bilateral deep brain stimulation vs best medical therapy for patients with advanced Parkinson disease: a randomized controlled trial. JAMA 301(1):63–73. https://doi.org/10.1001/jama.2008.929
76. Follett KA, Weaver FM, Stern M, Hur K, Harris CL, Luo P, Marks WJJ, Rothlind J, Sagher O, Moy C, Pahwa R, Burchiel K, Hogarth P, Lai EC, Duda JE, Holloway K, Samii A, Horn S, Bronstein JM, Stoner G, Starr PA, Simpson R, Baltuch G, De Salles A, Huang GD, Reda DJ

(2010) Pallidal versus subthalamic deep-brain stimulation for Parkinson's disease. N Engl J Med 362(22):2077–2091. https://doi.org/10.1056/NEJMoa0907083

77. Beric A, Kelly PJ, Rezai A, Sterio D, Mogilner A, Zonenshayn M, Kopell B (2001) Complications of deep brain stimulation surgery. Stereotact Funct Neurosurg 77(1–4):73–78. https://doi.org/10.1159/000064037
78. Oh MY, Abosch A, Kim SH, Lang AE, Lozano AM (2002) Long-term hardware-related complications of deep brain stimulation. Neurosurgery 50(6):1268–1276. https://doi.org/10.1093/neurosurgery/50.6.1268
79. Binder DK, Rau G, Starr PA (2003) Hemorrhagic complications of microelectrode-guided deep brain stimulation. Stereotactic and functional neurosurgery 80(1–4):28–31. https://doi.org/10.1159/000075147
80. Lyons KE, Wilkinson SB, Overman J, Pahwa R (2004) Surgical and hardware complications of subthalamic stimulation: a series of 160 procedures. Neurology 63(4):612–616. https://doi.org/10.1212/01.WNL.0000134653.52776.65
81. Uc EY, Follett KA (2007) Deep brain stimulation in movement disorders. Semin Neurol 27(2):170–182. https://doi.org/10.1055/s-2007-971175
82. Baizabal-Carvallo J, Jankovic J (2016) Movement disorders induced by deep brain stimulation. Parkinsonism Relat Disord 25:1–9. https://doi.org/10.1016/j.parkreldis.2016.01.014
83. Voon V, Fox SH (2007) Medication-related impulse control and repetitive behaviors in Parkinson disease. Arch Neurol 64(8):1089–1096
84. Eisenstein SA, Dewispelaere WB, Campbell MC, Lugar HM, Perlmutter JS, Black KJ, Hershey T (2014) Acute changes in mood induced by subthalamic deep brain stimulation in Parkinson disease are modulated by psychiatric diagnosis. Brain Stimul 7(5):701–708. https://doi.org/10.1016/j.brs.2014.06.002
85. Merola A, Romagnolo A, Rizzi L, Rizzone MG, Zibetti M, Lanotte M, Mandybur G, Duker AP, Espay AJ, Lopiano L (2017) Impulse control behaviors and subthalamic deep brain stimulation in Parkinson disease. J Neurol 264(1):40–48. https://doi.org/10.1007/s00415-016-8314-x
86. Healy S, Shepherd H, Mooney N, Costa AD, Osman-Farah J, Macerollo A (2022) The effect of deep brain stimulation on impulse control related disorders in Parkinson' s disease-a 10-year retrospective study of 137 patients. J Neurol Sci 440:120339. https://doi.org/10.1016/j.jns.2022.120339
87. Hammond C, Bergman H, Brown P (2007) Pathological synchronization in Parkinson's disease: networks, models and treatments. Trends Neurosci 30(7):357–364. https://doi.org/10.1016/j.tins.2007.05.004
88. Tass PA (1999) Phase resetting in medicine and biology: stochastic modelling and data analysis. Springer, Berlin. https://doi.org/10.1007/978-3-642-58676-3
89. Tass PA (2001) Desynchronizing double-pulse phase resetting and application to deep brain stimulation. Biol Cybern 85(5):343–354. https://doi.org/10.1007/s004220100268
90. Tass PA (2003) A model of desynchronizing deep brain stimulation with a demand-controlled coordinated reset of neural subpopulations. Biol Cybern 89(2):81–88. https://doi.org/10.1007/s00422-003-0425-7
91. Adamchic I, Hauptmann C, Barnikol UB, Pawelczyk N, Popovych O, Barnikol TT, Silchenko A, Volkmann J, Deuschl G, Meissner WG, Maarouf M, Sturm V, Freund HJ, Tass PA (2014) Coordinated reset neuromodulation for Parkinson's disease: proof-of-concept study. Mov Disord 29(13):1679–1684. https://doi.org/10.1002/mds.25923
92. Bouthour W, Mégevand P, Donoghue J, Lüscher C, Birbaumer N, Krack P (2019) Biomarkers for closed-loop deep brain stimulation in Parkinson disease and beyond. Nat Rev Neurol 15(6):343–352. https://doi.org/10.1038/s41582-019-0166-4
93. Hoang KB, Turner DA (2019) The emerging role of biomarkers in adaptive modulation of clinical brain stimulation. Neurosurgery 85(3):E430–E439. https://doi.org/10.1093/neuros/nyz096
94. Beudel M, Brown P (2016) Adaptive deep brain stimulation in Parkinson's disease. Parkinsonism Relat Disord 22 (Suppl 1), S123–6. https://doi.org/10.1016/j.parkreldis.2015.09.028

95. Tass PA, Majtanik M (2006) Long-term anti-kindling effects of desynchronizing brain stimulation: a theoretical study. Biol Cybern 94(1):58–66. https://doi.org/10.1007/s00422-005-0028-6
96. Madadi Asl M, Vahabie AH, Valizadeh A, Tass PA (2022) Spike-timing-dependent plasticity mediated by dopamine and its role in Parkinson's disease pathophysiology. Front Netw Physiol 2:817524. https://doi.org/10.3389/fnetp.2022.817524
97. Markram H, Lübke J, Frotscher M, Sakmann B (1997) Regulation of synaptic efficacy by coincidence of postsynaptic APs and EPSPs. Science 275(5297):213–215. https://doi.org/10.1126/science.275.5297.213
98. Abbott LF, Nelson SB (2000) Synaptic plasticity: taming the beast. Nat Neurosci 3:1178–1183. https://doi.org/10.1038/81453
99. Caporale N, Dan Y (2008) Spike timing-dependent plasticity: a hebbian learning rule. Annu Rev Neurosci 31:25. https://doi.org/10.1146/annurev.neuro.31.060407.125639
100. Feldman DE (2012) The spike-timing dependence of plasticity. Neuron 75:556–571. https://doi.org/10.1016/j.neuron.2012.08.001
101. Tass PA, Hauptmann C (2007) Therapeutic modulation of synaptic connectivity with desynchronizing brain stimulation. Int J Psychophysiol 64(1):53–61. https://doi.org/10.1016/j.ijpsycho.2006.07.013
102. Zeitler M, Tass PA (2015) Augmented brain function by coordinated reset stimulation with slowly varying sequences. Front Syst Neurosci 9:49. https://doi.org/10.3389/fnsys.2015.00049
103. Manos T, Zeitler M, Tass PA (2018) How stimulation frequency and intensity impact on the long-lasting effects of coordinated reset stimulation. PLoS Comput Biol 14(5):e1006113. https://doi.org/10.1371/journal.pcbi.1006113
104. Tyulmankov D, Tass PA, Bokil H (2018) Periodic flashing coordinated reset stimulation paradigm reduces sensitivity to on and off period durations. PLoS ONE 13(9):e0203782. https://doi.org/10.1371/journal.pone.0203782
105. Kromer JA, Tass PA (2020) Long-lasting desynchronization by decoupling stimulation. Phys Rev Res 2:033101. https://doi.org/10.1103/PhysRevResearch.2.033101
106. Khaledi-Nasab A, Kromer JA, Tass PA (2021) Long-lasting desynchronization of plastic neural networks by random reset stimulation. Front Physiol 11:622620. https://doi.org/10.3389/fphys.2020.622620
107. Kromer JA, Tass PA (2022) Synaptic reshaping of plastic neuronal networks by periodic multichannel stimulation with single-pulse and burst stimuli. PLoS Comput Biol 18(11):e1010568. https://doi.org/10.1371/journal.pcbi.1010568
108. Asl MM, Valizadeh A, Tass PA (2023) Decoupling of interacting neuronal populations by time-shifted stimulation through spike-timing-dependent plasticity. PLoS Comput Biol 19(2):e1010853. https://doi.org/10.1371/journal.pcbi.1010853
109. Hauptmann C, Tass PA (2009) Cumulative and after-effects of short and weak coordinated reset stimulation: a modeling study. J Neural Eng 6(1):016004. https://doi.org/10.1088/1741-2560/6/1/016004
110. Popovych OV, Xenakis MN, Tass PA (2015) The spacing principle for unlearning abnormal neuronal synchrony. PLoS ONE 10(2):e0117205. https://doi.org/10.1371/journal.pone.0117205
111. Popovych OV, Tass PA (2012) Desynchronizing electrical and sensory coordinated reset neuromodulation. Front Hum Neurosci 6:58. https://doi.org/10.3389/fnhum.2012.00058
112. Tass PA, Qin L, Hauptmann C, Dovero S, Bezard E, Boraud T, Meissner WG (2012) Coordinated reset has sustained aftereffects in parkinsonian monkeys. Ann Neurol 72(5):816–820. https://doi.org/10.1002/ana.23663
113. Butz M, Wörgötter F, van Ooyen A (2009) Activity-dependent structural plasticity. Brain Res Rev 60(2):287. https://doi.org/10.1016/j.brainresrev.2008.12.023
114. Butz M, van Ooyen A (2013) A simple rule for dendritic spine and axonal bouton formation can account for cortical reorganization after focal retinal lesions. PLoS Comput Biol 9:e1003259. https://doi.org/10.1371/journal.pcbi.1003259

115. Manos T, Diaz-Pier S, Tass PA (2021) Long-term desynchronization by coordinated reset stimulation in a neural network model with synaptic and structural plasticity. Front Physiol 12:716556. https://doi.org/10.3389/fphys.2021.716556
116. Chauhan K, Khaledi-Nasab A, Neiman AB, Tass PA (2022) Dynamics of phase oscillator networks with synaptic weight and structural plasticity. Sci Rep 12(1):15003. https://doi.org/10.1038/s41598-022-19445-3
117. Maistrenko YL, Lysyansky B, Hauptmann C, Burylko O, Tass PA (2007) Multistability in the Kuramoto model with synaptic plasticity. Phys Rev E 75(6):066207. https://doi.org/10.1103/PhysRevE.75.066207
118. Wang J, Nebeck S, Muralidharan A, Johnson MD, Vitek JL, Baker KB (2016) Coordinated reset deep brain stimulation of subthalamic nucleus produces long-lasting, dose-dependent motor improvements in the 1-methyl-4-phenyl-1,2,3,6-tetrahydropyridine non-human primate model of parkinsonism. Brain Stimul 9(4):609–617. https://doi.org/10.1016/j.brs.2016.03.014
119. Wang J, Fergus SP, Johnson LA, Nebeck SD, Zhang J, Kulkarni S, Bokil H, Molnar GF, Vitek JL (2022) Shuffling improves the acute and carryover effect of subthalamic coordinated reset deep brain stimulation. Front Neurol 13:716046. https://doi.org/10.3389/fneur.2022.716046
120. Chelangat Bore J, Campbell BA, Cho H, Pucci F, Gopalakrishnan R, Machado AG, Baker BK (2022) Long-lasting effects of subthalamic nucleus coordinated reset deep brain stimulation in the non-human primate model of Parkinsonism: a case report. Brain Stimul 15(3):598–600. https://doi.org/10.1016/j.brs.2022.04.005
121. Tass PA (2017) Vibrotactile coordinated reset stimulation for the treatment of neurological diseases: concepts and device specifications. Cureus 9(8):e1535. https://doi.org/10.7759/cureus.1535
122. Syrkin-Nikolau J, Neuville R, O'Day J, Anidi C, Miller Koop M, Martin T, Tass PA, Bronte-Stewart H (2018) Coordinated reset vibrotactile stimulation shows prolonged improvement in Parkinson's disease. Mov Disord 33(1):179–180. https://doi.org/10.1002/mds.27223
123. Pfeifer KJ, Kromer JA, Cook AJ, Hornbeck T, Lim EA, Mortimer BJP, Fogarty AS, Han SS, Dhall R, Halpern CH, Tass PA (2021) Coordinated reset vibrotactile stimulation induces sustained cumulative benefits in Parkinson's disease. Front Physiol 12:624317. https://doi.org/10.3389/fphys.2021.624317
124. Tass PA (2021) Vibrotactile coordinated reset stimulation for the treatment of Parkinson's disease. Neural Regen Res 17(7):1495–1497. https://doi.org/10.4103/1673-5374.329001
125. Esposito E, Di Matteo V, Di Giovanni G (2007) Death in the substantia nigra: a motor tragedy. Expert Rev Neurother 7(6):677–697. https://doi.org/10.1586/14737175.7.6.677
126. Haelterman NA, Yoon WH, Sandoval H, Jaiswal M, Shulman JM, Bellen HJ (2014) A mitocentric view of Parkinson's disease. Annu Rev Neurosci 37:137–159. https://doi.org/10.1146/annurev-neuro-071013-014317
127. McGregor MM, Nelson AB (2019) Circuit mechanisms of Parkinson's disease. Neuron 101(6):1042–1056. https://doi.org/10.1016/j.neuron.2019.03.004
128. Krizhevsky A, Sutskever I, Hinton G (2012) ImageNet classification with deep convolutional neural networks. In: Advances in neural information processing systems, vol 25. Curran Associates, Inc., pp 1097–1105. https://doi.org/10.5555/2999134.2999257
129. LeCun Y, Bengio Y, Hinton G (2015) Deep learning. Nature 521(7553):436–444. https://doi.org/10.1038/nature14539
130. Simonyan K, Zisserman A (2015) Very deep convolutional networks for large-scale image recognition. In: International conference on learning representations. ICLR
131. He K, Zhang X, Ren S, Sun J (2016) Deep residual learning for image recognition. In: Proceedings of the IEEE conference on computer vision and pattern recognition. IEEE, pp 770–778. https://doi.org/10.1109/CVPR.2016.90
132. Dosovitskiy A, Beyer L, Kolesnikov A, Weissenborn D, Zhai X, Unterthiner T, Dehghani M, Minderer M, Heigold G, Gelly S, Uszkoreit J, Houlsby N (2021) An image is worth 16x16 words: transformers for image recognition at scale. In: International conference on learning representations. Proceedings

133. Hinton G, Deng L, Yu D, Dahl GE, Mohamed AR, Jaitly N, Senior A, Vanhoucke V, Nguyen P, Sainath TN et al (2012) Deep neural networks for acoustic modeling in speech recognition: the shared views of four research groups. IEEE Signal Process Mag 29(6):82–97. https://doi.org/10.1109/MSP.2012.2205597
134. Sutskever I, Vinyals O, Le Q (2014) Sequence to sequence learning with neural networks. In: Ghahramani Z, Welling M, Cortes C, Lawrence ND, Weinberger KQ (eds) Advances in neural information processing systems, vol 27. Curran Associates, Inc., pp 3104–3112. https://doi.org/10.5555/2969033.2969173
135. Vaswani A, Shazeer N, Parmar N, Uszkoreit J, Jones L, Gomez AN, Kaiser Ł, Polosukhin I (2017) Attention is all you need. In: Advances in neural information processing systems, vol 30. Curran Associates, Inc., pp 5998–6008. https://doi.org/10.5555/3295222.3295349
136. Devlin J, Chang MW, Lee K, Toutanova K (2019) BERT: pre-training of deep bidirectional transformers for language understanding. In: Proceedings of NAACL-HLT. Proceedings, pp 4171–4186. http://arxiv.org/abs/1810.04805
137. Silver D, Huang A, Maddison CJ, Guez A, Sifre L, van den Driessche G, Schrittwieser J, Antonoglou I, Panneershelvam V, Lanctot M, Dieleman S, Grewe D, Nham J, Kalchbrenner N, Sutskever I, Lillicrap T, Leach M, Kavukcuoglu K, Graepel T, Hassabis D (2016) Mastering the game of Go with deep neural networks and tree search. Nature 529(7587):484–489. https://doi.org/10.1038/nature16961
138. Silver D, Schrittwieser J, Simonyan K, Antonoglou I, Huang A, Guez A, Hubert T, Baker L, Lai M, Bolton A, Chen Y, Lillicrap T, Hui F, Sifre L, van den Driessche G, Graepel T, Hassabis D (2017) Mastering the game of Go without human knowledge. Nature 550(7676):354–359. https://doi.org/10.1038/nature24270
139. Bowling M, Burch N, Johanson M, Tammelin O (2015) Heads-up limit hold'em poker is solved. Science 347(6218):145–149. https://doi.org/10.1126/science.1259433
140. Brown N, Sandholm T (2018) Superhuman AI for heads-up no-limit poker: Libratus beats top professionals. Science 359(6374):418–424. https://doi.org/10.1126/science.aao1733
141. Brown N, Sandholm T (2019) Superhuman ai for multiplayer poker. Science 365(6456):885–890. https://doi.org/10.1126/science.aay2400
142. Carleo G, Cirac I, Cranmer K, Daudet L, Schuld M, Tishby N, Vogt-Maranto L, Zdeborová L (2019) Machine Learning and the physical sciences. Rev Mod Phys 91:045002. https://doi.org/10.1103/RevModPhys.91.045002
143. Tunyasuvunakool K, Adler J, Wu Z, Green T, Zielinski M, Žídek A, Bridgland A, Cowie A, Meyer C, Laydon A, Velankar S, Kleywegt GJ, Bateman A, Evans R, Pritzel A, Figurnov M, Ronneberger O, Bates R, Kohl SAA, Potapenko A, Ballard AJ, Romera-Paredes B, Nikolov S, Jain R, Clancy E, Reiman D, Petersen S, Senior AW, Kavukcuoglu K, Birney E, Kohli P, Jumper J, Hassabis D (2021) Highly accurate protein structure prediction for the human proteome. Nature 596(7873):590–596. https://doi.org/10.1038/s41586-021-03828-1
144. Oliva A, Torralba A (2001) Modeling the shape of the scene: a holistic representation of the spatial envelope. Int J Comput Vis 42(3):145–175. https://doi.org/10.1023/A:1011139631724
145. Dalal N, Triggs B (2005) Histograms of oriented gradients for human detection. In: 2005 IEEE computer society conference on computer vision and pattern recognition (CVPR'05), vol 1. IEEE, Computer Vision and Pattern Recognition, pp 886–893. https://doi.org/10.1109/CVPR.2005.177
146. Bay H, Tuytelaars T, Van Gool L (2006) SURF: speeded up robust features. In: Leonardis A, Bischof H, Pinz A (eds) Computer vision-ECCV 2006. Springer, Berlin Heidelberg, Berlin, Heidelberg, pp 404–417. https://doi.org/10.1007/11744023_32
147. Goodfellow I, Bengio Y, Courville A (2016) Deep learning. MIT Press
148. MacKay DJ (2003) Information theory, inference and learning algorithms. Cambridge University Press
149. Hardt M, Recht B, Singer Y (2016) Train faster, generalize better: stability of stochastic gradient descent. In: Balcan MF, Weinberger KQ (eds) Proceedings of the 33rd international conference on machine learning. Proceedings of Machine Learning research, vol 48. PMLR, New York, New York, USA, pp 1225–1234

150. Wilson AC, Roelofs R, Stern M, Srebro N, Recht B (2017) The marginal value of adaptive gradient methods in machine learning. In: Guyon I, Luxburg UV, Bengio S, Wallach H, Fergus R, Vishwanathan S, Garnett R (eds) Advances in neural information processing systems, vol 30. Curran Associates, Inc
151. von Luxburg U, Bousquet O (2004) Distance-based classification with lipschitz functions. J Mach Learn Res 5:669–695
152. Bahri Y, Kadmon J, Pennington J, Schoenholz S, Sohl-Dickstein J, Ganguli S (2020) Statistical mechanics of deep learning. Annu Rev Condens Matter Phys 11(1):501–528. https://doi.org/10.1146/annurev-conmatphys-031119-050711
153. Zdeborová L (2020) Understanding Deep Learning is also a job for physicists. Nat Phys 16(6):602–604. https://doi.org/10.1038/s41567-020-0853-1
154. Geiger M, Petrini L, Wyart M (2021) Landscape and training regimes in deep learning. Phys Rep 924:1–18. https://doi.org/10.1016/j.physrep.2021.04.001
155. Grassberger P, Procaccia I (1983) Measuring the strangeness of strange attractors. Phys D Nonlinear Phenom 9(1–2):189–208. https://doi.org/10.1016/0167-2789(83)90252-4
156. Costa J, Hero A (2004) Learning intrinsic dimension and intrinsic entropy of high-dimensional datasets. In: 2004 12th European signal processing conference. Proceedings, pp 369–372
157. Levina E, Bickel P (2004) Maximum likelihood estimation of intrinsic dimension. In: Advances in neural information processing systems, vol 17. MIT Press
158. Spigler S, Geiger M, Wyart M (2020) Asymptotic learning curves of kernel methods: empirical data versus teacher-student paradigm. J Stat Mech Theory Exp 12:124001. https://doi.org/10.1088/1742-5468/abc61d
159. Pope P, Zhu C, Abdelkader A, Goldblum M, Goldstein T (2021) The intrinsic dimension of images and its impact on learning. In: International conference on learning representations. OpenReview.net
160. Bach F (2017) Breaking the curse of dimensionality with convex neural networks. J Mach Learn Res 18(1):629–681
161. Ghorbani B, Mei S, Misiakiewicz T, Montanari A (2019) Limitations of lazy training of two-layers neural network. In: Advances in neural information processing systems, vol 32. Proceedings, pp 9111–9121
162. Chizat L, Bach F (2020) Implicit bias of gradient descent for wide two-layer neural networks trained with the logistic loss. In: Conference on learning theory. PMLR, Proceedings of Machine Learning Research, pp 1305–1338
163. Ghorbani B, Mei S, Misiakiewicz T, Montanari A (2020) When do neural networks outperform kernel methods? In: Advances in neural information processing systems, vol 33. Proceedings
164. Geiger M, Spigler S, Jacot A, Wyart M (2020) Disentangling feature and lazy training in deep neural networks. J Stat Mech Theory Exp 11:113301
165. Daniely A, Malach E (2020) Learning parities with neural networks. In: Advances in neural information processing systems. Proceedings of the 34th international conference on neural information processing systems, vol 33 (2020)
166. Refinetti M, Goldt S, Krzakala F, Zdeborova L (2021) Classifying high-dimensional Gaussian mixtures: where kernel methods fail and neural networks succeed. In: Meila M, Zhang T (eds) Proceedings of the 38th international conference on machine learning. Proceedings of machine learning research, vol 139. PMLR, pp 8936–8947
167. Deng J, Dong W, Socher R, Li LJ, Li K, Fei-Fei L (2009) ImageNet: a large-scale hierarchical image database. In: 2009 IEEE conference on computer vision and pattern recognition. IEEE, Proceedings Computer Vision and Pattern Recognition, pp 248–255. https://doi.org/10.1109/CVPR.2009.5206848
168. Chung S, Lee D, Sompolinsky H (2018) Classification and geometry of general perceptual manifolds. Phys Rev X 8:031003. https://doi.org/10.1103/PhysRevX.8.031003
169. Goldt S, Mézard M, Krzakala F, Zdeborová L (2020) Modeling the influence of data structure on learning in neural networks: the hidden manifold model. Phys Rev X 10(4):041044. https://doi.org/10.1103/PhysRevX.10.041044
170. Goldt S, Loureiro B, Reeves G, Krzakala F, Mézard M, Zdeborová L (2022) The Gaussian equivalence of generative models for learning with shallow neural networks. In: Bruna J, Hesthaven J, Zdeborová L (eds) Proceedings of the 2nd mathematical and scientific machine learning conference. Proceedings of machine learning research, vol 145. PMLR, pp 426–471

171. Richards D, Mourtada J, Rosasco L (2021) Asymptotics of ridge(less) regression under general source condition. In: Banerjee A, Fukumizu K (eds) Proceedings of the 24th international conference on artificial intelligence and statistics. Proceedings of machine learning research, vol 130. PMLR, pp 3889–3897
172. Torlai G, Melko RG (2016) Learning thermodynamics with Boltzmann machines. Phys Rev B 94:165134. https://doi.org/10.1103/PhysRevB.94.165134
173. Carrasquilla J, Melko RG (2017) Machine learning phases of matter. Nat Phys 13(5):431–434. https://doi.org/10.1038/nphys4035
174. Van Nieuwenburg EP, Liu YH, Huber SD (2017) Learning phase transitions by confusion. Nat Phys 13(5):435–439. https://doi.org/10.1038/nphys4037
175. Wirnsberger P, Ballard AJ, Papamakarios G, Abercrombie S, Racanière S, Pritzel A, Jimenez Rezende D, Blundell C (2020) Targeted free energy estimation via learned mappings. J Chem Phys 153(14):144112. https://doi.org/10.1063/5.0027136
176. Seif A, Hafezi M, Jarzynski C (2021) Machine learning the thermodynamic arrow of time. Nat Phys 17(1):105–113. https://doi.org/10.1038/s41567-020-01154-0
177. Kim DK, Bae Y, Lee S, Jeong H (Oct2020) Learning entropy production via neural networks. Phys Rev Lett 125:140604. https://doi.org/10.1103/PhysRevLett.125.140604
178. Bengio Y, Simard P, Frasconi P (1994) Learning long-term dependencies with gradient descent is difficult. IEEE Trans Neural Netw 5(2):157–166. https://doi.org/10.1109/72.279181
179. Hochreiter S, Schmidhuber J (1997) Long short-term memory. Neural Comput 9(8):1735–1780. https://doi.org/10.1162/neco.1997.9.8.1735
180. Argun A, Thalheim T, Bo S, Cichos F, Volpe G (2020) Enhanced force-field calibration via machine learning. Appl Phys Rev 7(4):041404. https://doi.org/10.1063/5.0022280
181. Pathak J, Hunt B, Girvan M, Lu Z, Ott E (2018) Model-free prediction of large spatiotemporally chaotic systems from data: a reservoir computing approach. Phys Rev Lett 120(2):024102. https://doi.org/10.1103/PhysRevLett.120.024102
182. Kuramoto Y (1978) Diffusion-induced chaos in reaction systems. Prog Theoret Phys Suppl 64:346–367. https://doi.org/10.1143/PTPS.64.346
183. Sivashinsky GI (1977) Nonlinear analysis of hydrodynamic instability in laminar flames—i. derivation of basic equations. Acta Astronaut 4(11):1177–1206. https://doi.org/10.1016/0094-5765(77)90096-0
184. Sivashinsky GI (1980) On flame propagation under conditions of stoichiometry. SIAM J Appl Math 39(1):67–82. https://doi.org/10.1137/0139007
185. Lukoševičius M, Jaeger H (2009) Reservoir computing approaches to recurrent neural network training. Comput Sci Rev 3(3):127–149. https://doi.org/10.1016/j.cosrev.2009.03.005
186. Seif A, Loos SA, Tucci G, Roldán É, Goldt S (2022) The impact of memory on learning sequence-to-sequence tasks. arXiv. https://doi.org/1048550/arXiv.2205.14683
187. Krenn M, Pollice R, Guo SY, Aldeghi M, Cervera-Lierta A, Friederich P, dos Passos Gomes G, Häse F, Jinich A, Nigam A, Yao Z, Aspuru-Guzik A (2022) On scientific understanding with artificial intelligence. Nat Rev Phys 4(12):761–769. https://doi.org/10.1038/s42254-022-00518-3
188. Ott E, Grebogi C, Yorke JA (1990) Controlling chaos. Phys Rev Lett 64:1196–1199. https://doi.org/10.1103/PhysRevLett.64.1196
189. Romeiras F, Grebogi C, Ott E, Dayawansa W (1992) Controlling chaotic dynamical systems. Phys D 58(1):165–192. https://doi.org/10.1016/0167-2789(92)90107-X
190. Sarangapani J (2006) Neural network control of nonlinear discrete-time systems. Taylor & Francis, Boca Raton. https://doi.org/10.1201/9781420015454
191. Jaeger H, Haas H (2004) Harnessing nonlinearity: predicting chaotic systems and saving energy in wireless communication. Science 304:78–80. https://doi.org/10.1126/science.1091277
192. Waegeman T, Wyffels F, Schrauwen B (2012) Feedback control by online learning an inverse model. IEEE Trans Neural Netw Learn Syst 23:1637–1648. https://doi.org/10.1109/TNNLS.2012.2203285

193. Canaday D, Pomerance A, Gauthier DJ (2021) Model-free control of dynamical systems with deep reservoir computing. J Phys Complex 2:035025. https://doi.org/10.1088/2632-072X/ac24f3
194. Gauthier DJ, Bollt E, Griffith A, Barbosa WAS (2021) Next generation reservoir computing. Nat Commun 12:5564. https://doi.org/10.1038/s41467-021-25801-6
195. Kermack WO, McKendrick AG (1927) A contribution to the mathematical theory of epidemics. Proc R Soc A 115(772):700–721. https://doi.org/10.1098/rspa.1927.0118
196. Lehnert J, Hövel P, Selivanov AA, Fradkov AL, Schöll E (2014) Controlling cluster synchronization by adapting the topology. Phys Rev E 90(4):042914. https://doi.org/10.1103/physreve.90.042914
197. Balcan D, Hu H, Goncalves B, Bajardi P, Poletto C, Ramasco JJ, Paolotti D, Perra N, Tizzoni M, Van den Broeck W, Colizza V, Vespignani A (2009) Seasonal transmission potential and activity peaks of the new influenza A(H1N1): a Monte Carlo likelihood analysis based on human mobility. BMC Med 7(1):45. https://doi.org/10.1186/1741-7015-7-45
198. Chung LH (2015) Impact of pandemic control over airport economics: Reconciling public health with airport business through a streamlined approach in pandemic control. J Air Transp Manag 44:42–53. https://doi.org/10.1016/j.jairtraman.2015.02.003
199. Poletto C, Gomes M, y Piontti AP, Rossi L, Bioglio L, Chao DL, Longini Jr I, Halloran ME, Colizza V, Vespignani A (2014) Assessing the impact of travel restrictions on international spread of the 2014 West African Ebola epidemic. Eurosurveillance 19(42):20936. https://doi.org/10.2807/1560-7917.ES2014.19.42.20936
200. Kyrychko YN, Blyuss KB, Brovchenko I (2020) Mathematical modelling of the dynamics and containment of COVID-19 in Ukraine. Sci Rep 10(1):19662. https://doi.org/10.1038/s41598-020-76710-1
201. Maier BF, Brockmann D (2020) Effective containment explains subexponential growth in recent confirmed COVID-19 cases in china. Science 368(6492):742–746. https://doi.org/10.1126/science.abb4557
202. Prem K, Liu Y, Russell TW, Kucharski AJ, Eggo RM, Davies N, Flasche S, Clifford S, Pearson CA, Munday JD et al (2020) The effect of control strategies to reduce social mixing on outcomes of the COVID-19 epidemic in Wuhan, China: a modelling study. The Lancet Public Health 5(5):e261–e270. https://doi.org/10.1016/S2468-2667(20)30073-6
203. Humphries R, Spillane M, Mulchrone K, Wieczorek S, O'Riordain M, Hövel P (2021) A metapopulation network model for the spreading of SARS-CoV-2: case study for Ireland. Infect Dis Modell 6:420–437. https://doi.org/10.1016/j.idm.2021.07.002
204. Brockmann D, Helbing D (2013) The hidden geometry of complex, network-driven contagion phenomena. Science 342(6164):1337–1342. https://doi.org/10.1126/science.1245200
205. Iannelli F, Koher A, Brockmann D, Hövel P, Sokolov IM (2017) Effective distances for epidemics spreading on complex networks. Phys Rev E 95(1):012313. https://doi.org/10.1103/physreve.95.012313
206. Gold L, Balal E, Horak T, Cheu RL, Mehmetoglu T, Gurbuz O (2019) Health screening strategies for international air travelers during an epidemic or pandemic. J Air Transp Manag 75:27–38. https://doi.org/10.1016/j.jairtraman.2018.11.006
207. Cousins S (2020) New Zealand eliminates COVID-19. The Lancet 395(10235):1474. https://doi.org/10.1016/S0140-6736(20)31097-7
208. Czypionka T, Iftekhar EN, Prainsack B, Priesemann V, Bauer S, Valdez AC, Cuschieri S, Glaab E, Grill E, Krutzinna J et al (2022) The benefits, costs and feasibility of a low incidence COVID-19 strategy. Lancet Reg Health Eur 13:100294. https://doi.org/10.1016/j.lanepe.2022.100294
209. Gross T, D'Lima CJD, Blasius B (2006) Epidemic dynamics on an adaptive network. Phys Rev Lett 96(20):208701. https://doi.org/10.1103/physrevlett.96.208701
210. Gross T, Blasius B (2008) Adaptive coevolutionary networks: a review. J R Soc Interface 5(20):259–271. https://doi.org/10.1098/rsif.2007.1229
211. Watts DJ, Strogatz SH (1998) Collective dynamics of "small-world" networks. Nature 393:440–442. https://doi.org/10.1038/30918

212. Bajardi P, Barrat A, Savini L, Colizza V (2012) Optimizing surveillance for livestock disease spreading through animal movements. J R Soc Interface 9(76):2814–2825. https://doi.org/10.1098/rsif.2012.0157
213. Schöll E (2021) Partial synchronization patterns in brain networks. Europhys Lett 136:18001. https://doi.org/10.1209/0295-5075/ac3b97
214. Belik V, Mikolajczyk R, Hövel P (2016) Control of epidemics on hospital networks. In: Schöll E, Klapp SHL, Hövel P (eds) Control of self-organizing nonlinear systems, chap 22. Springer, Berlin, Heidelberg, pp 431–440. https://doi.org/10.1007/978-3-319-28028-8_22
215. Belik V, Karch A, Hövel P, Mikolajczyk R (2017) Leveraging topological and temporal structure of hospital referral networks for epidemic control. In: Masuda N, Holme P (eds) Temporal network epidemiology. Theoretical biology, chap 9. Springer, Singapore, pp 199–214.https://doi.org/10.1007/978-981-10-5287-3_9
216. Masuda N, Holme P (2017) Temporal network epidemiology. Springer, Singapore. https://doi.org/10.1007/978-981-10-5287-3
217. Holme P, Saramäki J (2019) Temporal network theory, vol 2. Springer. https://doi.org/10.1007/978-3-031-30399-9
218. Valdano E, Ferreri L, Poletto C, Colizza V (2015) Analytical computation of the epidemic threshold on temporal networks. Phys Rev X 5(2):021005. https://doi.org/10.1103/PhysRevX.5.021005
219. Frasca M, Sharkey KJ (2016) Discrete-time moment closure models for epidemic spreading in populations of interacting individuals. J Theor Biol 399:13–21. https://doi.org/10.1016/j.jtbi.2016.03.024
220. Koher A, Lentz HHK, Gleeson JP, Hövel P (2019) Contact-based model for epidemic spreading on temporal networks. Phys Rev X 9:031017. https://doi.org/10.1103/physrevx.9.031017
221. Humphries R, Mulchrone K, Tratalos J, More SJ, Hövel P (2021) A systematic framework of modelling epidemics on temporal networks. Appl Netw Sci 6(1):1–19. https://doi.org/10.1007/s41109-021-00400-9
222. Liu YY, Slotine JJ, Barabási AL (2011) Controllability of complex networks. Nature 473(7346):167–173. https://doi.org/10.1038/nature10011
223. Pósfai M, Hövel P (2014) Structural controllability of temporal networks. New J Phys 16(12):123055. https://doi.org/10.1088/1367-2630/16/12/123055
224. Pósfai M, Gao J, Cornelius SP, Barabási AL, D'Souza RM (2016) Controllability of multiplex, multi-time-scale networks. Phys Rev E 94(3):032316. https://doi.org/10.1103/PhysRevE.94.032316
225. Gross T, Blasius B (2008) Adaptive coevolutionary networks: a review. J R Soc Interface 5:259–271. https://doi.org/10.1098/rsif.2007.1229
226. Clifford P, Sudbury A (1973) A model for spatial conflict. Biometrika 60:581–588. https://doi.org/10.1093/biomet/60.3.581
227. Holley RA, Liggett TM (1975) Ergodic theorems for weakly interacting infinite systems and the voter model. Ann Probab 3:643–663. https://doi.org/10.1214/aop/1176996306
228. Vazquez F, Eguíluz VM, Miguel MS (2008) Generic absorbing transition in coevolution dynamics. Phys Rev Lett 100(108702). https://doi.org/10.1103/PhysRevLett.100.108702
229. Demirel G, Vazquez F, Böhme GA, Gross T (2014) Moment-closure approximations for discrete adaptive networks. Phys D 267:68–80. https://doi.org/10.1016/j.physd.2013.07.003
230. Durrett R, Gleeson JP, Lloyd AL, Mucha PJ, Shi F, Sivakoff D, Socolar JES, Varghese C (2012) Graph fission in an evolving voter model. Proc Natl Acad Sci USA 109:3682–3687. https://doi.org/10.1073/pnas.1200709109
231. Gross T, Dommar D'Lima CJ, Blasius B (2006) Epidemic dynamics on an adaptive network. Phys Rev Lett 96(208701). https://doi.org/10.1103/PhysRevLett.96.208701
232. Kuehn C, Bick C (2021) A universal route to explosive phenomena. Sci Adv 7(eabe3824). https://doi.org/10.1126/sciadv.abe3824
233. Zschaler G, Böhme GA, Seißinger M, Huepe C, Gross T (2012) Early fragmentation in the adaptive voter model on directed networks. Phys Rev E 85(046107). https://doi.org/10.1103/PhysRevE.85.046107

234. Böhme GA, Gross T (2011) Analytical calculation of fragmentation transitions in adaptive networks. Phys Rev E 83:035101. https://doi.org/10.1103/PhysRevE.83.035101
235. Kozma B, Barrat A (2008) Consensus formation on adaptive networks. Phys Rev E 77:016102. https://doi.org/10.1103/PhysRevE.77.016102
236. Yu Y, Xiao G, Li G, Tay WP, Teoh HF (2017) Opinion diversity and community formation in adaptive networks. Chaos 27(103115). https://doi.org/10.1063/1.4989668
237. Zhang X, Shan C, Jin Z, Zhu H (2019) Complex dynamics of epidemic models on adaptive networks. J Differ Equ 266:803–832. https://doi.org/10.1016/j.jde.2018.07.054
238. Ogura M, Preciado VM (2016) Epidemic processes over adaptive state-dependent networks. Phys Rev E 93(062316). https://doi.org/10.1103/PhysRevE.93.062316
239. Horstmeyer L, Kuehn C, Thurner S (2018) Network topology near criticality in adaptive epidemics. Phys Rev E 98(042313). https://doi.org/10.1103/PhysRevE.98.042313
240. Horstmeyer L, Kuehn C, Thurner S (2022) Balancing quarantine and self-distancing measures in adaptive epidemic networks. Bull Math Biol 84(79). https://doi.org/10.1007/s11538-022-01033-3
241. Avalos-Gaytán V, Almendral JA, Leyva I, Battiston F, Nicosia V, Latora V, Boccaletti S (2018) Emergent explosive synchronization in adaptive complex networks. Phys Rev E 97(042301). https://doi.org/10.1103/PhysRevE.97.042301
242. Berner R, Fialkowski J, Kasatkin DV, Nekorkin VI, Yanchuk S, Schöll E (2019) Hierarchical frequency clusters in adaptive networks of phase oscillators. Chaos 29(10):103134. https://doi.org/10.1063/1.5097835
243. Berner R, Schöll E, Yanchuk S (2019) Multiclusters in networks of adaptively coupled phase oscillators. SIAM J Appl Dyn Syst 18(4):2227–2266. https://doi.org/10.1137/18m1210150
244. Horstmeyer L, Kuehn C (2020) Adaptive voter model on simplicial complexes. Phys Rev E 101(022305). https://doi.org/10.1103/PhysRevE.101.022305
245. Papanikolaou N, Vaccario G, Hormann E, Lambiotte R, Schweitzer F (2022) Consensus from group interactions: an adaptive voter model on hypergraphs. Phys Rev E 105(054307). https://doi.org/10.1103/PhysRevE.105.054307
246. Alvarez LW, Alvarez W, Asaro F, Michel HV (1980) Extraterrestrial cause for the cretaceous-tertiary extinction. Science 208(4448):1095–1108. https://doi.org/10.1126/science.208.4448.1095
247. Brugger J, Feulner G, Petri S (2017) Baby, it's cold outside: climate model simulations of the effects of the asteroid impact at the end of the cretaceous. Geophys Res Lett 44(1):419–427. https://doi.org/10.1002/2016GL072241
248. Sparks R (1978) Gas release rates from pyroclastic flows: a assessment of the role of fluidisation in their emplacement. Bull Volcanol 41(1):1–13. https://doi.org/10.1007/BF02597679
249. Williams MA, Ambrose SH, van der Kaars S, Ruehlemann C, Chattopadhyaya U, Pal J, Chauhan PR (2009) Environmental impact of the 73 ka toba super-eruption in south asia. Palaeogeogr Palaeoclimatol Palaeoecol 284(3–4):295–314. https://doi.org/10.1016/j.palaeo.2009.10.009
250. Armstrong McKay DI, Staal A, Abrams JF, Winkelmann R, Sakschewski B, Loriani S, Fetzer I, Cornell SE, Rockström J, Lenton TM (2022) Exceeding 1.5 °c global warming could trigger multiple climate tipping points. Science 377(6611):eabn7950. https://doi.org/10.1126/science.abn7950
251. IPCC (2023) Climate Change 2021-the physical science basis: working group I contribution to the sixth assessment report of the intergovernmental panel on climate change. Cambridge University Press. https://doi.org/10.1017/9781009157896
252. Lenton TM, Held H, Kriegler E, Hall JW, Lucht W, Rahmstorf S, Schellnhuber HJ (2008) Tipping elements in the earth's climate system. Proc Natl Acad Sci 105(6):1786–1793. https://doi.org/10.1073/pnas.0705414105
253. Liu T, Chen D, Yang L, Meng J, Wang Z, Ludescher J, Fan J, Yang S, Chen D, Kurths J, Chen X, Havlin S, Schellnhuber HJ (2023) Teleconnections among tipping elements in the earth system. Nat Climate Change 13(1):67–74. https://doi.org/10.1038/s41558-022-01570-6

254. Wunderling N, Donges JF, Kurths J, Winkelmann R (2021) Interacting tipping elements increase risk of climate domino effects under global warming. Earth Syst Dyn 12(2):601–619. https://doi.org/10.5194/esd-12-601-2021
255. Sawicki J, Berner R, Löser T, Schöll E (2022) Modelling tumor disease and sepsis by networks of adaptively coupled phase oscillators. Front Netw Physiol 1:730385. https://doi.org/10.3389/fnetp.2021.730385
256. Berner R, Sawicki J, Thiele M, Löser T, Schöll E (2022) Critical parameters in dynamic network modeling of sepsis. Front Netw Physiol 2:904480. https://doi.org/10.3389/fnetp.2022.904480
257. Löser T (2020) Aspects of tumor progression. Med Hypotheses 144:110157. https://doi.org/10.1016/j.mehy.2020.110157
258. Bomans K, Schenz J, Sztwiertnia I, Schaack D, Weigand MA, Uhle F (2018) Sepsis induces a long-lasting state of trained immunity in bone marrow monocytes. Front Immunol 9:2685. https://doi.org/10.3389/fimmu.2018.02685
259. Male D, Brostoff J, Roth D, Roitt I (2012) Immunology, Saunders, Philadelphia, 8th edn
260. Rich RR, Chaplin DD (2019) The human immune response, chap 1. Elsevier, Amsterdam, pp 3–17. https://doi.org/10.1016/b978-0-7020-6896-6.00001-6
261. Weinberg RA (2014) The biology of cancer, 2nd edn. Garland Publishing Inc., Milton. https://doi.org/10.1201/9780429258794
262. Singer M, Deutschman CS, Seymour CW, Shankar-Hari M, Annane D, Bauer M, Bellomo R, Bernard GR, Chiche J, Coopersmith CM, Hotchkiss RS, Levy MM, Marshall JC, Martin GS, Opal SM, Rubenfeld GD, van der Poll T, Vincent JL, Angus DC (2016) The third international consensus definitions for sepsis and septic shock (sepsis-3). JAMA 315:801. https://doi.org/10.1001/jama.2016.0287
263. Longo DL (2011) Harrison's Hematology and Oncology. McGraw-Hill Companies, New York City
264. Zhang Y, Zuo C, Liu L, Hu Y, Yang B, Qiu S, Li Y, Cao D, Ju Z, Ge J, Wang Q, Wang T, Bai L, Yang Y, Li G, Shao Z, Gao Y, Li Y, Bian R, Miao H, Li L, Li X, Jiang C, Yan S, Wang Z, Wang Z, Cui X, Huang W, Xiang D, Wang C, Li Q, Wu X, Gong W, Liu Y, Shao R, Liu F, Li M, Chen L, Liu Y (2024) Single-cell RNA-sequencing atlas reveals an mdk-dependent immunosuppressive environment in erbb pathway-mutated gallbladder cancer. J Hepatol 75(5):1128–1141. https://doi.org/10.1016/j.jhep.2021.06.023
265. Altan-Bonnet G, Mukherjee R (2019) Cytokine-mediated communication: a quantitative appraisal of immune complexity. Nat Rev Immunol 19(4):205–217. https://doi.org/10.1038/s41577-019-0131-x
266. Morán GAG, Parra-Medina R, Cardona AG, Quintero-Ronderos P, Rodríguez ÉG (2013) Cytokines, chemokines and growth factors. In: Anaya JM, Shoenfeld Y, Rojas-Villarraga A, Levy RA, Cervera R (eds) Autoimmunity: from bench to bedside, chap 9. El Rosario University Press, Bogota, Colombia, pp 133–168
267. Schmidt K, Gensichen J, Fleischmann-Struzek C, Bahr V, Pausch C, Sakr Y, Reinhart K, Vollmar HC, Thiel P, Scherag A, Gantner J, Brunkhorst FM (2020) Long-term survival following sepsis: results of a single-center registry study with 4-year follow-up. Dtsch Arztebl Int 117(46):775. https://doi.org/10.3238/arztebl.2020.0775
268. Brunkhorst FM, Weigand MA, Pletz M, Gastmeier P, Lemmen SW, Meier-Hellmann A, Ragaller M, Weyland A, Marx G, Bucher M, Gerlach H, Salzberger B, Grabein BT, Werdan WK, Kluge S, Bone HG, Putensen C, Rossaint R, Quintel M, Spies C, Weiß B, John S, Oppert M, Jörres A, Brenner T, Elke G, Gründling M, Mayer K, Weimann A, Felbinger TW, Axer HDSGEV (2020) S3-Leitlinie Sepsis-Prävention, Diagnose, Therapie und Nachsorge. Medizinische Klinik-Intensivmedizin und Notfallmedizin 115(2):37–109. https://doi.org/10.1007/s00063-020-00685-0
269. Weis S, Pletz MW, Bauer M (2017) Sepsis–neue Definition, neue Kontroversen. In: Epidemiologisches bulletin, No 37, Robert Koch-Institut, Epidemiologie und Gesundheitsberichterstattung. https://doi.org/10.17886/epibull-2017-049.2

270. Vineis P, Schatzkin A, Potter JD (2010) Models of carcinogenesis: an overview. Carcinogenesis 31:1703. https://doi.org/10.1093/carcin/bgq087
271. Tracey KJ, Beutler B, Lowry SF, Merryweather J, Wolpe S, Milsark IW, Hariri RJ, Fahey TJR, Zentella A, Albert JD (1986) Shock and tissue injury induced by recombinant human cachectin. Sci 234(4775):470–474. https://doi.org/10.1126/science.3764421
272. Hotchkiss RS, Moldawer LL (2014) Parallels between cancer and infectious disease. N Engl J Med 371(4):380–383. https://doi.org/10.1056/nejmcibr1404664
273. Toussaint S, Gerlach H (2009) Activated protein C for sepsis. N Engl J Med 361(27):2646–2652. https://doi.org/10.1056/NEJMct0808063
274. Gerlach H (2016) Agents to reduce cytokine storm. F1000Res 5:2909. https://doi.org/10.12688/f1000research.9092.1
275. Pikovsky A, Rosenblum M, Kurths J (2001) Synchronization: a universal concept in nonlinear sciences, 1st edn. Cambridge University Press, Cambridge. https://doi.org/10.1017/CBO9780511755743
276. Strogatz SH (2001) Exploring complex networks. Nature 410:268–276. https://doi.org/10.1038/35065725
277. Albert R, Barabási AL (2002) Statistical mechanics of complex networks. Rev Mod Phys 74:47–97. https://doi.org/10.1103/revmodphys.74.47
278. Newman MEJ (2003) The structure and function of complex networks. SIAM Rev 45(2):167–256. https://doi.org/10.1137/s0036144503
279. Boccaletti S, Pisarchik AN, del Genio CI, Amann A (2018) Synchronization: from coupled systems to complex networks. Cambridge University Press, Cambridge. https://doi.org/10.1017/9781107297111
280. Berner R (2021) Patterns of synchrony in complex networks of adaptively coupled oscillators. Springer Theses, Springer, Cham. https://doi.org/10.1007/978-3-030-74938-5
281. Dahms T, Lehnert J, Schöll E (2012) Cluster and group synchronization in delay-coupled networks. Phys Rev E 86(1):016202. https://doi.org/10.1103/physreve.86.016202
282. Kuramoto Y, Battogtokh D (2002) Coexistence of coherence and incoherence in nonlocally coupled phase oscillators. Nonlin Phen Complex Syst 5(4):380–385. https://doi.org/10.1088/1468-9604/5/4/310
283. Abrams DM, Strogatz SH (2004) Chimera states for coupled oscillators. Phys Rev Lett 93(17):174102. https://doi.org/10.1103/physrevlett.93.174102
284. Panaggio MJ, Abrams DM (2015) Chimera states: coexistence of coherence and incoherence in networks of coupled oscillators. Nonlinearity 28:R67. https://doi.org/10.1088/0951-7715/28/3/r67
285. Sawicki J (2019) Delay controlled partial synchronization in complex networks. Springer Theses, Springer, Heidelberg. https://doi.org/10.1007/978-3-030-34076-6
286. Schöll E (2020) Chimeras in physics and biology: synchronization and desynchronization of rhythms. Nova Acta Leopold 425:67–95. https://doi.org/10.26164/leopoldina_10_00275
287. Schöll E, Zakharova A, Andrzejak RG (2020) Chimera states in complex networks. Front Appl Math Stat. Frontiers Media SA, Lausann, Research topic. https://doi.org/10.3389/978-2-88963-311-1
288. Seliger P, Young SC, Tsimring LS (2002) Plasticity and learning in a network of coupled phase oscillators. Phys Rev E 65(4):041906. https://doi.org/10.1103/physreve.65.041906
289. Aoki T, Aoyagi T (2009) Co-evolution of phases and connection strengths in a network of phase oscillators. Phys Rev Lett 102:034101. https://doi.org/10.1103/physrevlett.102.034101
290. Timms L, English LQ (2014) Synchronization in phase-coupled Kuramoto oscillator networks with axonal delay and synaptic plasticity. Phys Rev E 89(3):032906. https://doi.org/10.1103/physreve.89.032906
291. Kasatkin DV, Yanchuk S, Schöll E, Nekorkin VI (2017) Self-organized emergence of multilayer structure and chimera states in dynamical networks with adaptive couplings. Phys Rev E 96(6):062211. https://doi.org/10.1103/physreve.96.062211
292. Berner R, Vock S, Schöll E, Yanchuk S (2021) Desynchronization transitions in adaptive networks. Phys Rev Lett 126(2):028301. https://doi.org/10.1103/physrevlett.126.028301

293. Berner R, Yanchuk S, Schöll E (2021) What adaptive neuronal networks teach us about power grids. Phys Rev E 103:042315. https://doi.org/10.1103/physreve.103.042315
294. Kasatkin DV, Nekorkin VI (2018) Synchronization of chimera states in a multiplex system of phase oscillators with adaptive couplings. Chaos 28:093115. https://doi.org/10.1063/1.5031681
295. Berner R, Sawicki J, Schöll E (2020) Birth and stabilization of phase clusters by multiplexing of adaptive networks. Phys Rev Lett 124(8):088301. https://doi.org/10.1103/physrevlett.124.088301
296. Bashan A, Bartsch RP, Kantelhardt JW, Havlin S, Ivanov PC (2012) Network physiology reveals relations between network topology and physiological function. Nat Commun 31:702. https://doi.org/10.1038/ncomms1705
297. Ivanov PC, Bartsch RP (2014) Network physiology: mapping interactions between networks of physiologic networks, chap 10. In: Networks of networks: the last frontier of complexity. Springer, pp 203–222. https://doi.org/10.1007/978-3-319-37493-2_10
298. Bartsch RP, Liu KKL, Bashan A, Ivanov PC (2015) Network physiology: how organ systems dynamically interact. PLoS One 10:11. https://doi.org/10.1371/journal.pone.0142143
299. Moorman JR, Lake DE, Ivanov PC (2016) Early detection of sepsis-a role for network physiology? Crit Care Med 44:312. https://doi.org/10.1097/ccm.0000000000001548
300. Lin A, Liu KKL, Bartsch RP, Ivanov PC (2016) Delay-correlation landscape reveals characteristic time delays of brain rhythms and heart interactions. Phil Trans R Soc A 374:20150182. https://doi.org/10.1098/rsta.2015.0182
301. Costa F, Wiedenmann B, Schöll E, Tuszynski JA (2024) Emerging cancer therapies: targeting physiological networks and cellular bioelectrical differences with non-thermal systemic electromagnetic fields in the human body-a comprehensive review. Front Netw Physiol 4:1483401. https://doi.org/10.3389/fnetp.2023.1483401
302. Costa F, Tuszynski JA, Iemma AF, Trevizan WA, Wiedenmann B, Schöll E (2025) External low energy electromagnetic fields affect heart dynamics: surrogate for system synchronization, chaos control and cancer patient's health. Front Netw Physiol
303. Xu L, Chen Z, Hu K, Stanley HE, Ivanov PC (2006) Spurious detection of phase synchronization in coupled nonlinear oscillators. Phys Rev E 73:065201. https://doi.org/10.1103/physreve.73.065201
304. Chen Z, Hu K, Stanley HE, Novak V, Ivanov PC (2006) Cross-correlation of instantaneous phase increments in pressure-flow fluctuations: applications to cerebral autoregulation. Phys Rev E 73:031915. https://doi.org/10.1103/physreve.73.031915
305. Ivanov PC, Ma QDY, Bartsch RP (2009) Maternal-fetal heartbeat phase synchronization. Proc Natl Acad Sci USA 106:13641. https://doi.org/10.1073/pnas.0906987106
306. Bartsch RP, Schumann AY, Kantelhardt JW, Penzel T, Ivanov PC (2012) Phase transitions in physiologic coupling. Proc Natl Acad Sci USA 109:10181. https://doi.org/10.1073/pnas.1204568109
307. Chen L, Liu R, Liu ZP, Li M, Aihara K (2012) Detecting early-warning signals for sudden deterioration of complex diseases by dynamical network biomarkers. Sci Rep 2:342. https://doi.org/10.1038/srep00342
308. Liu R, Li M, Liu ZP, Wu J, Chen L, Aihara K (2012) Identifying critical transitions and their leading biomolecular networks in complex diseases. Sci Rep 2:813. https://doi.org/10.1038/srep00813
309. Liu R, Wang X, Aihara K, Chen L (May2014) Early diagnosis of complex diseases by molecular biomarkers, network biomarkers, and dynamical network biomarkers. Med Res Rev 34(3):455–478. https://doi.org/10.1002/med.21293
310. Liu R, Aihara K, Chen L (2013) Dynamical network biomarkers for identifying critical transitions and their driving networks of biologic processes. Quant Biol 1:105–114. https://doi.org/10.1007/s40484-013-0008-0
311. Shi J, Kirihara K, Tada M, Fujioka M, Usui K, Koshiyama D, Araki T, Chen L, Kasai K, Aihara K (2022) Criticality in the healthy brain. Front Netw Physiol 1:755685. https://doi.org/10.3389/fnetp.2021.755685

312. Jain S, Krishna S (2001) A model for the emergence of cooperation, interdependence, and structure in evolving networks. Proc Natl Acad Sci 98(2):543–547. https://doi.org/10.1073/pnas.98.2.543
313. Markram H, Lübke J, Frotscher M, Sakmann B (1997) Regulation of synaptic efficacy by coincidence of postsynaptic APs and EPSPs. Science 275(5297):213–215. https://doi.org/10.1126/science.275.5297.213
314. Abbott LF, Nelson S (2000) Synaptic plasticity: taming the beast. Nat Neurosci 3(11):1178–1183. https://doi.org/10.1038/81453
315. Meisel C, Gross T (2009) Adaptive self-organization in a realistic neural network model. Phys Rev E 80(6):061917. https://doi.org/10.1103/physreve.80.061917
316. Lücken L, Popovych OV, Tass PA, Yanchuk S (2016) Noise-enhanced coupling between two oscillators with long-term plasticity. Phys Rev E 93(3):032210. https://doi.org/10.1103/physreve.93.032210
317. Boccaletti S, Bianconi G, Criado R, del Genio CI, Gómez-Gardeñes J, Romance M, Sendiña Nadal I, Wang Z, Zanin M (2014) The structure and dynamics of multilayer networks. Phys Rep 544(1):1–122. https://doi.org/10.1016/j.physrep.2014.07.001
318. De Domenico M, Solé-Ribalta A, Cozzo E, Kivelä M, Moreno Y, Porter MA, Gómez S, Arenas A (2013) Mathematical formulation of multilayer networks. Phys Rev X 3(4):041022. https://doi.org/10.1103/physrevx.3.041022
319. De Domenico M, Nicosia V, Arenas A, Latora V (2015) Structural reducibility of multilayer networks. Nat Commun 6:6864. https://doi.org/10.1038/ncomms7864
320. Kivelä M, Arenas A, Barthélemy M, Gleeson JP, Moreno Y, Porter MA (2014) Multilayer networks. J Complex Netw 2(3):203–271. https://doi.org/10.1093/comnet/cnu016
321. Zhang X, Boccaletti S, Guan S, Liu Z (2015) Explosive synchronization in adaptive and multilayer networks. Phys Rev Lett 114(3):038701. https://doi.org/10.1103/physrevlett.114.038701
322. Maksimenko VA, Makarov VV, Bera BK, Ghosh D, Dana SK, Goremyko MV, Frolov NS, Koronovskii AA, Hramov AE (2016) Excitation and suppression of chimera states by multiplexing. Phys Rev E 94:052205. https://doi.org/10.1103/physreve.94.052205
323. Leyva I, Sevilla-Escoboza R, Sendiña-Nadal I, Gutiérrez R, Buldú JM, Boccaletti S (2017) Inter-layer synchronization in non-identical multi-layer networks. Sci Rep 7:45475. https://doi.org/10.1038/srep45475
324. Andrzejak RG, Ruzzene G, Malvestio I (2017) Generalized synchronization between chimera states. Chaos 27:053114. https://doi.org/10.1063/1.4983841
325. Sawicki J, Omelchenko I, Zakharova A, Schöll E (2018) Delay controls chimera relay synchronization in multiplex networks. Phys Rev E 98:062224. https://doi.org/10.1103/physreve.98.062224
326. Omelchenko I, Hülser T, Zakharova A, Schöll E (2019) Control of chimera states in multilayer networks. Front Appl Math Stat 4:67. https://doi.org/10.3389/fams.2018.00067
327. Rybalova E, Vadivasova T, Strelkova G, Anishchenko V, Zakharova A (2019) Forced synchronization of a multilayer heterogeneous network of chaotic maps in the chimera state mode. Chaos 29:033134. https://doi.org/10.1063/1.5090184
328. Nikitin D, Omelchenko I, Zakharova A, Avetyan M, Fradkov AL, Schöll E (2019) Complex partial synchronization patterns in networks of delay-coupled neurons. Phil Trans R Soc A 377:20180128. https://doi.org/10.1098/rsta.2018.0128
329. Drauschke F, Sawicki J, Berner R, Omelchenko I, Schöll E (2020) Effect of topology upon relay synchronization in triplex neuronal networks. Chaos 30:051104. https://doi.org/10.1063/5.0008341
330. Berner R, Mehrmann V, Schöll E, Yanchuk S (2021) The multiplex decomposition: an analytic framework for multilayer dynamical networks. SIAM J Appl Dyn Syst 20(4):1752–1772. https://doi.org/10.1137/21m1406180
331. Sawicki J, Koulen JM, Schöll E (2021) Synchronization scenarios in three-layer networks with a hub. Chaos 31:073131. https://doi.org/10.1063/5.0052443

332. Shepelev IA, Muni SS, Schöll E, Strelkova GI (2021) Repulsive inter-layer coupling induces anti-phase synchronization. Chaos 31:063116. https://doi.org/10.1063/5.0051050
333. Aoki T, Aoyagi T (2011) Self-organized network of phase oscillators coupled by activity-dependent interactions. Phys Rev E 84:066109. https://doi.org/10.1103/physreve.84.066109
334. Nekorkin VI, Kasatkin DV (2016) Dynamics of a network of phase oscillators with plastic couplings. AIP Conf Proc 1738(1):210010. https://doi.org/10.1063/1.4951993
335. Berner R, Sawicki J, Schöll E (2020) Birth and stabilization of phase clusters by multiplexing of adaptive networks. Phys Rev Lett 124(8):088301. https://doi.org/10.1103/physrevlett.124.088301
336. Berner R, Vock S, Schöll E, Yanchuk S (2021) Desynchronization transitions in adaptive networks. Phys Rev Lett 126(2):028301. https://doi.org/10.1103/physrevlett.126.028301
337. Leyva I, Sendiña-Nadal I, Sevilla-Escoboza R, Vera-Avila VP, Chholak P, Boccaletti S (2018) Relay synchronization in multiplex networks. Sci Rep 8(1):8629. https://doi.org/10.1038/s41598-018-26945-w
338. Franceschi C, Campisi J (2014) Chronic inflammation (inflammaging) and its potential contribution to age-associated diseases. J Gerontol A Biol Sci Med Sci 69:4. https://doi.org/10.1093/gerona/glu057
339. Calder PC, Bosco N, Bourdet-Sicard R, Capuron L, Delzenne N, Doré J, Franceschi C, Lehtinen MJ, Recker T, Salvioli S, Visioli F (2017) Health relevance of the modification of low grade inflammation in ageing (inflammageing) and the role of nutrition. Ageing Res Rev 40:95. https://doi.org/10.1016/j.arr.2017.09.001
340. Tragl KH (1999) Handbuch der internistischen Geriatrie. Springer. https://doi.org/10.1007/978-3-7091-8839-2
341. Fasshauer M, Klein J, Blüher M, Paschke R (2004) Adipokine: Mögliches Bindegliedzwischen Insulinresistenz und Adipositas. Dtsch Ärztebl Int 101:3491. https://doi.org/10.3238/arztebl.2004.3491
342. Gaillard RC (2007) Adipozyten: endokrine hochleistungsfabriken. Cardiovasc Med 10:163. https://doi.org/10.4414/cvm.2007.01248
343. Chovatiya R, Medzhitov R (2014) Stress, inflammation, and defense of homeostasis. Mol Cell 54:281. https://doi.org/10.1016/j.molcel.2014.03.030
344. Elisia I, Lam V, Hofs E, Li MY, Hay M, Cho B, Brooks-Wilson A, Rosin M, Bu L, Jia W, Krystal G (2017) Effect of age on chronic inflammation and responsiveness to bacterial and viral challenges. PLoS One 12:e0188881. https://doi.org/10.1371/journal.pone.0188881
345. Fulop T, Larbi A, Dupuis G, Le Page A, Frost EH, Cohen AA, Witkowski JM, Franceschi C (2018) Immunosenescence and inflamm-aging as two sides of the same coin: friends or foes? Front Immunol 8:1960. https://doi.org/10.3389/fimmu.2017.01960
346. Beneke G (1971) Altersabhängige Veränderung des Kollagens und der Bindegewebszellen, chap 1. Schattauer Verlag, Altern und Entwicklung, pp 1–37. https://doi.org/10.1007/978-3-642-61932-8_15
347. Nemetschek T (1971) Altersabhängige Abläufe am Kollagen, Altern und Entwicklung, vol 3, chap 3. Schattauer Verlag, pp 38–68
348. Virchow R (1978) Die krankhaften Geschwülste. Springer. https://doi.org/10.1007/978-3-642-66491-5_3
349. Brunkhorst FM, Weigand M, Pletz M, Gastmeier P, Lemmen SW, Meier-Hellmann A, Ragaller M, Weyland A, Marx G, Bucher M, Gerlach H, Salzberger B, Grabein B, Welte T, Werdan K, Kluge S, Bone HG, Putensen C, Rossaint R, Quintel M, Spies C, Weiß B, John S, Oppert M, Jörres A, Brenner T, Elke G, Gründling M, Mayer K, Weimann A, Felbinger TW, Axer H (2018) S3-leitlinie Sepsis-Prävention, Diagnose, Therapie und Nachsorge. https://doi.org/10.1007/s00063-020-00685-0
350. Storch V, Welsch U, Wink M (2013) Evolutionsbiologie, 3 edn. Springer Spektrum, Berlin, Heidelberg. https://doi.org/10.1007/978-3-642-32836-7
351. Delves PJ, Martin SJ, Burton DR, Roitt IM (2016) Roitt's essential immunology, 13th edn. Wiley

352. Rosendal C, Bouchon AI, Bierhaus A, Nawroth PP, Martin E, Bardenheuer HJ, Weigand MA (2004) Bedeutung der angeborenen immunantwort in der sepsis. Der Anaesthesist 53:10. https://doi.org/10.1007/s00101-003-0626-4
353. Egger G (2005) Die akute Entzündung. Springer, Wien New York. https://doi.org/10.1007/3-211-29899-1
354. Tisoncik JR, Korth MJ, Simmons CP, Farrar J, Martin TR, Katze MG (2012) Into the eye of the cytokine storm. Microbiol Mol Biol Rev 76:16–32. https://doi.org/10.1128/mmbr.05015-11
355. Schulte W, Bernhagen J, Bucala R (2013) Cytokines in sepsis: potent immunoregulators and potential therapeutic targets-an updated view. Mediators Inflamm 2013:165974. https://doi.org/10.1155/2016/842687
356. Hotchkiss RS, Moldawer LL, Opal SM, Reinhart K, Turnbull IR, Vincent JL (2016) Sepsis and septic shock. Nat Rev Dis Primers 2:16045. https://doi.org/10.1038/nrdp.2016.45
357. Xia S, Zhang X, Zheng S, Khanabdali R, Kalionis B, Wu J, Wan W, Tai X (2016) An update on inflamm-aging: mechanisms, prevention, and treatment. J Immunol Res 2016:8426874. https://doi.org/10.1155/2016/8426874
358. Thomas L (2020) Clinical laboratory diagnostics. https://www.clinical-laboratory-diagnostics-2020.com/
359. Löser T (2017) Process analysis of carcinogenesis: concept derivation of the tissue function "preservation of a homogeneous gene expression.". Theory Biosci 137:85. https://doi.org/10.1007/s12064-017-0256-z
360. Lamouille S, Xu J, Derynck R (2014) Molecular mechanisms of epithelial-mesenchymal transition. Nat Rev Mol Cell Biol 15:178. https://doi.org/10.1038/nrm3758
361. Chockley PJ, Keshamouni VG (2016) Immunological consequences of epithelial-mesenchymal transition in tumor progression. J Immunol 197:691. https://doi.org/10.4049/jimmunol.1600458
362. Wu Y, Zhou BP (2009) Inflammation: a driving force speeds cancer metastasis. Cell Cycle 8:3267. https://doi.org/10.4161/cc.8.20.9699
363. Karlsson MC, Gonzalez SF, Welin J, Fuxe J (2017) Epithelial-mesenchymal transition in cancer metastasis through the lymphatic system. Mol Oncol 11:781. https://doi.org/10.1002/1878-0261.12092
364. Mantovani A, Marchesi F, Malesci A, Laghi L, Allavena P (2017) Tumour-associated macrophages as treatment targets in oncology. Nat Rev Clin Oncol 14:399. https://doi.org/10.1038/nrclinonc.2016.217
365. Walther HE (1948) Krebsmetastasen. Benno Schwabe Verlag
366. Arends J, Bertz H, Bischoff SC, Fietkau R, Herrmann HJ, Holm E, Horneber M, Hütterer E, Körber J, Schmid I (2015) Klinische ernährung in der onkologie. Aktuel Ernahrungsmed 40:1. https://doi.org/10.1055/s-0035-1552741
367. Thomas L (1972) Germs. N Engl J Med 287:553. https://doi.org/10.1056/nejm197209142871109
368. Majetschak M, Schade VA (2001) Mechanismen der inflammatorischen Wirtsantwort bei schweren Infektionen. Intensivmedizin, Thieme Stuttgart, New York. https://doi.org/10.1055/b-002-7278
369. Seymour CW, Kennedy JN, Wang S, Chang CH, Elliott CF, Xu Z, Berry S, Clermont G, Cooper G, Gomez H, Huang DT, Kellum JA, Opal SM, Talisa V, van der Poll T, Visweswaran S, Vodovotz Y, Weiss JC, Yealy DM, Yende S, Angus DC (2019) Derivation, validation, and potential treatment implications of novel clinical phenotypes for sepsis. JAMA 321:20. https://doi.org/10.1001/jama.2019.5791
370. Franović I, Omel'chenko OE, Wolfrum M (2018) Phase-sensitive excitability of a limit cycle. Chaos 28(7):071105. https://doi.org/10.1063/1.5045179
371. Gastpar H (1982) Die Beeinflussung der Metastasierung über Gerinnungsmechanismen, chap unb. Thieme, pp 118–128. https://doi.org/10.1159/000406832
372. Eichinger S, Lechner K (2004) Hämorrhagische und thromboembolische Komplikationen bei malignen Erkrankungen, chap 34. Springer, pp 799–809. https://doi.org/10.1007/978-3-662-06670-6_34

373. Lippman ME (2016) Mammakarzinom, chap 39. ABW Wissenschaftsverlag, pp 633–649
374. Prescott HC, Osterholzer JJ, Langa KM, Angus DC, Iwashyna TJ (2016) Late mortality after sepsis: propensity matched cohort study. BMJ 353:i2375. https://doi.org/10.1136/bmj.i2375
375. Mostel Z, Perl A, Marck M, Mehdi SF, Lowell B, Bathija S, Santosh R, Pavlov VA, Chavan SS, Roth J (2019) Post-sepsis syndrome-an evolving entity that afflicts survivors of sepsis. Mol Med 26:6. https://doi.org/10.1186/s10020-019-0132-z
376. Junqueira LC, Carneiro J, Kelley RO (1995) Basic histology. McGraw Hill Education, New York City
377. Coussens LM, Werb Z (2002) Inflammation and cancer. Nature 420:860. https://doi.org/10.1038/nature01322
378. Heerboth S, Housman G, Leary M, Longacre M, Byler S, Lapinska K, Willbanks A, Sarkar S (2015) EMT and tumor metastasis. Clin Transl Med 4:6. https://doi.org/10.1186/s40169-015-0048-3
379. Porporato PE (2016) Understanding cachexia as a cancer metabolism syndrome. Oncogenesis 5:200. https://doi.org/10.1038/oncsis.2016.3
380. Razak NBA, Jones G, Bhandari M, Berndt MC, Metharon P (2018) Cancer-associated thrombosis: an overview of mechanisms, risk factors, and treatment. Cancers 10:380. https://doi.org/10.3390/cancers10100380
381. Greten FR, Grivennikov SI (2019) Inflammation and cancer: triggers, mechanisms, and consequences. Immunity 51:27. https://doi.org/10.1016/j.immuni.2019.06.025
382. Matsumoto H, Ogura H, Shimizu K, Ikeda M, Hirose T, Matsuura H, Takahashi K, Tanaka T, Shimazu T (2018) The clinical importance of a cytokine network in the acute phase of sepsis. Sci Rep 8:13995. https://doi.org/10.1038/s41598-018-32275-8
383. Sakaguchi H, Kuramoto Y (1986) A soluble active rotater model showing phase transitions via mutual entertainment. Prog Theor Phys 76(3):576–581. https://doi.org/10.1143/PTP.76.576
384. Madadi Asl M, Valizadeh A, Tass PA (2018) Dendritic and axonal propagation delays may shape neuronal networks with plastic synapses. Front Physiol 9:1849. https://doi.org/10.3389/fphys.2018.01849
385. Kuehn C (2015) Multiple time scale dynamics. Springer, Cham. https://doi.org/10.1007/978-3-319-12316-5
386. Desroches M, Guckenheimer J, Krauskopf B, Kuehn C, Osinga HM, Wechselberger M (2012) Mixed-mode oscillations with multiple time scales. SIAM Rev 54(2):211–288. https://doi.org/10.1137/100791233
387. Maistrenko Y, Lysyansky B, Hauptmann C, Burylko O, Tass PA (2007) Multistability in the kuramoto model with synaptic plasticity. Phys Rev E 75(6):066207. https://doi.org/10.1103/physreve.75.066207
388. Caporale N, Dan Y (2008) Spike timing-dependent plasticity: a Hebbian learning rule. Annu Rev Neurosci 31(1):25–46. https://doi.org/10.1146/annurev.neuro.31.060407.125639
389. Popovych OV, Yanchuk S, Tass PA (2013) Self-organized noise resistance of oscillatory neural networks with spike timing-dependent plasticity. Sci Rep 3:2926. https://doi.org/10.1038/srep02926
390. Röhr V, Berner R, Lameu EL, Popovych OV, Yanchuk S (2019) Frequency cluster formation and slow oscillations in neural populations with plasticity. PLoS ONE 14(11):e0225094. https://doi.org/10.1371/journal.pone.0225094
391. Berner R, Polanska A, Schöll E, Yanchuk S (2020) Solitary states in adaptive nonlocal oscillator networks. Eur Phys J Spec Top 229(12–13):2183–2203. https://doi.org/10.1140/epjst/e2020-900253-0
392. Acebrón JA, Bonilla LL, Pérez Vicente CJ, Ritort F, Spigler R (2005) The Kuramoto model: a simple paradigm for synchronization phenomena. Rev Mod Phys 77(1):137–185. https://doi.org/10.1103/revmodphys.77.137
393. Warburg O, Posener K, Negelein E (1924) Über den Stoffwechsel der Carcinomzelle. Biochem Z 152:309. https://doi.org/10.1007/BF01726151

394. Kuramoto Y (1984) Chemical oscillations, waves and turbulence. Springer, Berlin. https://doi.org/10.1007/978-3-642-69689-3
395. Daido H (1994) Generic scaling at the onset of macroscopic mutual entrainment in limit-cycle oscillators with uniform all-to-all coupling. Phys Rev Lett 73(5):760–763. https://doi.org/10.1103/physrevlett.73.760
396. Berner R, Yanchuk S, Maistrenko Y, Schöll E (2021) Generalized splay states in phase oscillator networks. Chaos 31(7):073128. https://doi.org/10.1063/5.0056664
397. Shannon CE (1948) A mathematical theory of communication. Bell Syst Tech J 27:379. https://doi.org/10.1002/j.1538-7305.1948.tb01338.x
398. Fleischmann C, Thomas-Rueddel DO, Hartmann M, Hartog CS, Welte T, Heublein S, Dennler U, Reinhart K (2016) Hospital incidence and mortality rates of sepsis: an analysis of hospital episode (DRG) statistics in Germany from 2007 to 2013. Dtsch Arztebl Int 113(10):159. https://doi.org/10.3238/arztebl.2016.0159
399. Omel'chenko OE, Wolfrum M, Maistrenko Y (2010) Chimera states as chaotic spatiotemporal patterns. Phys Rev E 81(6):065201(R). https://doi.org/10.1103/physreve.81.065201
400. Omelchenko I, Omel'chenko OE, Hövel P, Schöll E (2013) When nonlocal coupling between oscillators becomes stronger: patched synchrony or multichimera states. Phys Rev Lett 110:224101. https://doi.org/10.1103/physrevlett.110.224101
401. Omel'chenko OE (2018) The mathematics behind chimera states. Nonlinearity 31(5):R121. https://doi.org/10.1088/1261-6544/aaaa07
402. Omel'chenko OE, Knobloch E (2019) Chimerapedia: coherence-incoherence patterns in one, two and three dimensions. New J Phys 21(9):093034. https://doi.org/10.1088/1367-2630/ab3f6b
403. Gerster M, Berner R, Sawicki J, Zakharova A, Skoch A, Hlinka J, Lehnertz K, Schöll E (2020) FitzHugh-Nagumo oscillators on complex networks mimic epileptic-seizure-related synchronization phenomena. Chaos 30:123130. https://doi.org/10.1063/5.0021420
404. Scholz M, Witzenrath M, Loeffler M, Nouailles-Kursar G, Mueller-Redetzky H, Wienhold S, Ahnert P, Schirm S (2016) A biomathematical model of pneumococcal lung infection and antibiotic treatment in mice. PloS One e0156047. https://doi.org/10.1371/journal.pone.0156047
405. Vincent JL, Sakr Y, Reinhart K, Sprung CL, Gerlach H, Ranieri VM (2005) Is albumin administration in the acutely ill associated with increased mortality? Results of the soap study. Crit Care 9(6):R745-54. https://doi.org/10.1186/cc3895
406. Vincent JL, Sakr Y, Sprung CL, Ranieri VM, Reinhart K, Gerlach H, Moreno R, Carlet J, Le Gall JR, Payen D (2006) Sepsis in European intensive care units: results of the soap study. Crit Care Med 34(2):344–353. https://doi.org/10.1097/01.ccm.0000194725.48928.3a
407. Sprung CL, Sakr Y, Vincent JL, Le Gall JR, Reinhart K, Ranieri VM, Gerlach H, Fielden J, Groba CB, Payen D (2006) An evaluation of systemic inflammatory response syndrome signs in the sepsis occurrence in acutely ill patients (soap) study. Intensive Care Med 32(3):421–427. https://doi.org/10.1007/s00134-005-0039-8
408. Sakr Y, Reinhart K, Vincent JL, Sprung CL, Moreno R, Ranieri VM, De Backer D, Payen D (2006) Does dopamine administration in shock influence outcome? Results of the sepsis occurrence in acutely ill patients (soap) study. Crit Care Med 34(3):589–597. https://doi.org/10.1097/01.CCM.0000201896.45809.E3
409. Sakr Y, Vincent JL, Ruokonen E, Pizzamiglio M, Installe E, Reinhart K, Moreno R (2008) Sepsis and organ system failure are major determinants of post-intensive care unit mortality. J Crit Care 23(4):475–483. https://doi.org/10.1016/j.jcrc.2007.09.006
410. Taccone FS, Artigas AA, Sprung CL, Moreno R, Sakr Y, Vincent JL (2009) Characteristics and outcomes of cancer patients in European icus. Crit Care 13(1):R15. https://doi.org/10.1186/cc7713
411. Sakr Y, Lobo SM, Moreno RP, Gerlach H, Ranieri VM, Michalopoulos A, Vincent JL (2012) Patterns and early evolution of organ failure in the intensive care unit and their relation to outcome. Crit Care 16(6):R222. https://doi.org/10.1186/cc11868

412. Yiu HH, Graham AL, Stengel RF (2012) Dynamics of a cytokine storm. PLoS ONE 7(10):1–15. https://doi.org/10.1371/journal.pone.0045027
413. Ivanov PC (2021) The new field of network physiology: Building the human physiolome. Front Netw Physiol 1:711778. https://doi.org/10.3389/fnetp.2021.711778
414. Skoch A, Rehák Bučková B, Mareš J, Tintera J, Sanda P, Jajcay L, Horacek J, Spaniel F, Hlinka J (2022) Human brain structural connectivity matrices-ready for modelling. Sci Data 9(1), 486. https://doi.org/10.1038/s41597-022-01596-9
415. Ramlow L, Sawicki J, Zakharova A, Hlinka J, Claussen JC, Schöll E (2019) Partial synchronization in empirical brain networks as a model for unihemispheric sleep. Europhys Lett 126:50007. https://doi.org/10.1209/0295-5075/126/50007
416. Tzourio-Mazoyer N, Landeau B, Papathanassiou D, Crivello F, Etard O, Delcroix N, Mazoyer B, Joliot M (2002) Automated anatomical labeling of activations in SPM using a macroscopic anatomical parcellation of the MNI MRI single-subject brain. Neuroimage 15(1):273–289. https://doi.org/10.1006/nimg.2001.0978
417. Melicher T, Horacek J, Hlinka J, Spaniel F, Tintera J, Ibrahim I, Mikolas P, Novak T, Mohr P, Hoschl C (2015) White matter changes in first episode psychosis and their relation to the size of sample studied: a DTI study. Schizophr Res 162(1):22–28. https://doi.org/10.1016/j.schres.2015.01.029
418. Assaf Y, Pasternak O (2008) Diffusion tensor imaging (DTI)-based white matter mapping in brain research: a review. J Mol Neurosci 34(1):51–61. https://doi.org/10.1007/s12031-007-0029-0
419. Behrens TE, Berg HJ, Jbabdi S, Rushworth MF, Woolrich MW (2007) Probabilistic diffusion tractography with multiple fibre orientations: What can we gain? Neuroimage 34(1):144–155. https://doi.org/10.1016/j.neuroimage.2006.09.018
420. Chouzouris T, Omelchenko I, Zakharova A, Hlinka J, Jiruska P, Schöll E (2018) Chimera states in brain networks: empirical neural vs. modular fractal connectivity. Chaos 28(4):045112 (2018). https://doi.org/10.1063/1.5009812
421. Cabral J, Fernandes HM, Van Hartevelt TJ, James AC, Kringelbach ML (2013) Structural connectivity in schizophrenia and its impact on the dynamics of spontaneous functional networks. Chaos 23(4):046111. https://doi.org/10.1063/1.4851117
422. Schlottmann E, Schicke D, Krüger F, Lingnau B, Schneider C, Höfling S, Lüdge K, Porte X, Reitzenstein S (2019) Stochastic polarization switching induced by optical injection in bimodal quantum-dot micropillar lasers. Opt Express 27(20):28816–28831. https://doi.org/10.1364/oe.27.028816
423. Hlinka J, Coombes S (2012) Using computational models to relate structural and functional brain connectivity. Eur J Neurosci 36(2):2137–2145. https://doi.org/10.1111/j.1460-9568.2012.08081.x
424. Tomasi D, Volkow ND (2012) Laterality patterns of brain functional connectivity: gender effects. Cereb Cortex 22(6):1455–1462. https://doi.org/10.1093/cercor/bhr230
425. Kann S, Zhang S, Manza P, Leung HC, Li CSR (2016) Hemispheric lateralization of resting-state functional connectivity of the anterior insula: association with age, gender, and a novelty-seeking trait. Brain Connect 6(9):724–734. https://doi.org/10.1089/brain.2016.0443
426. Daianu M, Jahanshad N, Dennis EL, Toga AW, McMahon KL, de Zubicaray GI, Martin NG, Wright MJ, Hickie IB, Thompson PM (2012) Left versus right hemisphere differences in brain connectivity: 4-tesla hardi tractography in 569 twins. Proc IEEE Int Symp Biomed Imaging 2012:526–529. https://doi.org/10.1109/ISBI.2012.6235601
427. Schilling KG, Daducci A, Maier-Hein K, Poupon C, Houde JC, Nath V, Anderson AW, Landman BA, Descoteaux M (2019) Challenges in diffusion mri tractography-lessons learned from international benchmark competitions. Magn Res Imaging 57:194–209. https://doi.org/10.1111/j.1460-9568.2012.08081.x
428. Purpura DP (1959) Nature of electrocortical potentials and synaptic organizations in cerebral and cerebellar cortex. Int Rev Neurobiol 1:47. https://doi.org/10.1016/s0074-7742(08)60314-1

429. Creutzfeldt OD (1964) Experimenteller Nachweis von Beziehungen zwischen EEG-Wellen und Aktivität corticaler Nervenzellen. Naturwiss 51(7):166. https://doi.org/10.1007/bf00622299
430. Okamoto M, Dan H, Sakamoto K, Takeo K, Shimizu H, Kohno S, Oda I, Isobe S, Suzuki T, Kohyama K, Dan I (2004) Three-dimensional probabilistic anatomical cranio-cerebral correlation via the international 10–20 system oriented for transcranial functional brain mapping. Neuroimage 21(1):99–111. https://doi.org/10.1016/j.neuroimage.2003.08.026
431. Sharbrough F (1991) American electroencephalographic society guidelines for standard electrode position nomenclature. J Clin Neurophysiol 8:200–202. https://doi.org/10.1097/00004691-199104000-00002
432. Hodgkin AL, Huxley AF (1952) A quantitative description of membrane current and its application to conduction and excitation in nerve. J Physiol 117:500–544. https://doi.org/10.1113/jphysiol.1952.sp004764
433. Markram H, Gerstner W, Sjöström PJ (2011) A history of spike-timing-dependent plasticity. Front Synaptic Neurosci 3(4). https://doi.org/10.3389/fnsyn.2011.00004
434. Jonas E, Kording KP (2017) Could a neuroscientist understand a microprocessor? PLOS Comput Biol 13(1):1–24. https://doi.org/10.1371/journal.pcbi.1005268
435. Krakauer JW, Ghazanfar AA, Gomez-Marin A, MacIver MA, Poeppel D (2017) Neuroscience needs behavior: correcting a reductionist bias. Neuron 93(3):480–490. (2024/10/15 2017). https://doi.org/10.1016/j.neuron.2016.12.041
436. Eliasmith C, Stewart TC, Choo X, Bekolay T, DeWolf T, Tang Y, Rasmussen D (2012) A large-scale model of the functioning brain. Science 338:1202–1205. https://doi.org/10.1126/science.1225266
437. Sanz Leon P, Knock SA, Woodman MM, Domide L, Mersmann J, McIntosh AR, Jirsa V (2013) The virtual brain: a simulator of primate brain network dynamics. Front Neuroinform 7. https://doi.org/10.3389/fninf.2013.00010
438. Stiefel KM, Ermentrout GB (2016) Neurons as oscillators. J Neurophysiol 116(6):2950–2960
439. Dayan P, Abbott LF (2005) Theoretical neuroscience: computational and mathematical modeling of neural systems. J Cogn Neurosci 15:154. https://doi.org/10.1162/089892903321107891
440. FitzHugh R (1961) Impulses and physiological states in theoretical models of nerve membrane. Biophys J 1:445–466. https://doi.org/10.1016/s0006-3495(61)86902-6
441. Nagumo J, Arimoto S, Yoshizawa S (1962) An active pulse transmission line simulating nerve axon. Proc IRE 50:2061–2070. https://doi.org/10.1109/JRPROC.1962.288235
442. Bassett DS, Zurn P, Gold JI (2018) On the nature and use of models in network neuroscience. Nat Rev Neurosci 19:566–578. https://doi.org/10.1038/s41583-018-0038-8
443. Chernihovskyi A, Mormann F, Müller M, Elger CE, Baier G, Lehnertz K (2005) EEG analysis with nonlinear excitable media. J Clin Neurophysiol 22:314–329. https://doi.org/10.1097/01.wnp.0000179968.14838.e7
444. Chernihovskyi A, Lehnertz K (2007) Measuring synchronization with nonlinear excitable media. Int J Bifurc Chaos 17(10):3425–3429. https://doi.org/10.1142/s0218127407019159
445. Kiskowski MA, Alber MS, Thomas GL, Glazier JA, Bronstein NB, Pu J, Newman SA (2004) Interplay between activator-inhibitor coupling and cell-matrix adhesion in a cellular automaton model for chondrogenic patterning. Dev Biol 271:372. https://doi.org/10.1016/j.ydbio.2004.03.038
446. Pereda AE (2014) Electrical synapses and their functional interactions with chemical synapses. Nat Rev Neurosci 15:250–263. https://doi.org/10.1038/nrn3708
447. Massobrio P, de Arcangelis L, Pasquale V, Jensen HJ, Plenz D (2015) Criticality as a signature of healthy neural systems. Front Syst Neurosci 9:22. https://doi.org/10.3389/fnsys.2015.00022
448. Petkoski S, Iatsenko D, Basnarkov L, Stefanovska A (2013) Mean-field and mean-ensemble frequencies of a system of coupled oscillators. Phys Rev E 87(3):032908. https://doi.org/10.1103/PhysRevE.87.032908
449. Omelchenko I, Provata A, Hizanidis J, Schöll E, Hövel P (2015) Robustness of chimera states for coupled FitzHugh-Nagumo oscillators. Phys Rev E 91:022917. https://doi.org/10.1103/physreve.91.022917

450. Tsigkri-DeSmedt ND, Hizanidis J, Schöll E, Hövel P, Provata A (2017) Chimeras in leaky integrate-and-fire neural networks: effects of reflecting connectivities. Eur Phys J B 90(7):139. https://doi.org/10.1140/epjb/e2017-80162-0
451. Hoke M, Lehnertz K, Pantev C, Lütkenhöner B (1989) Spatiotemporal aspects of synergetic processes in the auditory cortex as revealed by the magnetoencephalogram. Brain dynamics. Springer, Berlin, pp 84–105. https://doi.org/10.1007/978-3-642-74557-7_7
452. Mormann F, Lehnertz K, David P, Elger CE (2000) Mean phase coherence as a measure for phase synchronization and its application to the EEG of epilepsy patients. Phys D 144(3–4):358–369. https://doi.org/10.1016/S0167-2789(00)00087-7
453. Chernihovskyi A, Elger CE, Lehnertz K (2009) Analysis of synchronization phenomena in broadband signals with nonlinear excitable media. EURASIP J Adv Sig Pr 2009:1–7. https://doi.org/10.1155/2009/134974
454. Sadleir L, Farrell K, Smith S, Connolly M, Scheffer I (2006) Electroclinical features of absence seizures in childhood absence epilepsy. Neurology 67(3):413–418. https://doi.org/10.1212/01.wnl.0000228251.93545.bf
455. Gastaut H (1970) Clinical and electroencephalographical classification of epileptic seizures. Epilepsia 11(1):102–112. https://doi.org/10.1111/j.1528-1157.1970.tb03871.x
456. Pecora LM, Carroll TL (1998) Master Stability Functions for Synchronized Coupled Systems. Phys Rev Lett 80(10):2109–2112. https://doi.org/10.1103/physrevlett.80.2109
457. Choe CU, Dahms T, Hövel P, Schöll E (2010) Controlling synchrony by delay coupling in networks: from in-phase to splay and cluster states. Phys Rev E 81(2):025205(R). https://doi.org/10.1103/physreve.81.025205
458. Lehnert J (2016) Controlling synchronization patterns in complex networks. Springer Theses, Springer, Heidelberg. https://doi.org/10.1007/978-3-319-25115-8
459. von Luxburg U (2007) A tutorial on spectral clustering. Stat Comput 17(4):395–416. https://doi.org/10.1007/s11222-007-9033-z
460. Jain AK, Murty MN, Flynn PJ (1999) Data clustering: a review. ACM Comput Surv 31(3):264–323
461. MacQueen JB (1967) Some methods for classification and analysis of multivariate observations. In: Cam LML, Neyman J (eds) Proceedings of the fifth Berkeley symposium on mathematical statistics and probability, vol 1. University of California Press, pp 281–297 (1967)
462. Atev SE (2011) Using asymmetry in the spectral clustering of trajectories. PhD thesis, University of Minnesota
463. Scikit-learn developers: clustering (2024). https://scikit-learn.org/stable/modules/clustering.html
464. Mykietyshyn M (2022) Topology and dynamics of an empirical brain connectivity. Technische Universität Berlin, MSc
465. Sawicki J, Mykietyshyn M, Rico B, Sprekeler H, Schöll E (2025) Pacemaker in modeling synchronization patterns in the brain. Chaos
466. Rattenborg NC, Horacio O, Kempenaers B, Lesku JA, Meerlo P, Scriba MF (2017) Sleep research goes wild: new methods and approaches to investigate the ecology, evolution and functions of sleep. Phil Trans R Soc B 372(1734):20160251. https://doi.org/10.1098/rstb.2016.0251
467. Wright KP (2009) Encyclopedia of neuroscience, chap EEG in S. Springer, p 85. https://doi.org/10.1007/978-3-540-29678-2_2892
468. Bergner A, Frasca M, Sciuto G, Buscarino A, Ngamga EJ, Fortuna L, Kurths J (2012) Remote synchronization in star networks. Phys Rev E 85:026208. https://doi.org/10.1103/physreve.85.026208
469. Malhotra RK, Avidan AY (2014) Sleep stages and scoring technique, chap 3. W.B. Saunders, pp 77–99. https://doi.org/10.1016/B978-1-4557-1267-0.00003-5
470. Abhang PA, Gawali BW (2015) Correlation of EEG images and speech signals for emotion analysis. Br J Appl Sci Technol 10:1. https://doi.org/10.9734/BJAST/2015/19000

471. Steriade M, McCormick DA, Sejnowski TJ (1993) Thalamocortical oscillations in the sleeping and aroused brain. Science 262(5134):679–685. https://doi.org/10.1126/science.8235588
472. Dang-Vu TT, Schabus M, Desseilles M, Albouy G, Boly M, Darsaud A, Gais S, Rauchs G, Sterpenich V, Vandewalle G, Carrier J, Moonen G, Balteau E, Degueldre C, Luxen A, Phillips C, Maquet P (2008) Spontaneous neural activity during human slow wave sleep. Proc Natl Acad Sci USA 105(39):15160–15165. https://doi.org/10.1073/pnas.0801819105
473. Sejnowski TJ, Destexhe A (2000) Why do we sleep? Brain Res 886(1–2):208–223. https://doi.org/10.1016/S0006-8993(00)03007-9
474. Schwartz JRL, Roth T (2008) Neurophysiology of sleep and wakefulness: basic science and clinical implications. Curr Neuropharmacol 6(4):367–378. https://doi.org/10.2174/157015908787386050
475. Mukhametov LM, Supin AY, Polyakova IG (1977) Interhemispheric asymmetry of the electroencephalographic sleep patterns in dolphins. Brain Res 134(3):581–584. https://doi.org/10.1016/0006-8993(77)90835-6
476. Niedernostheide FJ, Arps M, Dohmen R, Willebrand H, Purwins HG (1992) Spatial and spatio temporal patterns in pnpn semiconductor devices. Phys Status Solidi B 172:249. https://doi.org/10.1002/pssb.2221720123
477. Tarpley RJ, Ridgway SH (1994) Corpus callosum size in delphinid cetaceans. Brain Behav Evol 44(3):156. https://doi.org/10.1159/000113587
478. Rattenborg NC, Amlaner CJ, Lima SL (2000) Behavioral, neurophysiological and evolutionary perspectives on unihemispheric sleep. Neurosci Biobehav Rev 24:817–842. https://doi.org/10.1016/s0149-7634(00)00039-7
479. Mascetti GG (2016) Unihemispheric sleep and asymmetrical sleep: behavioral, neurophysiological, and functional perspectives. Nat Sci Sleep 8:221. https://doi.org/10.2147/NSS.S71995
480. Tamaki M, Bang JW, Watanabe T, Sasaki Y (2016) Night watch in one brain hemisphere during sleep associated with the first-night effect in humans. Curr Biol 26(9):1190–1194. https://doi.org/10.1016/j.cub.2016.02.063
481. Abrams DM, Mirollo RE, Strogatz SH, Wiley DA (2008) Solvable model for chimera states of coupled oscillators. Phys Rev Lett 101(8):084103. https://doi.org/10.1103/physrevlett.101.084103
482. Motter AE, Albert R (2012) Networks in motion. Phys Today 65:43–48. https://doi.org/10.1063/PT.3.1514
483. Shima S, Kuramoto Y (2004) Rotating spiral waves with phase-randomized core in nonlocally coupled oscillators. Phys Rev E 69(3):036213. https://doi.org/10.1103/physreve.69.036213
484. Bal T, Debay D, Destexhe A (2000) Cortical feedback controls the frequency and synchrony of oscillations in the visual thalamus. J Neurosci 20(19):7478–7488. https://doi.org/10.1523/JNEUROSCI.20-19-07478.2000
485. Blumenfeld H, McCormick DA (2000) Corticothalamic inputs control the pattern of activity generated in thalamocortical networks. J Neurosci 20(13):5153–5162. https://doi.org/10.1523/JNEUROSCI.20-13-05153.2000
486. Singer W (2018) Neuronal oscillations: unavoidable and useful? Eur J Neurosci 48(7):2389–2398. https://doi.org/10.1111/ejn.13796
487. Lehnertz K, Bialonski S, Horstmann MT, Krug D, Rothkegel A, Staniek M, Wagner T (2009) Synchronization phenomena in human epileptic brain networks. J Neurosci Methods 183(1):42–48. https://doi.org/10.1016/j.jneumeth.2009.06.004
488. Jiruska P, de Curtis M, Jefferys JGR, Schevon CA, Schiff SJ, Schindler K (2013) Synchronization and desynchronization in epilepsy: controversies and hypotheses. J Physiol 591(4):787–797. https://doi.org/10.1113/jphysiol.2012.239590
489. Jirsa VK, Stacey WC, Quilichini PP, Ivanov AI, Bernard C (2014) On the nature of seizure dynamics. Brain 137:2210. https://doi.org/10.1093/brain/awu133
490. Lehnertz K, Ansmann G, Bialonski S, Dickten H, Geier C, Porz S (2014) Evolving networks in the human epileptic brain. Physica D 267:7–15. https://doi.org/10.1016/j.physd.2013.06.009

491. Ngugi AK, Kariuki S, Bottomley C, Kleinschmidt I, Sander JW, Newton CR (2011) Incidence of epilepsy: a systematic review and meta-analysis. Neurology 77:1005–1012. https://doi.org/10.1212/WNL.0b013e31822cfc90
492. Fisher RS, van Emde Boas W, Blume W, Elger CE, Genton P, Lee P, Engel J Jr (2005) Epileptic seizures and epilepsy: definitions proposed by the International League Against Epilepsy (ILAE) and the International Bureau for Epilepsy (IBE). Epilepsia 46:470–472. https://doi.org/10.1111/j.0013-9580.2005.66104.x
493. Spencer SS (2002) Neural networks in human epilepsy: evidence of and implications for treatment. Epilepsia 43:219–227. https://doi.org/10.1046/j.1528-1157.2002.26901.x
494. Richardson MP (2012) Large scale brain models of epilepsy: dynamics meets connectomics. J Neurol Neurosurg Psychiatry 83(12):1238–1248. https://doi.org/10.1136/jnnp-2011-301944
495. Kuhlmann L, Lehnertz K, Richardson MP, Schelter B, Zaveri HP (2018) Seizure prediction-ready for a new era. Nat Rev Neurol 14:618–630. https://doi.org/10.1038/s41582-018-0055-2
496. Boccaletti S, Latora V, Moreno Y, Chavez M, Hwang DU (2006) Complex networks: structure and dynamics. Phys Rep 424(4–5):175–308. https://doi.org/10.1016/j.physrep.2005.10.009
497. Schöll E, Klapp SH, Hövel P (2016) Control of self-organizing nonlinear systems. Springer
498. Bick C, Goodfellow M, Laing CR, Martens EA (2020) Understanding the dynamics of biological and neural oscillator networks through exact mean-field reductions: a review. J Math Neurosci 10:9. https://doi.org/10.1186/s13408-020-00086-9
499. Wolf P, Beniczky S (2014) Understanding ictogenesis in generalized epilepsies. Exp Rev Neurother 14(7):787–798. https://doi.org/10.1586/14737175.2014.925803
500. Lopes Da Silva F, Blanes W, Kalitzin S, Parra J, Suffczynski P, Velis DN (2003) Epilepsies as dynamical diseases of brain systems: basic models of the transition between normal and epileptic activity. Epilepsia 44:72–83. https://doi.org/10.1111/j.0013-9580.2003.12005.x
501. Ferlazzo E, Zifkin BG, Andermann E, Andermann F (2005) Cortical triggers in generalized reflex seizures and epilepsies. Brain 128(4):700–710. https://doi.org/10.1093/brain/awh446
502. Ponten SC, Douw L, Bartolomei F, Reijneveld JC, Stam CJ (2009) Indications for network regularization during absence seizures: weighted and unweighted graph theoretical analysis. Exp Neurol 217:197–204. https://doi.org/10.1016/j.expneurol.2009.02.001
503. Chavez M, Valencia M, Navarro V, Latora V, Martinerie J (2010) Functional modularity of background activities in normal and epileptic brain networks. Phys Rev Lett 104:118701. https://doi.org/10.1103/PhysRevLett.104.118701
504. Baier G, Goodfellow M, Taylor PN, Wang Y, Garry DJ (2012) The importance of modeling epileptic seizure dynamics as spatio-temporal patterns. Front Physiol 3:281. https://doi.org/10.3389/fphys.2012.00281
505. Benjamin O, Fitzgerald TH, Ashwin P, Tsaneva-Atanasova K, Chowdhury F, Richardson MP, Terry JR (2012) A phenomenological model of seizure initiation suggests network structure may explain seizure frequency in idiopathic generalised epilepsy. J Math Neurosci 2:1. https://doi.org/10.1186/2190-8567-2-1
506. Terry JR, Benjamin O, Richardson MP (2012) Seizure generation: the role of nodes and networks. Epilepsia 53:e166–e169. https://doi.org/10.1111/j.1528-1167.2012.03560.x
507. Petkov G, Goodfellow M, Richardson MP, Terry JR (2014) A critical role for network structure in seizure onset: a computational modeling approach. Front Neurol 5:261. https://doi.org/10.3389/fneur.2014.00261
508. Goodfellow M, Schindler K, Baier G (2011) Intermittent spike-wave dynamics in a heterogeneous, spatially extended neural mass model. Neuroimage 55(3):920–932. https://doi.org/10.1016/j.neuroimage.2010.12.074
509. Goodfellow M, Glendinning P (2013) Mechanisms of intermittent state transitions in a coupled heterogeneous oscillator model of epilepsy. J Math Neurosci 3(1):17. https://doi.org/10.1186/2190-8567-3-17
510. Lopes M, Richardson MP, Abela E, Rummel C, Schindler K, Goodfellow M, Terry JR (2017) An optimal strategy for epilepsy surgery: Disruption of the rich-club? PLoS Comp Biol 13(8):e1005637. https://doi.org/10.1371/journal.pcbi.1005637

511. Rothkegel A, Lehnertz K (2014) Synchronization in populations of sparsely connected pulse-coupled oscillators. Europhys Lett 105(3):30003. https://doi.org/10.1209/0295-5075/105/30003
512. Poel W, Zakharova A, Schöll E (2015) Partial synchronization and partial amplitude death in mesoscale network motifs. Phys Rev E 91:022915. https://doi.org/10.1103/physreve.91.022915
513. Zakharova A (2020) Chimera patterns in networks: interplay between dynamics, structure, noise, and delay. In: Understanding complex systems. Springer, Cham. https://doi.org/10.1007/978-3-030-21714-3
514. Andrzejak RG, Rummel C, Mormann F, Schindler K (2016) All together now: Analogies between chimera state collapses and epileptic seizures. Sci Rep 6:23000. https://doi.org/10.1038/srep23000
515. Rothkegel A, Lehnertz K (2014) Irregular macroscopic dynamics due to chimera states in small-world networks of pulse-coupled oscillators. New J Phys 16:055006. https://doi.org/10.1088/1367-2630/16/5/055006
516. Mormann F, Kreuz T, Andrzejak RG, David P, Lehnertz K, Elger CE (2003) Epileptic seizures are preceded by a decrease in synchronization. Epilepsy Res 53(3):173–185. https://doi.org/10.1016/s0920-1211(03)00002-0
517. Feldt S, Osterhage H, Mormann F, Lehnertz K, Zochowski M (2007) Internetwork and intranetwork communications during bursting dynamics: applications to seizure prediction. Phys Rev E 76(2):021920. https://doi.org/10.1103/physreve.76.021920
518. Lüttjohann A, Schoffelen JM, van Luijtelaar G (2013) Peri-ictal network dynamics of spike-wave discharges: Phase and spectral characteristics. Exp Neurol 239:235. https://doi.org/10.1016/j.expneurol.2012.10.021
519. Barrat A, Barthélemy M, Pastor-Satorras R, Vespignani A (2004) The architecture of complex weighted networks. Proc Natl Acad Sci USA 101(11):3747–3752. https://doi.org/10.1073/pnas.0400087101
520. Boccaletti S, Bragard J (2006) Controlling spatio-temporal chaos in the scenario of the one-dimensional complex Ginzburg-Landau equation. Philos Trans R Soc Lond Ser A 364(1846):2383–2395. https://doi.org/10.1098/rsta.2006.1830
521. Enright MB, Leitner DM (2005) Mass fractal dimension and the compactness of proteins. Phys Rev E 71(1):011912. https://doi.org/10.1103/PhysRevE.71.011912
522. Hahn HK, Georg M, Peitgen HO (2005) Fractal aspects of three-dimensional vascular constructive optimization, vol 4. Springer. https://doi.org/10.1007/3-7643-7412-8_5
523. Katsaloulis P, Verganelakis DA, Provata A (2009) Fractal dimension and lacunarity of tractography images of the human brain. Fractals 17(02):181–189. https://doi.org/10.1142/S0218348X09004284
524. Katsaloulis P, Ghosh A, Philippe AC, Provata A, Deriche R (2012) Fractality in the neuron axonal topography of the human brain based on 3-D diffusion MRI. Eur Phys J B 85(5):1–7. https://doi.org/10.1140/epjb/e2012-30045-y
525. Plotnikov SA, Lehnert J, Fradkov AL, Schöll E (2016) Synchronization in heterogeneous FitzHugh-Nagumo networks with hierarchical architecture. Phys Rev E 94:012203. https://doi.org/10.1103/physreve.94.012203
526. Sawicki J, Omelchenko I, Zakharova A, Schöll E (2019) Delay-induced chimeras in neural networks with fractal topology. Eur Phys J B 92:54. https://doi.org/10.1140/epjb/e2019-90309-6
527. Krishnagopal S, Lehnert J, Poel W, Zakharova A, Schöll E (2017) Synchronization patterns: from network motifs to hierarchical networks. Phil Trans R Soc A 375:20160216. https://doi.org/10.1098/rsta.2016.0216
528. Hizanidis J, Panagakou E, Omelchenko I, Schöll E, Hövel P, Provata A (2015) Chimera states in population dynamics: networks with fragmented and hierarchical connectivities. Phys Rev E 92:012915. https://doi.org/10.1103/physreve.92.012915
529. Ulonska S, Omelchenko I, Zakharova A, Schöll E (2016) Chimera states in networks of Van der Pol oscillators with hierarchical connectivities. Chaos 26:094825. https://doi.org/10.1063/1.4960191

530. Tsigkri-DeSmedt ND, Hizanidis J, Hövel P, Provata A (2016) Multi-chimera states and transitions in the leaky integrate-and-fire model with excitatory coupling and hierarchical connectivity. Eur Phys J ST 225(6):1149. https://doi.org/10.1140/epjst/e2016-02661-4
531. Sawicki J, Omelchenko I, Zakharova A, Schöll E (2017) Chimera states in complex networks: interplay of fractal topology and delay. Eur Phys J Spec Top 226(9):1883–1892. https://doi.org/10.1140/epjst/e2017-70036-8
532. zur Bonsen A, Omelchenko I, Zakharova A, Schöll E (2018) Chimera states in networks of logistic maps with hierarchical connectivities. Eur Phys J B 91:65. https://doi.org/10.1140/epjb/e2018-80630-y
533. Erdös P, Rényi A (1960) On the evolution of random graphs. Publ Math Inst Hung Acad Sci 5:17–61. https://doi.org/10.1515/9781400841356.38
534. Karnatak R, Ansmann G, Feudel U, Lehnertz K (2014) Route to extreme events in excitable systems. Phys Rev E 90:022917. https://doi.org/10.1103/physreve.90.022917
535. Ansmann G, Lehnertz K, Feudel U (2016) Self-induced switchings between multiple space-time patterns on complex networks of excitable units. Phys Rev X 6(1):011030. https://doi.org/10.1103/physrevx.6.011030
536. Hilgetag CC, Goulas A (2016) Is the brain really a small-world network? Brain Struct Func 221(4):2361–2366. https://doi.org/10.1007/s00429-015-1035-6
537. Papo D, Zanin M, Martínez JH, Buldú JM (2016) Beware of the small-world neuroscientist! Front Hum Neurosci 10:96. https://doi.org/10.3389/fnhum.2016.00096
538. Gastner MT, Ódor G (2016) The topology of large open connectome networks for the human brain. Sci Rep 6:27249. https://doi.org/10.1038/srep27249
539. Bialonski S, Horstmann MT, Lehnertz K (2010) From brain to earth and climate systems: small-world interaction networks or not? Chaos 20(1):013134. https://doi.org/10.1063/1.3290992
540. Hlinka J, Hartman DH, Paluš M (2012) Small-world topology of functional connectivity in randomly connected dynamical systems. Chaos 22(3):033107. https://doi.org/10.1063/1.4729608
541. Hlinka J, Hartman DH, Jajcay N, Tomeček D, Tintěra J, Paluš M (2017) Small-world bias of correlation networks: from brain to climate. Chaos 27(3):035812. https://doi.org/10.1063/1.4984217
542. Horstmann MT, Bialonski S, Noennig N, Mai H, Prusseit J, Wellmer J, Hinrichs H, Lehnertz K (2010) State dependent properties of epileptic brain networks: comparative graph-theoretical analyses of simultaneously recorded EEG and MEG. Clin Neurophysiol 121(2):172–185. https://doi.org/10.1016/j.clinph.2009.10.021
543. Ansmann G, Lehnertz K (2012) Surrogate-assisted analysis of weighted functional brain networks. J Neurosci Methods 208(2):165–172. https://doi.org/10.1016/j.jneumeth.2012.05.019
544. Xia M, Wang J, He Y (2013) Brainnet viewer: A network visualization tool for human brain connectomics. PLoS ONE 8(7):e68910-15. https://doi.org/10.1371/journal.pone.0068910
545. Guo J, Kim R, Chen Y, Negishi M, Jhun S, Weiss S, Ryu JH, Bai X, Xiao W, Feeney E, Rodriguez-Fernandez J, Mistry H, Crunelli V, Crowley MJ, Mayes LC, Constable RT, Blumenfeld H (2016) Impaired consciousness in patients with absence seizures investigated by functional mri, eeg, and behavioural measures: a cross-sectional study. Lancet Neurol 15(13):1336–1345. https://doi.org/10.1016/s1474-4422(16)30295-2
546. Kilpatrick CJ, Davis SM, Tress BM, Rossiter SC, Hopper JL, Vandendriesen ML (1990) Epileptic seizures in acute stroke. Arch Neurol 47(2):157–160. https://doi.org/10.1001/archneur.1990.00530020069015
547. Goodfellow M, Rummel C, Abela E, Richardson MP, Schindler K, Terry JR (2016) Estimation of brain network ictogenicity predicts outcome from epilepsy surgery. Sci Rep 6:29215. https://doi.org/10.1038/srep29215
548. Sinha N, Dauwels J, Kaiser M, Cash SS, Brandon Westover M, Wang Y, Taylor PN (2016) Predicting neurosurgical outcomes in focal epilepsy patients using computational modelling. Brain 140(2):319–332. https://doi.org/10.1093/brain/aww299
549. Olmi S, Petkoski S, Guye M, Bartolomei F, Jirsa VK (2019) Controlling seizure propagation in large-scale brain networks. PLoS Comp Biol. 15:e1006805. https://doi.org/10.1371/journal.pcbi.1006805

Chapter 5
Modeling Musical Perception

In the preceding chapter, we have demonstrated the significant utility of adaptation and synchronization as essential modeling tools for physiological systems, with their successful application serving as evidence of their universal relevance. Building on this framework, the current chapter explores how these two phenomena are equally critical to understanding perceptual physiology and its modeling.

The opening section of this chapter delves into the pivotal role of adaptive mechanisms within physiological systems, emphasizing their contributions to perceptual processes. Adaptation is a fundamental property of biological systems, enabling organisms to respond dynamically and efficiently to ever-changing environments. Evolution frequently converges on analogous adaptive strategies to address similar challenges, underscoring the universality and robustness of these mechanisms. Section 5.1 introduces key regulatory principles, develops foundational aspects of a physical culture theory, and elaborates on mechanisms and models of perception. It further discusses practical applications, including the utilization of neural networks to replicate adaptive processes.

This chapter also extends the discussion to partial synchronization patterns in networks with empirically derived structural connectivity. Synchronization and coherence are widely recognized as essential to brain network function, underpinning a range of processes in both health and disease. For example, during auditory perception, an observable enhancement in coherence between global brain dynamics and auditory input highlights the importance of synchronization in sensory processing.

In Sect. 5.2, we explore the influence of an external sound source on a network of FitzHugh-Nagumo oscillators, leveraging structural connectivity data obtained through diffusion-weighted magnetic resonance imaging (MRI) from healthy human subjects. This investigation focuses on the emergence of synchronization patterns in response to auditory stimuli, examining how variations in the frequency and amplitude of the input modulate network dynamics. By systematically altering these parameters, we reveal that synchronization levels can be dynamically adjusted, providing insights into mechanisms that regulate neural coherence.

J. Sawicki, *Music Psychology—Balance of Relations*,
Current Research in Systematic Musicology 13,
https://doi.org/10.1007/978-3-031-97003-0_5

Section 5.3 expands on these findings by analyzing how synchronization phenomena are influenced by the interplay of network topology, input spatial location, and the frequency characteristics of cortical input signals. This analysis highlights the intricate dependencies that govern synchronization within neural networks.

Finally, in Sect. 5.4, we investigate the specific effects of music on the dynamics of such networks. Simulations demonstrate a notable increase in coherence between the network's global dynamics and auditory input when stimulated with particular musical pieces. This enhanced coherence illustrates the brain's remarkable capacity to align its activity with external auditory stimuli. The findings emphasize the critical role of specific frequency bands in shaping the interaction between external inputs and intrinsic neural dynamics.

By integrating adaptation and synchronization within perceptual models, this chapter underscores their significance in explaining both fundamental physiological processes and the complex interplay between neural systems and external stimuli. These insights pave the way for deeper understanding and more refined computational models of perception, with implications for sensory processing, cognition, and brain-computer interfaces.

5.1 Perception Phenomena and Neural Adaptivity

The following section examines the important role of adaptive mechanisms in physiological systems, with a focus on perceptual processes. Adaptation represents a fundamental feature of biological systems, enabling organisms to respond effectively to dynamic environments. Evolution frequently converges on analogous solutions to related challenges, suggesting the universality of certain adaptive strategies. This section outlines key regulatory principles, develops foundational concepts for a physical culture theory, elaborates on mechanisms and models of perception, and discusses practical medical applications involving neural networks. Further details can be found in the comprehensive work [1, 2], to which this section refers.

Organisms, including their nervous systems, have evolved mechanisms to adapt across varying timescales. These range from rapid responses to transient changes in sensory inputs to gradual adjustments over evolutionary periods. Learning and memory formation exemplify adaptive processes, where neuronal circuits undergo modifications driven by short- and long-term synaptic plasticity (Sect. 5.1.1). Long-term potentiation (LTP) and depression (LTD), key forms of synaptic plasticity, serve as neural substrates for experience-dependent adaptation, reinforcing connections that underlie learning.

Experimental studies reveal that repetitive exposure to identical sensory stimuli at constant intervals reduces the neural response magnitude in the auditory cortex, a phenomenon termed sensory adaptation. This decrement in response is hypothesized to result from dynamic modulation of synaptic coupling between neurons, reflecting an efficient neural coding strategy to conserve resources while prioritizing novel stimuli (Sect. 5.1.1). Adaptation mechanisms thus enhance perceptual salience by filtering redundant information and reallocating cognitive resources.

Music exemplifies an active adaptation process, engaging neural and emotional systems to achieve dynamic interaction with sound. Adaptations in music involve not only physiological changes but also behavioral and emotional adjustments. These include the development of new strategies for engaging with music, eliciting emotional responses, or honing novel skills such as rhythm perception and motor coordination. From a theoretical perspective, music can be conceptualized as an adaptive system represented by spatiotemporal electrical fields in the brain. These fields comprise sequences of impulses and energy bursts, which propagate and dissipate, inducing new neural activity through feedback mechanisms (Sect. 5.1.2). Such feedback loops highlight the brain's capacity to integrate past experiences into present processing, providing a framework for understanding the brain-music interaction.

The study of adaptive mechanisms has profound implications for modeling perception and developing medical applications. Neural network models incorporating synaptic plasticity and sensory adaptation principles offer valuable insights into how sensory systems process dynamic inputs. For example, computational models that simulate sensory adaptation can elucidate mechanisms of auditory scene analysis and attention modulation. Additionally, these models aid in designing therapeutic interventions, such as cochlear implants or brain-computer interfaces, by mimicking natural adaptation processes to improve functionality and user experience.

The emerging field of physical culture theory in Sect. 5.1.3 considers music and other human activities as adaptive systems rooted in physical and biological processes. In this framework, the brain's response to music is envisioned as a complex interplay of energy transfer and feedback loops, characterized by dynamic electric fields that guide neural activity. This perspective integrates insights from neuroscience, physics, and psychology, offering a holistic understanding of how humans interact with their environments through music and movement. By emphasizing the energetic and temporal dimensions of brain dynamics, physical culture theory provides a robust model for exploring the adaptive nature of perception and cognition.

5.1.1 Adaptation and Neuronal Coding

"To live is to adapt to the world around us" [3]. The environment of an organism is subject to changes that span vastly different timescales, from immediate alterations such as fluctuations in lighting conditions to gradual processes like climate change. To cope with this variability, organisms–and consequently their brains–have evolved sophisticated strategies to adapt across these diverse timescales. These adaptive mechanisms range from rapid responses to transient sensory stimuli to slower, long-term adjustments shaped by evolutionary processes [1].

In this context, the brain exhibits remarkable flexibility, enabling it to respond dynamically to environmental changes. Adaptive processes on shorter timescales are particularly critical for real-time sensory processing and survival. These include rapid synaptic adjustments, transient modulation of neural excitability, and short-term plasticity mechanisms that fine-tune neural responses to immediate demands. The following sections delve into key examples of such short-timescale adaptive mechanisms, which highlight the brain's capacity to respond and recalibrate in the face of sudden environmental changes.

Single neurons adapt to environmental changes through two primary strategies: intrinsic modifications and extrinsic adjustments. Intrinsic changes involve alterations in a neuron's internal properties, such as variations in its excitability, which can modulate how readily the neuron generates action potentials in response to input stimuli [4]. In contrast, extrinsic changes pertain to modifications in the strength of synaptic connections onto the neuron, which dynamically adjust the efficacy of information transfer between neurons.

A key extrinsic mechanism associated with short-term adaptation (on the order of tens to hundreds of milliseconds) is short-term synaptic plasticity. This phenomenon enables rapid, transient adjustments in synaptic efficacy based on the temporal pattern of input spikes. Specifically, input spikes arriving within short intervals can induce either short-term depression (a temporary reduction in synaptic strength) or short-term facilitation (a transient enhancement of synaptic strength) [5] (see Sect. 4.1.1.2). These transient effects play a crucial role in shaping neuronal responses to rapidly fluctuating stimuli and optimizing signal transmission within neural circuits.

On longer timescales, lasting modifications in synaptic strength arise from long-term synaptic plasticity, which forms the basis for learning and memory. Experimentally, long-term changes are commonly induced using a pairing protocol, such as spike-timing-dependent plasticity (STDP) [6]. In STDP, the relative timing of pre- and postsynaptic spikes determines whether synaptic strength is enhanced or diminished. When a presynaptic spike precedes a postsynaptic spike by approximately 10 milliseconds, the synaptic connection undergoes long-term potentiation (LTP), leading to a lasting increase in efficacy. Conversely, when the presynaptic spike follows the postsynaptic spike within a window of 10-100 milliseconds, the connection exhibits long-term depression (LTD), resulting in a sustained decrease in strength [7, 8].

Both short- and long-term synaptic plasticity have been observed not only at synapses between excitatory neurons but also at inhibitory-to-excitatory synapses, highlighting their universality across diverse synaptic types. These inhibitory synaptic plasticity mechanisms are critical for maintaining the balance of excitation and inhibition within neural networks, a key determinant of proper brain function [9, 10]. Together, these adaptive mechanisms enable neurons to fine-tune their responses to environmental changes across various temporal scales, supporting both rapid sensory processing and long-term behavioral adaptations.

A well-established experimental framework for investigating short-term adaptation mechanisms is the oddball paradigm [11]. In this paradigm, a standard stimulus–typically visual or auditory–is presented repeatedly, creating a familiar and predictable pattern. Intermittently, a second, less frequent stimulus, referred to as the deviant or novel stimulus, is introduced, which represents an unpredictable event. At the whole-brain level, electroencephalogram (EEG) recordings demonstrate that

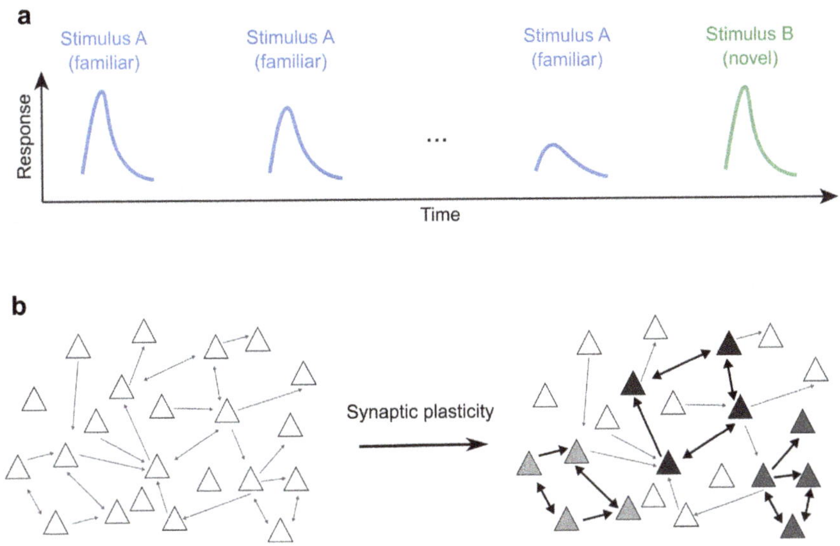

Fig. 5.1 **a** Oddball paradigm. Presenting a stimulus repeatedly (stimulus a) leads to decrease of the neuronal response while the deviant stimulus (stimulus b) leads to high neuronal response. **b** Synaptic plasticity leading to strongly recurrently connected structures (assemblies). Figure taken from [1]

the presentation of the deviant stimulus elicits a pronounced negative deflection in the EEG signal compared to responses elicited by the standard stimulus. This phenomenon is known as mismatch negativity (MMN) [12, 13].

At the level of single neurons or neuronal populations in sensory cortices, experiments consistently reveal enhanced neuronal responses to deviant stimuli relative to standard stimuli [14–16] (Fig. 5.1a). These findings underscore the neural encoding of novelty and predictability, which are key features of adaptive sensory processing. Computational models have proven instrumental in elucidating the underlying mechanisms of such short-term adaptation. Several studies highlight the role of short-term synaptic plasticity as a crucial mechanism mediating the differential responses to familiar and novel stimuli [17–19]. Notably, short-term plasticity at inhibitory synapses is particularly important for modulating neuronal responses in a context-dependent manner, adapting dynamically to the temporal properties of sensory input [20, 21].

In addition to short-term mechanisms, evidence suggests that long-term synaptic plasticity also contributes to the distinction between responses to standard and deviant stimuli. Specifically, long-term plasticity at inhibitory-to-excitatory synapses has been proposed as a mechanism for reducing excitatory responses to familiar stimuli while preserving heightened responses to novel stimuli [22]. In this framework, increased inhibitory-to-excitatory synaptic strength leads to suppression of neural activity for frequently encountered inputs, effectively encoding familiarity, whereas novel stimuli bypass this inhibition, resulting in elevated responses. These findings collectively illustrate the interplay between short-term and long-term plasticity in shaping adaptive neural responses to environmental changes, offering valuable insights into the fundamental principles of sensory adaptation and novelty detection in the brain.

Numerous functional implications have been proposed to explain the differential neuronal activity observed for familiar and novel stimuli, where reduced activity for familiar stimuli and heightened activity for novel stimuli serve distinct adaptive purposes. These implications span diverse concepts, including efficient coding to reduce redundancy in sensory processing, rapid detection of unexpected or significant events, and mechanisms aligned with Bayesian inference frameworks [3, 11].

A particularly influential theory in this context is predictive coding. Within this framework, the brain is conceptualized as a predictive machine, aiming to minimize the discrepancy between its internal model of the external world and incoming sensory information [23]. Novel stimuli, which deviate from internal predictions, elicit heightened neuronal responses that are interpreted as prediction errors. These errors signal the need to update the internal model to better align with the observed sensory environment.

Despite the appeal and explanatory power of predictive coding, the precise neural mechanisms underlying these computations remain an area of active investigation. A key question pertains to how predictive processes are implemented at the synaptic and circuit levels, particularly in relation to short- and long-term plasticity mechanisms. Short-term plasticity may facilitate rapid adjustments to transient changes in sensory input, while long-term plasticity may underpin more enduring adaptations that refine predictive models over time. However, the exact interplay between these forms of plasticity, and their roles in encoding prediction errors or updating neural representations, is not yet fully understood.

Future research integrating experimental, computational, and theoretical approaches is essential to unravel how predictive coding frameworks map onto the biophysical processes of the brain. Understanding these mechanisms has the potential to bridge gaps between high-level cognitive theories and the molecular and circuit-level adaptations that enable perception and learning.

Neuronal circuits must exhibit robustness against perturbations to ensure reliable function under varying conditions. Experimental studies investigating this robustness often involve sensory deprivation paradigms, such as occluding one eye during the development of the brain. These manipulations lead to homeostatic adjustments in the affected neuronal circuits, compensating for the altered sensory input and restoring functional balance [24]. A related question in neurobiology pertains to the regulation of intrinsic neuronal properties, such as ion channel conductance densities, which are crucial for maintaining proper circuit dynamics. Studies employing computational models and machine learning techniques have revealed that similar neuronal circuit behaviors can arise despite significant variability in these conductance densities. This degeneracy enables neuronal circuits to dynamically adjust and compensate for perturbations on rapid timescales [25–27].

Neuromodulators such as serotonin and dopamine play an essential role in regulating these intrinsic properties, acting as chemical signals that modulate neuronal excitability and synaptic strength [28]. Computational investigations have begun to explore the interplay between intrinsic and extrinsic neuronal properties, examining their combined effects on neuronal activity and their role in stabilizing critical behavioral states, such as transitions between sleep and wakefulness [29].

Learning and memory formation are additional adaptive processes that rely on mechanisms of plasticity. Short-term synaptic plasticity is hypothesized to support working memory by temporarily modulating synaptic efficacy [30], while long-term synaptic plasticity forms the basis for long-term memory storage by persistently altering synaptic connections [31]. One influential theory posits that groups of highly interconnected neurons, known as neuronal assemblies, serve as the fundamental units of representation in the brain. Computational models have demonstrated that long-term plasticity is instrumental in establishing and maintaining these connectivity structures [32, 33] (Fig. 5.1b).

Neuronal circuits also face the challenge of the stability-flexibility tradeoff. While synaptic connectivity must remain stable to preserve long-term memories and ensure robustness against perturbations, circuits must also retain sufficient flexibility to accommodate re-learning or the acquisition of new representations [34]. Computational models have proposed several mechanisms to address this tradeoff, including reverberatory neuronal activity [35], inhibitory-to-excitatory plasticity [36], and the integration of multiple synaptic plasticity and homeostatic mechanisms [37].

Despite significant advances, our understanding of the interplay among intrinsic neuronal properties, short-term plasticity, and long-term plasticity in sensory adaptation remains incomplete. This area of research is particularly important due to its implications for neuropsychiatric disorders. Deficits in excitatory and inhibitory circuit balance have been implicated in conditions such as schizophrenia and autism spectrum disorders [38, 39], while the mismatch negativity response has been proposed as a biomarker for psychotic disorders [40]. Advancing our knowledge of these cellular mechanisms may pave the way for novel therapeutic interventions (see Sect. 4.1.1.3).

5.1.2 Adaptation in Auditory Cortex

Many auditory stimuli, including speech and music, evolve over time, yet the brain perceives them as cohesive, continuous entities (see also Sect. 5.1.3). This phenomenon relies on the brain's ability to perform temporal integration, a process where incoming stimuli are represented and merged with traces of preceding stimuli [1]. Temporal integration is thought to occur primarily in the auditory cortex, while subcortical regions are suggested to be involved in source localization and spectral analysis [41]. Intracranial and extracranial measurements have demonstrated that neural responses in the auditory cortex are highly context-sensitive [42, 43]. The response to a given stimulus is modulated by the temporal sequence and spectral characteristics of preceding stimuli, as shown in various studies [44–46].

The simplest form of such context sensitivity is observed when a stimulus is presented repeatedly at a constant interval. In this case, the neural response magnitude decreases progressively, a phenomenon termed adaptation. Adaptation is stimulus-specific and depends on the inter-stimulus interval [47]. The specificity of adaptation has been studied extensively using oddball paradigms, where a frequently presented standard stimulus is occasionally replaced by a deviant stimulus. Neural responses to the deviant are consistently stronger than those to the standard, a phenomenon termed stimulus-specific adaptation or mismatch response [42, 45] (see also Fig. 5.1 in Sect. 5.1.1). Despite its importance, the precise mechanisms underlying adaptation in the auditory cortex remain unclear.

Single neurons exhibit adaptation due to intrinsic properties, a phenomenon known as spike frequency adaptation (see also Sect. 4.1.1.2) [47]. Adaptation is also observed in subcortical structures, including the auditory nerve fibers, inferior colliculus, and thalamus, which relay information to the auditory cortex. However, adaptation in the auditory cortex is not merely inherited from these regions. The timescales of adaptation differ significantly along the auditory pathway, with cortical adaptation aligning more closely with the temporal structure of stimuli [47, 48]. Moreover, adaptation is not present in all subdivisions of the inferior colliculus and thalamus. For instance, it is largely absent in the lemniscal pathway, which projects to the core area of the primary auditory cortex [45, 47].

Neurons in the auditory cortex function within networks, interconnected by highly plastic synapses. A prevailing hypothesis is that adaptation arises from modulations in synaptic coupling between neurons. However, the precise mechanisms driving these modulations remain a topic of ongoing investigation [49, 50]. Short-term synaptic depression has been proposed as a key physiological mechanism underlying adaptation [21, 51–53] (see also Sects. 4.1.1.2 and 5.1.1). This form of plasticity, triggered by repetitive stimulation of presynaptic neurons, involves vesicle depletion and desensitization of calcium channels at presynaptic terminals [5]. Short-term synaptic depression operates on timescales consistent with context-sensitive responses and plays critical roles in temporal filtering [54], gain control [55], and efficient information transfer [56].

In recent research, we incorporated short-term synaptic depression dynamics into a computational model of the auditory cortex, using a network architecture based on the hierarchical core-belt-parabelt structure observed in mammals [57–59]. This model features mean-field excitatory and inhibitory cell populations, interconnected with synapses modulated by short-term synaptic depression. These dynamics depend on the spectrotemporal characteristics of auditory stimuli. By linearizing the state equations and applying a slow-fast approximation to synaptic depression dynamics, we analyzed the system as a set of damped harmonic oscillators, or normal modes [58, 59].

The findings reveal that the properties of these normal modes–such as frequency, phase, spatial wave patterns, and decay rates–are determined by both the macrostructure (e.g., anatomical organization) and microstructure (e.g., synaptic weights) of the auditory cortex, as well as by the stimulus properties. Activity elicited by external stimuli propagates spatiotemporally across the auditory cortex, with synaptic depression shaping the local and global dynamics. This framework contrasts with traditional views of electromagnetic brain activity, such as magnetoencephalography (MEG) signals, which are often attributed to the summation of localized neural generators. Instead, our normal-mode approach suggests that adaptation in the auditory cortex arises from modulations in normal mode properties driven by synaptic plasticity, with reductions in response magnitude being a secondary consequence [59].

This normal-mode perspective offers a comprehensive framework for understanding how local synaptic processes contribute to the emergent dynamics of the auditory cortex. By linking synaptic mechanisms to observable neural responses, it advances our understanding of sensory adaptation and opens new avenues for investigating auditory processing in health and disease.

5.1.3 Adaptation and Impulse Pattern Formulation

The study of music represents a fundamentally interdisciplinary endeavor, integrating diverse scientific domains to achieve a comprehensive understanding [60]. Musical instruments are designed for auditory perception, active performance, social interaction, and individual application, making it necessary to bridge fields such as the physics of musical instruments, music psychology, neuromusicology, music sociology, and political science. Moreover, understanding music requires consideration of its roles within culture, technology, economy, and ethnicity, as well as its connections to natural resources, such as the use of wood or alternative materials in instrument construction.

Music itself embodies a continuous process of adaptation [1]. Listeners adjust to novel compositions, while musicians adapt to audience preferences, newly available instruments, and innovative compositional techniques (see also Sect. 3.1.5). Instrument makers respond to contemporary performance demands, emerging materials, and advancing technologies. Similarly, societies adapt to new musical works, genres, and presentation formats, including mass media and streaming platforms. These adaptive processes encompass changes in strategies, emotional responses, and the development of new skills. Participants in these processes may embrace and integrate new developments or resist and oppose them, reflecting the dynamic interplay of tradition and innovation.

Contemporary research in music employs methodologies tailored to the specific needs of each discipline [60]. Music psychology often relies on statistical and Bayesian approaches to analyze perception and cognition. Musical acoustics uses analytical equations and numerical discretization methods such as finite-element and finite-difference techniques to investigate the physics of sound and instruments. Ethnomusicology traditionally employs heuristic and historical methodologies, though computational and analytical approaches, including mathematical modeling of tonal systems, are increasingly incorporated. Across disciplines, machine learning methods are becoming indispensable. For example, connectionist models are widely used in computational composition [61], while self-organizing Kohonen maps [62] are frequently applied for analytical purposes [63–65]. This interdisciplinary framework underscores the complexity of musical phenomena and the necessity for collaborative approaches in understanding music as a dynamic and adaptive cultural artifact.

The methodologies used, therefore, strongly depend on the subfields, but some also intertwine, e.g., in the field of psycho-acoustics, relating physics to perception, using algorithms calculating loudness, brightness, pitch, spatial audio, or the like. Still, to arrive at a common, robust, suitable algorithm able to model music in a global, holistic way in the future, also including extra-musical players like ecology, economy, or politics, a common ground is needed, not debatable among the very diverse disciplines involved. For example, a Physical Culture Theory suggests music as an adaptive system to consist of impulses, physical energy bursts sent out, returning with a specific damping, thereby causing new impulses [66]. In its most general form, the Impulse Pattern Formulation can be written as a system parameter g representing an impulse sent out by one subsystem. This impulse is reflected at n other subsystems with damping parameters α and β_k for each reflection point k like [67, 68]:

$$g_+ = g - \ln\left(\frac{1}{\alpha}\left(g - \sum_{k=1}^{n} \beta_k e^{g-g_{k-}}\right)\right). \tag{5.1}$$

The system parameter g is updated at each iteration step to g_+, taking the most recent g and the previous g_{k-} into consideration. The logarithm reflects the exponential damping found in most systems. Adaptation is present for $g_+ = g$. E.g. with musical instruments, g can be taken as a periodicity of a musical tone. During the initial transient phase, $g_+ \,!= g$, and the system struggles, leading to a complex initial transient sound. After the initial phase, a stable periodicity is reached, and a musical pitch is heard. For example, with a guitar, two subsystems are present, the string and the guitar body, both with their own eigenfrequencies. Still, when playing a note, the string's vibration takes over the guitar body's vibration, i.e., the body adapts to the string's pitch. Thereby, the Impulse Pattern Formulation is able to model the guitar tone very precisely, which is especially reflected in the length and complexity of the initial sound phase [67].

Such an Impulse Pattern Formulation algorithm is scale-free and therefore able to model and predict very small networks as well as overall or general behavior fast and precise in musical acoustics [68, 69] or music perception and action [70]. Such a self-organizing system is found as a basis for all musical instrument families. Moreover, it is the basis of brain dynamics [71] and all interactions in society or politics.

To accommodate aesthetic and artistic phenomena, systems must integrate consciousness and its experiential content, such as perceiving sound, vision, emotions, or cognition. The Physical Culture Theory posits that conscious content corresponds to spatio-temporal electric fields within the brain, sufficiently complex to produce diverse experiences. These spatio-temporal fields are conceptualized as intricate impulse patterns, framing brain dynamics as a self-organizing system rather than a linear interplay of bottom-up and top-down processes. While this view acknowledges localized brain functions–such as the auditory cortex processing sound from the auditory pathway–it also emphasizes the circular nature of neural processing, where regions like the cortex and cochlea engage in bidirectional communication. Consequently, the brain's adaptation to sensory input emerges as an active, whole-brain process, despite the localized nature of sensory input pathways, as e.g. in the auditory oddball paradigm, as introduced in Sects. 5.1.1 and 5.1.2 (see Fig. 5.1).

In this framework, global neural networks in the brain exhibit dynamics ranging from stability and bistability to bifurcations, complexity, and chaos [67]. This mirrors the behavior of musical instrument sounds, where a stable pitch emerges only after an intricate transient phase. Similarly, each new tone in a melody undergoes such transformations. Brain activity follows a parallel pattern [72], as does ensemble performance, where musicians continuously adapt to the interplay of their co-performers' outputs. This dynamic system represents a perpetual interaction between novelty and adaptation in response to shifting contexts.

Adaptation, while often successful in achieving a steady state, can also fail, resulting in extended periods of chaos, noise, or bifurcations. This duality of adaptation and disruption is fundamental and recurrent across all levels of music, whether in sound production, the development of musical compositions, genre evolution, or the broader trajectory of music history. Understanding this interplay sheds light on how both adaptation and disruption are integral to the musical experience.

5.1.4 Summary

This section has explored the critical role of adaptive mechanisms within physiological systems, emphasizing their importance in perceptual processes. Adaptation constitutes a fundamental characteristic of biological systems, facilitating efficient responses to dynamically changing environmental conditions. Neuronal adaptation to environmental changes involves intrinsic modifications, such as changes in excitability affecting action potential generation, and extrinsic adjustments, which alter synaptic strength to regulate inter-neuronal information transfer.

We can distinguish between two types of neuronal adaptation: Short-term synaptic plasticity (milliseconds) provides transient changes in synaptic efficacy through mechanisms like depression and facilitation, optimizing neural responses to dynamic stimuli. Long-term plasticity supports learning and memory via durable synaptic strength changes, exemplified by spike-timing-dependent plasticity (STDP), where precise spike timing determines synaptic potentiation or depression. Both excitatory and inhibitory synapses employ these mechanisms to maintain circuit stability, with short-term plasticity modulating sensory input responses and long-term plasticity encoding novelty and familiarity. These processes underlie sensory adaptation and predictive coding, demonstrated in paradigms like the oddball framework. Short-term plasticity contributes to working memory, while long-term plasticity forms neuronal assemblies crucial for durable memory storage.

Perception of dynamic auditory stimuli, including speech and music, relies on temporal integration, which merges incoming stimuli with preceding traces for cohesive representation. This occurs primarily in the auditory cortex, with subcortical regions aiding in source localization and spectral analysis. Neural responses in the auditory cortex are context-sensitive, influenced by the temporal and spectral characteristics of prior stimuli.

Repeated stimuli induce adaptation, where response magnitudes decrease progressively, as seen in stimulus-specific adaptation (SSA). SSA, studied in oddball paradigms, shows heightened neural responses to rare, deviant stimuli and involves intrinsic mechanisms (e.g., spike frequency adaptation) and synaptic processes like short-term synaptic depression (STSD). STSD, driven by repetitive presynaptic activity, includes vesicle depletion and calcium channel desensitization, facilitating temporal filtering, gain control, and efficient information transfer.

A computational model of the auditory cortex incorporating short-term synaptic depression (STSD) demonstrates that adaptation arises from modulations in normal modes (damped harmonic oscillators), determined by cortical anatomy, synaptic microstructure, and stimulus properties. These normal modes govern spatiotemporal activity patterns and response decay rates, revealing that adaptation reflects synaptic plasticity-driven modulations rather than localized signal summation. This framework links local synaptic processes to global cortical dynamics, advancing understanding of auditory processing and its pathological disruptions.

Contemporary music research employs discipline-specific methodologies while fostering interdisciplinary integration. Music psychology utilizes statistical and Bayesian techniques to study perception and cognition, whereas musical acoustics focuses on the physics of sound and instruments using analytical equations and numerical methods, such as finite-element and finite-difference approaches. Ethnomusicology combines heuristic and historical methods with emerging computational tools like tonal system modeling. Machine learning bridges these fields, with connectionist models enhancing computational composition and self-organizing Kohonen maps aiding in analytical studies.

Interdisciplinary areas, such as psychoacoustics, connect physics and perception through algorithms addressing parameters like loudness and pitch. Future progress in developing holistic, robust algorithms for music modeling requires unified frameworks transcending disciplinary boundaries. A proposed Impulse Pattern Formulation, grounded in Physical Culture Theory, models music as an adaptive system of impulses and energy bursts with damping effects, represented by system parameters. This scale-free algorithm has shown promise in modeling musical acoustics, perception, brain dynamics, and sociopolitical interactions, offering a universal foundation for understanding music and its broader implications.

In summary, the study of adaptive mechanisms in physiological systems, particularly within the contexts of perception and music, underscores their universal importance in biological function and offers pathways for innovative applications in neural modeling and medicine. These adaptive processes illustrate the brain's remarkable capacity to learn, adapt, and thrive in ever-changing environments.

5.2 Influence of Sound on Empirical Brain Networks

In this section, we investigate the impact of an external sound source on a network of FitzHugh-Nagumo oscillators, utilizing empirically derived structural connectivity from diffusion-weighted magnetic resonance imaging (MRI) data obtained from healthy human subjects as introduced in Sect. 4.3.1. Our analysis focuses on the synchronization patterns that emerge in response to auditory input, with particular attention to how the frequency and amplitude of the sound source influence these patterns. By systematically varying these parameters, we demonstrate that synchronization levels can be significantly modulated, providing a mechanism to control network dynamics.

This section highlights the role of frequency as a critical determinant of synchronization patterns. Specifically, synchronization is observed to intensify when the input frequency approaches the natural frequencies of the uncoupled oscillators or the collective resonance frequency of the coupled network. Similarly, the amplitude of the sound source serves as a secondary control parameter, influencing the degree and spatial extent of synchronization within the network. These findings underscore the sensitivity of neural dynamics to auditory stimuli, suggesting potential pathways through which sound and music might exert their effects on the brain.

To elucidate the modalities of music's influence on the human brain, we employ a minimal model that abstracts key features of auditory input and network dynamics. This approach allows us to isolate fundamental mechanisms underlying synchronization and explore their implications for auditory perception. By leveraging the FitzHugh-Nagumo model–a paradigmatic representation of excitable neuronal dynamics–we gain insights into how external perturbations can drive neural systems into partially synchronized states. These states are thought to be critical for integrating sensory input with cognitive processes, such as attention and memory formation.

Furthermore, our work builds on and extends prior studies that have linked synchronization phenomena to musical perception and neural dynamics. For instance, we explore the potential of using auditory inputs as a means to probe the functional connectivity of the brain, particularly in regions associated with auditory processing. This is of particular relevance for understanding the neural basis of music perception, where synchronization has been implicated in the segmentation of musical structure and the anticipation of rhythmic patterns.

Importantly, this section serves as a foundation for further investigations into how sound and music influence the brain at different scales. Future work could expand this framework by incorporating additional complexities, such as heterogeneous network topologies, variable transmission delays, or more detailed auditory input models that capture the full spectrum of musical properties. Such extensions would enable a more comprehensive understanding of how structural connectivity and auditory stimuli interact to shape brain dynamics.

For readers interested in a more detailed exploration of the methodologies and findings discussed here, we refer to the publications [73, 74], which are the basis of this section and provide an in-depth analysis of the mechanisms and implications of auditory-driven synchronization in neural networks. This foundational work not only advances our understanding of the interplay between sound and brain dynamics but also opens avenues for novel applications, such as neurorehabilitation and the design of auditory-based interventions for neurological disorders. This section is structured as follows: Section 5.2.1 provides a concise overview of synchronization phenomena in neural dynamics. Section 5.2.2 introduces the neural network model, which is based on empirically derived connectivity data and includes external inputs to the auditory cortex. In Sect. 5.2.3, we examine synchronization regions within this framework, with particular focus on synchronization tongues and their influence on the system's dynamics. Section 5.2.4 investigates the transition to synchronization within this neural architecture. Finally, Sect. 5.2.5 presents concluding remarks and discusses the broader implications of the results described in this section.

5.2.1 Synchronization Phenomena in Neural Dynamics

Synchronization phenomena are fundamental to the dynamical activities of the brain, underpinning both physiological and pathological states. For example, high synchronization levels characterize slow-wave sleep [75, 76] and transitions from wakefulness to sleep [77, 78]. Partial synchronization has been identified as a key mechanism explaining phenomena such as the first-night effect [79] and unihemispheric sleep, where one hemisphere exhibits sleep-like dynamics while the other remains active [76, 80–82], as discussed in Sect. 4.4. In pathological contexts, excessive synchronization, such as that observed in epileptic seizures, can lead to adverse outcomes [83], as explored in Sect. 4.5.

In contrast, synchronization is also central to cognitive functions, such as the development and perception of syntax [84–86], and serves as a theoretical framework for understanding musical acoustics and music psychology [87–91]. During music listening, event-related potentials (ERPs) recorded via electroencephalography (EEG) reveal synchronized activity across various brain regions [92, 93]. This synchronization correlates with the perception of large-scale musical structures, suggesting that neuronal dynamics associated with music perception engage multiple brain areas [86]. These findings underscore the role of synchronization in integrating auditory inputs into cohesive perceptual experiences.

To investigate these dynamics, we use the FitzHugh-Nagumo (FHN) model, a paradigmatic framework for neuronal spiking dynamics, to explore partial synchronization patterns induced by external sound sources. A notable feature of musical perception, tonal fusion, involves the emergence of a single perceived pitch from a spectrum of overtones [94]. To simplify analysis, we focus on an external sound source with fixed amplitude and single frequency, abstracting from the complexity of music while examining its influence on brain oscillations. This minimalistic approach enables us to isolate the fundamental effects of periodic perturbations on neural synchronization.

A particularly intriguing phenomenon in complex networks is relay synchronization, where non-adjacent layers synchronize through an intermediate relay layer [95]. A basic model of this phenomenon involves a triplex network, with the central layer mediating synchronization between outer layers. Remote synchronization is heavily influenced by network symmetries [96–100]. Extensions of this concept have demonstrated partial synchronization patterns in three-layer multiplex networks, with the relay layer facilitating synchronization of chimera states in the outer layers [101–106].

Incorporating time delays into neural models further enhances their realism. Delays arise from finite communication speeds along white-matter tracts and depend on the spatial distance between regions. These delays critically influence whether stimulation effects remain localized or propagate across the network [107]. In delayed systems, synchronization dynamics are organized by the interplay between time delays and the oscillatory properties of network nodes [108], emphasizing the significance of temporal factors in shaping brain network behavior.

In the present section, we focus on the minimal requirements for a simple model to exhibit partial synchronization patterns, as observed in experimental studies [92, 93]. While the inclusion of time delays represents an important extension for future research, we defer this consideration to maintain focus on foundational mechanisms. This approach provides a tractable framework for understanding the core principles underlying synchronization phenomena and their potential implications for auditory perception and broader neural dynamics.

5.2.2 Brain Network and the Auditory Cortex

We consider an empirical structural brain network shown in Fig. 4.29 where every region of interest is modeled by a single FitzHugh-Nagumo (FHN) oscillator. The auditory cortex is the part of the temporal lobe that processes auditory information in humans. It is a part of the auditory system, performing basic and higher functions in hearing and is located bilaterally, roughly at the upper sides of the temporal lobes, i.e., corresponding to the AAL indexing $k = 41, 86$ (temporal sup L/R) in Table 4.3. The auditory cortex takes part in the spectrotemporal analysis of the inputs passed on from the ear.

Each node corresponding to a brain region is modeled by the FitzHugh-Nagumo (FHN) model with external stimulus, a paradigmatic model for neuronal spiking [109–111]. Note that while the FitzHugh-Nagumo model is a simplified model of a single neuron, it is also often used as a generic model for excitable media on a coarse-grained level [112, 113]. Thus the dynamics of the network reads:

$$
\begin{aligned}
\epsilon \dot{u}_k = {} & u_k - \frac{u_k^3}{3} - v_k \\
& + \sigma \sum_{j=1}^{N} A_{kj} \left[B_{uu}(u_j - u_k) + B_{uv}(v_j - v_k) \right] \\
& + C_k \gamma \cos \omega t
\end{aligned}
\tag{5.2a}
$$

$$
\begin{aligned}
\dot{v}_k = {} & u_k + a \\
& + \sigma \sum_{j=1}^{N} A_{kj} \left[B_{vu}(u_j - u_k) + B_{vv}(v_j - v_k) \right],
\end{aligned}
\tag{5.2b}
$$

where $\epsilon = 0.05$ describes the timescale separation between the fast activator variable (neuron membrane potential) u and the slow inhibitor (recovery variable) v [109]. Depending on the threshold parameter a, the FHN model may exhibit excitable behavior ($|a| > 1$) or self-sustained oscillations ($|a| < 1$). We use the FHN model in the oscillatory regime and thus fix the threshold parameter at $a = 0.5$ sufficiently far from the Hopf bifurcation point. The external stimulus is modeled by a trigonometric function with frequency ω and amplitude γ and is applied to the brain areas $k = 41, 86$ associated with the auditory cortex, i.e. $C_k = 1$ if $k = 41$ or 86 and zero otherwise. The coupling between the single regions is given by the coupling strength σ. As we are looking for partial synchronization patterns we fix $\sigma = 0.6$ similar to numerical studies of synchronization phenomena during unihemispheric sleep in Sect. 4.4 [82] and epileptic seizures in Sect. 4.5 [83] where partial synchronization patterns have been observed. The interaction scheme between nodes is characterized by a rotational coupling matrix:

$$
\mathbf{B} = \begin{pmatrix} B_{uu} & B_{uv} \\ B_{vu} & B_{vv} \end{pmatrix} = \begin{pmatrix} \cos\phi & \sin\phi \\ -\sin\phi & \cos\phi \end{pmatrix},
\tag{5.3}
$$

with coupling phase $\phi = \frac{\pi}{2} - 0.1$, causing primarily an activator-inhibitor cross-coupling. This particular scheme was shown to be crucial for the occurrence of partial synchronization patterns in ring topologies [114] as it reduces the stability of the completely synchronized state. Also in the modeling of epileptic-seizure-related synchronization phenomena [83], where a part of the brain synchronizes, it turned out that such a cross-coupling is important. The subtle interplay of excitatory and inhibitory interaction is typical of the critical state at the edge of different dynamical regimes in which the brain operates [115], and gives rise to partial synchronization patterns which are not found otherwise.

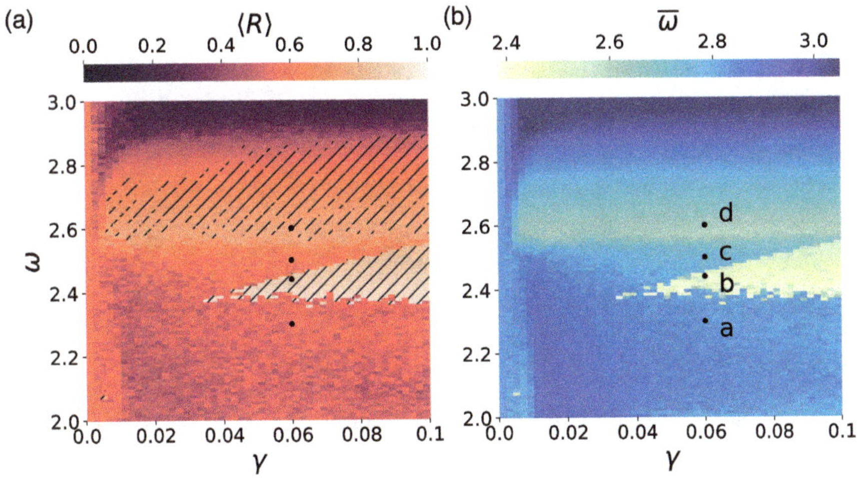

Fig. 5.2 Synchronization tongues in brain network with external stimulus: **a** The temporal mean of the Kuramoto order parameter $\langle R \rangle$ for simulation time $\Delta T = 10\,000$ and **b** the spatially averaged mean phase velocity $\overline{\omega}$ in the parameter plane of the frequency ω of the external stimulus and its amplitude γ. The light color in panel (**a**) stands for synchronization and the darker color for desynchronization. In the hatched region the standard deviation of $\langle R \rangle$ is less than 0.1, which indicates the absence of strong fluctuations of R in time. The dynamics of the four marked dots in each panel are shown in Figs. 5.3a–d and 5.4a–d. Other parameters are given by $\sigma = 0.6, \epsilon = 0.05$, $a = 0.5$, and $\phi = \frac{\pi}{2} - 0.1$. Figure taken from [73]

5.2.3 *Synchronization Regions*

We investigate synchronization scenarios emerging from an external periodic stimulus in the auditory cortices of both hemispheres ($k = 41, 86$). Figure 5.2 shows synchronization scenarios of an empirical structural brain network in dependence of the frequency ω and amplitude γ of the external stimulus. The light colored regions in Fig. 5.2a indicates synchronized dynamics, whereas the darker colors indicate desynchronized dynamics. There is a light colored stripe for $\omega = 2.6$ which indicates a Kuramoto order parameter $\langle R \rangle \approx 0.8$ and a light colored tongue starting at $\omega = 2.4, \gamma = 0.04$. The hatched region in Fig. 5.2a stands for a low standard deviation < 0.1 of the temporal mean of the Kuramoto order parameter $\langle R \rangle$. It indicates the absence of strong fluctuations of $R(t)$ and therefore a constant high level of synchrony in time. Figure 5.2b shows the drop of the spatially averaged mean phase velocity $\overline{\omega}$ in case of coherent dynamics in the synchronization regions of Fig. 5.2a. In the upper region, $\overline{\omega}$ takes over the value of the frequency ω of the external stimulus, whereas in the synchronization tongue $\overline{\omega}$ keeps its value of $\overline{\omega} = 2.4$.

It turns out that by taking the frequency ω of the external stimulus as a control parameter, one can change the level of synchrony of the system. Figure 5.3 depicts the details of the transition to synchronization for increasing values of the frequency ω of the external stimulus. Fixing the amplitude $\gamma = 0.06$, we take a closer look

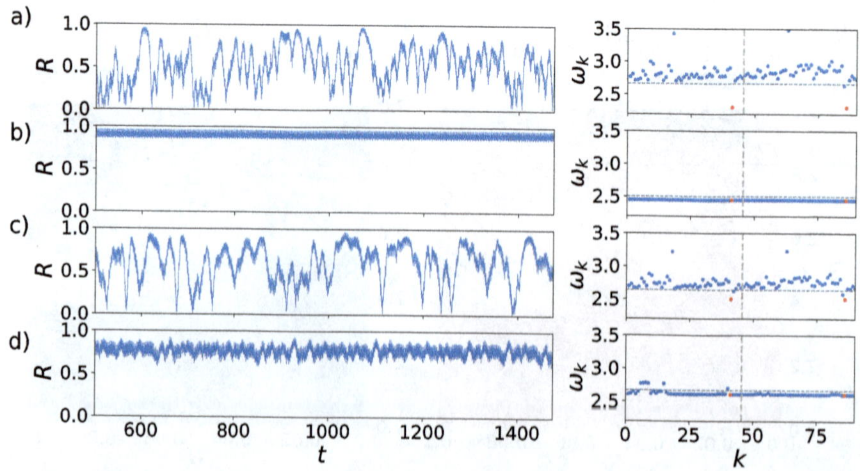

Fig. 5.3 Dynamical scenarios: dynamics inside and outside the synchronization regions (marked as black dots in Fig. 5.2) by the Kuramoto order parameter R (left column) and the mean phase velocities ω_k (right column) for increasing values of the frequency ω of the external stimulus $\omega = 2.30$ (**a**), $\omega = 2.44$ (**b**), $\omega = 2.50$ (**c**), and $\omega = 2.60$ (**d**) for fixed amplitude $\gamma = 0.06$. The vertical dashed line in the right column separates the left and right hemisphere; the horizontal grey dotted line indicates the temporal average of the mean-field frequency Ω. The red dots mark the nodes of the auditory cortical regions ($k = 41, 86$). Other parameters are as in Fig. 5.2. Figure taken from [73]

on the temporal evolution of R and the mean phase velocities in the system for different regions in Fig. 5.2. In Fig. 5.3a, the temporal evolution of the Kuramoto order parameter is similar to the system behavior without external stimulus, i.e., it exhibits large temporal fluctuations. In the right column the phase velocities of all nodes are plotted, the horizontal grey dotted line indicates the temporal average of the collective mean-field frequency Ω. Only the phase velocity of the auditory cortex follows the frequency of the external driving stimulus $\omega = 2.3$ and therefore is lower than the frequency of the other nodes $\omega_k \approx 2.8$. Increasing the external frequency to $\omega = 2.4$ yields an abrupt transition to a synchronized state. In Fig. 5.3b, the Kuramoto order parameter $R \approx 0.95$ and the mean phase velocities indicate a synchronous dynamical behavior, which agrees with the collective frequency Ω of the mean-field (grey dotted horizontal line). With a further increment to $\omega = 2.5$, the system loses synchrony (see Fig. 5.3c) and enters the region between the two synchronization regions in Fig. 5.2a. For $\omega = 2.6$ in Fig. 5.3d, which corresponds to the natural frequency of the uncoupled oscillators, the system regains synchronization, though the Kuramoto order parameter with $R \approx 0.8$ is lower than in the synchronization tongue. Remarkable is the fact of a dynamical asymmetry shown by the mean phase velocities. While the nodes of the right hemisphere exhibit an equal mean phase velocity, a part of the left hemisphere exhibits a faster dynamic similar to dynamics of unihemispheric sleep studied in [82]. In such states one hemisphere is synchronized, whereas the other hemisphere is partly desynchronized.

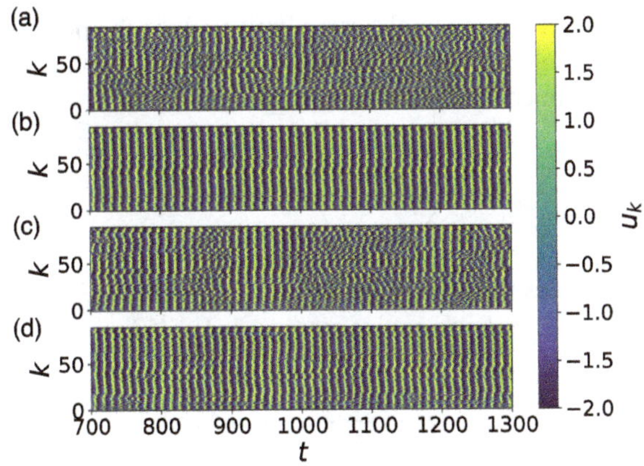

Fig. 5.4 Synchronized and desynchronized dynamics: Shown are the space-time plot of the variable u_k inside and outside the synchronization regions (marked as black dots in Fig. 5.2) for increasing values of the frequency ω of the external stimulus $\omega = 2.30$ (**a**), $\omega = 2.44$ (**b**), $\omega = 2.50$ (**c**), and $\omega = 2.60$ (**d**) for fixed amplitude $\gamma = 0.06$. The panels correspond to the panels in Fig. 5.3. Other parameters are as in Fig. 5.2. Figure taken from [73]

For a better insight, Fig. 5.4 shows the space-time plot of the variable u_k for the corresponding parameter values in Fig. 5.3. In Fig. 5.4b, d, the dynamics inside the two synchronization regions is depicted. The perturbation in the mean phase velocity profile in the right panel of Fig. 5.3d, can be detected also in the corresponding perturbations in Fig. 5.4d. Comparing Fig. 5.4a and c, we can see an increase of synchronized time segments. This increase will be analyzed quantitatively in more detail in the inset of Fig. 5.5.

5.2.4 Transition to Synchronization

There are two frequencies which play an important role for the dynamics of the system. On the one hand, in Fig. 5.2a a broad synchronization region is located at a frequency $\omega \approx 2.6$, which is the frequency of the uncoupled FHN oscillator ω_{FHN}. Although the external stimulus effects only the two auditory nodes ($k = 41, 86$), we can observe a transition to synchronization of the whole system approaching $\omega \approx 2.6$ already for small values of the amplitude $\gamma > 0.004$. On the other hand, we can detect a synchronization tongue with a lower boundary at $\omega \approx 2.4$ and an upper boundary increasing linearly with the amplitude γ. In contrast to the first, smooth transition, we can find here a sharp transition to synchronized dynamics, similar to a first order transition, depicted by the high contrast of the boarder of the synchronization tongue in Fig. 5.2a. In this synchronization tongue, the nodes oscillate with an equal mean phase velocity (see Fig. 5.3b), but there are phase differences between them, as indicated by $0.95 < R(t) < 1$ and shown in the phase-time plot in Fig. 5.4b. Using the fact that u_j/v_j and u_k/v_k are on the same limit cycle in the phase space and have the same mean phase velocity, the phase differences in the coupling term of Eq. (5.2) can be effectively summed up in following way:

$$\sum_j A_{kj}\mathbf{B}\begin{pmatrix} u_j - u_k \\ v_j - v_k \end{pmatrix} \approx \Delta t_{\text{eff}}\,\mathbf{B}\begin{pmatrix} \dot{u}_k \\ \dot{v}_k \end{pmatrix}, \tag{5.4}$$

where $\Delta t_{\text{eff}} \ll 1$ denotes the effective sum of the time intervals of all phase differences. Neglecting $\cos\phi \ll 1$ and setting $\sin\phi \approx 1$, Eq. (5.2) reads for $k \neq 41, 86$:

$$\begin{aligned} \varepsilon\dot{u}_k &= u_k - \frac{u_k^3}{3} - v_k - \sigma\,\Delta t_{\text{eff}}\,\dot{v}_k \\ \dot{v}_k &= u_k + a + \sigma\,\Delta t_{\text{eff}}\,\dot{u}_k \end{aligned} \tag{5.5}$$

The local dynamics of Eq. (5.2) is governed by a slow-fast system (FitzHugh-Nagumo oscillator), where the slow part essentially determines the period of the oscillations. Hence, considering the slow motion on the falling branches of the u-nullcline ($\dot{u}_k = 0$) by inserting the second equation into the first one

$$v_k = u_k - \frac{u_k^3}{3} - \sigma\,\Delta t_{\text{eff}}\,(u_k + a), \tag{5.6}$$

the time derivative of the falling branches yields with $\dot{v}_k$ from Eq. (5.5)

$$\dot{u}_k = \frac{u_k + a}{1 - u_k^2 - 2\sigma\,\Delta t_{\text{eff}}}. \tag{5.7}$$

The separation of the variables gives

$$\mathrm{d}t = \frac{1 - u_k^2 - 2\sigma\,\Delta t_{\text{eff}}}{u_k + a}\,\mathrm{d}u_k, \tag{5.8}$$

where $\mathrm{d}t$ can be integrated over one oscillation period T. As shown in [116], this leads in case of synchronization to a linear dependence of the oscillation period $T_{\text{sync}} = \int_0^T \mathrm{d}t$ on the effective sum of the phase differences proportional to Δt_{eff}. For incoherent distribution of the phases of each node k (see Fig. 5.4d), the phase differences between the single nodes are also strongly distributed and thus $\Delta t_{\text{eff}} \approx 0$. In this case, the natural frequency of the uncoupled system plays an important role, provided that the mean phase velocity of all oscillators is still almost equal as in case of Fig. 5.2d.

This could explain on one side the fact that we observe a synchronization tongue at $\omega \approx 2.4$ (which is smaller than the frequency of an uncoupled oscillator $\omega_{\text{FHN}} \approx 2.6$), and on the other side, the linear boundaries of the synchronization tongue for increasing amplitude γ. The increase of γ yields an increase of the sum of the phase

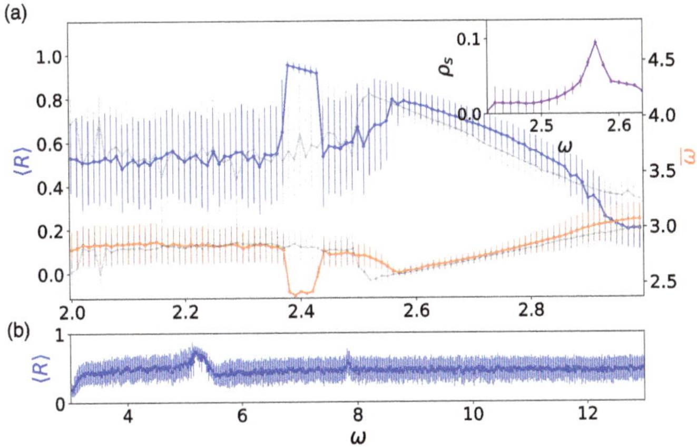

Fig. 5.5 Transition scenarios: **a** temporal mean of the Kuramoto order parameter $\langle R\rangle$ (dark blue) and the spatially averaged mean phase velocities $\overline{\omega}$ (light orange) in dependence on the frequency ω of the external stimulus for a fixed amplitude $\gamma = 0.052$. The vertical bars indicate the standard deviation of the temporal mean of the Kuramoto order parameter and the spatially averaged mean phase velocities, respectively. As input nodes, the auditory cortices $k = 41, 86$ are chosen. In case of a different input ($k = 1, 45$) the corresponding light grey curves are shown in panel (**a**). The inset in panel (**a**) depicts $\rho_s = \frac{N_s}{\Delta T_L}$, the number N_s of synchronized time intervals ($R(t) > 0.8\ \forall t$) divided by a simulation time of $\Delta T_L = 30\,000$ for values of the frequency ω between the two synchronization regions. The vertical bars denote the standard deviation of the length of these synchronized time intervals. **b** $\langle R\rangle$ for a larger range of driving frequencies ω, showing higher resonance tonges. Other parameters are as in Fig. 5.2. Figure taken from [73]

differences in the coupling term of Eq. (5.2) and therefore an increase of the effective sum of the time intervals Δt_{eff}.

In Fig. 5.5a, both transitions are depicted in dependence on the frequency ω for a fixed amplitude $\gamma = 0.052$. We can see an abrupt increase and decrease of the temporal mean of the Kuramoto order parameter $\langle R\rangle$ before and after $\omega \approx 2.4$, respectively. In contrast, in approaching the upper synchronization region starting from $\omega \approx 2.6$, $\langle R\rangle$ increases more slowly than at the transition to the synchronization tongue ($\omega \approx 2.4$). In case of synchronization the standard deviation of $\langle R\rangle$, displayed by the vertical bars, is smaller than in case of desynchronized dynamics. That holds also for the spatially averaged mean phase velocities $\overline{\omega}$, which in case of synchronization takes over the lower value of the frequency ω of the external stimulus. Also for $\omega > 2.6$, $\overline{\omega}$ is equal to ω, whereas the standard deviation of $\overline{\omega}$ increases linearly with ω. In contrast, there is no effect on the system for $\omega < 2.4$. Neither $\langle R\rangle$ nor $\overline{\omega}$ show a different behavior for such values of ω. The high value of the standard deviation of $\langle R\rangle$ stands for dynamics as shown in Fig. 5.3a, where the Kuramoto order parameter $R(t)$ is fluctuating over its whole bandwidth $R \in [0, 1]$. Simulations show that for $\omega > 3.0$ the dynamical behavior of the system becomes similar to that with $\omega < 2.3$. For both parameter intervals of ω, there is no effect on the system. Simulations show also that a similar transition to synchronization at $\omega = 2.6$ can be found for higher

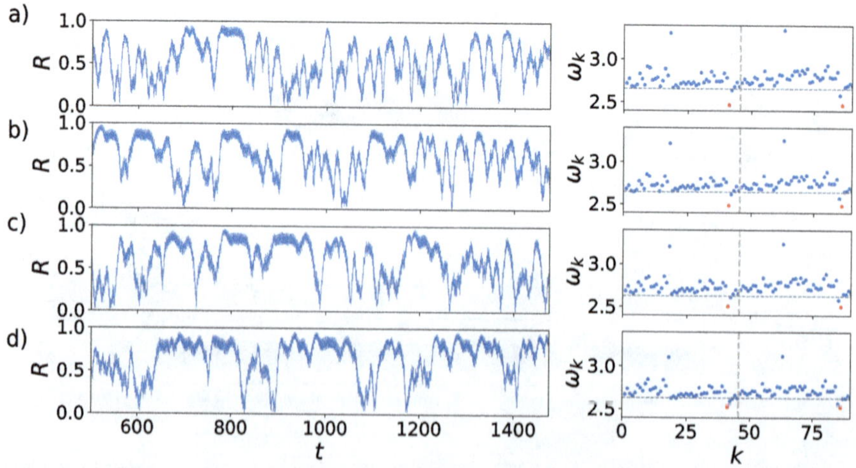

Fig. 5.6 Transition scenarios: The increase of the regularity and duration of synchronized time intervals shown by the temporal evolution of the Kuramoto order parameter R (left column) and the mean phase velocities ω_k (right column) for increasing values of the frequency ω of the external stimulus $\omega = 2.47$ (**a**), $\omega = 2.49$ (**b**), $\omega = 2.51$ (**c**), and $\omega = 2.53$ (**d**) for fixed amplitude $\gamma = 0.06$. The vertical dashed line in the right column separates the left and right hemisphere, and the horizontal grey dotted line indicates the temporal average of the mean-field frequency Ω. The red colored dots indicate the nodes of the auditory cortical regions ($k = 41, 86$). Other parameters as in Fig. 5.2. Figure taken from [73]

harmonics, i.e., multiple values of $\omega = 2.6$. In Fig. 5.5b, we can identify synchronization regions for $\omega = 5.2, 7.8$ and 10.4 becoming less pronounced for increasing ω, i.e., having a smaller extension in the plane of ω and γ. In contrast, we could not detect repeated synchronization tongues of ω for multiple values of $\omega = 2.4$. This indicates the existence of two different synchronization mechanisms.

The existence of two synchronization regions depends on the choice to which nodes the external stimulus is supplied. In case of a different input, for instance $k = 1, 45$ in contrast to $k = 41, 86$, the light grey curves in Fig. 5.5a depict the corresponding dependence of the Kuramoto order parameter $\langle R \rangle$ and the spatially averaged mean phase velocities $\overline{\omega}$ upon the frequency ω of the external stimulus. The synchronization region at $\omega \approx 2.4$ is missing here and only one synchronization region remains ($\omega > 2.5$).

In the following, we analyze the region between the two synchronization areas in more detail. Figure 5.6 depicts the dynamical behavior when we approach the synchronization region by increasing the frequency ω of the external stimulus in the neighborhood of the synchronization region at $\omega = 2.6$. For $\omega = 2.47$ in Fig. 5.6a, the time series of the Kuramoto order parameter shows familiar temporal fluctuations with only short episodes of synchronization ($R(t) > 0.8$). In [83] the authors define the threshold $R = 0.8$ as the onset of an epileptic seizure. By increasing the frequency ω, one can increase the quantity of these episodes, as well as their duration. Figure 5.6d with $\omega = 2.51$ features much longer duration of synchronized episodes, moreover the duration of the single episodes are comparable in length. This transition in Fig. 5.6a-d can be also seen in Fig. 5.5a. The inset of Fig. 5.5a confirms the increasing regularity between the two synchronization regions by depicting $\rho_s = \frac{N_s}{\Delta T_L}$ versus ω, where N_s is the number of synchronized time intervals ($R(t) > 0.8\,\forall\, t$) and $\Delta T_L = 30\,000$ is the simulation time. The vertical bars denote the standard deviation of the length of these synchronized time intervals. With increasing ω not only the number of synchronized time intervals is increasing, but the standard deviation of their duration is decreasing. For $\omega > 2.6$ we enter the synchronization region, where the value of ρ_s drops due to the nearly consistently synchronized dynamics.

Finally, the mean phase velocities in the right column of Fig. 5.6 display the transition to frequency synchronization. While the frequency of the two driven nodes ($k = 41, 86$) converges to the frequency of an uncoupled FHN oscillator $\omega_{\text{FHN}} \approx 2.6$, also the frequencies of all the other nodes are adjusted, especially those with a much higher frequency ($k = 18, 63$).

5.2.5 *Summary*

In this section, we have explored the impact of an external sound source on the dynamics of a network with empirically derived structural connectivity. Our findings reveal that the synchronization dynamics of the system are strongly influenced by the frequency and amplitude of the sound source. Specifically, synchronization is observed to emerge when the frequency of the external stimulus aligns with certain characteristic frequencies of the system, particularly the natural frequency of the uncoupled oscillator and the collective frequency of the coupled system. We demonstrate that the degree of synchronization increases progressively as the input frequency approaches these key values or their harmonics.

Additionally, we have analyzed the relationship between the synchronization boundaries and the amplitude of the external sound, which can be interpreted as the sound's volume. Our results indicate that increasing the amplitude of the input sound expands the region of synchronization, thereby enhancing the stability of synchronous states. This highlights the role of amplitude as a control parameter in modulating brain network dynamics. Furthermore, we examined the system's behavior during transitions to synchronization, showing that fine-tuning the external sound frequency can significantly enhance the level of synchrony in the network.

These results align with experimental observations from Bader's group [92, 93], which demonstrate that music induces measurable synchrony in the human brain. Their research indicates that listening to music leads to a periodic alternation between synchronized and desynchronized states, a phenomenon that reflects the system's inherent variability. Such variability can be understood as indicative of a critical state, where the brain operates at the boundary of different dynamical regimes [115]. This criticality is characterized by features such as hysteresis and avalanche phenomena, commonly observed in phase transitions and other critical phenomena [117–119]. By carefully selecting the parameters of the external sound source, we have observed intriguing dynamical behaviors near synchronization transitions, including an induced alternation between high and low levels of synchrony.

Moreover, these results have broader implications for understanding the effect of auditory stimuli on brain dynamics. The periodic alternation between synchronization and desynchronization may be fundamental to cognitive processing, as it reflects the brain's ability to dynamically adjust to external stimuli. Such alternations could underlie mechanisms of attention and expectation while listening to music, highlighting the interplay between external inputs and intrinsic neural variability.

Our findings also underscore the potential for external sound sources to serve as models for the influence of music on the human brain. Specifically, our results provide a theoretical framework for understanding how external auditory inputs can drive synchronization patterns across neural networks. Future work may extend this approach by incorporating more complex input signals, such as real musical stimuli, and investigating their effects on empirical brain network dynamics in more detail (see Sect. 5.4). This could offer deeper insights into how music shapes neural processes and contributes to phenomena such as emotional engagement, memory, and cognitive enhancement.

5.3 Interplay of Synchronization and Cortical Input

Synchronization patterns and coherence are widely recognized as fundamental to the operation of brain networks, playing pivotal roles in both healthy and pathological states. Notably, during auditory perception, there is an observable enhancement in coherence between the network's global dynamics and the auditory input. In this section, we demonstrate that synchronization phenomena are governed by a delicate interplay involving the network's topology, the spatial location of the input, and the frequency characteristics of the cortical input signals. To explore these dynamics, we investigate the effects of external stimulation on a network of FitzHugh-Nagumo oscillators configured with empirically derived structural connectivity. Furthermore, we examine the impact of cortical stimulation applied to various regions, with a particular focus on the auditory cortex. More detailed information can be find in the publications [74, 120], this section is based on. This section is organized as follows: Section 5.3.1 provides a brief overview of synchronization processes in cortical and subcortical brain regions. In Sect. 5.3.2, we present the neural network model, which is grounded in empirically derived connectivity patterns and incorporates region-specific neural inputs to various cortical areas. Section 5.3.3 explores the distinct functional contributions of different cortical regions, with particular emphasis on the role of the auditory cortex. Finally, Sect. 5.3.4 offers concluding remarks and discusses the broader implications of the findings presented in this section.

5.3.1 Synchronization in Cortical and Subcortical Regions

The human brain is a highly intricate organ, characterized by the distinct yet interconnected roles of the cortical and subcortical regions, which are fundamental to cognition, behavior, and physiological regulation. The cortex, or cerebral cortex, constitutes the brain's outermost layer and governs higher-order functions, including perception, reasoning, decision-making, language, and voluntary motor control. It is anatomically divided into the frontal, parietal, temporal, and occipital lobes, each specializing in distinct processes such as sensory interpretation, memory, and planning, as discussed in Sect. 4.3.1.1.

The brain's exceptional capacity for complex functionality emerges from the dynamic interaction between the cortex and the subcortex, a collection of structures that includes the basal ganglia, thalamus, and limbic system. These subcortical components are essential for motor control, emotional regulation, memory formation, and autonomic processes such as maintaining heart rate and respiration. Together, the cortical and subcortical regions form an integrated neural network that synthesizes sensory inputs with internal states to produce coordinated, adaptive, and context-appropriate behaviors.

Neuronal Synchronization is a fundamental mechanism underpinning this integration. It plays a crucial role in both physiological processes, such as cognition and learning [121], and pathological conditions like Parkinson's disease in Sect. 4.1.1.3 [122] and epilepsy in Sect. 4.5 [83]. Synchronization enables efficient communication and coordination across spatially distributed brain regions, facilitating complex cognitive and motor tasks.

In the context of motor coordination, functional Magnetic Resonance Imaging (fMRI) studies reveal that task complexity dictates the degree and extent of brain region involvement [123]. Simple tasks, such as synchronization to a steady rhythm, primarily recruit contralateral sensorimotor, supplementary motor cortices, and the ipsilateral cerebellum. However, more complex tasks like syncopation require the recruitment of additional regions, including the basal ganglia, dorsolateral premotor cortex, prefrontal regions, and temporal association cortices. These findings underscore the brain's capacity for adaptive reorganization, engaging broader networks to meet task-specific demands.

Similar principles apply to other specialized functions, such as language processing in autism [124]. Altered synchronization patterns in high-functioning autistic individuals reflect a reliance on alternative or additional neural resources to compensate for underconnectivity between cortical regions. Reduced interregional coordination challenges large-scale information integration, necessitating the recruitment of distinct neural circuits to perform complex tasks.

Temporal coding in neural systems further highlights the critical role of oscillatory synchronization in facilitating communication across spatially segregated brain regions [125, 126]. Rhythms in the beta/gamma ranges (20–80 Hz) are particularly vital for precise neuronal responses and context-dependent processing, engaging cortico-cortical and cortico-subcortical interactions to support motor planning, attention, and perceptual grouping. Structures such as the thalamus, basal ganglia, and long-range cortical connections contribute significantly to maintaining this synchronization. Developmental plasticity further enhances synchronization by optimizing conduction velocities and refining cortico-cortical connections [127].

These studies collectively emphasize that for specialized and demanding tasks, the brain recruits a more extensive and distributed network of cortical and subcortical regions, leveraging synchronization as a dynamic and adaptive mechanism to meet complex functional demands. Additionally, studies indicate that neural dynamics during music perception involve widespread brain regions [86]. Electroencephalography (EEG)-based investigations have demonstrated increased coherence between global neural dynamics and auditory inputs triggered by specific musical stimuli [128]. This highlights the importance of examining how auditory inputs influence brain networks, particularly in relation to the auditory cortex.

Among the diverse modeling approaches available [111], the FitzHugh-Nagumo (FHN) system serves as a paradigmatic model for excitable media of the activator-inhibitor type. It provides a coarse-grained representation of neuronal spiking dynamics and synchronization, making it well-suited for studying partial synchronization patterns induced by external auditory stimuli. Here, we use the FHN model to explore synchronization patterns elicited by an external sound source targeting the auditory cortex.

A notable characteristic of musical sound perception is the phenomenon of sound fusion [94]. Despite the complexity of a sound's overtone spectrum, listeners perceive a single pitch, which represents a fusion of the spectral components. Understanding this fusion in the context of synchronization provides valuable insights into the neurophysiological basis of auditory perception.

In this section, we consider an external sound source with fixed amplitude and a single frequency, intentionally simplifying the complexities of music and its differential effects across various frequency bands in brain oscillations. To isolate the effects of periodic perturbations in Sect. 5.2, we adopt a minimal model that omits node-specific behaviors. This approach enables a clear demonstration of synchronization phenomena using networks of coupled neural oscillators.

The FitzHugh-Nagumo (FHN) model is employed to capture the nonlinear dynamics of individual neurons or entire brain regions, representing these dynamics through a fast excitatory and a slow inhibitory variable. Interactions between neurons or brain regions are governed by a coupling matrix, which can either be constructed using standard network science techniques or derived from empirical structural connectivity data, such as those obtained from diffusion-weighted magnetic resonance imaging (MRI) of human subjects [129, 130].

The significance of distance-dependent transmission time delays in large-scale brain synchronization has been highlighted in previous studies [108, 131]. In the present section, we focus on the interplay between synchronization and cortical input across various brain regions, as defined by the 90 areas of the human brain according to the Automated Anatomical Labeling (AAL) atlas. This approach allows for a detailed examination of how external perturbations influence synchronization patterns within large-scale brain networks.

5.3.2 Brain Network and Cortical Regions

We consider an empirical structural brain network (see Fig. 4.29) where every brain area is modeled by a single FitzHugh-Nagumo (FHN) oscillator [109–113]. The FHN dynamics of the network with external stimulus at a specific pair of cortical regions reads:

$$\begin{aligned}\epsilon \dot{u}_k &= u_k - \frac{u_k^3}{3} - v_k \\ &+ \sigma \sum_{j \in N_{\mathrm{H}}} A_{kj} \left[B_{uu}(u_j - u_k) + B_{uv}(v_j - v_k) \right] \\ &+ \varsigma \sum_{j \notin N_{\mathrm{H}}} A_{kj} \left[B_{uu}(u_j - u_k) + B_{uv}(v_j - v_k) \right], \\ &+ C_k^I \gamma \cos \omega t\end{aligned} \tag{5.9a}$$

$$\begin{aligned}\dot{v}_k &= u_k + a \\ &+ \sigma \sum_{j \in N_{\mathrm{H}}} A_{kj} \left[B_{vu}(u_j - u_k) + B_{vv}(v_j - v_k) \right] \\ &+ \varsigma \sum_{j \notin N_{\mathrm{H}}} A_{kj} \left[B_{vu}(u_j - u_k) + B_{vv}(v_j - v_k) \right],\end{aligned} \tag{5.9b}$$

with $k \in N_{\mathrm{H}}$ where N_{H} denotes either the set of nodes k belonging to the left (N_L) or the right (N_R) hemisphere, and $\epsilon = 0.05$ describes the timescale separation between the fast activator variable (neuron membrane potential) u and the slow inhibitor (recovery variable) v [109]. Depending on the threshold parameter a, the FHN model may exhibit excitable behavior ($|a| \geq 1$); or self-sustained oscillations ($|a| < 1$). For all nodes we use the oscillatory regime in order to obtain clear insight into the topological effect of different locations of the input, and we fix $a = 0.5$ sufficiently far from the Hopf bifurcation point. The external input stimulus is modeled by a trigonometric function with frequency ω and amplitude γ and is applied to a specific pair of cortical regions $k = I_0$ and $k = I_0 + 45$, where the index $I = I_0$ denotes the stimulated area. For instance, $k = 41$ and $k = 86$ are associated with the auditory cortex, i.e. $C_k^I = 1$ if $k = 41$ or $k = 86$ and zero otherwise. The intra-hemispheric coupling between the single regions is given by the coupling strength σ, and the inter-hemispheric coupling is given by ς. As we are looking for partial synchronization patterns we fix $\sigma = 0.7$ and $\varsigma = 0.15$ similar to numerical studies of synchronization phenomena during unihemispheric sleep in Sect. 4.4 [132] where partial synchronization patterns have been observed. Given the uncertainty in the empirical connectivity data [130], note that the precise choice of the interhemispheric coupling is not crucial since the cortical input is chosen symmetrically in the corresponding areas of both hemispheres. The interaction scheme between nodes is characterized by a rotational coupling matrix:

$$\mathbf{B} = \begin{pmatrix} B_{uu} & B_{uv} \\ B_{vu} & B_{vv} \end{pmatrix} = \begin{pmatrix} \cos\phi & \sin\phi \\ -\sin\phi & \cos\phi \end{pmatrix}, \tag{5.10}$$

with coupling phase $\phi = \frac{\pi}{2} - 0.1$, causing dominant activator-inhibitor cross-coupling [114], i.e., the prevalence of excitatory versus inhibitory couplings. Also in the modeling of epileptic-seizure-related synchronization phenomena in Sect. 4.5 [83], where a part of the brain synchronizes, it turned out that such a cross-coupling is important. Mathematically, this means that signals of other neuronal areas are coupled via a coupling phase, which introduces a phase lag or time delay. The subtle interplay of excitatory and inhibitory interaction enables intermittent periods of either high or low synchronization. This is typical of the critical state at the edge of different dynamical regimes in which the brain operates [115, 133, 134].

5.3.3 *Role of Auditory Cortex*

In this section, we will investigate the role of the auditory cortex in the collective dynamics of the human brain. For this purpose, we feed a periodic external input into specific areas of our neural network, using the regions in pairs as described in the AAL atlas (Table 5.1). Depending on the selected cortical regions I, a different influence on the degree of synchronization of the overall network can be observed as shown for three different input regions in Fig. 5.7. The light colored regions in Fig. 5.7 indicate synchronized dynamics, whereas the darker colors indicate desynchronized dynamics. For $k = 14$, $I = 1$ in Fig. 5.7a, there is a slightly light colored stripe around the natural uncoupled frequency $\omega = 2.6$ ($\langle R \rangle \approx 0.7$). For $k = 34$, $I = 45$ in Fig. 5.7b, there is a pronounced light colored synchronization region ($\langle R \rangle \approx 1$) starting at $\omega = 2.4$. For $k = 41$, $I = 15$ in Fig. 5.7c, a triangular synchronization tongue splits off from the bottom left of the broad synchronization region, starting at $\omega = 2.4$. This is shown in more detail in the close-up in Fig. 5.7d.

To further elaborate the different influence of different input regions, the global order parameter $\langle R \rangle$ is shown in Fig. 5.8 in dependence of the frequency ω of the external stimulus and its cortical input region I for four values of the input strength γ. In the case of the auditory cortex ($I = 15$), a distinct influence on the neuronal network can be observed, which is not so pronounced for other regions. There is a threshold for global synchronization at input frequencies of $\omega = 2.4$. Even for very small input strengths $\gamma = 0.11$ (see Fig. 5.8a), synchronization of the entire network can be achieved for certain input regions I. On the other hand, even for very large input strengths $\gamma = 7.0$ (see Fig. 5.8c), some input regions never induce synchronization of the entire system. For better visibility, the input region indices $I = \mathcal{P}(1, ..., N)$ (modulo 45) are permuted according to their synchronizability by re-arranging them in ascending order of the row sum of the temporal mean of the Kuramoto order parameters $\langle R \rangle$ averaged over all 4 panels of Fig. 5.8. Increasing the

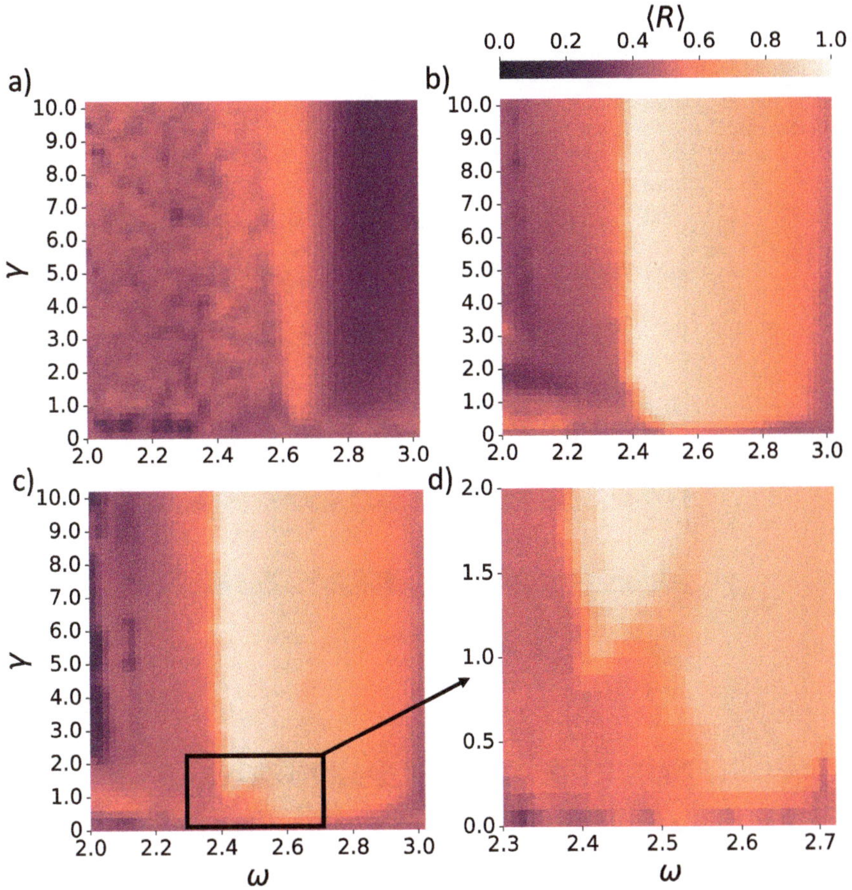

Fig. 5.7 Synchronization regions in brain network with external stimulus: $\langle R \rangle$ in the parameter plane of the frequency ω and the input strength γ of the external stimulus for cortical input regions **a** $k = 14, I = 1$, **b** $k = 34, I = 45$, and **c** $k = 41, I = 15$. Panel (**d**) shows a blow-up of (**c**). Other parameters: $\sigma = 0.7$, $\varsigma = 0.15$, $\epsilon = 0.05$, $a = 0.5$, and $\phi = \frac{\pi}{2} - 0.1$. Figure taken from [120]

input strength further to $\gamma = 11.0$ (see Fig. 5.8d) does not lead to any quantitative change of the synchronization tongue.

Interestingly, the auditory cortex $I = 15$ (relabeled, marked yellow in Table 5.1) is an input region that allows for the synchronization of the entire system at sufficiently large γ, but not at very small γ (Fig. 5.8a). In this sense it plays an intermediate role between cortical areas which can easily synchronize the brain (marked pink in Table 5.1) and those which never synchronize (marked blue in Table 5.1).

Figure 5.9 shows details of the synchronization dynamics for the three different groups of input regions in terms of the time evolution of R (left) and the mean phase velocities (right). Fixing the input amplitude γ and frequency ω beyond the threshold of the synchronization tongue, we find three different scenarios: In Fig. 5.9a, the time

Table 5.1 Cortical and subcortical regions [135] as introduced in Table 4.3. I denotes the relabeled index of the regions ordered according to increasing influence on global synchronization. A small input signal ($\gamma = 0.11$, see Fig. 5.8a) to the pink shaded regions has a big impact on the synchronization of the whole system ($\langle R \rangle > 0.8$), whereas even strong inputs ($\gamma = 11.0$, see Fig. 5.8d) to the blue shaded regions have no impact on synchronization ($\langle R \rangle < 0.5$). The brain areas $k = 41, 86$ (yellow shaded, $I = 15$) are associated with the auditory cortex

k: L/R	I	Region	Lobe
1/46	35	Precentral	Central region
2/47	18	Frontal Sup	Frontal lobe
3/48	7	Frontal Sup Orb	Frontal lobe
4/49	31	Frontal Mid	Frontal lobe
5/50	8	Frontal Mid Orb	Frontal lobe
6/51	34	Frontal Inf Oper	Frontal lobe
7/52	29	Frontal Inf Tri	Frontal lobe
8/53	3	Frontal Inf Orb	Frontal lobe
9/54	11	Rolandic Oper	Central Region
10/55	30	Supp Motor Area	Frontal lobe
11/56	5	Olfactory	Frontal lobe
12/57	33	Frontal Sup Medial	Frontal lobe
13/58	21	Frontal Med Orb	Frontal lobe
14/59	1	Rectus	Frontal lobe
15/60	22	Insula	Insula
16/61	20	Cingulum Ant	Limbic lobe
17/62	25	Cingulum Mid	Limbic lobe
18/63	26	Cingulum Post	Limbic lobe
19/64	13	Hippocampus	Limbic lobe
20/65	16	ParaHippocampal	Limbic lobe
21/66	2	Amygdala	Sub cort gray nuc
22/67	39	Calcarine	Occipital lobe
23/68	43	Cuneus	Occipital lobe
24/69	37	Lingual	Occipital lobe
25/70	42	Occipital Sup	Occipital lobe
26/71	38	Occipital Mid	Occipital lobe
27/72	32	Occipital Inf	Occipital lobe
28/73	23	Fusiform	Occipital lobe
29/74	41	Postcentral	Central region
30/75	28	Parietal Sup	Parietal lobe
31/76	36	Parietal Inf	Parietal lobe
32/77	44	Supramarginal	Parietal lobe
33/78	40	Angular	Parietal lobe
34/79	45	Precuneus	Parietal lobe
35/80	14	Paracentral Lobule	Frontal lobe
36/81	9	Caudate	Sub cort gray nuc
37/82	19	Putamen	Sub cort gray nuc
38/83	12	Pallidum	Sub cort gray nuc
39/84	17	Thalamus	Sub cort gray nuc
40/85	10	Heschl	Temporal lobe
41/86	15	Temporal Sup	Temporal lobe
42/87	4	Temporal Pole Sup	Limbic lobe
43/88	24	Temporal Mid	Temporal lobe
44/89	6	Temporal Pole Mid	Limbic lobe
45/90	27	Temporal Inf	Temporal lobe

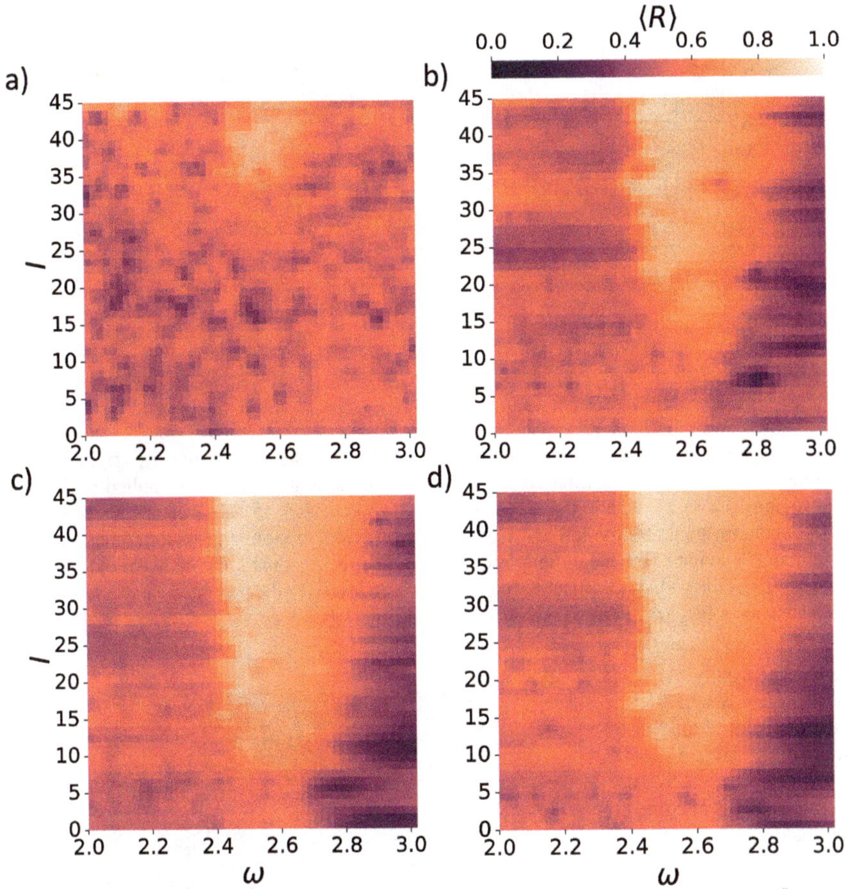

Fig. 5.8 Same as in Fig. 5.7 in the parameter plane of the frequency ω of the external stimulus and its cortical input region I (see Table 5.1) for input strengths **a** $\gamma = 0.11$, **b** $\gamma = 1.1$, **c** $\gamma = 7.0$, and **d** $\gamma = 11.0$. Other parameters as in Fig. 5.7. Figure taken from [120]

evolution of the Kuramoto order parameter is similar to the system behavior without external stimulus, i.e., it exhibits large time variations and no global synchronization. The mean phase velocities of all nodes, in particular in the left hemisphere, show no frequency synchrony, except for some local areas, and stronger frequency synchrony in the right hemisphere, which displays a remarkable asymmetry. The horizontal gray dotted line shows the time average of the collective mean field frequency Ω, which in the right hemisphere agrees well with the individual mean phase velocities Ω_k, however it is clearly distinct from the frequency of the two stimulated areas (red dots). Fig. 5.9b shows input to the auditory cortex, leading to episodes of strong global synchrony interrupted by intervals of desynchronization (left panel), and overall mean phase velocities close to the collective mean frequency Ω (right panel). Figure 5.9c shows constantly high global synchronization equal to the input

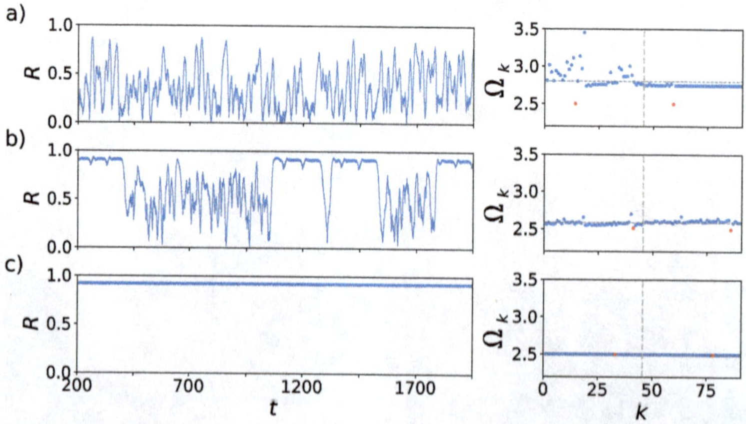

Fig. 5.9 Dynamical scenarios: $R(t)$ (left column) and mean phase velocities Ω_k (right column) for frequency $\omega = 2.5$ of the external stimulus and fixed amplitude $\gamma = 1.1$ applied to different input regions **a** $k = 14$, $I = 1$, **b** $k = 41$, $I = 15$, **c** $k = 34$, $I = 45$. The vertical dashed line in the right column separates the left and right hemisphere; the horizontal grey dotted line indicates the temporal average of the mean-field frequency Ω. The red dots mark the nodes of (**a**) the rectus $k = 14, 59$, (**b**) the auditory cortical regions ($k = 41, 86$), and (c) the precuneus $k = 34, 79$. Other parameters as in Fig. 5.8. Figure taken from [120]

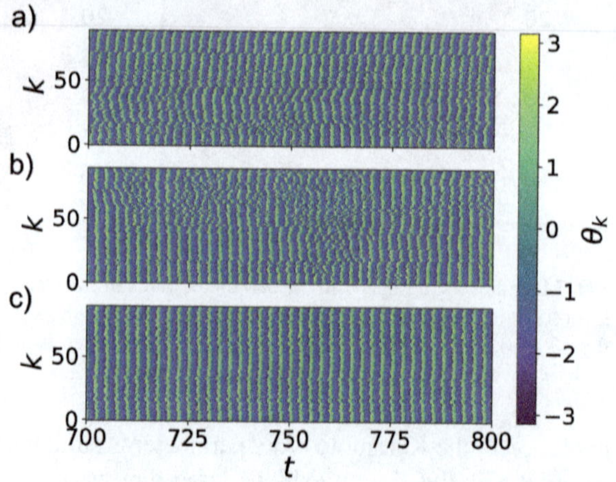

Fig. 5.10 Partially synchronized dynamics: space-time plot of the phase variable $\theta_k = \arctan(u_k, v_k)$ for frequency $\omega = 2.5$ of the external stimulus and fixed amplitude $\gamma = 1.1$ applied to different input regions **a** $I = 1$, **b** $I = 15$, **c** $I = 45$, corresponding to Fig. 5.9a,b,c, respectively. Other parameters are as in Fig. 5.8. Figure taken from [120]

frequency. Thus the auditory cortex plays an intermediate role between brain areas which cannot synchronize the whole system and others which do so strongly in a very stable way.

For further insight into the local synchronization, Fig. 5.10 shows the space-time representation of the phase variable θ_k for the corresponding parameters in Fig. 5.9. Figure 5.10a shows quite irregular dynamics independent of the cortical input, while in Panel (c) strong global synchronization is induced by the cortical input. In Panel (b) the alternating synchronization and desynchronization dynamics can be clearly seen. This delicate balance of synchrony and asynchrony for auditory stimulation is probably typical for the critical state and the high sensitivity of the brain operating between different dynamic regimes.

5.3.4 Summary

The interaction between synchronization and cortical input is fundamental to understanding how the brain perceives music. Here, we present computational simulations of dynamic networks of nonlinear oscillators to explore brain function. By employing simple oscillator models at network nodes combined with complex brain network architectures and targeted cortical inputs, we gain insights into network responses shaped by the structural connectivity of the stimulated nodes across varying input locations. Notably, our results reveal distinct synchronization dynamics when periodic input is applied to nodes outside the auditory cortex.

The simulations use empirical structural network connectivities derived from diffusion-weighted magnetic resonance imaging (MRI) of humans, alongside the paradigmatic FitzHugh-Nagumo (FHN) model. This approach demonstrates significant instances of partial brain synchronization and supports the prevailing view that the brain operates near a critical state, positioned at the boundary of different dynamical regimes [115, 133, 134].

In particular, our investigation highlights the interplay between network topology and cortical input signals across various brain areas. We observe episodes of heightened synchrony that depend on the input frequency, amplitude, and cortical location. Two frequencies are especially critical for synchronization: the natural frequency of the uncoupled oscillator and the collective frequency of the coupled network. Synchronization increases as the input frequency approaches either the oscillator's natural frequency or its harmonics. Furthermore, the cortical input location significantly impacts network synchronizability, with the auditory cortex being a key region for inducing variability in synchrony. Consistent with previous findings [136], we note marked differences between stimulations targeting specific network modules and those distributed across the network. Extending this analysis to inputs into various sensory cortices, possibly using alternative brain atlases [137, 138], would be an interesting avenue for future research.

These findings align with experimental evidence [93, 128] showing that music elicits synchrony in the human brain. For example, music listening induces periodic alternations between synchronization and desynchronization, reflecting the brain's dynamic variability. This alternation can be interpreted as a critical state situated between fully synchronized and desynchronized regimes. It is well-documented that in critical states, the brain exhibits characteristics such as hysteresis and avalanche phenomena, akin to behaviors observed in critical phenomena and phase transitions [117–119, 139, 140]. By adjusting model parameters, we observe intriguing dynamical transitions involving intermittent shifts between high and low synchronization levels [73]. Thus, simulating the effects of an external sound source provides a framework for modeling the influence of music on brain synchronization dynamics.

In a more detailed analysis of music perception using a cochlear model for auditory input [128], the impact of real music on a simulated network of FHN oscillators with empirical structural connectivity is compared to experimental EEG data in Sect. 5.4. These simulations reveal increased coherence between global network dynamics and musical input signals, with sensitivity to specific frequency bands. The results correspond to experimental observations of global neural synchronization in the gamma-band range, particularly its enhancement before transitions between different musical form segments. This synchronization increase before musical boundaries and decrease afterward characterizes large-scale musical form perception. Together, these findings emphasize the intricate role of synchronization in translating musical stimuli into dynamic brain activity.

5.4 Modeling the Perception of Music

In this section, we investigate the influence of music on the dynamics of a network of FitzHugh-Nagumo oscillators, leveraging empirical structural connectivity data derived from healthy human subjects. Our simulations demonstrate an increase in coherence between the global dynamics of the network and the input signal when stimulated by a specific piece of music. This coherence reflects the brain's capacity to synchronize its activity with external auditory stimuli, highlighting the critical role of frequency bands in shaping this interaction.

We show that the level of coherence is highly dependent on the frequency band of the input signal. High-frequency bands, such as those in the gamma range, are particularly effective in driving synchronization within the network, consistent with their known association with neocortical activity. Our findings are supported by comparisons with experimental studies, which report global neural synchronization in the gamma-band range across different brain regions. These studies also observe an increase in synchronization just before transitions between distinct parts of the musical structure, often referred to as musical high-level events. Such transitions likely represent moments of heightened cognitive engagement and expectation in response to the evolving musical form.

Furthermore, our results suggest a separation in musical form-related brain synchronization across different frequency domains. High frequencies, associated with neocortical activity, are primarily linked to cognitive processing and the perception of complex musical structures. In contrast, low-frequency bands, such as those related to dance movements, appear to facilitate interaction between cortical and subcortical regions. This dichotomy underscores the multifaceted impact of music on brain dynamics, engaging distinct neural circuits depending on the frequency characteristics of the auditory input.

In addition to these findings, we explore how the interplay between cortical and subcortical regions contributes to the perception of rhythm and motion in music.

Low-frequency synchronization likely reflects the brain's ability to align with the rhythmic components of music, which are often connected to motor activity and dance-like movements. This resonates with the idea that music not only stimulates auditory regions but also engages motor systems, fostering a holistic neural response.

Our analysis of coherence patterns induced by music aligns with a broader understanding of brain dynamics during auditory processing. The observed synchronization patterns emphasize the brain's adaptability and its ability to coordinate activity across widespread networks in response to auditory stimuli. This insight has implications for the study of music therapy, where rhythmic and melodic elements of music are used to enhance neural synchronization in clinical populations.

By comparing our simulation results with experimental electroencephalography (EEG) and functional magnetic resonance imaging (fMRI) data, we establish a robust framework for understanding how music modulates brain activity. Specifically, the observed increase in synchronization in the gamma-band range and its association with transitions in musical form provide a valuable model for interpreting the neural basis of musical perception and engagement.

These findings pave the way for future research into the neural mechanisms underlying the interaction between music and brain dynamics. By incorporating more complex models of auditory input and examining their effects on brain networks, researchers can gain deeper insights into how music influences cognition, emotion, and motor functions. For more detailed information and a comprehensive discussion of these results, we refer readers to the publications [74, 128].

This section is structured as follows: Section 5.4.1 gives a short introduction into synchronization and its role for perception. In Sect. 5.4.2, we detail the transformation of musical input into a neural signal using a comprehensive cochlea model. Section 5.4.3 introduces the neural network model, which is based on empirical connectivity patterns and incorporates neural input to the auditory cortex derived from musical stimuli. In Sect. 5.4.4, we describe the methodologies employed to characterize the neural output. Section 5.4.5 presents the results of computational simulations, highlighting the observed dynamical scenarios. In Sect. 5.4.6, we compare these findings with experimental data obtained from human subjects. Finally, Sect. 5.4.7 provides concluding remarks and discusses the broader implications of this section.

5.4.1 Synchronization and Perception

Synchronization is a central concept in the dynamics of neural networks, playing a key role across a range of physiological, pathological, and cognitive processes. Elevated synchronization levels are typically observed during slow-wave sleep [75, 76] and the transition from wakefulness to sleep [77, 78]. Partial synchronization has recently been proposed as a framework for explaining phenomena such as the first-night effect [79] and unihemispheric sleep [76, 80–82], investigated in Sect. 4.4. Furthermore, synchronized brain activity in Sect. 4.5 is implicated in epileptic seizures, where excessive synchronization in localized brain regions can result in severe outcomes [83]. Conversely, synchronization also facilitates cognitive functions, including the development and perception of syntax [84–86].

Synchronization theory has become a valuable tool in understanding musical acoustics and music psychology [87–91]. While the neurophysiological mechanisms involved in music perception remain under investigation, synchronization is widely believed to play a critical role in attention modulation and expectation formation. Event-related potentials (ERPs) measured through electroencephalography (EEG) during music listening demonstrate synchronized dynamics across multiple brain regions [92, 93]. These findings indicate that increased synchronization is associated with the perception of musical structures, particularly in the beta (13–30 Hz) and gamma (30–120 Hz) frequency bands (see Table 4.5) [73, 88].

Synchronization is also employed as a theoretical framework for elucidating brain processes related to syntax development and perception [1, 84–86]. Notably, synchronization is integral to music perception, where it modulates attentional focus and supports the formation of auditory expectations [87–89, 91, 141]. EEG studies have shown that synchronized dynamics across brain regions during music listening reflect large-scale musical perception, where specific sound frequencies elicit distinct synchronization patterns in the brain [73, 88].

Sensory-guided actions rely on the integration of sensory information, perceptual decision-making, and motor execution. Research underscores the critical role of interareal phase synchronization in coordinating these processes across sensory, frontoparietal, and motor brain regions [142]. Using magnetoencephalography (MEG), studies have demonstrated that perceiving and reporting weak somatosensory stimuli are linked to enhanced phase synchronization in delta/theta (3–7 Hz) and gamma (40–60 Hz) frequency bands. This dynamic synchronization connects task-relevant neural systems, such as sensory, motor, and frontoparietal attentional networks. The strength of synchronization is also correlated with response times, highlighting its role in sensory-guided actions.

Further exploration of the temporal coupling of neural activity underscores its significance in efficient information processing during both perception and action [143]. Temporal alignment of neural oscillations reduces randomness, increases mutual information, and enhances robust sensory and cognitive processing. Computational and clinical research supports these findings, emphasizing the importance of synchronized oscillatory rhythms in integrating distributed neural activities.

Broader frameworks also extend these insights to memory and sensory processing. The concept of "synchrony through synaptic plasticity" suggests that synchronized input timing, regardless of its source, is essential for the functionality of neural networks, including signal compression and sparsening. Studies on auditory perception, particularly in cochlear implants, have demonstrated the crucial role of restoring neural synchrony through advanced sound-coding strategies. These approaches improve pitch perception and speech comprehension, further illustrating the importance of precise neural timing.

In Sects. 5.2 and 5.3, investigations into the influence of sound on brain networks have shown that external auditory stimuli, when linked to the auditory cortex, can induce partial synchronization patterns [73]. However, this approach simplifies the complexity of music by not considering its differential effects on various frequency bands of brain oscillations. Numerous modeling approaches have been proposed for neural systems in general [108, 111, 144–146] (see also Sect. 4.1.1.2), as well as for music-related phenomena in particular [147]. In this context, we utilize the FitzHugh-Nagumo model to simulate the spiking dynamics of neurons and investigate the coherence between brain network dynamics and an external music source. Our network model combines simple nodal dynamics with complex structural connectivity derived from empirical measurements. The model is validated by reproducing experimental data, demonstrating its ability to capture neural responses to musical stimuli.

By focusing on minimal models that exhibit partial synchronization patterns, researchers can isolate and better understand the direct impact of auditory stimuli on neural dynamics [92, 93]. This approach advances theoretical understanding while also facilitating the development of interventions aimed at leveraging neural synchronization for therapeutic and cognitive enhancement.

5.4.2 From Sound to Neural Spikes

The transformation of sound into neural spikes is the subject of much current research [148–153]. Music, speech, or any sound enters through the outer and middle ear as sound pressure, then acting on the oval window of the cochlea, as described in Sect. 2.6.1. The movement of the oval window is then transferred to a pressure in the lymph liquid of the cochlea surrounding the basilar membrane, which again acts on the basilar membrane, causing traveling waves there. Due to spatial differences in stiffness and damping on the membrane, sinusoidal waves with a single frequency show an increase in amplitude up to a point with maximum amplitude, the position of the so-called best-frequency, with a fast decay afterwards. Therefore, different positions on the basilar membrane represent different frequencies, making the cochlea a Fourier analyzer. The stereocilia on the basilar membrane at the position of respective best-frequency are then transferring the mechanical energy into neural spikes. The frequency distribution on the basilar membrane is logarithmic. Movements of neighboring frequencies lead to interactions, causing roughness perception up to a frequency band of a musical major third. These bands are called critical bands, and the basilar membrane consists of 24 such bands. The spikes leaving the respective bands are fed into the auditory pathway, consisting of several neural nuclei, where the nucleus cochlearis or the trapezoid body are the first two. The interaction between these neural nuclei is manifold with several feedback loops and binaural connections [154] ending at the auditory cortex on both hemispheres. Still up to the A1 region of the auditory cortex, the critical bands are maintained, where neural connections of higher nuclei are connected to bands on the basilar membrane, which is called tonotopy.

Many auditory features are present, extracted, or perceived already in this pathway, like sound localization, pitch, or timbre [155] (see also Sect. 2.3), although research has not concluded on further processing in the cortex [156]. Music perception of larger temporal content, like song or sonata form, are not part of processing in the auditory pathway up to the cortex, as far as we know. Still the feedback loops within the pathway are both directions, up and down, afferent and efferent (see Sect. 2.6.2), so e.g. there is one connection down from the cortex to the cochlea with only one nucleus in between, tuning the basilar membrane tension through efferent nerves, according to cortex activity [154].

Up to now, no model of the whole auditory pathway exists on a detailed neural level. The model used in this section therefore concentrates on main findings, i.e., the transition from sound to neural spikes, the tonotopy of neural connections up to the cortex, as well as partial synchronization of phases in the cochlea, which are also present as coincidence detection in the auditory pathway. A Finite-Difference Time Domain (FDTD) physical model of the cochlea is used [150]. The basilar membrane is about 3.5 cm long and only between 0.1 and 0.12 cm wide, so it is more a rod than a membrane. Therefore, the present model assumes a differential equation of a membrane like

$$\frac{K(x)}{\mu(x)}\frac{\partial^2 u}{\partial x^2} - d(x)\frac{\partial u}{\partial t} = \frac{\partial^2 u}{\partial t^2} + f(t)\,, \tag{5.11}$$

with basilar membrane displacement u along a one-dimensional axis x, basilar membrane stiffness $K(x) = 2 \times 10^9 e^{-3.4x}$ dyn/cm^3 changing along x, and linear mass density $\mu(x) = m/A(x)$ with mass m over cross section A again changing along the basilar membrane and $A(x) = 0.1\,\text{cm} \times (0.1\,\text{cm} + 0.02\,\text{cm} \times x/l)$ with basilar membrane length $l = 3.5$ cm taking into account the slight widening of the basilar membrane over its length. The boundary conditions of the basilar membrane are homogeneous Dirichlet boundary conditions which do not allow for displacements on the boundaries, but any derivative is allowed in accordance with the physiological conditions. Comparison between a membrane and a rod model shows no considerable differences, therefore a rod model is used. Here d is damping, and $f(t)$ is the driving force of the lymph fluid which drives the basilar membrane.

To calculate the spikes omitted by the cochlea, the recording of the musical piece used is fed into the cochlea model. Here the amplitudes of the digital musical sound file are taken as sound pressures acting on the oval window of the cochlea and therefore immediately on the peri- and endolymph around the basilar membrane. As the speed of sound in the lymph (~1 500 m/s) is much larger than the speed of waves on the basilar membrane which is between ~ 100 m/s at the oval window and down to ~10 m/s at the helicotrema, an instantaneous action of the pressure at the oval window on the basilar membrane is reasonable and known as long-wave approximation [157]. This holds for frequencies up to ~4 kHz, where pitch perception stops and humans only hear a very high sound. This approximation is used in the model. It leads to the force $f(t)$ in Eq. (5.11) which represents the amplitudes of the digital musical sound file acting instantaneously on all points of the basilar membrane at each time point respectively. It is interesting to see that the traveling wave on the basilar membrane is therefore not caused by an external input slowly traveling through the cochlea but by the intrinsic solution of the inhomogeneous differential equation of the basilar membrane driven by a periodic force over its whole length instantaneously.

Depending on the brain region, neurological measurements reveal different time scales [158]. In this section we choose 50 ms as a time integration step as this is consistent with a characteristic time scale in music as well as in visual perception. In music, 50 ms correspond to the second integration time, below which two events cannot be distinguished one from another. This leads to a threshold of 20 Hz, above which musical pitches are perceived and below which adjacent events are heard as rhythms. In vision, 18-24 frames per second lead to a continuous visual perception, again corresponding to about 50 ms time intervals. Therefore, in terms of hearing and seeing, the brain seems to update perceptional input on a time-scale of 50 ms [88].

The transition between mechanical displacement and electrical spike is performed using two conditions according to literature [159]. A neural spike at one point X on the basilar membrane at time τ is excited if two conditions hold

$$u(X,\tau) > u(X-1,\tau), u(X+1,\tau) \tag{5.12a}$$

$$u(X,\tau) > u(X,\tau-1), u(X,\tau+1). \tag{5.12b}$$

The condition in Eq. (5.12a) means a maximum shearing of two nervous fibers as a necessary condition to an opening of the ion channels at the fibers. This only happens with a positive slope, as only then the stereocilia are driven away from each other. With a negative slope the cilia are getting closer and therefore no stress appears at the tip links between them. This corresponds to the rectification process in gammatone filter banks. The condition in Eq. (5.12b) is a temporal maximum positive peak of the basilar membrane displacement. It is the temporal equivalent to the spatial condition of a maximum acceleration, where the tip link between the cell and its neighboring cells is most active.

To calculate the spikes omitted by the cochlea, the recording of the musical piece used is fed into the cochlea model. Therefore, the original piece, available as a digital recording of 44.1 kHz sample rate (CD-Quality) is upsampled to 192 kHz to meet Finite-Difference Time Domain (FDTD) stability criteria. The cochlea model is then run with a time-discretization step of $\Delta t = 1/192\,000$ s. Each time when a neural spike appears, the time point, strength, and critical band of the spike is stored. Therefore, after processing, a time series $I(t)$ of all spikes leaving the cochlea is obtained.

Figure 5.11a displays an example of an artificially generated so-called *tone complex* with $f_0 = 475$ Hz and ten partial tones (harmonics) with amplitudes $1/m$ where $m = 1, 2, 3, \ldots, 10$. The respective spike output of the basilar membrane model is shown in Fig. 5.11b. Each time when the sound wave has a maximum amplitude, a pressure pulse is traveling over the basilar membrane, which emits electrical spikes at respective best-frequency positions on the membrane in accordance with the frequencies in the activating sound. As traveling waves on the membrane start at the basal

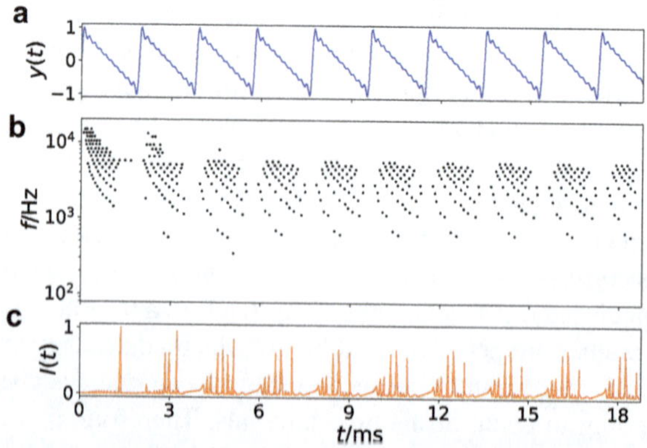

Fig. 5.11 Example of transformation of a sound wave into a spike pattern of the cochlea model. **a** Time series of an artificially generated tone complex $y(t)$ versus time t in ms with $f_0 = 475$ Hz and ten partial tones (harmonics) with amplitudes $1/m$ where $m = 1, 2, 3, \ldots, 10$. **b** Spikes (black dots) leaving the cochlea as calculated from the model [150], where the vertical axis represents the cochlea position with best-frequency f in Hz indicated, i.e., categorized into 24 so-called critical bands. **c** Time series $I(t)$ of the sum of all spike weights leaving the cochlea at a certain time t. Note that the first 5 ms are transients

end, next to the oval window, where high frequencies have their best-frequency location, and travel down the membrane towards the upper end, the helicotrema, where low frequencies are located, low frequencies show a time-delay with respect to higher frequencies. If the spikes of all critical bands are summed up for a certain point in time, a time series $I(t)$ of all neural spikes leaving the cochlea can be generated, as exemplarily shown in Fig. 5.11c.

5.4.3 *Neural Network Model and Music*

The auditory cortex is the part of the temporal lobe that processes auditory information in humans. It is a part of the auditory system, performing basic and higher functions in hearing and is located bilaterally, roughly at the upper sides of the temporal lobes, i.e., corresponding to the AAL indexing $k = 41, 86$ (temporal sup L/R) in Table 4.3. The auditory cortex takes part in the spectrotemporal analysis of the input passed on from the ear. Figure 5.12 displays the time-series of impulses which are supplied to the brain by means of the auditory cortex. These neural impulses were obtained by the method of Bader described in Sect. 5.4.2 [150–152]. Here, in contrast to Fig. 5.11, a real piece of music was used, namely the hip hop music song *One Mic*, composed by the American rapper *Nas* and released in 2002. During the transition from acoustic mechanical to electrical excitation within the cochlea, synchronization appears to improve perception of pitch, speech, or localization. The sampling rate of these impulses obtained by Bader's method is $f_s = 192$ kHz.

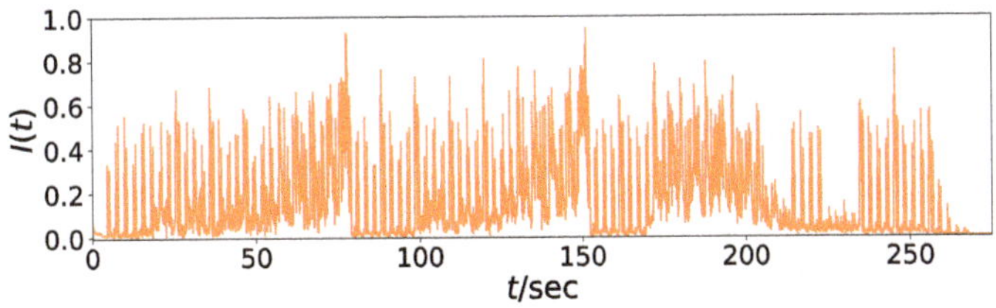

Fig. 5.12 Time-series of the neural input signal $I(t)$ obtained from the music song *One Mic* transformed by a method developed by Bader [86]. The song has a length of about 270 s and was released in 2002 by American rapper *Nas*. Figure taken from [128]

Each node corresponding to a brain region is modeled by the FitzHugh-Nagumo (FHN) model with external stimulus, a paradigmatic model for neural spiking [109–111]. Note that while the FitzHugh-Nagumo model is a simplified model of a single neuron, it is also often used as a generic model for excitable media on a coarse-grained level [112, 113]. Thus the dynamics of the network reads:

$$\begin{aligned}\epsilon\dot{u}_k &= u_k - \frac{u_k^3}{3} - v_k \\ &\quad + \sigma \sum_{j\in N_\mathrm{H}} A_{kj}\left[B_{uu}(u_j - u_k) + B_{uv}(v_j - v_k)\right] \\ &\quad + \varsigma \sum_{j\notin N_\mathrm{H}} A_{kj}\left[B_{uu}(u_j - u_k) + B_{uv}(v_j - v_k)\right], \\ &\quad + C_k I(t)\end{aligned} \tag{5.13a}$$

$$\begin{aligned}\dot{v}_k &= u_k + a \\ &\quad + \sigma \sum_{j\in N_\mathrm{H}} A_{kj}\left[B_{vu}(u_j - u_k) + B_{vv}(v_j - v_k)\right] \\ &\quad + \varsigma \sum_{j\notin N_\mathrm{H}} A_{kj}\left[B_{vu}(u_j - u_k) + B_{vv}(v_j - v_k)\right],\end{aligned} \tag{5.13b}$$

with $k \in N_\mathrm{H}$ where N_H denotes either the set of nodes k belonging to the left (N_L) or the right (N_R) hemisphere. Parameter $\epsilon = 0.05$ describes the timescale separation between the fast activator variable (neuron membrane potential) u and the slow inhibitor (recovery variable) v [109]. Depending on the threshold parameter a, the FitzHugh-Nagumo (FHN) model may exhibit excitable behavior ($|a| > 1$) or self-sustained oscillations ($|a| < 1$). We use the FHN model in the oscillatory regime and thus fix the threshold parameter at $a = 0.5$ sufficiently far from the Hopf bifurcation point. The coupling within the hemispheres is given by the coupling strength σ while the coupling between the hemispheres is given by the inter-hemispheric coupling strength ς. As we are looking for partial synchronization patterns we fix $\sigma = 0.7$ and $\varsigma = 0.15$ similar to numerical studies of synchronization phenomena during unihemispheric sleep in Sect. 4.4 [82] where partial synchronization patterns have been observed. The interaction scheme between nodes is characterized by a rotational coupling matrix:

$$\mathbf{B} = \begin{pmatrix} B_{uu} & B_{uv} \\ B_{vu} & B_{vv} \end{pmatrix} = \begin{pmatrix} \cos\phi & \sin\phi \\ -\sin\phi & \cos\phi \end{pmatrix}, \tag{5.14}$$

with coupling phase $\phi = \frac{\pi}{2} - 0.1$, causing primarily an activator-inhibitor cross-coupling. This particular scheme was shown to be crucial for the occurrence of partial synchronization patterns in ring topologies [114] as it reduces the stability of the completely synchronized state. Also in the modeling of epileptic-seizure-related synchronization phenomena [83], where a part of the brain synchronizes, it turned out that such a cross-coupling is important. The subtle interplay of excitatory and inhibitory interaction is typical of the critical state at the edge of different dynamical regimes in which the brain operates [115, 134], and gives rise to partial synchronization patterns which are not found otherwise.

The external stimulus $I(t)$ describes the impulses evoked by the music piece *One Mic* by *Nas* and is applied to the brain areas $k = 41, 86$ associated with the auditory cortex, i.e., $C_k = 1$ if $k = 41$ or 86 and zero otherwise. Since $I(t)$ is a time series which is calculated from a real piece of music, see Sect. 5.4.2, it has a physical dimension in seconds. On the other hand, the FitzHugh-Nagumo model uses dimensionless time. In order to compare our simulations with real data, we must transform the dimensionless time units of the FHN oscillator model to real time units by comparing the FHN oscillation period of a single FHN oscillator $T \approx 2.5$ (corresponding to dimensionless frequency $f_{\mathrm{FHN}} \approx 0.4$) to the characteristic frequencies n_b in Hz of an empirical time series. Depending upon the frequency band n_b (in Hz) chosen, the simulation time is converted to real time by $1\,\mathrm{s} = 2.5 n_b$ simulation time units, or the simulated frequency (in Hz) is

$$f_b = n_b / f_{\mathrm{FHN}}. \tag{5.15}$$

5.4.4 *Neural Dynamics and Coherence*

Additionally, we introduce a new measure which specifies the coherence between the Kuramoto order parameter and the input signal by using the time average of the Kuramoto order parameter weighted with the input signal

$$\gamma = \frac{1}{\Delta T} \int_0^{\Delta T} R(t) I(t)\, \mathrm{d}t \tag{5.16}$$

to quantify the overlap of coherent episodes (R large) with large input signals, averaged over time. The coherence γ is maximum if the synchronization is large whenever the signal is large. It is small if the overall synchronization is low, or if the modulation of the synchronization in time is not in phase with the modulation of the input signal amplitude. For $\gamma = 0$ the Kuramoto order parameter and the input signal do not overlap at any time point. An increased value of $\gamma \in [0, 1]$ means increased overlap between the Kuramoto order parameter and the input signal. The motivation for introducing the measure γ lies in the fact that in the human brain the increase and decrease of synchronization follows the large-scale form of the listened music in a coherent way [92, 93].

The input signal $I(t)$ is obtained from the original music song *One Mic* by the cochlea model described in Sect. 5.4.2 (see Fig. 5.11). The song has a length of about 4.5 min and the sampling rate of the obtained input signal is given by $f_s = 192$ kHz. Sampling is the reduction of a continuous-time signal to a discrete-time signal, e.g., the conversion of a sound wave (a continuous signal) to a sequence of samples (a discrete-time signal). The sampling rate f_s is then the average number of samples obtained in one second. According to the Nyquist criterion, the frequency information of $I(t)$ is then band-limited to $f_b < \frac{1}{2} f_s$.

The local dynamics of our model, the FitzHugh-Nagumo oscillator, has no explicit time scale. Its intrinsic angular frequency is dimensionless and given by $\omega_k = \omega_{\text{FHN}} = 2\pi f_{\text{FHN}} \approx 2.51$. To include the time signal $I(t)$ correctly in our dimensionless model, we scale the dimensional neural input obtained in Sect. 5.4.2 by the frequency n_b (in Hz), or equivalently, convert the dimensionless output of our simulation into physical dimensions f_b according to Eq. (5.15). This parameter n_b effectively removes the time scale from the input, but on the other hand, it can also be seen as creating a link between our dimensionless model and the input signal $I(t)$.

5.4.5 *Frequency Bands and Coherence*

Next, we investigate dynamical scenarios emerging from an external stimulus in the auditory cortices of both hemispheres ($k = 41, 86$). In order to compare our simulations with the empirical analysis of the influence of music upon the brain [92, 93], we may choose different frequency bands n_b, and hence a different scaling of the time in the external stimulus. This can be visualized by plotting the coherence measure γ in dependence on the characteristic frequency n_b (in Hz), see Fig. 5.13. We find a strong non-monotonic behavior of $\gamma(n_b)$ and it turns out that by taking the frequency band n_b of the external stimulus as a control parameter, one can change the level of coherence between the system dynamics and the external stimulus. Although the standard deviation of the coherence measure is relatively large for an ensemble size of 200 simulations (indicated by the vertical bars), we find a pronounced maximum of the coherence γ for $n_b = 12 - 48$ Hz corresponding to the gamma-band of brain waves ($f_b \approx 30 - 120$ Hz) shown in Fig. 5.13 by purple shading. This means that for that frequency n_b the level of synchronization follows the external signal most closely. It is in agreement with what has been observed in empirical brain analysis of the perception of music [92, 93].

Figure 5.14a–c depicts the details of the change of the time series of the Kuramoto order parameter $R(t)$ with increasing values of the frequency band n_b of the external stimulus $I(t)$, which is shown in Fig. 5.14d. It represents a part of the neural input

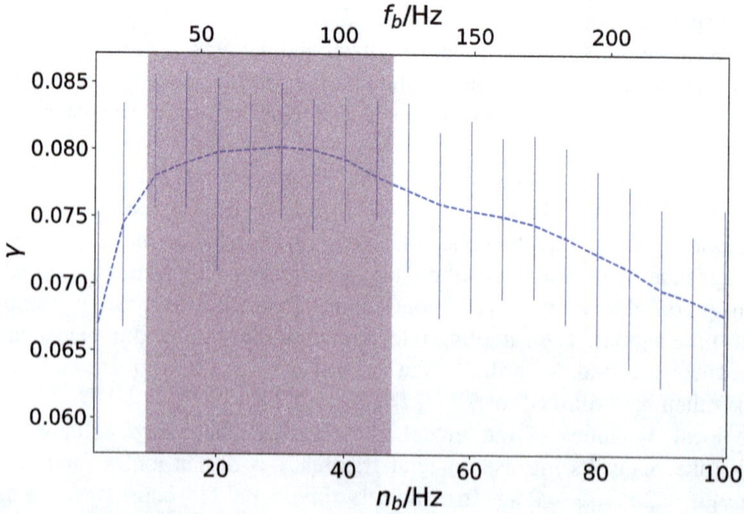

Fig. 5.13 Coherence between network dynamics and external stimulus: coherence measure γ in dependence on the characteristic music frequency n_b (in Hz). The labeling on the upper x-axis denotes the corresponding frequency $f_b = n_b/f_{\mathrm{FHN}}$ in the brain, where $f_{\mathrm{FHN}} \approx 0.4$ is the dimensionless frequency of the FitzHugh-Nagumo (FHN) model, and the purple shaded region indicates the gamma-band ($f_b \approx 30 - 120$ Hz). The vertical bars indicate the standard deviation of the coherence measure for an ensemble of 200 simulations. The dashed line is obtained by a Savitzky-Golay filter. Other parameters are given by $\sigma = 0.7$, $\varsigma = 0.15$, $\epsilon = 0.05$, $a = 0.5$, and $\phi = \frac{\pi}{2} - 0.1$. Figure taken from [128]

signal $I(t)$ constructed from the music song *One Mic* and shown in Fig. 5.12b. We take a closer look at the temporal evolution of R and the mean phase velocities ω_k in the system for different values of n_b chosen from three different regimes in Fig. 5.13. With increasing value of n_b in panels (a)–(c), the time scale of the simulated neural output in Hz changes from lower to higher frequencies f_b which is also seen in the temporal fluctuations of $R(t)$. Furthermore we observe on the one hand an increasing amplitude of the temporal fluctuations of R. On the other hand, the temporal average of the Kuramoto order parameter R decreases with increasing n_b, marked by a horizontal grey dotted line in the left column: While for a small value of $n_b = 5$ Hz in Fig. 5.14a the Kuramoto order parameter R assumes rather large values, and small values $R < 0.2$ are not reached, for high values of $n_b = 90$ Hz in Fig. 5.14c rather small values of R are measured. This trend can be seen by means of the temporal average of the Kuramoto order parameter R. For $n_b = 30$ Hz in Fig. 5.14b, the temporal average of R takes a value ≈ 0.5 and the time evolution shows regular oscillations between low ($R < 0.2$) and high values ($R > 0.8$). This aspect will be further discussed in the next section, since it can also be observed in experiments.

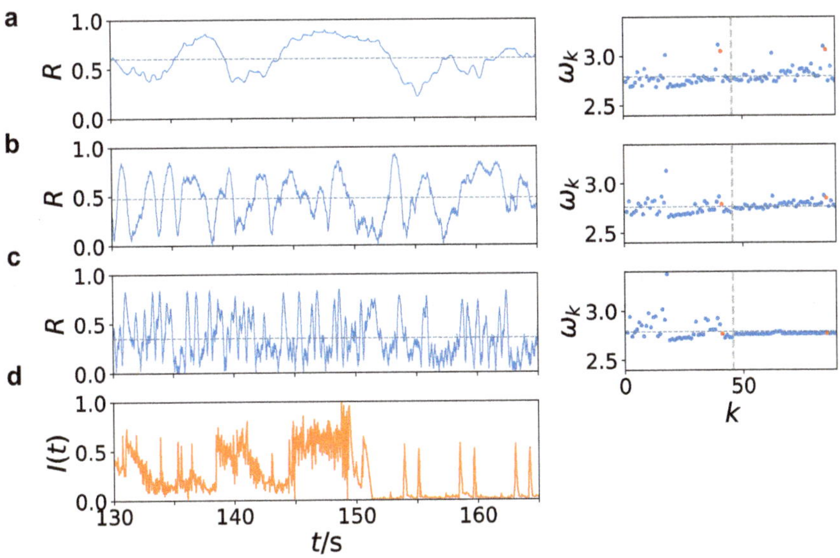

Fig. 5.14 Dynamical scenarios: network dynamics for low and high values of coherence γ. Kuramoto order parameter R versus time in s (left column) and dimensionless mean phase velocity profile $\omega_k = 2\pi f_k$ versus k (right column) for increasing values of the frequency n_b of the external stimulus $I(t)$ **a** $n_b = 5$ Hz, **b** $n_b = 30$ Hz and **c** $n_b = 90$ Hz. In panel (**d**) the corresponding external stimulus $I(t)$ is plotted, which is a blowup of a part of Fig. 5.12B. The vertical dashed line in the right column separates the left and right brain hemisphere; the red dots mark the nodes of the auditory regions ($k = 41, 86$). The horizontal grey dotted line indicates the temporal average of the Kuramoto order parameter R in the left column, and the spatial average of the mean-field frequency $\overline{\omega}$ in the right column. Other parameters are as in Fig. 5.13. Figure taken from [128]

As shown in Fig. 5.13, in the case of $n_b = 30$ Hz the coherence γ is maximum. Even though a higher value of the temporal average of $R(t)$, as observed in Fig. 5.14a for $n_b = 5$, might imply a higher value of γ according to Eq. (5.16), Fig. 5.14b shows that it is more important that $R(t)$ and $I(t)$ show a similar temporal modulation, as in Fig. 5.14b for $n_b = 30$. Despite the averaging over 250 simulations over the whole simulation time in Fig. 5.13, the time segment in Fig. 5.14b shows such a similarity in the modulation: We can see simultaneous drops of $R(t) < 0.1$ and $I(t) < 0.1$ for example at $t \approx 138, 140, 150$, whereas the values in between are higher, even if they fluctuate.

In the right column of Fig. 5.14 the dimensionless mean phase velocities ω_k of all nodes are plotted, the horizontal grey dotted line indicates the spatial average, i.e., the collective mean-field frequency $\overline{\omega}$, which does not change for different n_b since it is determined by the intrinsic collective dynamics. In contrast, the node dynamics of the auditory regions ($k = 41, 86$), indicated by red dots, depends on n_b since it receives the external input signal which has a higher frequency in dimensionless units if the time is scaled in larger units $1/n_b$. For $n_b = 5$ Hz in Fig. 5.14a, the mean phase velocity of the auditory cortex is higher compared to the spatial average of the collective mean-field frequency $\overline{\omega}$. For $n_b = 30$ Hz in Fig. 5.14b, the mean phase velocity of the auditory cortex approaches $\overline{\omega}$ having a bigger impact on the dynamics of the whole system than in Fig. 5.14a for $n_b = 5$ Hz. Remarkable is the fact of a dynamical asymmetry shown by the mean phase velocities in Fig. 5.14c.

While the nodes of the right hemisphere exhibit equal mean phase velocity, i.e., they are frequency synchronized, the left hemisphere remains desynchronized and exhibits on average faster dynamics. This may indicate that regardless of the input $I(t)$ the system can exhibit partial synchronization. Such behavior is similar to the dynamics of unihemispheric sleep studied in [82], where no external input has been applied to the dynamical system. In such states one hemisphere is synchronized, whereas the other hemisphere is partially desynchronized.

5.4.6 Comparison with Experiments

The coupling of oscillatory neural signals within the usual frequency bands has been thought to be a mechanism that is related to a broad range of perceptual, sensorimotor, and cognitive processes, such as Gestalt perception and binding [160–166], timing and expectation [167, 168], attention [169–171], consciousness [172–175], or motor functions [176] as well as in music perception [177, 178]. According to [174], oscillatory brain activity is usually clustered into several frequency bands: delta (0.5 – 3.5 Hz), theta (4 – 7 Hz), alpha (8 – 12 Hz), beta (13 – 30 Hz) and gamma (> 30 Hz) as shown in Table 4.5. Since the gamma-band is the 'youngest' frequency band which has become of interest (from about the late 1990s), the ranges and definitions vary from source to source. Here, we refer to the classification of [179], who speak of a low gamma range for frequencies above 30 Hz up to 60 Hz, and high gamma for frequencies above 60 Hz up to about 120 Hz. For everything above 120 Hz, we use the term 'fast oscillations' as employed by [180, p.114]. The gamma-band frequency range is of particular interest in the context of large-scale synchronization since it is thought to be a mechanism that integrates information from different parts of the cortex.

5.4.6.1 Musical Form

The musical form as the hierarchically highest level of musical structure and its perception is related to some of the mentioned processes above [93, 181]. Perceptually, notes, bars, and phrases are grouped and integrated into a high-level part of the form by the Gestalt laws [182–185]. The contrast of the form's parts, such as the concatenation of verse and chorus in a song, the sonata form of classical music, or the continuous night-long tension build-up and decay in Techno, House or Electronic Dance Music, characterize the musical form and the learned knowledge about the underlying structures leads to the build-up of expectation and their fulfillment as well as to modulated attention. On an emotional level, this can be expressed in the terms of tension and relaxation [186, 187]. Also, the transition from "potential energy" (expectations) into "kinetic energy" (dancing) as proposed by [188] can be related to the processing of musical form in the sense of entrainment of neurons in the motor cortex by neurons from the auditory cortex [176].

The characteristic of contrasting parts can be revealed not only by music analysis using pen and paper but also by different computational methods by the music information retrieval discipline, like the amplitude of a piece of music that corresponds to the subjective perception of loudness. Also other properties of the stimulus, such as the spectral centroid that corresponds to the perceived brightness of a sound, or the fractal correlation dimension [189, 190] corresponding to the perceived density and thereby representing the complexity of a piece of music, are drivers of the musical form [88, 93, 156, 191, 192].

Based on the aforementioned correlations between the processes associated with the perception of musical form and neural synchronization we expect the dynamics of neural synchronization to correspond to the dynamics of the amplitude of the stimulus. Again, the musical amplitude corresponds to perceived loudness, and is calculated as integration of energy over time intervals. Then synchronization between different brain regions is high when the amplitude of the musical piece is high, and synchronization is low when the amplitude of the piece is low. We expect such brain synchronization to be strong due to the prominence of the gamma-band in perception of musical parameters.

5.4.6.2 Experiment Procedure, Artefact Correction and Calculation of the Grand Average

In an experiment, we have recorded the electroencephalogram (EEG) from human scalps to examine the perception of music large-scale form (Fig. 5.15). We have taken into account the usual guidelines regarding ethical procedure (informed consent). The subjects were mainly found and recruited through the Institute of Systematic Musicology Hamburg and had instrumental lessons on at least one instrument (mean duration 10.0 years, standard deviation 4.6 years) or corresponding experience as DJ (Discjockey). They participated in accordance with local ethics committee guidelines. 25 musically skilled subjects listened to the song *One Mic* from the artist *Nas* three times each. The song was released in 2001 on his Album *Stillmatic* on Columbia Records. The electroencephalogram (EEG) signals were recorded with a sample rate of 500 Hz from 32 electrodes, positioned following the 10-20 method of placement [193]. After artifact correction, recorded data for each channel has been averaged over subjects and trials to obtain a grand average of 75 trials for each channel to increase the signal to noise ratio and enhance event-related potentials. This type of averaging reveals *evoked potentials* (in contrast to *induced potentials*) and is related to the presented stimulus in a classical event-related potential manner [163, 178, 194]. Since the positions of the electrodes do not differ over measurement time, the differences in correlation strength between different electrodes cannot be

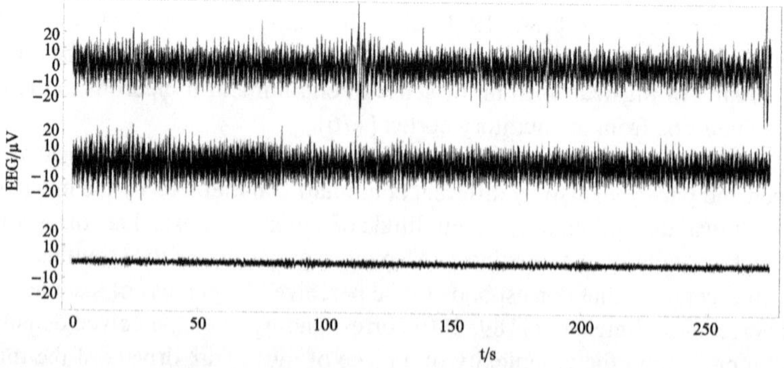

Fig. 5.15 Recorded and averaged electroencephalogram (EEG) data: top and middle plot show recorded EEG time series after pre-processing for one electrode (Fp1) from two different participants. The bottom plot shows the time series of the same electrode averaged over 25 subjects and 3 trials. Figure taken from [128]

explained by spurious synchrony [195–197]. For a more detailed description of the experimental procedure, technical details and pre-processing, see [93].

5.4.6.3 Wavelet Frequency Decomposition

In Fig. 5.16, all channels have been decomposed into nine independent frequency bands that correspond approximately to the frequency bands mentioned above by using a continuous wavelet transformation with a Mexican Hat wavelet [179]. In contrast to a bandpass filter with a subsequent Hilbert transform, using a Mexican hat wavelet for filtering is fast and efficient since one can decompose the recorded EEG data into the desired frequency-bands in one step by defining the number of octaves. The continuous wavelet transform of a uniformly sampled sequence $\{x_1, x_2, \ldots x_n\} = \{x(t_0), x(t_0 + \Delta t), \ldots, x(t_0 + (n-1)\Delta t)\}$ is given by

$$w(u, s) = \frac{1}{\sqrt{s}} \sum_{k=1}^{n} x_k \, \psi\left(\frac{(k-u)\Delta t}{s}\right), \tag{5.17}$$

where $s \in \mathbb{R}$ corresponds to the frequency of the EEG band and $u = 1, \ldots, n$ labels the wavelet coefficients with the number n of analyzed sample points defining the time window of observation. As wavelet function ψ a Mexican Hat wavelet is used, given by

$$\psi(x) = \frac{-2}{\sqrt[4]{\pi}\sqrt{3\sigma}} \left(\frac{x^2}{\sigma^2} - 1\right) \exp\left(-\frac{x^2}{2\sigma^2}\right), \tag{5.18}$$

where σ is the width of the wavelet. The EEG bands used align very well with a musical scale, where each higher band doubles the frequency of its respective lower

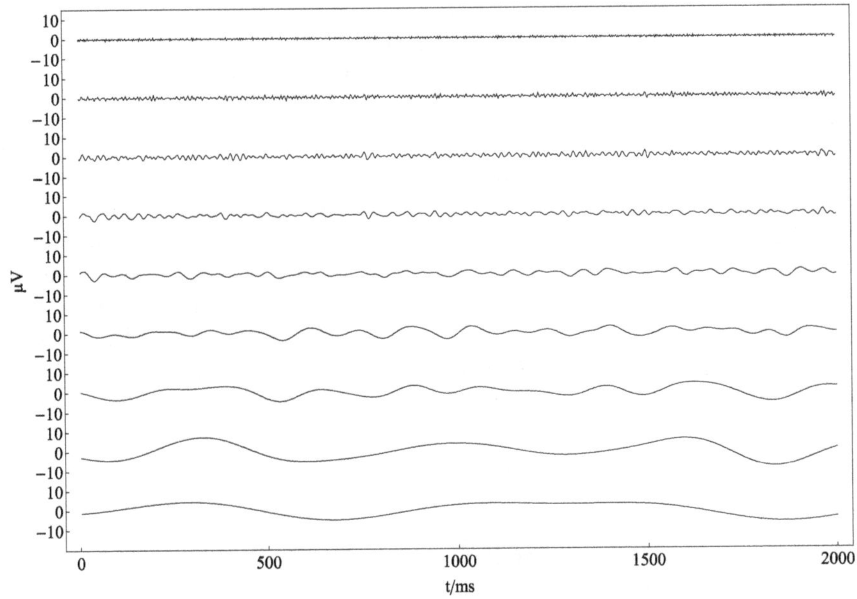

Fig. 5.16 Nine frequency bands (FB) after wavelet transformation: Result of the continuous wavelet transform for the first two seconds of the averaged time series in Fig. 5.15. From top to bottom frequency bands correspond to FB 1: 125 – 250 Hz, FB 2: 62.5 – 125 Hz, FB 3: 31.25 – 62.5 Hz, FB 4: 15.63 – 31.25 Hz, FB 5: 7.81 – 15.63 Hz, FB 6: 3.91 – 7.81 Hz, FB 7: 1.95 – 3.91 Hz, FB 8: 0.98 – 1.95 Hz, FB 9: 0.49 – 0.98 Hz. Figure taken from [128]

band, corresponding to a musical octave. Please note that this relation might only be at chance, still it may also relate to the fact that all human senses relate physics to perception in a logarithmic way [198]. It is therefore convenient to scale s in the wavelet transform in the same mathematical way as an equal-tempered musical scale like $s_{\text{oct}} = \alpha\, 2^{\,\text{oct}-1}$, where oct $\in \{1, 2, \ldots, 9\}$ is the octave number related to the nine frequency bands shown in Fig. 5.16 and α is the smallest wavelet scale.

5.4.6.4 Analysis of Synchronization

For each electrode pair of these nine data sets filtered in this way, the synchronization is calculated in the next step. Thus, we can analyze the synchronization dynamics as a function of the frequency bands. As a synchronization measure the linear cross-correlation, for simplicity taken without time delays, is used, which in fact results in the Pearson correlation coefficient r. This is widely used as a non-directed measure of the strength of the correlation between two variables or sequences $\{x_1, x_2, \ldots, x_n\}$ and $\{y_1, y_2, \ldots, y_n\}$ [199–201]:

$$r = r_{x,y} = \frac{\frac{1}{n}\sum_{i=1}^{n}(x_i - \overline{x})(y_i - \overline{y})}{\sqrt{\frac{1}{n}\sum_{i=1}^{n}(x_i - \overline{x})^2}\sqrt{\frac{1}{n}\sum_{i=1}^{n}(y_i - \overline{y})^2}}, \tag{5.19}$$

where $\overline{x}, \overline{y}$ denotes the mean of x, y, respectively. Since we aim to reveal synchronization dynamics on the level of musical form, we calculate the correlation within successive 1-second time windows for each possible pair of electrodes of each wavelet-filtered dataset, which results in $32 \times 31/2 \times 9 = 4464$ time series with a resolution of 1 s, and each time series has a length of 270 s corresponding to the stimulus length (see Fig. 5.17). In recent decades, various methods for measuring synchronization have been introduced [200, 202]. The advantage of the Pearson correlation coefficient r is that it allows for easy and efficient calculation of the linear correlations between two variables or time series, and the results are very similar to those obtained by other common methods such as the phase-locking value [203]. For a comparison of the different synchronization measures see [204].

5.4.6.5 Analysis of Correlation Between Synchronization Dynamics and Stimulus

In order to relate this huge number of time series to the amplitude modulation of the stimulus, we first average the amplitude of the stimulus and the correlation coefficients calculated for the 496 electrode pairs and 9 frequency bands within successive 4-second windows to avoid minor amplitude fluctuations and obtain a scaling corresponding to about two musical bars that fits to changes related to the musical form (Fig. 5.17). In the second step, we correlate all 4464 time series from the synchronization analysis with the amplitude time series of the stimulus. In the third step, we select the 25 time series per frequency band that correlate most strongly with the time series of the amplitude of the stimulus, shown in Fig. 5.18. Now, we average these 25 time series per frequency band, which results in a single time series of 270 s length for each frequency band, respectively. These averaged time series for each frequency band are correlated over the whole recorded time with the amplitude

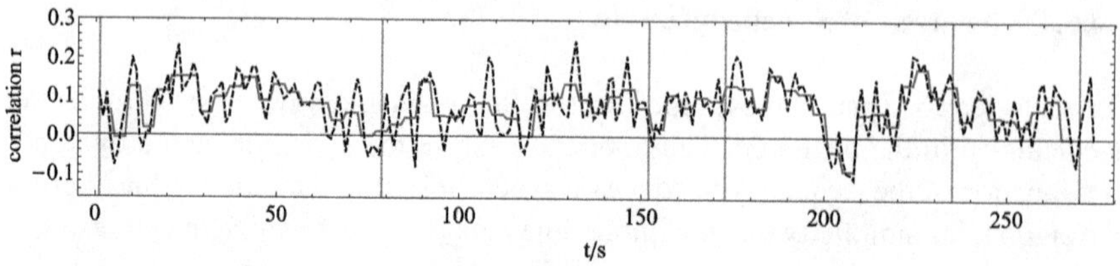

Fig. 5.17 Example of the synchronization dynamics between two electrodes. Dashed black line: Time series of the Pearson correlation coefficient r calculated for successive 1-second time windows ($n = 500$ in Eq. (5.19)) between averaged EEG recordings of electrode Fp1 (lower plot in Fig. 5.15) and electrode T7. Blue line: Pearson correlation coefficient averaged over 4 consecutive 1-second time windows of the dashed black line. Figure taken from [128]

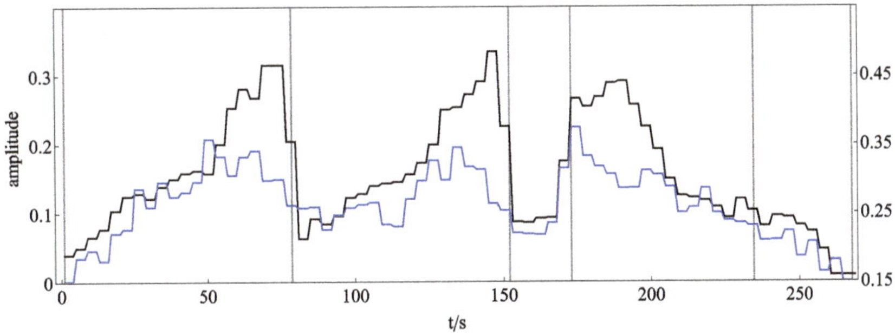

Fig. 5.18 Comparison of whole brain synchronization dynamics and representation of the musical form of the stimulus. The black line shows the amplitude dynamics of the stimulus as a representation of the musical form, averaged over each of 4 consecutive seconds. The blue line shows the average of the 25 correlation time series between two electrodes from each frequency band that correlates most strongly with the amplitude dynamics of the stimulus. Figure taken from [128]

dynamics of the stimulus (see Fig. 5.19a). In can be shown that the low and the high gamma-band (FB 2 and FB 3) correlate strongly with the stimulus as expected, but also the slow oscillations (FB 7–9) correlate very well (see discussion below). By this, we can reveal how good the synchronization dynamics in each frequency band corresponds to the amplitude dynamics of the stimulus on the level of musical form. In the next step, we average these time series over all frequency bands and correlate the resulting time series with the amplitude dynamics of the stimulus as well. These two time series correlate with a Pearson coefficient of $r = 0.76$. Therefore, we can conclude that the higher the amplitude of the stimulus, the higher the synchronization between the most correlated time series of the different frequency bands. According to [205], this is a strong effect.

Note that the correlation between sound amplitude (perceived loudness) or other parameters like brightness or fractal correlation dimension (see inset of Fig. 5.19a) and brain synchronization is not trivial. First, brain synchronization appears at frequencies much lower than most musical frequencies. Secondly, synchronization appears with multiple perceptual parameters. Thirdly, increasing, e.g., the sound amplitude might lead to an increase of the network amplitude, but here it leads to an enhanced synchronization, pointing to a highly nonlinear process in the network, caused by the activity of the brain when perceiving sound.

It is interesting to note that the correlation with the stimulus is highest when the time series from all frequency bands are averaged. The correlation coefficient of the averages of the 25 most correlated time-series as a function of the individual frequency bands is shown in Fig. 5.19a. It shows two regimes of high correlation, separated by a frequency band (FB 5) with low correlation. Here, the central nervous system in the spinal cord and its relation to the locomotor system are expected to be responsible for the dynamics in the frequency bands FB 6–9 due to their frequency range close to walking and dancing [206]. Note that the electroencephalogram (EEG) recordings are performed on the skull, and therefore represent the brain dynamics

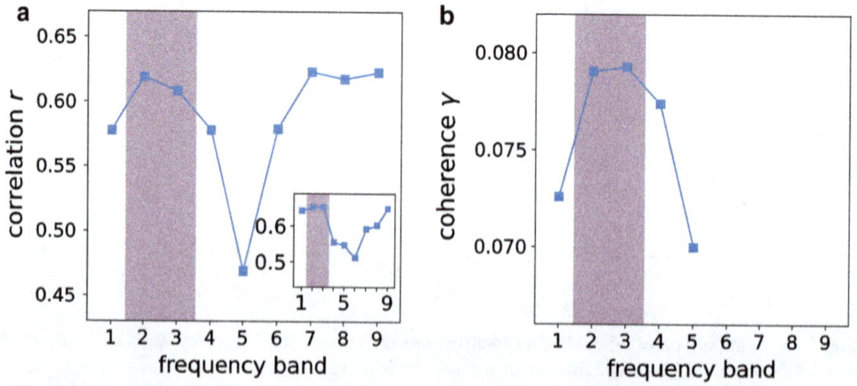

Fig. 5.19 Comparison between experimental and numerical results: **a** Experimentally recorded correlation r of the individual averages of the amplitude time series for each frequency band most strongly correlated with the stimulus as a function of frequency band (FB) FB 1: $125 - 250$ Hz, FB 2: $62.5 - 125$ Hz, FB 3: $31.25 - 62.5$ Hz, FB 4: $15.63 - 31.25$ Hz, FB 5: $7.81 - 15.63$ Hz, FB 6: $3.91 - 7.81$ Hz, FB 7: $1.95 - 3.91$ Hz, FB 8: $0.98 - 1.95$ Hz, FB 9: $0.49 - 0.98$ Hz. The inset depicts the Pearson correlation coefficient r as a function of frequency band where instead of the amplitude the fractal dimension [189, 190] has been used for the calculation of r. **b** Numerically simulated coherence γ between network dynamics and external stimulus, where the corresponding frequency bands are averaged from Fig. 5.13. As in Fig. 5.13, the purple shaded regions in both panels indicate the gamma-band ($f_b \approx 30{-}120$ Hz), respectively. Figure taken from [128]

of the neocortex which is interacting with the brain stem. Therefore, the high correlations between synchronization and musical form in frequency bands FB 6–9 can be interpreted as caused by the interaction of the neocortex with subcortical brain regions. Likewise, the high correlations in frequency bands FB 2–3 are interpreted as activity of the neocortex solely, as expected. The results therefore also suggest a separation of musical form-related synchronization between cortical (frequency bands FB 2–3) and subcortical (frequency bands FB 6–9) regions.

The high correlations observed in frequency bands 2 and 3 for the sound amplitude (see Fig. 5.19a) as well as for the fractal correlation dimension (see inset of Fig. 5.19a) correspond to a frequency range of $31.25 - 125$ Hz (gamma-band). On the other hand in Fig. 5.13, the strongest coherence between the Kuramoto order parameter (measure for global neural synchronization) and the external input can be found for $n_b = 10 - 40$ Hz. Taking into account that the natural frequency of each node is $f_{\text{FHN}} \approx 0.4$, we can calculate the corresponding frequency band $f_b = n_b / f_{\text{FHN}}$. As shown by the upper x-axis in Fig. 5.13, the strongest coherence in our model can be observed for a frequency band of $f_b = 40 - 100$ Hz, which agrees with the gamma-band in the brain. For comparison with the experiment, we show the corresponding numerically simulated results in Fig. 5.19b, where the respective frequency bands are averaged from Fig. 5.13. Both experimental and numerical results show a pronounced maximum of correlation between stimulus and brain dynamics for the gamma-band (bands 2 and 3) in Fig. 5.19. Note that the second maximum in the experimental data (panel a), which is due to the interaction of the neocortex with subcortical brain regions as discussed above, is absent in the simulated data (panel b) since the computer simulation is only performed for the neocortex, using a cochlea input, but neglecting brain stem activity.

5.4.7 Summary

We have systematically investigated the influence of music on a simulated network of FitzHugh-Nagumo oscillators, using empirical structural connectivity derived from healthy human subjects, and compared our findings with electroencephalogram (EEG) measurements. This section reveals an increase in coherence between the global network dynamics and the input signal induced by a specific musical piece. Notably, we demonstrated that the level of coherence is critically dependent on the frequency band. Comparisons with experimental data indicate that global neural synchronization between distinct brain regions in the gamma-band range is heightened just before transitions between different parts of a musical form, which we term as musical high-level events. This synchronization increase, observed prior to large-scale musical form boundaries and diminishing afterward, underscores its significance in representing musical large-scale form perception.

The transformation of auditory stimuli into neural spikes occurs in the cochlea, a sensory organ connected directly to the auditory cortex. Through the basilar membrane, the cochlea processes different frequencies organized into critical bands, enabling nuanced auditory perception. To simulate this process, we applied a cochlear model to transform a specific musical piece into an input signal that mimics neural spikes elicited by the music. This input was fed into a network of FitzHugh-Nagumo oscillators, structurally connected based on empirical data. Using a scaling approach to convert dimensionless time units in the oscillator model into real-world temporal measurements, we explored the dynamics of the system across varying frequency bands. To assess the degree of alignment between the input signal and the network's response, we introduced a coherence measure. Our findings reveal that this coherence is highly frequency-dependent, peaking within the gamma-band, thereby highlighting the critical role of this frequency range in driving coherence between input signals and network dynamics.

These results align with prior work and independent experimental studies [92, 93], which similarly identify music as a powerful inducer of synchronization within the human brain. Our analyses further reveal that listening to music induces a periodic alternation between synchronized and desynchronized states, intricately linked to the structure of the perceived music. Detailed experimental investigations into the effects of real music on neural activity, focusing on established brain frequency bands, reaffirmed the prominence of the gamma-band in musical form perception. Using measures such as the Pearson correlation coefficient of sound amplitude and fractal correlation dimension, we observed a pronounced maximum in the gamma-band, corroborating the simulation results. Furthermore, our findings suggest a functional differentiation between high-frequency neural activity, associated with neocortical processes, and low-frequency activity, connected to interactive dynamics between cortical and subcortical regions, including dance and rhythm-associated movements.

The periodic alternation between synchronization and desynchronization observed in both experimental and simulated contexts reflects a fundamental property of brain dynamics: variability indicative of a critical state poised at the boundary between synchronized and desynchronized regimes. Such criticality is characteristic of brain activity, associated with phenomena like hysteresis and avalanches observed in phase transitions and other critical systems [115, 117–119, 134].

By selecting appropriate model parameters and quantification measures, we identified compelling dynamic behaviors across frequency bands, highlighting the induced coherence between musical input and brain dynamics in both simulated and experimental settings. In summary, musical input to the brain facilitates significant coherence and synchronization, particularly within the gamma-band. These insights provide a framework to further investigate the fundamental mechanisms underlying the influence of music on human brain dynamics and offer potential applications in music therapy and cognitive neuroscience.

References

1. Sawicki J, Berner R, Loos SAM, Anvari M, Bader R, Barfuss W, Botta N, Brede N, Franović I, Gauthier DJ, Goldt S, Hajizadeh A, Hövel P, Karin O, Lorenz-Spreen P, Miehl C, Mölter J, Olmi S, Schöll E, Seif A, Tass PA, Volpe G, Yanchuk S, Kurths J (2023) Perspectives on adaptive dynamical systems. Chaos Interdiscip J Nonlinear Sci 33(7):071501. https://doi.org/10.1063/5.0147231
2. Johnson-Groh M (2023) A look at adaptive systems from biology to machine learning. Scilight 2023(30):301105. https://doi.org/10.1063/10.0020288
3. Whitmire CJ, Stanley GB (2016) Rapid sensory adaptation redux: a circuit perspective. Neuron 92:298–315. https://doi.org/10.1016/j.neuron.2016.09.046
4. Debanne D, Inglebert Y, Russier M (2019) Plasticity of intrinsic neuronal excitability. Current Opin Neurobiol 54:73–82. https://doi.org/10.1016/j.conb.2018.09.001
5. Zucker RS, Regehr WG (2002) Short-term synaptic plasticity. Annu Rev Physiol 64:355–405
6. Feldman DE (2012) The spike-timing dependence of plasticity. Neuron 75:556–571. https://doi.org/10.1016/j.neuron.2012.08.001
7. Markram H, Lübke J, Frotscher M, Sakmann B (1997) Regulation of synaptic efficacy by coincidence of postsynaptic APs and EPSPs. Science 275(5297):213–215. https://doi.org/10.1126/science.275.5297.213
8. Bi GG, Poo MM (1998) Synaptic modifications in cultured hippocampal neurons: dependence on spike timing, synaptic strength, and postsynaptic cell type. J Neurosci 18(24):10464–10472. https://doi.org/10.1523/JNEUROSCI.18-24-10464.1998
9. Motanis H, Seay MJ, Buonomano DV (2018) Short-term synaptic plasticity as a mechanism for sensory timing. Trends Neurosci 41(10):701–711. https://doi.org/10.1016/j.tins.2018.08.001
10. Wu YK, Miehl C, Gjorgjieva J (2022) Regulation of circuit organization and function through inhibitory synaptic plasticity. Trends Neurosci 45(12):884–898. https://doi.org/10.1016/j.tins.2022.10.006
11. Weber AI, Krishnamurthy K, Fairhall AL (2019) Coding Principles in Adaptation. Annu Rev Vis Sci 5:427–449. https://doi.org/10.1146/annurev-vision-091718-014818

12. Näätänen R, Simpson M, Loveless NE (1982) Stimulus deviance and evoked potentials. Biol Psychol 14:53–98. https://doi.org/10.1016/0301-0511(82)90017-5
13. Ross JM, Hamm JP (2020) Cortical microcircuit mechanisms of mismatch negativity and its underlying subcomponents. Front Neural Circuits 14(13) (2020). https://doi.org/10.3389/fncir.2020.00013
14. Ulanovsky N, Las L, Nelken I (2003) Processing of low-probability sounds by cortical neurons. Nature Neuroscience 6(4):391–398. https://doi.org/10.1038/nn1032
15. Natan RG, Briguglio JJ, Mwilambwe-Tshilobo L, Jones SI, Aizenberg M, Goldberg EM, Geffen MN (2015) Complementary control of sensory adaptation by two types of cortical interneurons. eLife 4:e09868. https://doi.org/10.7554/eLife.09868
16. Homann J, Ann S, Chen KS, Tank DW, Berry MJ (2022) Novel stimuli evoke excess activity in the mouse primary visual cortex. Proc Natl Acad Sci USA 119(5):e2108882119. https://doi.org/10.1073/pnas.2108882119
17. Mill R, Coath M, Wennekers T, Denham SL (2011) Abstract stimulus-specific adaptation models. Neural Comput 23:435–476. https://doi.org/10.1162/NECO_a_00077
18. Mill R, Coath M, Wennekers T, Denham SL (2012) Characterising stimulus-specific adaptation using a multi-layer field model. Brain Res 1434:178–188. https://doi.org/10.1016/j.brainres.2011.08.063
19. Hershenhoren I, Taaseh N, Antunes FM, Nelken I (2014) Intracellular correlates of stimulus-specific adaptation. J Neurosci 34(9):3303–3319. https://doi.org/10.1523/JNEUROSCI.2166-13.2014
20. Park Y, Geffen MN (2020) A circuit model of auditory cortex. PLoS Comput Biol 16(7):e1008016. https://doi.org/10.1371/journal.pcbi.1008016
21. Seay MJ, Natan RG, Geffen MN, Buonomano DV (2020) Differential short-term plasticity of PV and SST neurons accounts for adaptation and facilitation of cortical neurons to auditory tones. J Neurosci 40(48):9224–9235. https://doi.org/10.1523/JNEUROSCI.0686-20.2020
22. Schulz A, Miehl C, Berry II MJ, Gjorgjieva J (2021) The generation of cortical novelty responses through inhibitory plasticity. eLife 10:e65309. https://doi.org/10.7554/eLife.65309
23. Rao RPN, Ballard DH (1999) Predictive coding in the visual cortex: a functional interpretation of some extra-classical receptive-field effects. Nat Neurosci 2(1):79–87. https://doi.org/10.1038/4580
24. Turrigiano GG, Nelson SB (2004) Homeostatic plasticity in the developing nervous system. Nat Rev Neurosci 5:97–107. https://doi.org/10.1038/nrn1327
25. Onasch S, Gjorgjieva J (2020) Circuit stability to perturbations reveals hidden variability in the balance of intrinsic and synaptic conductances. J Neurosci 40(16):3186–3202. https://doi.org/10.1523/JNEUROSCI.0985-19.2020
26. Gonçalves PJ, Lueckmann JM, Deistler M, Nonnenmacher M, Öcal K, Bassetto G, Chintaluri C, Podlaski WF, Haddad SA, Vogels TP, Greenberg DS, Macke JH (2020) Training deep neural density estimators to identify mechanistic models of neural dynamics. eLife 9:1–46. https://doi.org/10.7554/ELIFE.56261
27. Deistler M, Macke JH, Gonçalves PJ (2022) Energy efficient network activity from disparate circuit parameters. Proc Natl Acad Sci 119(44):e2207632119. https://doi.org/10.1073/pnas.2207632119
28. Marder E, O'Leary T, Shruti S (2014) Neuromodulation of circuits with variable parameters: single neurons and small circuits reveal principles of state-dependent and robust neuromodulation. Annu Rev Neurosci 37:329–346. https://doi.org/10.1146/annurev-neuro-071013-013958
29. Jacquerie K, Drion G (2021) Robust switches in thalamic network activity require a timescale separation between sodium and T-type calcium channel activations. PLoS Comput Biol 17(5):e1008997. https://doi.org/10.1371/journal.pcbi.1008997
30. Mongillo G, Barak O, Tsodyks M (2008) Synaptic theory of working memory. Science 319(5869):1543–1546. https://doi.org/10.1126/science.1150769
31. Abbott LF, Nelson SB (2000) Synaptic plasticity: taming the beast. Nat Neurosci 3:1178–1183. https://doi.org/10.1038/81453

32. Yuste R (2015) From the neuron doctrine to neural networks. Nat Rev Neurosci 16:487–497. https://doi.org/10.1038/nrn3962
33. Miehl C, Onasch S, Festa D, Gjorgjieva J (2023) Formation and computational implications of assemblies in neural circuits. J Physiol 601(15):3071–3090. https://doi.org/10.1113/JP282750
34. Fusi S (2017) Computational models of long term plasticity and memory. https://doi.org/10.48550/arXiv.1706.04946
35. Litwin-Kumar A, Doiron B (2014) Formation and maintenance of neuronal assemblies through synaptic plasticity. Nat Commun 5(5319). https://doi.org/10.1038/ncomms6319
36. Miehl C, Gjorgjieva J (2022) Stability and learning in excitatory synapses by nonlinear inhibitory plasticity. PLoS Comput Biol 18(12):e1010682. https://doi.org/10.1371/journal.pcbi.1010682
37. Zenke F, Agnes EJ, Gerstner W (2015) Diverse synaptic plasticity mechanisms orchestrated to form and retrieve memories in spiking neural networks. Nat Commun 6(1). https://doi.org/10.1038/ncomms7922
38. Hamm JP, Peterka DS, Gogos JA, Yuste R (2017) Altered cortical ensembles in mouse models of schizophrenia. Neuron 94:153–167. https://doi.org/10.1016/j.neuron.2017.03.019
39. Batista-Brito R, Zagha E, Ratliff JM, Vinck M (2018) Modulation of cortical circuits by top-down processing and arousal state in health and disease. Current Opin Neurobiol 52:172–181. https://doi.org/10.1016/j.conb.2018.06.008
40. Light GA, Näätänen R (2013) Mismatch negativity is a breakthrough biomarker for understanding and treating psychotic disorders. Proc Natl Acad Sci USA 110(38):15175–15176. https://doi.org/10.1073/pnas.1313287110
41. Nelken I (2004) Processing of complex stimuli and natural scenes in the auditory cortex. Current Opin Neurobiol 14:474–480. https://doi.org/10.1016/j.conb.2004.06.005
42. Butler RA (1968) Effect of changes in stimulus frequency and intensity on habituation of the human vertex potential. J Acoust Soc Am 44:945–950. https://doi.org/10.1121/1.1911233
43. Brosch M, Schreiner CE (2000) Sequence sensitivity of neurons in cat primary auditory cortex. Cerebral Cortex 10:1155–1167. https://doi.org/10.1093/cercor/10.12.1155
44. Näätänen R, Gaillard AW, Mäntysalo S (1978) Early selective-attention effect on evoked potential reinterpreted. Acta Psychol 42:313–329. https://doi.org/10.1016/0001-6918(78)90006-9
45. Ulanovsky N, Las L, Farkas D, Nelken I (2004) Multiple time scales of adaptation in auditory cortex neurons. J Neurosci 24:10440–10453. https://doi.org/10.1523/JNEUROSCI.1905-04.2004
46. Zacharias N, König R, Heil P (2012) Stimulation-history effects on the M100 revealed by its differential dependence on the stimulus onset interval. Psychophysiology 49:909–919. https://doi.org/10.1111/j.1469-8986.2012.01370.x
47. Pérez-González D, Malmierca MS (2014) Adaptation in the auditory system: an overview. Front Integr Neurosci 8:1–10. https://doi.org/10.3389/fnint.2014.00019
48. Malmierca MS, Anderson LA, Antunes FM (2015) The cortical modulation of stimulus-specific adaptation in the auditory midbrain and thalamus: a potential neuronal correlate for predictive coding. Front Syst Neurosci 9(19). https://doi.org/10.3389/fnsys.2015.00019
49. Friston K (2005) A theory of cortical responses. Philos Trans R Soc 360:815–836. https://doi.org/10.1098/rstb.2005.1622
50. May PJC (2021) The adaptation model offers a challenge for the predictive coding account of mismatch negativity. Front Hum Neurosci 15:721574. https://doi.org/10.3389/fnhum.2021.721574
51. May PJC, Tiitinen H (2010) Mismatch negativity (MMN), the deviance-elicited auditory deflection, explained. Psychophysiology 47:66–122. https://doi.org/10.1111/j.1469-8986.2009.00867.x
52. Wang P, Knösche TR (2013) A realistic neural mass model of the cortex with laminar-specific connections and synaptic plasticity-evaluation with auditory habituation. PLoS One 8:e77876. https://doi.org/10.1371/journal.pone.0077876

53. Kudela P, Boatman-Reich D, Beeman D, Anderson WS (2018) Modeling Neural Adaptation in Auditory Cortex. Front Neural Circuits 12:72. https://doi.org/10.3389/fncir.2018.00072
54. Fortune ES, Rose GJ (2001) Short-term synaptic plasticity as a temporal filter. Trends in Neurosciences 24:381–385. https://doi.org/10.1016/s0166-2236(00)01835-x
55. Abbott LF, Varela J, Sen K, Nelson S (1997) Synaptic depression and cortical gain control. Science 275(5297):221–224. https://doi.org/10.1126/science.275.5297.221
56. Salmasi M, Loebel A, Glasauer S, Stemmler M (2019) Short-term synaptic depression can increase the rate of information transfer at a release site. PLoS Computational Biology 15:1–21. https://doi.org/10.1371/journal.pcbi.1006666
57. May PJC, Tiitinen H (2013) Temporal binding of sound emerges out of anatomical structure and synaptic dynamics of auditory cortex. Front Comput Neurosci 7(152). https://doi.org/10.3389/fncom.2013.00152
58. Hajizadeh A, Matysiak A, May PJC, König R (2019) Explaining event-related fields by a mechanistic model encapsulating the anatomical structure of auditory cortex. Biol Cybern 113:321–345. https://doi.org/10.1007/s00422-019-00795-9
59. Hajizadeh A, Matysiak A, Wolfrum M, May PJC, König R (2022) Auditory cortex modelled as a dynamical network of oscillators: understanding event-related fields and their adaptation. Biol Cybern 116:475–499. https://doi.org/10.1007/s00422-022-00936-7
60. Bader R (ed) (2019) Springer handbook of systematic musicology. Springer Handbooks, Springer, Berlin, Germany. https://doi.org/10.1007/978-3-319-51790-5
61. Briot JP, Hadjeres G, Pachet FD (2020) Deep learning techniques for music generation. Springer International Publishing. https://doi.org/10.1007/978-3-319-70163-9
62. Kohonen T (1995) Self-organizing maps. Springer, Berlin Heidelberg. https://doi.org/10.1007/978-3-642-97610-0
63. Blaß M, Fischer JL, Plath N (2020) Computational phonogram archiving. Phys Today 73(12):50–55. https://doi.org/10.1063/pt.3.4636
64. Bader R (ed) (2019) Computational phonogram archiving. Current research in systematic musicology, 1st edn. Springer Nature, Cham, Switzerland. https://doi.org/10.1007/978-3-030-00993-5
65. Leman M, Carreras F (1997) Schema and gestalt: testing the hypothesis of psychoneural isomorphism by computer simulation. In: Music, gestalt, and computing. Springer Berlin Heidelberg, pp 144–168. https://doi.org/10.1007/bfb0034112
66. Bader R (2021) How music works. Springer International Publishing. https://doi.org/10.1007/978-3-030-67155-6
67. Bader R (2013) Nonlinearities and synchronization in musical acoustics and music psychology. Springer, Berlin Heidelberg. https://doi.org/10.1007/978-3-642-36098-5
68. Linke S, Bader R, Mores R (2019) The impulse pattern formulation (IPF) as a model of musical instruments–investigation of stability and limits. Chaos Interdiscip J Nonlinear Sci 29(10):103109. https://doi.org/10.1063/1.5092511
69. Linke S, Bader R, Mores R (2021) Influence of the supporting table on initial transients of the fretted zither: an impulse pattern formulation model. In: Proceedings of meetings on acoustics ASA. https://doi.org/10.1121/2.0001494
70. Linke S, Bader R, Mores R (2021) Modeling synchronization in human musical rhythms using impulse pattern formulation (ipf). https://doi.org/10.48550/ARXIV.2112.03218
71. Bader R (2022) Impulse pattern formulation (ipf) brain model. https://doi.org/10.48550/ARXIV.2212.11021
72. Kozma R, Freeman WJ (2016) Cognitive phase transitions in the cerebral cortex-enhancing the neuron doctrine by modeling neural fields. Springer International Publishing. https://doi.org/10.1007/978-3-319-24406-8
73. Sawicki J, Schöll E (2021) Influence of sound on empirical brain networks. Front Appl Math Stat 7:662221. https://doi.org/10.3389/fams.2021.662221
74. Sawicki J (2025) From empirical brain networks towards modeling music perception–a perspective. https://doi.org/10.48550/arXiv.2504.07721

75. Steriade M, McCormick DA, Sejnowski TJ (1993) Thalamocortical oscillations in the sleeping and aroused brain. Science 262(5134):679–685. https://doi.org/10.1126/science.8235588
76. Rattenborg NC, Amlaner CJ, Lima SL (2000) Behavioral, neurophysiological and evolutionary perspectives on unihemispheric sleep. Neurosci Biobehav Rev 24:817–842. https://doi.org/10.1016/S0149-7634(00)00062-5
77. Schwartz JRL, Roth T (2008) Neurophysiology of sleep and wakefulness: basic science and clinical implications. Curr Neuropharmacol 6(4):367–378. https://doi.org/10.2174/157015908787386104
78. Moroni F, Nobili L, De Carli F, Massimini M, Francione S, Marzano C, Proserpio P, Cipolli C, De Gennaro L, Ferrara M (2012) Slow eeg rhythms and inter-hemispheric synchronization across sleep and wakefulness in the human hippocampus. Neuroimage 60(1):497–504. https://doi.org/10.1016/j.neuroimage.2011.12.053
79. Tamaki M, Bang JW, Watanabe T, Sasaki Y (2016) Night watch in one brain hemisphere during sleep associated with the first-night effect in humans. Curr Biol 26(9):1190–1194. https://doi.org/10.1016/j.cub.2016.02.063
80. Rattenborg NC, Voirin B, Cruz SM, Tisdale R, Dell'Omo G, Lipp HP, Wikelski M, Vyssotski AL (2016) Evidence that birds sleep in mid-flight. Nat Commun 7:12468. https://doi.org/10.1038/ncomms12468
81. Mascetti GG (2016) Unihemispheric sleep and asymmetrical sleep: behavioral, neurophysiological, and functional perspectives. Nat Sci Sleep 8:221. https://doi.org/10.2147/NSS.S71970
82. Ramlow L, Sawicki J, Zakharova A, Hlinka J, Claussen JC, Schöll E (2019) Partial synchronization in empirical brain networks as a model for unihemispheric sleep. Europhys Lett 126:50007. https://doi.org/10.1209/0295-5075/126/50007
83. Gerster M, Berner R, Sawicki J, Zakharova A, Skoch A, Hlinka J, Lehnertz K, Schöll E (2020) FitzHugh-Nagumo oscillators on complex networks mimic epileptic-seizure-related synchronization phenomena. Chaos 30:123130. https://doi.org/10.1063/5.0021420
84. Koelsch S, Rohrmeier M, Torrecuso R, Jentschke S (2013) Processing of hierarchical syntactic structure in music. Proc Natl Acad Sci USA 110:15443. https://doi.org/10.1073/pnas.1300272110
85. Large EW, Herrera JA, Velasco MJ (2015) Neural Networks for Beat Perception in Musical Rhythm. Front Neurosci 9:159. https://doi.org/10.3389/fnsys.2015.00159
86. Bader R (2020) Neural coincidence detection strategies during perception of multi-pitch musical tones. https://doi.org/10.48550/arXiv.2001.06212
87. Joris PX, Carney LH, Smith PH, Yin TCT (1994) Enhancement of neural synchronization in the anteroventral cochlear nucleus. I. Responses to tones at the characteristic frequency. J Neurophysiol 71:1022. https://doi.org/10.1152/jn.1994.71.3.1022
88. Bader R (2013) Nonlinearities and synchronization in musical acoustics and music psychology. Springer, Berlin. https://doi.org/10.1007/978-3-642-27640-5
89. Sawicki J, Abel M, Schöll E (2018) Synchronization of organ pipes. Eur Phys J B 91(2):24. https://doi.org/10.1140/epjb/e2017-80485-8
90. Hou YS, Xia GQ, Jayaprasath E, Yue DZ, Wu ZM (2020) Parallel information processing using a reservoir computing system based on mutually coupled semiconductor lasers. Appl Phys B 126(3):40. https://doi.org/10.1007/s00340-019-7351-4
91. Shainline JM (2020) Fluxonic processing of photonic synapse events. IEEE J Sel Top Quantum Electron 26(1):7700315. https://doi.org/10.1109/JSTQE.2019.2923188
92. Hartmann L, Bader R (2014) Neuronal synchronization of musical large scale form: an eeg-study. In: Proceedings meetings on acoustic, 168th meeting of the acoustic society of America 22:1
93. Hartmann L, Bader R (2020) Neural synchronization of music large-scale form. https://doi.org/10.48550/arXiv.2005.06938
94. Schneider A (2018) Pitch and pitch perception. Springer Berlin Heidelberg, Berlin, Heidelberg, pp 605–685. https://doi.org/10.1007/978-3-662-55004-5_31

95. Leyva I, Sendiña-Nadal I, Sevilla-Escoboza R, Vera-Avila VP, Chholak P, Boccaletti S (2018) Relay synchronization in multiplex networks. Sci Rep 8(1):8629. https://doi.org/10.1038/s41598-018-26945-w
96. Bergner A, Frasca M, Sciuto G, Buscarino A, Ngamga EJ, Fortuna L, Kurths J (2012) Remote synchronization in star networks. Phys Rev E 85:026208. https://doi.org/10.1103/physreve.85.026208
97. Nicosia V, Valencia M, Chavez M, Díaz-Guilera A, Latora V (2013) Remote synchronization reveals network symmetries and functional modules. Phys Rev Lett 110:174102. https://doi.org/10.1103/physrevlett.110.174102
98. Gambuzza LV, Cardillo A, Fiasconaro A, Fortuna L, Gómez-Gardeñes J, Frasca M (2013) Analysis of remote synchronization in complex networks. Chaos 23:043103. https://doi.org/10.1063/1.4821862
99. Zhang L, Motter AE, Nishikawa T (2017) Incoherence-mediated remote synchronization. Phys Rev Lett 118(17):174102. https://doi.org/10.1103/physrevlett.118.174102
100. Zhang Y, Nishikawa T, Motter AE (2017) Asymmetry-induced synchronization in oscillator networks. Phys Rev E 95(6):062215. https://doi.org/10.1103/physreve.95.062215
101. Sawicki J, Omelchenko I, Zakharova A, Schöll E (2018) Delay controls chimera relay synchronization in multiplex networks. Phys Rev E 98:062224. https://doi.org/10.1103/physreve.98.062224
102. Sawicki J, Omelchenko I, Zakharova A, Schöll E (2018) Synchronization scenarios of chimeras in multiplex networks. Eur Phys J Spec Top 227:1161. https://doi.org/10.1140/epjst/e2018-800039-y
103. Sawicki J (2019) Delay controlled partial synchronization in complex networks. Springer Theses, Springer, Heidelberg. https://doi.org/10.1007/978-3-030-34076-6
104. Winkler M, Sawicki J, Omelchenko I, Zakharova A, Anishchenko V, Schöll E (2019) Relay synchronization in multiplex networks of discrete maps. EPL 126:50004. https://doi.org/10.1209/0295-5075/126/50004
105. Drauschke F, Sawicki J, Berner R, Omelchenko I, Schöll E (2020) Effect of topology upon relay synchronization in triplex neuronal networks. Chaos 30:051104. https://doi.org/10.1063/5.0008341
106. Sawicki J, Koulen JM, Schöll E (2021) Synchronization scenarios in three-layer networks with a hub. Chaos 31:073131. https://doi.org/10.1063/5.0050595
107. Muldoon SF, Pasqualetti F, Gu S, Cieslak M, Grafton ST, Vettel JM, Bassett DS (2016) Stimulation-based control of dynamic brain networks. PLoS Comput Biol 12(9):e1005076. https://doi.org/10.1371/journal.pcbi.1005076
108. Petkoski S, Jirsa VK (2019) Transmission time delays organize the brain network synchronization. Phil Trans R Soc A 377:20180132. https://doi.org/10.1098/rsta.2018.0132
109. FitzHugh R (1961) Impulses and physiological states in theoretical models of nerve membrane. Biophys J 1:445–466. https://doi.org/10.1016/s0006-3495(61)86902-6
110. Nagumo J, Arimoto S, Yoshizawa S (1962) An active pulse transmission line simulating nerve axon. Proc IRE 50:2061–2070. https://doi.org/10.1109/JRPROC.1962.288235
111. Bassett DS, Zurn P, Gold JI (2018) On the nature and use of models in network neuroscience. Nat Rev Neurosci 19:566–578. https://doi.org/10.1038/s41583-018-0038-8
112. Chernihovskyi A, Mormann F, Müller M, Elger CE, Baier G, Lehnertz K (2005) EEG analysis with nonlinear excitable media. J Clin Neurophysiol 22:314–329. https://doi.org/10.1097/01.wnp.0000179968.14838.e7
113. Chernihovskyi A, Lehnertz K (2007) Measuring synchronization with nonlinear excitable media. Int J Bifurc Chaos 17(10):3425–3429. https://doi.org/10.1142/s0218127407019159
114. Omelchenko I, Omel'chenko OE, Hövel P, Schöll E (2013) When nonlocal coupling between oscillators becomes stronger: patched synchrony or multichimera states. Phys Rev Lett 110:224101. https://doi.org/10.1103/physrevlett.110.224101
115. Massobrio P, de Arcangelis L, Pasquale V, Jensen HJ, Plenz D (2015) Criticality as a signature of healthy neural systems. Front Syst Neurosci 9:22. https://doi.org/10.3389/fnsys.2015.00022

116. Sawicki J, Omelchenko I, Zakharova A, Schöll E (2019) Delay-induced chimeras in neural networks with fractal topology. Eur Phys J B 92:54. https://doi.org/10.1140/epjb/e2019-90303-9
117. Kim H, Moon JY, Mashour GA, Lee U (2018) Mechanisms of hysteresis in human brain networks during transitions of consciousness and unconsciousness: theoretical principles and empirical evidence. PLoS Comput Biol 14(8):e1006424. https://doi.org/10.1371/journal.pcbi.1006424
118. Ribeiro TL, Copelli M, Caixeta F, Belchior H, Chialvo DR, Nicolelis MA, Ribeiro S (2010) Spike avalanches exhibit universal dynamics across the sleep-wake cycle. PloS One 5(11). https://doi.org/10.1371/journal.pone.0014129
119. Steyn-Ross A, Steyn-Ross M (2010) Modeling phase transitions in the brain, vol 509. Springer, Berlin. https://doi.org/10.1007/978-1-4419-0796-7
120. Sawicki J, Schöll E (2024) Interplay of synchronization and cortical input in models of brain networks. Europhys Lett 146(4):41001. https://doi.org/10.1209/0295-5075/ad438e
121. Singer W (2018) Neuronal oscillations: unavoidable and useful? Eur J Neurosci 48(7):2389–2398. https://doi.org/10.1111/ejn.13796
122. Tass PA, Rosenblum M, Weule J, Kurths J, Pikovsky A, Volkmann J, Schnitzler A, Freund HJ (1998) Detection of n:m phase locking from noisy data: application to magnetoencephalography. Phys Rev Lett 81:3291. https://doi.org/10.1103/PhysRevLett.81.3291
123. Mayville JM, Jantzen KJ, Fuchs A, Steinberg FL, Kelso JAS (2002) Cortical and subcortical networks underlying syncopated and synchronized coordination revealed using fmri. functional magnetic resonance imaging. Hum Brain Mapp 17(4):214–229. https://doi.org/10.1002/hbm.10065
124. Just MA, Cherkassky VL, Keller TA, Minshew NJ (2004) Cortical activation and synchronization during sentence comprehension in high-functioning autism: evidence of underconnectivity. Brain 127(8):1811–1821. https://doi.org/10.1093/brain/awh199
125. Singer W (1993) Synchronization of cortical activity and its putative role in information processing and learning. Annu Rev Physiol 55:349–374. https://doi.org/10.1146/annurev.ph.55.030193.002025
126. Uhlhaas P, Pipa G, Lima B, Melloni L, Neuenschwander S, Nikolic D, Singer W (2009) Neural synchrony in cortical networks: history, concept and current status. Front Integr Neurosci 3:17. https://doi.org/10.3389/neuro.07.017.2009
127. Löwel S, Singer W (1992) Selection of intrinsic horizontal connections in the visual cortex by correlated neuronal activity. Science 255(5041):209–212. https://doi.org/10.1126/science.1372754
128. Sawicki J, Hartmann L, Bader R, Schöll E (2022) Modelling the perception of music in brain network dynamics. Front Netw Physiol 2:910920. https://doi.org/10.3389/fnetp.2022.910920
129. Sanz-Leon P, Knock SA, Spiegler A, Jirsa VK (2015) Mathematical framework for large-scale brain network modeling in The Virtual Brain. Neuroimage 111:385–430. https://doi.org/10.1016/j.neuroimage.2015.01.002
130. Skoch A, Rehák Bučková B, Mareš J, Tintera J, Sanda P, Jajcay L, Horacek J, Spaniel F, Hlinka J (2022) Human brain structural connectivity matrices-ready for modelling. Sci Data 9(1):486. https://doi.org/10.1038/s41597-022-01596-9
131. Schöll E, Hiller G, Hövel P, Dahlem MA (2009) Time-delayed feedback in neurosystems. Phil Trans R Soc A 367:1079–1096. https://doi.org/10.1098/rsta.2008.0258
132. Schöll E (2021) Partial synchronization patterns in brain networks. Europhys Lett 136:18001. https://doi.org/10.1209/0295-5075/ac3b97
133. Wilting J, Priesemann V (2019) 25 years of criticality in neuroscience-established results, open controversies, novel concepts. Curr Opin Neurobiol 58:105–111. https://doi.org/10.1016/j.conb.2019.08.003
134. Shi J, Kirihara K, Tada M, Fujioka M, Usui K, Koshiyama D, Araki T, Chen L, Kasai K, Aihara K (2022) Criticality in the healthy brain. Front Netw Physiol 1:755685. https://doi.org/10.3389/fnetp.2021.755685

135. Tzourio-Mazoyer N, Landeau B, Papathanassiou D, Crivello F, Etard O, Delcroix N, Mazoyer B, Joliot M (2002) Automated anatomical labeling of activations in SPM using a macroscopic anatomical parcellation of the MNI MRI single-subject brain. Neuroimage 15(1):273–289. https://doi.org/10.1006/nimg.2001.0978
136. Franović I, Klinshov V (2018) Stimulus-evoked activity in clustered networks of stochastic rate-based neurons. Eur Phys J Special Top 227(10):1063–1076. https://doi.org/10.1140/epjst/e2018-800080-6
137. Bohland JW, Allen CB, Mitra PP (2009) The brain atlas concordance problem: quantitative comparison of anatomical parcellations. PLOS ONE 4(9):1–18. https://doi.org/10.1371/journal.pone.0007200
138. Evans AC, Janke AL, Collins DL, Baillet S (2012) Brain templates and atlases. Neuroimage 62(2):911–922. https://doi.org/10.1016/j.neuroimage.2012.01.024
139. Scarpetta S, Apicella I, Minati L, de Candia A (2018) Hysteresis, neural avalanches, and critical behavior near a first-order transition of a spiking neural network. Phys Rev E 97:062305. https://doi.org/10.1103/PhysRevE.97.062305
140. Contreras M, Medeiros ES, Zakharova A, Hövel P, Franović I (2023) Scale-free avalanches in arrays of FitzHugh-Nagumo oscillators. Chaos Interdiscip J Nonlinear Sci 33(9):093106. https://doi.org/10.1063/5.0165778
141. Hou Y, Song B, Hu Y, Pan Y, Hu Y (2020) The averaged inter-brain coherence between the audience and a violinist predicts the popularity of violin performance. NeuroImage 211:116655. https://doi.org/10.1016/j.neuroimage.2020.116655
142. Hirvonen J, Monto S, Wang SH, Palva JM, Palva S (2018) Dynamic large-scale network synchronization from perception to action. Netw Neurosci 2(4):442–463. https://doi.org/10.1162/netn_a_00039
143. Bahmer A, Rimmele JM, Gupta DS (2024) Editorial: understanding the role of oscillations, mutual information and synchronization in perception and action. Front Comput Neurosci 18. https://doi.org/10.3389/fncom.2024.1452001
144. Kacprzyk J, Pedrycz W (2015) Springer handbook of computational intelligence. Springer Handbooks, Springer, Berlin Heidelberg. https://doi.org/10.1007/978-3-662-43505-2
145. Bassett DS, Sporns O (2017) Network neuroscience. Nat Neurosci 20:353 EP. https://doi.org/10.1038/nn.4502
146. Petkoski S, Palva JM, Jirsa VK (2018) Phase-lags in large scale brain synchronization: Methodological considerations and in-silico analysis. PLoS Comput Biol 14:e1006160. https://doi.org/10.1371/journal.pcbi.1006160
147. Friston KJ, Friston DA (2013) A free energy formulation of music generation and perception: helmholtz revisited. Springer International Publishing, Heidelberg, pp 43–69. https://doi.org/10.1007/978-3-319-00107-4_2
148. Tritsch NX, Rodríguez-Contreras A, Crins TT, Wang HC, Borst JG, Bergles DE (2010) Calcium action potentials in hair cells pattern auditory neuron activity before hearing onset. Nat Neurosci 13(9):1050–1052. https://doi.org/10.1038/nn.2604
149. Mizrahi A, Shalev A, Nelken I (2014) Single neuron and population coding of natural sounds in auditory cortex. Curr Opin Neurobiol 24(1):103–110. https://doi.org/10.1016/j.conb.2013.09.007
150. Bader R (2015) Phase synchronization in the cochlea at transition from mechanical waves to electrical spikes. Chaos 25:103124. https://doi.org/10.1063/1.4927206
151. Bader R (2017) Pitch and timbre discrimination at wave-to-spike transition in the cochlea. https://doi.org/10.48550/arXiv.1711.05596
152. Bader R (2018) Cochlear spike synchronization and neuron coincidence detection model. Chaos 28:023105. https://doi.org/10.1063/1.5011450
153. Guo XX, Xiang SY, Qu Y, Han YN, Wen AJ, Hao Y (2021) Enhanced prediction performance of a neuromorphic reservoir computing system using a semiconductor nanolaser with double phase conjugate feedbacks. J Light Technol 39(1):129–135. https://doi.org/10.1109/jlt.2020.3023451

154. Ryugo D, Fay R, Popper A (2010) Auditory and vestibular efferents. Springer Handbook of Auditory Research, Springer New York. https://doi.org/10.1007/978-1-4419-0993-0
155. Lyon R, Shamma S (1996) Auditory representations of timbre and pitch. In: Hawkins HL, McMullen TA, Popper AN, Fay RR (eds) Auditory computation, chap 6. Springer, New York, pp 221–270. https://doi.org/10.1007/978-1-4612-4070-9_6
156. Bader R (2021) How music works. Springer, Cham. https://doi.org/10.1007/978-3-030-67155-6
157. de Boer E (1991) Auditory physics. Physical principles in hearing theory. iii. Phys Rep 203(3):125–231. https://doi.org/10.1016/0370-1573(91)90068-w
158. Spitmaan M, Seo H, Lee D, Soltani A (2020) Multiple timescales of neural dynamics and integration of task-relevant signals across cortex. Proc Natl Acad Sci USA 117(36):22522–22531. https://doi.org/10.1073/pnas.2005993117
159. Hubbard AE, Mountain DC (1996) Analysis and synthesis of cochlear mechanical function using models. In: Hawkins HL, McMullen TA, Popper AN, Fay RR (eds) Auditory Computation, chap 3. Springer, New York, pp 62–120. https://doi.org/10.1007/978-1-4612-4070-9_3
160. Gray CM, Singer W (1989) Stimulus-specific neuronal oscillations in orientation columns of cat visual cortex-1989. Proc Natl Acad Sci 86(5):1698–1702. https://doi.org/10.1073/pnas.86.5.1698
161. Tallon C, Bertrand O, Bouchet P, Pernier J (1995) Gamma-range activity evoked by coherent visual stimuli in humans. Eur J Neurosci 7(6):1285–1291. https://doi.org/10.1111/j.1460-9568.1995.tb01118.x
162. Keil A, Müller MM, Ray WJ, Gruber T, Elbert T (1999) Human gamma band activity and perception of a gestalt. J Neurosci 19(16):7152–7161. https://doi.org/10.1523/JNEUROSCI.19-16-07152.1999
163. Tallon-Baudry C, Bertrand O (1999) Oscillatory gamma activity in humans and its role in object representation. Trends Cogn Sci 3(4):151–162. https://doi.org/10.1016/s1364-6613(99)01299-1
164. Rodriguez E, George N, Lachaux JP, Martinerie J, Renault B, Varela FJ (1999) Perception's shadow: long-distance synchronization of human brain activity. Nature 397(6718):430–433. https://doi.org/10.1038/17120
165. Engel AK, Fries P, Singer W (2001) Dynamic predictions: oscillations and synchrony in top-down processing. Nat Rev Neurosci 2(10):704–716. https://doi.org/10.1038/35094565
166. Engel AK, Singer W (2001) Temporal binding and the neural correlates of sensory awareness. Trends Cognit Sci 5(1):16–25. https://doi.org/10.1016/s1364-6613(00)01568-0
167. Buhusi C, Meck W (2005) What makes us tick? Functional and neural mechanisms of interval timing. Nat Rev Neurosci 6:755. https://doi.org/10.1038/nrn1764
168. Buhusi C, Meck W (2009) Relativity theory and time perception: single or multiple clocks? PloS one 4(7):e6268. https://doi.org/10.1371/journal.pone.0006268
169. Womelsdorf T, Fries P (2007) The role of neuronal synchronization in selective attention. Curr Opin Neurobiol 17(2):154–160. https://doi.org/10.1016/j.conb.2007.02.002
170. Fries P (2009) Neuronal gamma-band synchronization as a fundamental process in cortical computation. Annu Rev Neurosci 32:209–224. https://doi.org/10.1146/annurev.neuro.051508.135603
171. Nikolić D, Fries P, Singer W (2013) Gamma oscillations: precise temporal coordination without a metronome. Trends Cogn Sci 17(2):54–55. https://doi.org/10.1016/j.tics.2012.12.003
172. Baars BJ (2006) Global workspace theory of consciousness: toward a cognitive neuroscience of human experience. In: Laureys S (ed) The boundaries of consciousness, progress in brain research, vol 150. Elsevier professional, s.l., pp 45–53. https://doi.org/10.1016/s0079-6123(05)50004-9
173. Dehaene S, Changeux JP, Naccache L (2011) The global neuronal workspace model of conscious access: from neuronal architectures to clinical applications. In: Dehaene S, Christen Y (eds) Characterizing consciousness: from cognition to the clinic?. Research and Perspectives

in Neurosciences, Springer Berlin Heidelberg, Berlin, Heidelberg, pp 55–84. https://doi.org/10.1007/978-3-642-18015-6_4
174. Engelhardt G, Schaller G, Brandes T (2016) Bosonic Josephson effect in the Fano-Anderson model. Phys Rev A 94:013608. https://doi.org/10.1103/physreva.94.013608
175. Owen M, Guta MP (2019) Physically sufficient neural mechanisms of consciousness. Front Syst Neurosci 13:24. https://doi.org/10.3389/fnsys.2019.00024
176. Thaut MH, McIntosh GC, Hoemberg V (2015) Neurobiological foundations of neurologic music therapy: rhythmic entrainment and the motor system. Front Psychol 5. https://doi.org/10.3389/fpsyg.2014.01185
177. Bhattacharya J, Petsche H, Pereda E (2001) Long-range synchrony in the gamma band: role in music perception. J Neurosci 21(16):6329–6337. https://doi.org/10.1523/JNEUROSCI.21-16-06329.2001
178. Zanto T, Large EW, Fuchs A, Kelso JAS (2005) Gamma-band responses to perturbed auditory sequences: evidence for synchronization of perceptual processes. Music Percept Interdiscip J 22(3):531–547. https://doi.org/10.1525/mp.2005.22.3.531
179. Freeman WJ, Quian Quiroga R (2013) Imaging brain function with EEG: advanced temporal and spatial analysis of electroencephalographic signals. Springer, New York. https://doi.org/10.1007/978-1-4614-4984-3
180. Buzsáki G (2006) Rhythms of the brain. Oxford University Press. https://doi.org/10.1093/acprof:oso/9780195301069.001.0001
181. Lerdahl F, Jackendoff R (1990) A generative theory of tonal music, 4, print. MIT Press, Cambridge, Massachusetts
182. Leman M (1997) Music, gestalt, and computing: studies in cognitive and systematic musicology, SpringerLink Bücher, vol 1317. Springer, Berlin Heidelberg, Berlin, Heidelberg. https://doi.org/10.1007/bfb0034102
183. Deutsch D (2013) The psychology of music. Academic Press series in cognition and perception, 3rd edn. Academic, Oxford
184. Neuhaus C (2013) Processing musical form: behavioural and neurocognitive approaches. Musicae Sci 17(1):109–127. https://doi.org/10.1177/1029864912468998
185. Deliége I, Melen M (2014) Cue abstraction in the representation musical form. In: Deliége I, Sloboda J (eds) Perception and cognition of music. Psychology Press, Hove, pp 387–412. https://doi.org/10.4324/9780203105184
186. Koelsch S (2014) Brain correlates of music-evoked emotions. Nat Rev Neurosci 15:170. https://doi.org/10.1038/nrn3666
187. Lehne M, Koelsch S (2015) Toward a general psychological model of tension and suspense. Front Psychol 6:79. https://doi.org/10.3389/fpsyg.2015.00079
188. Kurth E (1931) Musikpsychologie. Hesse, Berlin
189. Grassberger P, Procaccia I (1983) Characterization of strange attractors. Phys Rev Lett 50:346. https://doi.org/10.1103/PhysRevLett.50.346
190. Grassberger P, Procaccia I (1983) Measuring the strangeness of strange attractors. Phys D 9:189. https://doi.org/10.1016/0167-2789(83)90298-1
191. Linke S, Bader R, Mores R (2021) Modeling synchronization in human musical rhythms using impulse pattern formulation (ipf). https://doi.org/10.48550/arXiv.2112.03218
192. Bader R, Zielke A, Franke J (2021) Timbre-based machine learning of clustering Chinese and Western Hip Hop music. Audio Engineering Society. https://doi.org/10.31235/osf.io/8ef7g
193. Jasper HH (1958) The ten-twenty electrode system of the international federation. Electroencephalogr Clin Neurophysiol 10:371–375
194. Tallon-Baudry C, Bertrand O, Delpuech C, Pernier J (1996) Stimulus specificity of phase-locked and non-phase-locked 40 hz visual responses in human. J Neurosci 16(13):4240–4249. https://doi.org/10.1523/JNEUROSCI.16-13-04240.1996
195. Holsheimer J, Feenstra BWA (1977) Volume conduction and eeg measurements within the brain: a quantitative approach to the influence of electrical spread on the linear relationship of activity measured at different locations. Electroencephalogr Clin Neurophysiol 43(1):52–58. https://doi.org/10.1016/0013-4694(77)90194-8

196. Kayser J, Tenke CE (2006) Principal components analysis of laplacian waveforms as a generic method for identifying erp generator patterns: I. evaluation with auditory oddball tasks. Clin Neurophysiol 117(2):348–368. https://doi.org/10.1016/j.clinph.2005.08.034
197. Bhat WA (2018) Is a data-capacity gap inevitable in big data storage? Computer 51(9):54–62. https://doi.org/10.1109/mc.2018.3620975
198. Schneider A (2018) Fundamentals. In: Bader R (ed) Springer handbook of systematic musicology. Springer Handbooks, Springer, Berlin and Heidelberg, pp 559–603. https://doi.org/10.1007/978-3-662-55004-5_30
199. Glantz SA (2002) Primer of biostatistics, 5th edn. McGraw-Hill, New York
200. Bastos AM, Schoffelen JM (2015) A tutorial review of functional connectivity analysis methods and their interpretational pitfalls. Front Syst Neurosci 9:175. https://doi.org/10.3389/fnsys.2015.00175
201. Guevara Erra R, Perez Velazquez JL, Rosenblum M (2017) Neural synchronization from the perspective of non-linear dynamics. Front Comput Neurosci 11. https://doi.org/10.3389/fncom.2017.00098
202. Blinowska KJ (2011) Review of the methods of determination of directed connectivity from multichannel data. Med Biol Eng Comput 49(5):521–529. https://doi.org/10.1007/s11517-011-0739-x
203. Lachaux JP, Rodriguez E, Martinerie J, Varela FJ (1999) Measuring phase synchrony in brain signals. Hum Brain Map 8(4):194–208. https://doi.org/10.1002/(SICI)1097-0193(1999)8:4<194::AID-HBM4>3.0.CO;2-C
204. Jalili M, Barzegaran E, Knyazeva MG (2014) Synchronization of eeg: bivariate and multivariate measures. IEEE Trans Neural Syst Rehabil Eng Publ IEEE Eng Med Biol Soc 22(2):212–221. https://doi.org/10.1109/tnsre.2013.2289899
205. Cohen J (1992) A power primer. Psycholog Bull 112(1):155–159. https://doi.org/10.1037/0033-2909.112.1.155
206. van Noorden L, Moelants D (1999) Resonance in the perception of musical pulse. J New Music Res 28(1):43–66. https://doi.org/10.1076/jnmr.28.1.43.3122

Chapter 6
Conclusion

This monograph has examined music psychology through the lens of complex systems theory, presenting an interdisciplinary approach that integrates theoretical, computational, and empirical perspectives. By conceptualizing music perception as an emergent phenomenon driven by dynamic interactions within neural networks, it has elucidated the mechanisms that enable the brain to process and respond to auditory stimuli. Fundamental to this analysis are the principles of adaptation and synchronization, which offer a cohesive framework for understanding the interaction between individual components and their collective behavior within complex systems.

At the core of this perspective lies the principle of the balance of relations, which underscores the necessity of understanding entities not in isolation but through the dynamic relationships that define them. This paradigm extends across multiple scientific disciplines, including philosophy, psychology, music, and complex systems theory, emphasizing the interdependencies and interactions that shape system dynamics. From systems theory to cybernetics and synergetics, it provides a unified framework for analyzing how balance within interconnected components regulates stability and adaptation across diverse domains, whether in neural networks, biological systems, or the physical world.

The interdisciplinary nature of this work–drawing from musicology, neuroscience, physics, mathematics, biology, and computer science–demonstrates the power of collaborative research to bridge the growing fragmentation of scientific disciplines. This integrative approach reflects the broader movement towards interdisciplinarity in contemporary science, addressing the limitations of overspecialization by fostering cohesive frameworks that connect disparate fields. This monograph has shown how music psychology benefits from this synthesis, enabling novel insights into phenomena such as neural synchronization with auditory inputs, hemispheric asymmetries in brain dynamics, and the modulation of perception through external stimuli like music.

At its core, this work emphasizes that music perception is not simply a product of localized brain activity but a complex interplay between structural and dynamic factors. The adaptive capacity of neural networks allows the brain to respond flexibly

J. Sawicki, *Music Psychology—Balance of Relations*,
Current Research in Systematic Musicology 13,
https://doi.org/10.1007/978-3-031-97003-0_6

to ever-changing auditory environments, while synchronization facilitates coherence and coordination across different brain regions. These mechanisms extend beyond music perception, offering explanatory power for broader physiological and cognitive processes.

Through computational models, such as FitzHugh-Nagumo oscillators on empirically derived brain networks, this monograph has demonstrated how specific neural and network properties contribute to music-induced synchronization. Simulations have highlighted the role of network topology, input parameters, and intrinsic dynamics in shaping these processes, revealing the brain's remarkable capacity to align with musical stimuli. By identifying the critical frequency bands and structural features that influence neural coherence, the findings offer pathways for future research into sensory processing, cognition, and therapeutic applications.

In conclusion, the intersection of music psychology and complex systems theory represents a fertile ground for advancing our understanding of the mind and brain. By leveraging the principles of adaptation and synchronization, this monograph has not only deepened our comprehension of music perception but also underscored the value of interdisciplinary research in tackling complex scientific questions. The convergence of these fields holds promise for further breakthroughs, from enhancing artificial intelligence models to developing innovative approaches for treating neurological disorders and refining brain-computer interfaces. As such, the study of music psychology through the perspective of complex systems exemplifies the transformative potential of integrating diverse scientific disciplines into a coherent and holistic framework.

The methodologies explored in this monograph offer novel perspectives for advancing research in music psychology, outlined in Sect. 6.2. Section 6.1 provides a comprehensive summary of the findings presented across the individual chapters.

6.1 Summary

This monograph is organized into three parts: The first part (Chaps. 1 and 2) has explored the shared foundations of music psychology and complex systems research. The second part (Chap. 3) has presented key mechanisms central to both disciplines. The third part (Chaps. 4 and 5) has examined the modeling of physiological, neuronal, and perceptual processes, providing insights into their functioning within associated systems.

Chapter 1 discusses the increasing importance of interdisciplinarity in scientific research, particularly in fields addressing fragmented knowledge frameworks, and positions this approach as essential for fostering integrated understanding across disciplines. This monograph applies complex systems theory to the study of music psychology, integrating insights from musicology, physics, neuroscience, and computer science to investigate music perception. Central to this work is the principle of the balance of relations, which emphasizes the importance of understanding entities through the dynamic relationships that define them, rather than in isolation. This paradigm is prevalent across various scientific fields, including philosophy, psychology, music, and complex systems theory, underscoring the interdependencies and interactions within systems. The primary focus of this investigation is on understanding the mechanisms that underlie music perception, rather than exploring its specific neural correlates. By employing the tools of complex systems theory, this study aims to uncover fundamental principles that can simplify complex, chaotic phenomena into underlying, universal rules.

Chapter 2 explores the intersection of music psychology and complex systems theory, emphasizing their shared focus on emergent patterns, dynamic interactions, and self-organization. It traces their historical development, from philosophical inquiries into musical harmony to modern network theory and dynamical systems, establishing a theoretical foundation for their integration. The chapter examines psychoacoustic principles underlying music perception, highlighting the interplay of pitch, timbre, loudness, and rhythm as interconnected perceptual dimensions. Additionally, it provides a comprehensive overview of complex systems theory, demonstrating how mathematical formalism unifies analyses across neuroscience, physics, and computational modeling. The role of computational tools in bridging neuroscience and complex systems is also explored, with discussions on machine learning, connectomics, and neural modeling in music neuroscience.

The chapter further investigates the auditory system as a complex network, describing the transformation of sound waves into neural signals and the hierarchical processing of auditory information. A network-based approach in music psychology is presented as a paradigm shift from traditional localization models, emphasizing the distributed and dynamic nature of neural circuits involved in music perception. This perspective integrates sensory, motor, cognitive, and emotional processes, providing a comprehensive framework for understanding music cognition.

Chapter 3 examines adaptation and synchronization as core mechanisms in complex systems, exploring their mathematical underpinnings and emergent behaviors. Complex systems are defined by the interplay of individual components and network architecture, producing collective phenomena such as self-organization. Adaptation enables components to modify responses dynamically, while synchronization fosters coherence across systems, both of which are integral to understanding network dynamics.

Despite advances in modeling these mechanisms through differential equations, graph theory, and stochastic processes, a unified mathematical framework remains a challenge. Applications of these concepts span from multilayered networks with partial synchronization patterns to neural networks where adaptive control strategies and machine learning reveal insights into the behavior of high-dimensional systems. Synchronization studies, particularly in neuroscience and the field of music performance, elucidate mechanisms underlying cognitive functions and disorders like epilepsy. By analyzing specific patterns such as chimera states and relay synchronization, the chapter outlines critical research challenges and potential directions for understanding and optimizing complex networked systems.

Chapter 4 applies complex systems theory to model physiological and neuronal systems, emphasizing the roles of adaptation and synchronization as universal principles. Examples include neural synchronization in cognitive processes and disease dynamics, such as epilepsy. A novel functional model examines tumor disease and sepsis through network interactions of cytokine activity. FitzHugh-Nagumo oscillators applied to brain networks reveal partial synchronization patterns mediated by structural connectivity.

Moreover, we demonstrate the potential of complex systems frameworks to bridge empirical data and theoretical models, advancing understanding in domains such as artificial intelligence, socio-economic dynamics, and medical applications. Synchronization phenomena derived from empirical data provide insights into disease states and therapeutic strategies, linking structural and functional aspects of neural activity.

Building on the framework of adaptation and synchronization, Chap. 5 explores their application in modeling perceptual physiology. Adaptation enables dynamic responses to environmental changes, highlighting evolutionarily conserved strategies. Synchronization is critical to sensory processing, as seen in auditory perception, where coherence between global brain dynamics and auditory stimuli enhances processing.

Using network data from diffusion-weighted magnetic resonance imaging (MRI), simulations with neuronal oscillators reveal how auditory stimuli modulate synchronization patterns in neural networks. Music-specific analyses demonstrate increased coherence between network dynamics and auditory input, illustrating how frequency bands shape the interaction of neural systems with external stimuli. These findings underscore the significance of adaptation [1] and synchronization [2] in sensory processing, paving the way for advanced computational models and applications in cognition, sensory perception, and brain-computer interfaces.

6.2 Outlook

Music psychology, viewed through the framework of complex systems, offers a unique interdisciplinary perspective that bridges the humanities and natural sciences. By leveraging concepts like adaptation and synchronization, this approach elucidates the intricate mechanisms underlying music perception and emphasizes their broader implications for understanding cognitive and neural processes.

The principle of the balance of relations offers a fundamental perspective for understanding complex systems by focusing on the dynamic interdependencies that shape them. As a unifying paradigm spanning disciplines such as philosophy, psychology, music, and complex systems theory, it provides a structured approach to analyzing stability, adaptation, and emergent behavior in interconnected systems. From biological networks to artificial intelligence, this principle continues to inform research by emphasizing relational dynamics over isolated entities.

While artificial intelligence (AI) is currently at the forefront of technological advancements, it is unlikely to replace well-founded theoretical modeling once it becomes an established tool. Instead, AI can complement existing frameworks by enhancing computational modeling and analysis, but the fundamental understanding of system interactions will remain rooted in conceptual and scientific principles.

In music research, the principle of the balance of relations has promising applications, particularly in the development of digital musical instruments and the modeling of acoustic phenomena. Moreover, by integrating this paradigm with emerging digital music formats and online music platforms, it can also inspire new approaches in music sociology, shedding light on how technological and social dynamics shape contemporary musical experiences. Ultimately, this principle will continue to drive interdisciplinary innovation, fostering deeper insights into complex relational structures across diverse scientific and artistic domains.

The exploration of music psychology as a complex system reflects the growing need for interdisciplinary frameworks to address the fragmentation of modern scientific disciplines. Combining insights from musicology, neuroscience, physics, mathematics, biology, and computer science, this approach embodies the integrative spirit of interdisciplinarity. This synthesis not only bridges gaps within the multidisciplinarity of specialized fields but also promotes the development of cohesive models capable of addressing complex and multifaceted phenomena.

Music perception transcends isolated neural processes, emerging from dynamic interactions within neural networks. Adaptation, the ability of components to dynamically adjust to environmental changes, and synchronization, the alignment of temporal dynamics across systems, are central to this understanding. Together, these mechanisms illuminate how the brain achieves coherence and coordination in response to auditory stimuli. Methods and tools from complex systems science provide a powerful opportunity to holistically understand the dynamic interactions of the brain.

Adaptation ensures the flexibility of neural networks to evolving auditory environments, while synchronization aligns neural oscillations, enabling seamless integration of musical information. These principles resonate with broader physiological and cognitive processes, emphasizing the universality of complex system dynamics. In this spirit, for example, the development of new machine learning (ML) and artificial intelligence (AI) algorithms may potentially profit from a deeper understanding of the brain provided by the field of neuroscience, and vice versa.

Synchronization in neural systems occurs in varied forms, from zero-lag phase synchronization to more complex states such as partial or clustered synchronization. These patterns, observable in models like FitzHugh-Nagumo oscillators, underscore the brain's ability to dynamically adapt to music. Empirical data from neuroimaging studies have revealed that music-specific stimuli enhance coherence within neural networks, particularly within critical frequency bands that shape auditory perception. Investigating synchronization mechanisms offers critical insights into developing interventions aimed at enhancing mental health and optimizing strategies for music performance. Within this context, music psychology plays a pivotal role, with music-based therapies undergoing rigorous research to evaluate their effectiveness in addressing mental health disorders. These therapeutic approaches leverage the principles of complex systems to elucidate the dynamic processes underlying treatment outcomes, thereby advancing our understanding of the interplay between neural synchronization, psychological well-being, and musical engagement.

The adaptability of network topology–whether through directed relay pathways, feedback loops, or asymmetrical interactions–further demonstrates the brain's capacity to organize and process information dynamically. Such insights contribute to understanding not only music perception but also conditions like epilepsy and disorders involving disrupted neural synchronization. Indeed, it has long been recognized that "The adaptiveness of the human organism, the facility with which it acquires new representations and strategies and becomes adept in dealing with highly specialized environments, makes it an elusive and fascinating target of our scientific inquiries and the very prototype of the artificial" [3].

The principles of adaptation and synchronization extend beyond music psychology, offering tools for advancing fields like artificial intelligence, social dynamics, and medical applications. For instance, the study of synchronization patterns in neural networks informs therapeutic strategies for neurological disorders, while adaptation provides models for machine learning algorithms that emulate human cognition.

Future research may focus on refining models by integrating structural and functional aspects of neural activity, incorporating thresholds or discreteness into oscillatory dynamics, and exploring dynamical and topological asymmetries. These efforts could deepen our understanding of neural processing, from sensory perception to complex cognitive functions.

Music psychology, as studied through complex systems, exemplifies the transformative potential of interdisciplinary research. By embracing the concepts of adaptation and synchronization, this perspective has illuminated the emergent nature of music perception, offering new avenues for understanding the brain and its dynamic interaction with the environment. This synthesis not only advances scientific knowledge but also underscores the broader applicability of complex systems frameworks in addressing fundamental questions across disciplines.

References

1. Sawicki J, Berner R, Loos SAM, Anvari M, Bader R, Barfuss W, Botta N, Brede N, Franović I, Gauthier DJ. Goldt S, Hajizadeh A, Hövel P, Karin O, Lorenz-Spreen P, Miehl C, Mölter J, Olmi S, Schöll E, Seif A, Tass PA, Volpe G, Yanchuk S, Kurths J (2023) Perspectives on adaptive dynamical systems. Chaos Interdiscip J Nonlinear Sci 33(7):071501. https://doi.org/10.1063/5.0147231
2. Sawicki J (2019) Delay controlled partial synchronization in complex networks. Springer Theses, Springer, Heidelberg. https://doi.org/10.1007/978-3-030-34076-6
3. Simon HA (1969) The sciences of the artificial. Karl Taylor Compton lectures. M.I.T. Press, Cambridge

www.ingramcontent.com/pod-product-compliance
Lightning Source LLC
Chambersburg PA
CBHW050239250825
31606CB00003B/17
9783031970023